Understanding Biplots

Understanding Biplots

John Gower
The Open University, UK

Sugnet Lubbe
University of Cape Town, South Africa

Niël le Roux
University of Stellenbosch, South Africa

WILEY
A John Wiley and Sons, Ltd., Publication

This edition first published 2011

Registered office
John Wiley & Sons Ltd, The Atrium, Southern Gate, Chichester, West Sussex, PO19 8SQ, United Kingdom

For details of our global editorial offices, for customer services and for information about how to apply for permission to reuse the copyright material in this book please see our website at www.wiley.com.

Library of Congress Cataloguing-in-Publication Data

Gower, John.
Understanding biplots / John Gower, Sugnet Lubbe, Niel le Roux.
p. cm.
Includes bibliographical references and index.
ISBN 978-0-470-01255-0 (cloth)
1. Multivariate analysis–Graphic methods. 2. Graphical modeling (Statistics) I. Lubbe, Sugnet, 1973- II. le Roux, Niel. III. Title.
QA278.G685 2010
519.5′35–dc22

2010024555

A catalogue record for this book is available from the British Library.

Print ISBN: 978-0-470-01255-0
ePDF ISBN: 978-0-470-97320-2
oBook ISBN: 978-0-470-97319-6

Set in 10/12pt Times by Laserwords Private Limited, Chennai, India
Printed and bound in Singapore by Markono Print Media Pte Ltd

Contents

Preface

This book grew from an earlier book, *Biplots* (Gower and Hand, 1996), the first monograph on the subject of biplots, written in a fairly concentrated and not easily understood style. Colleagues tactfully suggested that there was a need for a friendlier book on biplots. This book is our response. Although it covers similar ground to the Gower and Hand (1996) book, it omits some topics and adds others. No attempt has been made to be encyclopedic and many biplot methods, especially those concerned with three-way tables, are totally ignored.

Our aims in writing this book have been threefold: first, to provide the geometric background, which is essential for understanding, together with its algebraic manifestations, which are essential for writing computer programs; second, to provide a wealth of illustrative examples drawn from a wide variety of fields of application, illustrating different representatives of the biplot family; and third, to provide computer functions written in R that allow routine multivariate descriptive methods to be easily used, together with their associated biplots. It also provides additional tools for those wishing to work interactively and to develop their own extensions.

We hope that research workers in the applied sciences will find the book a useful introduction to the possibilities for presenting certain types of data in informative ways and give them the background to make valid interpretations. Statisticians may find it of interest both as a source of potential research projects and useful examples.

This project has taken longer than we had planned and we are keenly aware that some topics remain less friendly than we might have hoped. We thank Kathryn Sharples, Susan Barclay, Richard Davies, Heather Kay and Prachi Sinha-Sahay at Wiley for both their forbearance and support. We also thank our long-suffering spouses, Janet, Pieter and Magda, if not for their active support, then at least for their forbearance.

John Gower
Sugnet Lubbe
Niël le Roux
www.wiley.com/go/biplots

1 *Introduction*

Biplots have been with us at least since Descartes, if not from the time of Ptolemy who had a method for fixing the map positions of cities in the ancient world. The essential ingredients are coordinate axes that give the positions of points. From the very beginning, the concept of distance was central to the Cartesian system, a point being fixed according to its distance from two orthogonal axes; distance remains central to much of what follows. Descartes was concerned with how the points moved in a smooth way as parameters changed, so describing straight lines, conics and so on. In statistics, we are interested also in isolated points presented in the form of a scatter diagram where, typically, the coordinate axes represent variables and the points represent samples or cases. Cartesian geometry soon developed three-dimensional and then multidimensional forms in which there are many coordinate axes. Although two-dimensional scatter diagrams are invaluable for showing data, multidimensional scatter diagrams are not. Therefore, statisticians have developed methods for approximating multidimensional scatter in two, or perhaps three, dimensions. It turns out that the original coordinate axes can also be displayed as part of the approximation, although inevitably they lose their orthogonality. The essential property of all biplots is the two modes, such as variables and samples. For obvious reasons, we shall be concerned mainly with two-dimensional approximations but should stress at the outset that the *bi-* of biplots refers to the two modes and not the usual two dimensions used for display.

Biplots, not necessarily referred to by name, have been used in one form or another for many years, especially since computer graphics have become readily available. The term 'biplot' is due to Gabriel (1971) who popularized versions in which the variables are represented by directed vectors. Gower and Hand (1996) particularly stressed the advantages of presenting biplots with calibrated axes, in much the same way as for conventional coordinate representations. A feature of this book is the wealth of examples of different kinds of biplots. Although there are many novel ideas in this book, we acknowledge our debts to many others whose work is cited either in the current text or in the bibliography of Gower and Hand (1996).

Understanding Biplots John Gower, Sugnet Lubbe and Niël le Roux

1.1 Types of biplots

We may distinguish two main types of biplot:

- *asymmetric* (biplots giving information on sample units and variables of a data matrix);
- *symmetric* (biplots giving information on rows and columns of a two-way table).

In symmetric biplots, rows and columns may be interchanged without loss of information, while in asymmetric biplots variables and sample units are different kinds of object that may not be interchanged.

Consider the data on four variables measured on 21 aircraft in Table 1.1. The corresponding biplot in Figure 1.1 represents the 21 aircraft as sample points and the four variables as biplot axes. It will not be sensible to exchange the two sets, representing the aircraft as continuous axes and the variables as points. Next, consider the two-way table in Table 1.2. Exchanging the rows and columns of this table will have no effect on the information contained therein. For such a symmetric data set, both the rows and columns are represented as points as shown in Figure 1.2. Details on the construction of these biplots are deferred to later chapters.

Table 1.1 Values of four variables, ***SPR*** (specific power, proportional to power per unit weight), ***RGF*** (flight range factor), ***PLF*** (payload as a fraction of gross weight of aircraft) and ***SLF*** (sustained load factor), for 21 aircraft labelled in column 2. From Cook and Weisberg (1982, Table 2.3.1), derived from 1979 RAND Corporation report.

	Aircraft	SPR	RGF	PLF	SLF
A	FH-1	1.468	3.30	0.166	0.10
B	FJ-1	1.605	3.64	0.154	0.10
C	F-86A	2.168	4.87	0.177	2.90
D	F9F-2	2.054	4.72	0.275	1.10
E	F-94A	2.467	4.11	0.298	1.00
F	F3D-1	1.294	3.75	0.150	0.90
G	F-89A	2.183	3.97	0.000	2.40
H	XF10F-1	2.426	4.65	0.117	1.80
I	F9F-6	2.607	3.84	0.155	2.30
J	F100-A	4.567	4.92	0.138	3.20
K	F4D-1	4.588	3.82	0.249	3.50
M	F11F-1	3.618	4.32	0.143	2.80
N	F-101A	5.855	4.53	0.172	2.50
P	F3H-2	2.898	4.48	0.178	3.00
Q	F102-A	3.880	5.39	0.101	3.00
R	F-8A	0.455	4.99	0.008	2.64
S	F-104A	8.088	4.50	0.251	2.70
T	F-105B	6.502	5.20	0.366	2.90
U	YF-107A	6.081	5.65	0.106	2.90
V	F-106A	7.105	5.40	0.089	3.20
W	F-4B	8.548	4.20	0.222	2.90

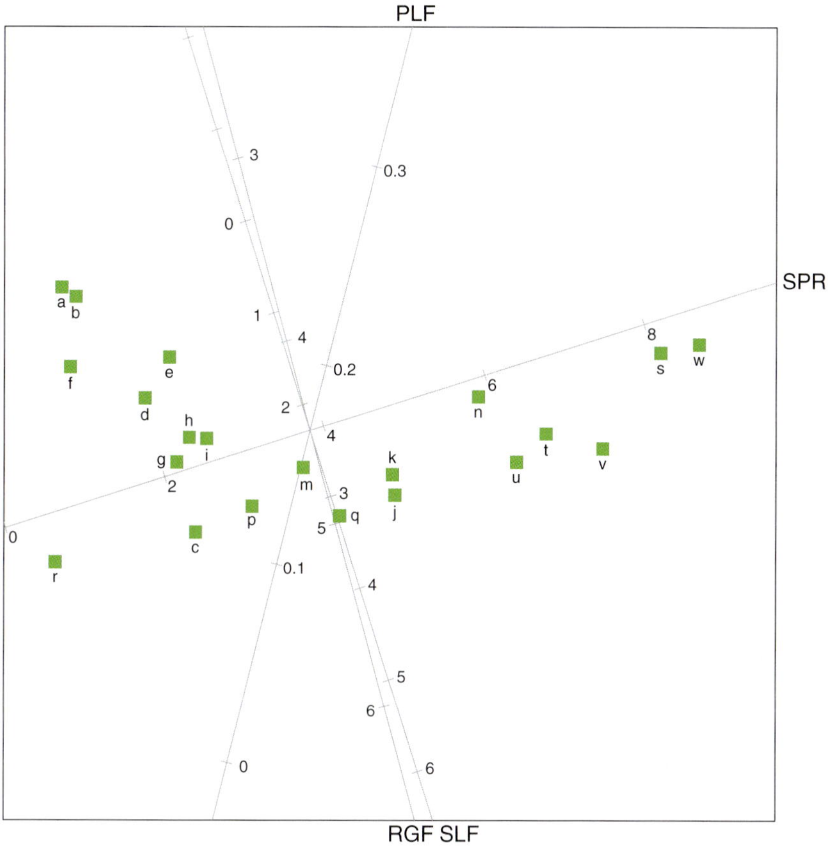

Figure 1.1 Principal component analysis biplot according to the Gower and Hand (1996) representation.

Table 1.2 Species × Temperature two-way table of percentage cellulose measured in wood pulp from four species after a hot water wash.

Temperature (°C)	Species			
	Amea	Edun	Egran	Emac
90	47.12	40.61	46.36	45.15
130	48.59	46.57	45.96	45.76
140	59.49	49.73	55.71	49.95
150	63.59	68.18	70.94	56.32
160	71.18	69.50	65.13	71.18
170	67.12	65.30	69.85	67.58

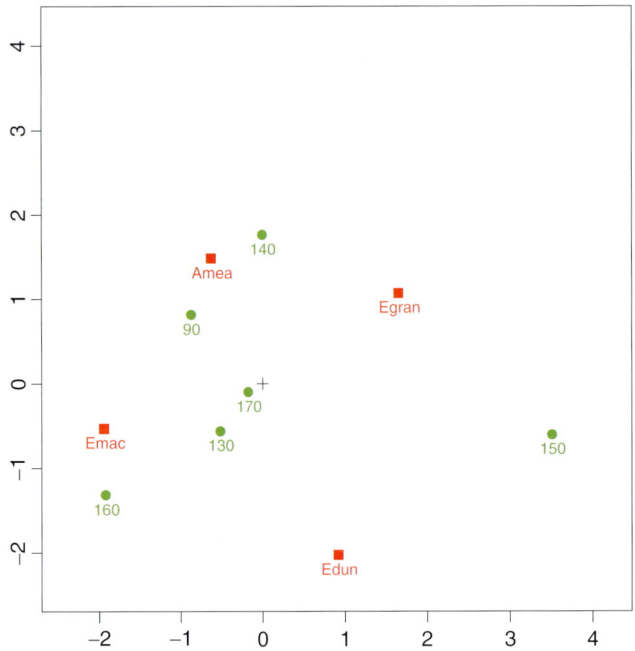

Figure 1.2 Biplot for a two-way table representing Species × Temperature.

We shall see that this distinction between symmetric and asymmetric biplots affects what is permissible in the construction of a biplot. Within this broad classification, other major considerations are:

- the types of variable (quantitative, qualitative, ordinal, etc.);
- the method used for displaying samples (multidimensional scaling and related methods);
- what the biplot display is to be used for (especially for prediction or for interpolation).

The following can be represented in an asymmetric biplot:

- distances between samples;
- relationships between variables;
- inner products between samples and variables.

However, only two of these characteristics can be optimally represented in a single biplot. In the simple biplot in Figure 1.1 all the calibration scales are linear with evenly spaced calibration points. Other types of scale are possible and we shall meet them later in other types of biplots. Figure 1.3 shows the main possibilities.

Figure 1.3(a) is the familiar equally spaced calibration of a linear axis that we have already met in Figure 1.1. Figure 1.3(b) shows logarithmic calibration of a linear axis;

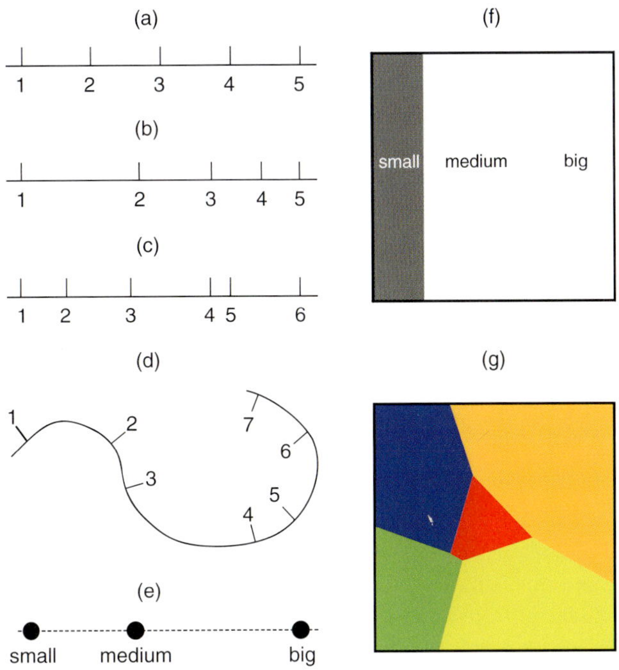

Figure 1.3 Different types of scale. (a) A linear scale with equally spaced calibration as used in principal component analysis. (b) A linear scale with logarithmic calibration. (c) A linear scale with irregular calibration. (d) A curvilinear scale with irregular calibration. (e) A linear scale for an ordered categorical variable. (f) Linear regions for ordered categorical variables (g) A categorical variable, colour, defined over convex regions.

this is an example of regular but unequally spaced calibration. In Figure 1.3(c) the axis remains linear but the calibrations are irregularly spaced. In Figure 1.3(d) the axis is nonlinear and calibrations are irregularly spaced; in principle, nonlinear axes could have equally spaced calibrations or regularly space calibrations, but in practice such combinations are unlikely. Figure 1.3(e) shows an ordered categorical variable, ***size***, not recorded numerically but only as *small*, *medium* and *big*. The calibration is indicated as a set of correctly ordered markers on a linear axis, but this is shown as a dotted line to indicate that intermediate markers are undefined (i.e. interpolation is not permitted). In Figure 1.3(f) the ordered categorical variable ***size*** is represented by linear regions; all samples in a region are associated with that level of ***size***. Figure 1.3(g) shows an unordered categorical variable, ***colour***, with five levels: *blue*, *green*, *yellow*, *orange* and *red*. These levels label convex regions. In general, the levels of unordered categorical variables may be represented by convex regions in many dimensions. Examples of these calibrations occur throughout the book.

1.2 Overview of the book

The basic steps for constructing many asymmetric biplots are summarized in Figure 1.4. Starting from a data matrix $\mathbf{X}$, first we calculate a distance matrix $\mathbf{D}$: $n \times n$. The essence

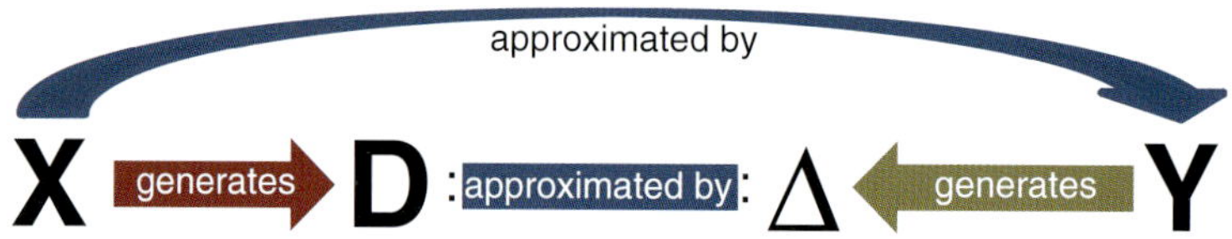

Figure 1.4 Construction of an asymmetric biplot.

of the methodology is approximating the distance matrix **D** by a matrix of Pythagorean distances $\mathbf{\Delta}$: $n \times n$. Operationally, this is achieved iteratively by updating r-dimensional coordinates **Y**, that generate $\mathbf{\Delta}$, to improve the approximation to **D**. It is hoped that a small choice of r (hopefully 2) will give a good approximation. Finally, the curved arrow represents two ideas: (i) in principal component analysis (PCA) **Y** approximates **X**; and (ii) more generally, information on **X** can be represented in the map of **Y** (the essence of biplots). These are the basic steps of multidimensional scaling (see Cox and Cox, 2001).

In general, the points given by **Y** generate distances in $\mathbf{\Delta}$ that approximate the values in **D**. In addition, and this is the special contribution of biplots, approximations to the true values **X** may be deduced from **Y**. In the simplest case, the PCA biplot, this approximation is made by projecting the orthogonal axes of **X** onto a subspace occupied by **Y**. In the subsequent chapters, we will discuss more general forms of asymmetric biplots. The most general of these, appropriately named the generalized biplot, has as special case the PCA biplot when all variables in **X** are continuous and the matrix **D** consists of Pythagorean distances. When restricting the variables in **X** to be continuous only, the rows of **X** represent the samples as points in p-dimensional space with an associated coordinate system. In the biplot, we represent the samples as points whose coordinates are given by the rows of **Y** and the coordinate system of **X** by appropriately defined biplot axes. These axes become nonlinear biplot trajectories when the definition of distance in the matrix **D** necessitates a nonlinear transformation from **X** to **Y**. The methodology outlined by Figure 1.4 allows us to also include categorical variables. Even though a categorical variable cannot be represented in the space of **X** by a linear coordinate axis, we can calculate the matrix **D** and proceed from there.

Thus, a biplot adds to **Y** information on the variables given in **X**. In multidimensional scaling, **D** may be observed directly and not derived from **X**, and then biplots cannot be constructed. The different types of asymmetric biplots discussed above depend on the properties of the variables in the matrix **X** and the distance metric producing the matrix **D**. Many special cases of importance fall within this general framework and are illustrated by applications in the following chapters. Several definitions of distance used in constructing **D** occur using both quantitative and qualitative variables (or mixtures of the two). For symmetric biplots, the position is simpler as we have only two main possibilities: (i) a quantitative variable classified in a two-way table and (ii) a two-way table of counts.

In Figure 1.5 the biplots to be discussed in the designated chapters are represented diagrammatically. The distances associated with the matrix **D** in Figure 1.4 is divided into subsets for the different types of biplots. The matrix $\mathbf{\Delta}$ always consists of Pythagorean distances to allow intuitive interpretation of the rows of **Y**.

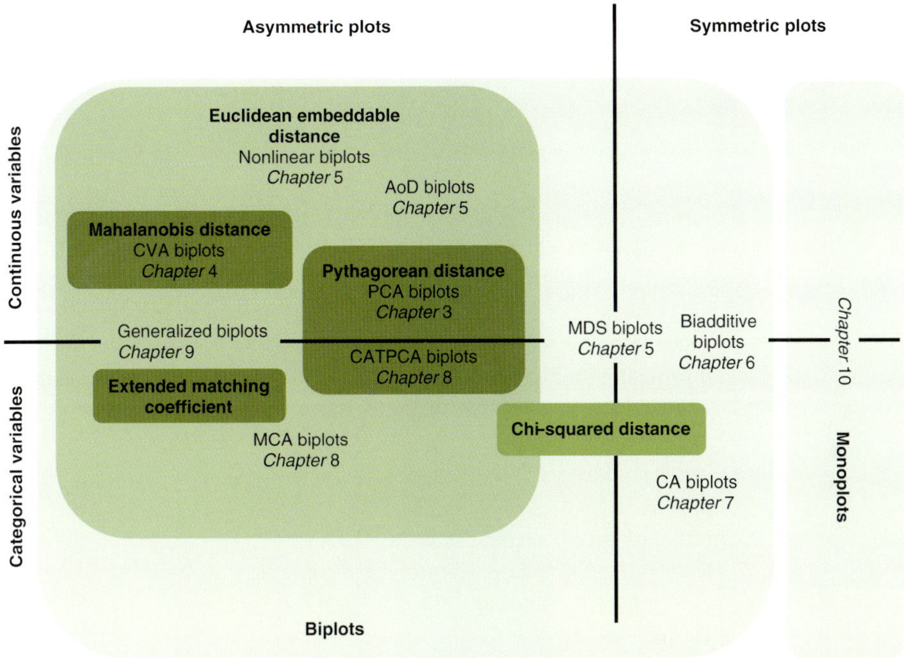

Figure 1.5 Summary of the different types of biplots discussed in subsequent chapters.

In a symmetric biplot, rows and columns have equal status and we aim to find two sets of coordinates $\mathbf{A}$ and $\mathbf{B}$, one for the rows and one for the columns respectively. Now, the main interest is in the inner product $\mathbf{AB}'$ and there is less interest in distance interpretations. A popular version of correspondence analysis (CA) approximates chi-squared distance, treating either the rows or columns as if they were 'variables' and thus giving two asymmetric biplots, not linked by a useful inner product. This form of CA is not a biplot and is sometimes referred to as a joint plot (see also Figure 10.4); other forms of CA do treat $\mathbf{X}$ symmetrically.

1.3 Software

A library of functions has been developed in the R language (R Development Core Team, 2009) and is available on the website www.wiley.com/go/biplots. Throughout this book reference will be made to the functions associated with the biplots being discussed. Examples of the commands to reproduce the figures in this book are given in the text. Sections are also included with specific information about the core functions needed for the different types of biplots.

1.4 Notation

Matrices are used extensively to enable the mathematically inclined reader to understand the algebra behind the different biplots. Bold upper-case letters indicate matrices and bold lower-case letters indicate vectors. Any column vector **x**: $p \times 1$ when presented as a row vector will be denoted by $\mathbf{x}' : 1 \times p$. The following symbols are used extensively throughout the text:

n	number of samples
p	number of variables
K	number of groups or classes into which the samples are divided
m	$\min(p, K-1)$
$\mathbf{X} : n \times p$	a data matrix with n samples measured on p variables. Unless stated otherwise, the matrix $\mathbf{X}$ is assumed to be centred to have column means equal to zero.
$\mathbf{G}$	an indicator matrix, usually with n rows, where each row consists of zeros except for a one in the column associated with that particular sample
$\mathbf{N}$	diagonal matrix of the group sizes, $\mathbf{N} = (\mathbf{G}'\mathbf{G})^{-1}$
$\mathbf{n}$	$\text{diag}(\mathbf{N})$
$\overline{\mathbf{X}} : K \times p$	matrix of group means, $\overline{\mathbf{X}} = \mathbf{N}^{-1}\mathbf{G}'\mathbf{X}$
$\mathbf{I}$	identity matrix, size determined by context
$\mathbf{J} : p \times p$	$\begin{bmatrix} \mathbf{I}_r & \mathbf{0} : r \times (p-r) \\ \mathbf{0} : (p-r) \times r & \mathbf{0} : (p-r) \times (p-r) \end{bmatrix}$
$\mathbf{1}$	column vector of ones, size determined by context
d_{ij}	the distance between sample i and sample j
δ_{ij}	the fitted distance between sample i and sample j
$\mathbf{D} : n \times n$	a matrix derived from the pairwise distances of all n samples with ijth element $-\frac{1}{2}d_{ij}^2$. The latter quantities are termed *ddistances*.
$\text{diag}(\mathbf{A} : p \times p)$	the $p \times p$ diagonal matrix formed by replacing all the off-diagonal elements of $\mathbf{A}$ with zeros; or, depending on the context, the p-vector consisting of the diagonal elements of $\mathbf{A}$
$\text{diag}(\mathbf{a})$	a diagonal matrix with the elements of the vector $\mathbf{a}$ on the diagonal
$\mathbf{R}$	diagonal matrix of row totals
$\mathbf{C}$	diagonal matrix of column totals
$\mathbf{E}$	$\mathbf{R}\mathbf{1}\mathbf{1}'\mathbf{C}/n$
$\|\mathbf{A}\|^2$	$\text{tr}(\mathbf{A}\mathbf{A}')$
$\mathbf{A}^*\mathbf{B}$	elementwise multiplication
$\mathbf{A}/\mathbf{B}$	elementwise division

The notion of *distance* is discussed in Chapter 5. Here we mention two concepts which the reader will need throughout the book. *Pythagorean distance* is the ordinary Euclidean distance between two samples $\mathbf{x}_i$ and $\mathbf{x}_j$ with

$$d_{ij}^2 = \sum_{k=1}^{p} (x_{ik} - x_{jk})^2.$$

Any distance metric that can be embedded in a Euclidean space is termed *Euclidean embeddable*.

1.4.1 Acronyms

AoD	analysis of distance
CA	correspondence analysis
CVA	canonical variate analysis
EMC	extended matching coefficient
JCA	joint correspondence analysis
MCA	multiple correspondence analysis
MDS	multidimensional scaling
PCA	principal component analysis

2 Biplot basics

In accordance with our aim of understanding biplots, the focus in this chapter is to look at biplot basics from the viewpoint of an ordinary scatterplot.

The chapter begins by introducing two- and three-dimensional biplots as ordinary scatterplots of two or three variables. In Section 2.2 biplots are considered as extensions of the ordinary scatterplot by providing for more than three variables. Generalizing, a biplot provides for a graphical display, in at most three dimensions, of data that typically exist in a higher-dimensional space. The concept of approximating a data matrix is thus crucial in biplot methodology. Subsequent sections explore how to represent multidimensional sample points in a biplot, how to equip the biplot with calibrated axes representing the variables and how to refine the biplot display. Emphasis is placed on how to use biplot axes analogously to axes in a scatterplot, that is, for adding new samples to the plot (interpolation) and reading off for any sample point its values for the different variables (prediction). It is then shown how to use a regression method for adding new variables to the plot. Various enhancements to configurations of sample points in a biplot, including how to describe large data sets, are discussed next. Finally, some examples are given, together with the R code for constructing all the graphical displays shown in the chapter. We strongly suggest that readers work through these examples for a thorough understanding of the basics of biplot construction. In later chapters, we provide only the function calls to more elaborate R functions for fine-tuning the various types of biplot.

2.1 A simple example revisited

The data of Table 1.1 are available in the accompanying R package `UBbipl` in the form of the dataframe `aircraft.data`. We first convert columns 3 to 6 to a data matrix, `aircraft.mat`, with row names the first column of Table 1.1 and column names the

Understanding Biplots John Gower, Sugnet Lubbe and Niël le Roux

abbreviations used for the variables in Table 1.1. This is done by issuing the following instructions from the R prompt:

```
> aircraft.mat <- aircraft.data[, 2:5]
> aircraft.mat
    SPR   RGF   PLF   SLF
  a 1.468 3.30 0.166 0.10
  b 1.605 3.64 0.154 0.10
  .......................
  v 7.105 5.40 0.089 3.20
  w 8.548 4.20 0.222 2.90
```

Next, we construct a scatterplot of the two variables ***SPR*** and ***RGF*** with the instructions:

```
> plot(x = aircraft.mat[,1], y = aircraft.mat[,2], xlab = "",
  ylab = "", xlim = c(0,10), ylim = c(2,6), pch = 15,
  col = "green", yaxp = c(2,6,4), bty = "n")
> text(x = aircraft.mat[,1], y = aircraft.mat[,2],
  labels = dimnames(aircraft.mat)[[1]], pos = 1)
> mtext("RGF", side = 2, at = 6.4, line = -0.35)
> mtext("SPR", side = 1, at = 10.4, line = -0.50)
```

The scatterplot in Figure 2.1 is an example of what is probably the simplest form of an asymmetric biplot. It shows a plot of the columns ***SPR*** and ***RGF***, giving performance figures for power and range of the 21 types of aircraft introduced in Table 1.1. It is a scatterplot of two variables referred to orthogonal axes. The familiar elements of Figure 2.1 are:

- points representing the aircraft;
- a directed line for each of the variables, known as a coordinate axis, with its label;
- scales marked on the axes giving the values of the variables.

Note also the convention followed of labelling the axes at the end where the calibrations are at their highest values. It is an asymmetric biplot because it gives information of two types, (i) concerning the 21 aircraft and (ii) concerning the two variables, which cannot be interchanged. When a point representing an aircraft is projected orthogonally onto an axis, one may read off the value of the corresponding variable and this will agree precisely with the value given in Table 1.1. Indeed, this is not surprising, because the values of the variables were those used in the first place to construct the coordinate positions of the points. Notice the difference between the top and bottom panels of Figure 2.1. Which of $\boldsymbol{k}$ and $\boldsymbol{n}$ is nearest to $\boldsymbol{j}$? From the top panel, it appears to be $\boldsymbol{n}$, but a simple calculation shows the true distances to be

$$\text{dist}(\boldsymbol{j}, \boldsymbol{k}) = \sqrt{0.021^2 + 1.1^2} = 1.10,$$

$$\text{dist}(\boldsymbol{j}, \boldsymbol{n}) = \sqrt{1.288^2 + 0.39^2} = 1.34,$$

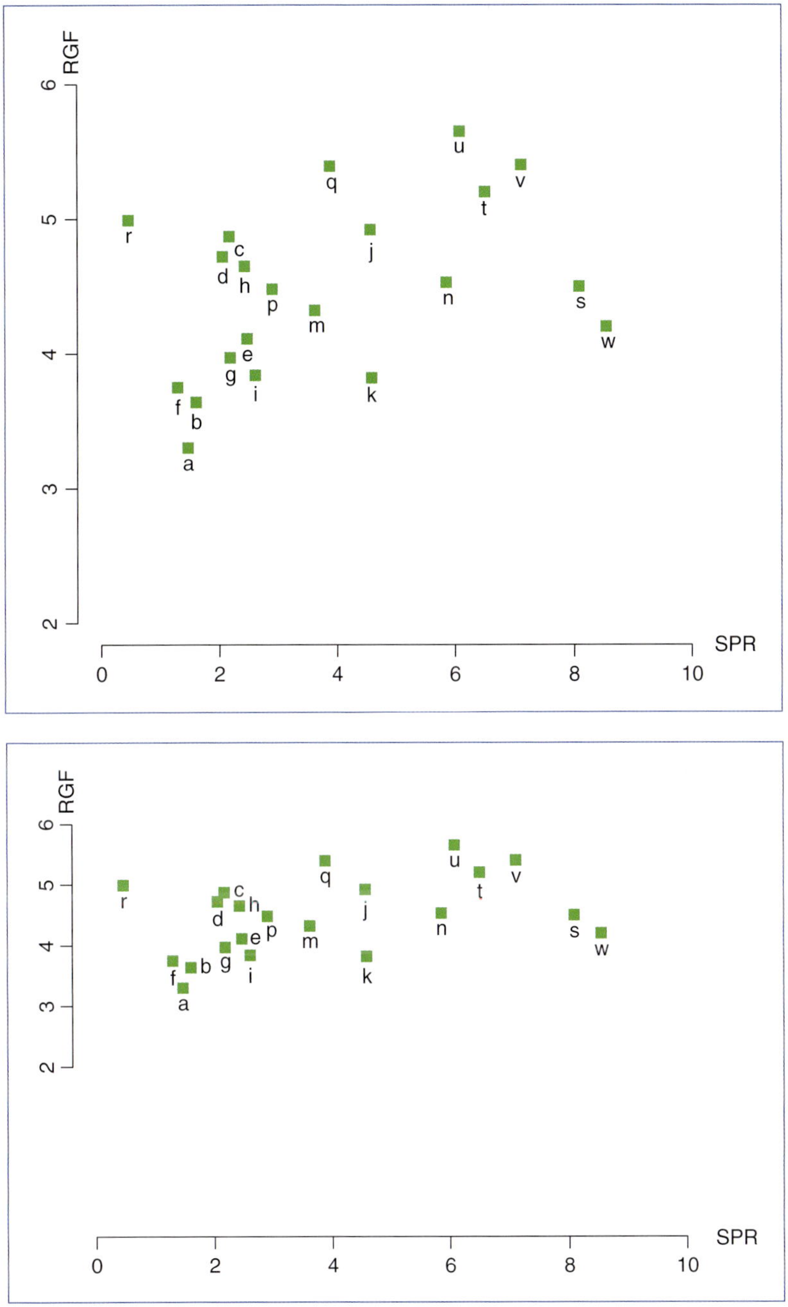

Figure 2.1 Scatterplot of variables ***SPR*** and ***RGF*** from the aircraft data in Table 1.1: (top) constructed with default settings; (bottom) constructed with an aspect ratio of unity.

so that $\boldsymbol{k}$ is nearer to $\boldsymbol{n}$, as is correctly displayed in the bottom panel. This example clearly demonstrates how one can go seriously wrong by constructing biplots that do not respect the aspect ratio. An aspect ratio of unity is not necessary for the validity of reading the scales by projection but, in much of what follows, we shall see that the relative scaling (or aspect ratio) of axes is crucial. The scatterplot in the bottom panel of Figure 2.1 has an aspect ratio of one. The call to the `plot` function to reproduce this scatterplot requires `asp = 1` instead of the `asp` default. The window for plotting is then set up so that one data unit in the x direction is equal in length to one data unit in the y direction. If this precaution is not taken when constructing biplots the inter-point distances in the biplot are distorted.

Figure 2.1 happens to be in two dimensions, but this is not necessary for a biplot. Indeed, if we make a three-dimensional Cartesian plot of the first three variables, this too would be a biplot (see Figure 2.2). The three-dimensional biplot in Figure 2.2 can be obtained by first using the following code and then interactively rotating and zooming the biplot to the desired view by using the left and right mouse buttons, respectively.

```
> library(rgl)
> open3d()
> view3d(theta = 180, phi = 45, fov = 40, zoom = 0.8)
> points3d(aircraft.mat, size = 10, col = "green", box = FALSE,
  xlim = c(3,6), ylim = c(1,9), zlim = c(0,0.5))
> text3d(aircraft.mat, texts = dimnames(aircraft.data)[[1]],
  adj = c(0.25, 1.2), cex = 0.75)
> axes3d(c("y","x","z-+"), cex = 0.75)
> aspect3d(1, 1, 0.5)
> title3d("","","SPR","RGF","PLF")
```

It is also possible to construct one-dimensional biplots, and although we consider such biplots as well as three-dimensional biplots in later chapters; for the remainder of this chapter we restrict ourselves to two-dimensional biplots.

2.2 The biplot as a multidimensional scatterplot

Although the plots in Figures 2.1 and 2.2 are commonly known as scatterplots, they are simple examples of biplots. Suppose now that we wish to show all four variables of Table 1.1. A perfect Cartesian representation would require four dimensions, so we would find it convenient if we could approximate the information in a two-dimensional (say) display. There are many ways of representing the aircraft by points in two dimensions so that their actual inter-point distances in the four dimensions are approximated. This is the concern of multidimensional scaling (MDS). We shall meet several methods of MDS in later chapters, but here we use one of the simplest methods by expressing the data matrix in terms of its singular value decomposition (SVD). We shall see that many of the ideas introduced in this chapter carry over easily into various forms of biplot discussed in later chapters.

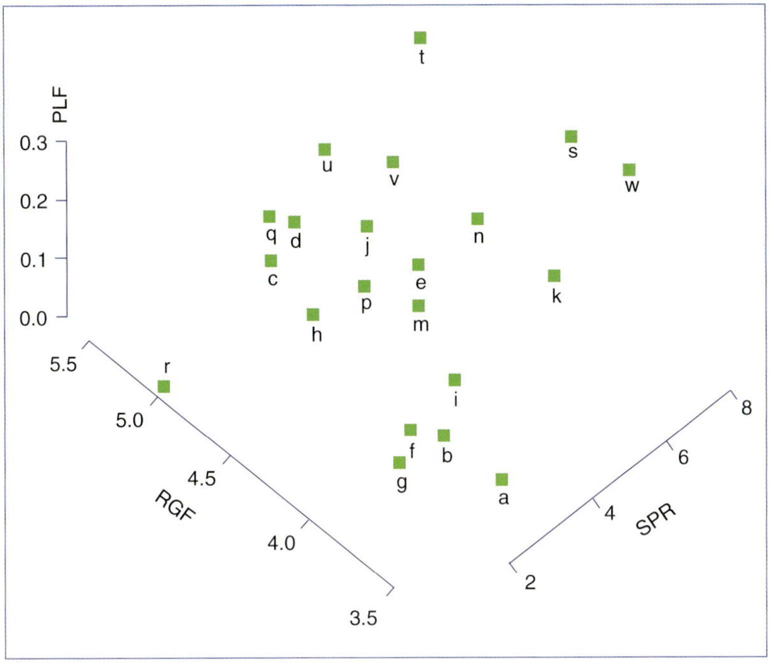

Figure 2.2 Three-dimensional scatterplot of variables ***SPR***, ***RGF*** and ***PLF*** of the aircraft data in Table 1.1.

Figure 2.3 shows the resulting plot where we have first subtracted the means of the individual variables from each aircraft's measurements. The same plot appears in both panels of Figure 2.3, the only difference being that the axes have been translated to pass through the point (0, 0) in the bottom panel. The orthogonal axes give the directions of what are known as the two principal axes. These mathematical constructs do not necessarily have any substantive interpretation. Nevertheless, attempts at interpretation in terms of *latent variables* are commonplace and sometimes successful. Any two oblique axes may determine the two-dimensional space, so there is an extensive literature on the search for interpretable oblique coordinate axes. Rather than dealing with latent variables, biplots offer the complementary approach of representing the original variables. Clearly, it is not possible to show four sets of orthogonal axes in two dimensions, so we are forced to use oblique representations. The axes representing the latent variables will generally not be shown; they form only what may be regarded as one-, two- or three-dimensional *scaffolding axes* on which the biplot is built.

How is Figure 2.3 constructed? The usual way of proceeding (Gabriel, 1971) is based on the SVD,

$$\mathbf{X} : n \times p = \mathbf{U}^* \mathbf{\Sigma}^* (\mathbf{V}^*)', \tag{2.1}$$

where, assuming that $n \geq p$, $\mathbf{U}^*$ is an $n \times n$ orthogonal matrix with columns known as the left singular vectors of $\mathbf{X}$, the matrix $\mathbf{V}^*$ is a $p \times p$ orthogonal matrix with columns

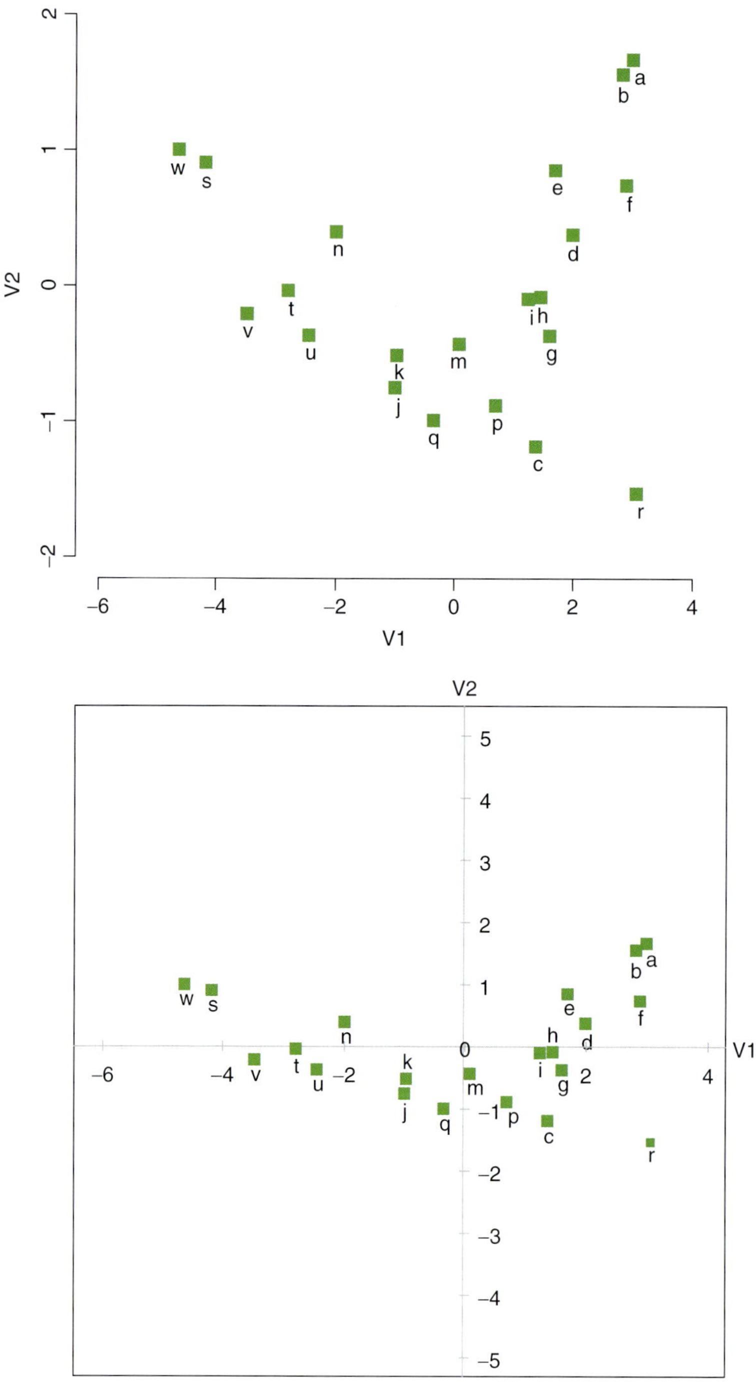

Figure 2.3 Principal axes ordination resulting from an SVD of the data matrix giving a two-dimensional scatterplot of the four-dimensional aircraft data. The bottom panel is similar to the top panel, except for the translation of the axes to pass through zero and an aspect ratio of unity.

known as the right singular vectors of $\mathbf{X}$, while the matrix $\mathbf{\Sigma}^*$ is of the form

$$\mathbf{\Sigma}^* : n \times p = \underset{\quad k \quad p-k}{\begin{bmatrix} \mathbf{\Sigma} & \mathbf{0} \\ \mathbf{0} & \mathbf{0} \end{bmatrix}} \begin{matrix} k \\ n-k \end{matrix}. \tag{2.2}$$

In (2.2), k denotes the rank of $\mathbf{X}$ while $\mathbf{\Sigma}$ is a $k \times k$ diagonal matrix with diagonal elements the nonzero singular values of $\mathbf{X}$, assumed to be presented in nonincreasing order. It follows that (2.1) can also be written as

$$\mathbf{X} : n \times p = \mathbf{U\Sigma V}', \tag{2.3}$$

where $\boldsymbol{U} : n \times k$ and $\boldsymbol{V} : p \times k$ consist of the first k columns of $\mathbf{U}^*$ and $\mathbf{V}^*$, respectively. The matrices $\mathbf{U}$ and $\mathbf{V}$ are both orthonormal.

An r-dimensional approximation of $\mathbf{X}$ is given by

$$\hat{\mathbf{X}}_{[r]} = \mathbf{U\Sigma}_{[r]}\mathbf{V}',$$

where $\mathbf{\Sigma}_{[r]}$ replaces the $p - r$ smallest diagonal values of $\mathbf{\Sigma}$ by zero. In the remainder of this chapter we discuss approximation, axes, interpolation, prediction, projection, and the like, from the viewpoint of extending scatter diagrams to more than two or three dimensions. We use mainly a simple type of biplot, the principal component analysis (PCA) biplot, as the instrument for introducing these concepts. In Chapter 3 we shall consider the PCA biplot as a distinct type of biplot in more detail while in subsequent chapters we shall show how the basic concepts generalize to more complicated data structures. Underpinning PCA is a result, proved by Eckart and Young (1936), that the r-dimensional approximation of $\mathbf{X}$ given by $\hat{\mathbf{X}}_{[r]} = \mathbf{U\Sigma}_{[r]}\mathbf{V}'$ is optimal in the least-squares sense that

$$\left\| \mathbf{X} - \hat{\mathbf{X}} \right\|^2 = \mathrm{tr}\{(\mathbf{X} - \hat{\mathbf{X}})(\mathbf{X} - \hat{\mathbf{X}})'\} \tag{2.4}$$

is minimized for all matrices $\hat{\mathbf{X}}_{[r]}$ of rank not larger than r.

It turns out to be convenient to express these results in terms of what we term $\mathbf{J}$-notation. Here the $p \times p$ matrix $\mathbf{J}$ is defined by

$$\underset{p \times p}{\mathbf{J}} = \begin{bmatrix} \mathbf{I}_r & \mathbf{0} : r \times (p-r) \\ \mathbf{0} : (p-r) \times r & \mathbf{0} : (p-r) \times (p-r) \end{bmatrix}. \tag{2.5}$$

Note that $\mathbf{J}^2 = \mathbf{J}$ and $(\mathbf{I} - \mathbf{J})^2 = \mathbf{I} - \mathbf{J}$ and recall that diagonal matrices commute. With this notation we can write the above as

$$\hat{\mathbf{X}}_{[r]} = \mathbf{U\Sigma JV}' = \mathbf{UJ\Sigma V}' = \mathbf{UJ\Sigma JV}'.$$

Of course, the final $p - r$ columns of $\mathbf{UJ}$ and $\mathbf{VJ}$ vanish but the matrices $\mathbf{UJ}$ and $\mathbf{VJ}$ remain $p \times p$. In some instances, it is more convenient to use the notation $\mathbf{U}_r$ and $\mathbf{V}_r$ to denote the first r columns of $\mathbf{U}$ and $\mathbf{V}$, respectively.

In the biplot, we want to represent the approximated rows and columns of our data matrix $\mathbf{X}$, that is, we want to represent the rows and columns of $\hat{\mathbf{X}}_{[r]}$. A standard result is that the orthogonal projections of all the rows of $\mathbf{X}$ onto the two dimensions $\mathbf{v}_1$ and $\mathbf{v}_2$, given by the first two columns of $\mathbf{V}$, are given by the rows of

$$\mathbf{X}\mathbf{V}_2\mathbf{V}_2'. \tag{2.6}$$

The projections (2.6) are points expressed in terms of the coordinates of the original p dimensions. When they are referred to the coordinates of the orthogonal vectors $\mathbf{v}_1$ and $\mathbf{v}_2$ they become

$$\mathbf{X}\mathbf{V}_2. \tag{2.7}$$

We can now construct a scatterplot of the two-dimensional approximation of $\mathbf{X}$ by plotting the samples as the rows of (2.7) as is shown in Figure 2.3. The R code for obtaining these scatterplots is as follows:

```
> aircraft.mat.centered <- scale(aircraft.mat, center = TRUE,
     scale = FALSE)
> svd.X.centered <- svd(aircraft.mat.centered)
> x <- aircraft.mat.centered %*% svd.X.centered$v[,1]
> y <- aircraft.mat.centered %*% svd.X.centered$v[,2]
> plot(x = x, y = y, xlim = c(-6,4), ylim = c(-2,2), pch = 15,
  col = "green", cex = 1.2, xlab = "V1", ylab = "V2",
  frame.plot = FALSE)
> text(x = x, y = y, label = dimnames(aircraft.mat)[[1]],
  pos = 1)
> windows()
> PCAbipl(cbind(x,y), colours = c("green",rep("black",8)),
  pch.samples = 15, exp.factor = 14, n.int = c(5,3),
  offset = c(0, 0, 0.5, 0.5), pos.m = c(1,4),
  offset.m = c(-0.25, -0.25))
```

The scatterplot in the bottom panel of Figure 2.3 is similar to that appearing in the top panel except for the translation of the ordination axes to pass through the origin and for the aspect ratio of unity. The effect of the difference in aspect ratios is clear. The R function `PCAbipl` is discussed in detail in Chapter 3.

Figure 2.3 is not yet a biplot because only the rows of $\mathbf{X}$ have a representation, and no representation of the columns (variables) is given. Chapter 3 gives the detailed algebraic and geometrical justifications of how to provide for the variables. Here, the following outline suffices, writing $\mathbf{X} = \mathbf{AB}$, then each element of $\mathbf{X}$ is given by $x_{ij} = \mathbf{a}_i'\mathbf{b}_j$, the inner product of a row marker (rows of $\mathbf{A}$) and a column marker (columns of $\mathbf{B}$). From (2.3) we have $\mathbf{X} = \mathbf{U\Sigma V}'$, which implies that $\mathbf{XV} = \mathbf{U\Sigma}$. Since (2.7) approximates the row markers, we set $\mathbf{A} = \mathbf{U\Sigma}$ and it follows that $\mathbf{B} = \mathbf{V}'$. Therefore the columns of $\mathbf{X}$ are approximated by the first two rows of $\mathbf{V}$.

An r-dimensional approximation of $\mathbf{X}$ is shown in Figure 2.4 for $r = 2$. In the top panel the rows are represented by green markers as in Figure 2.3, together with red markers for the columns (the variables). Therefore Figure 2.4 is a two-dimensional *biplot* of $\mathbf{X}$. In the bottom panel the variables are represented by vectors as suggested by

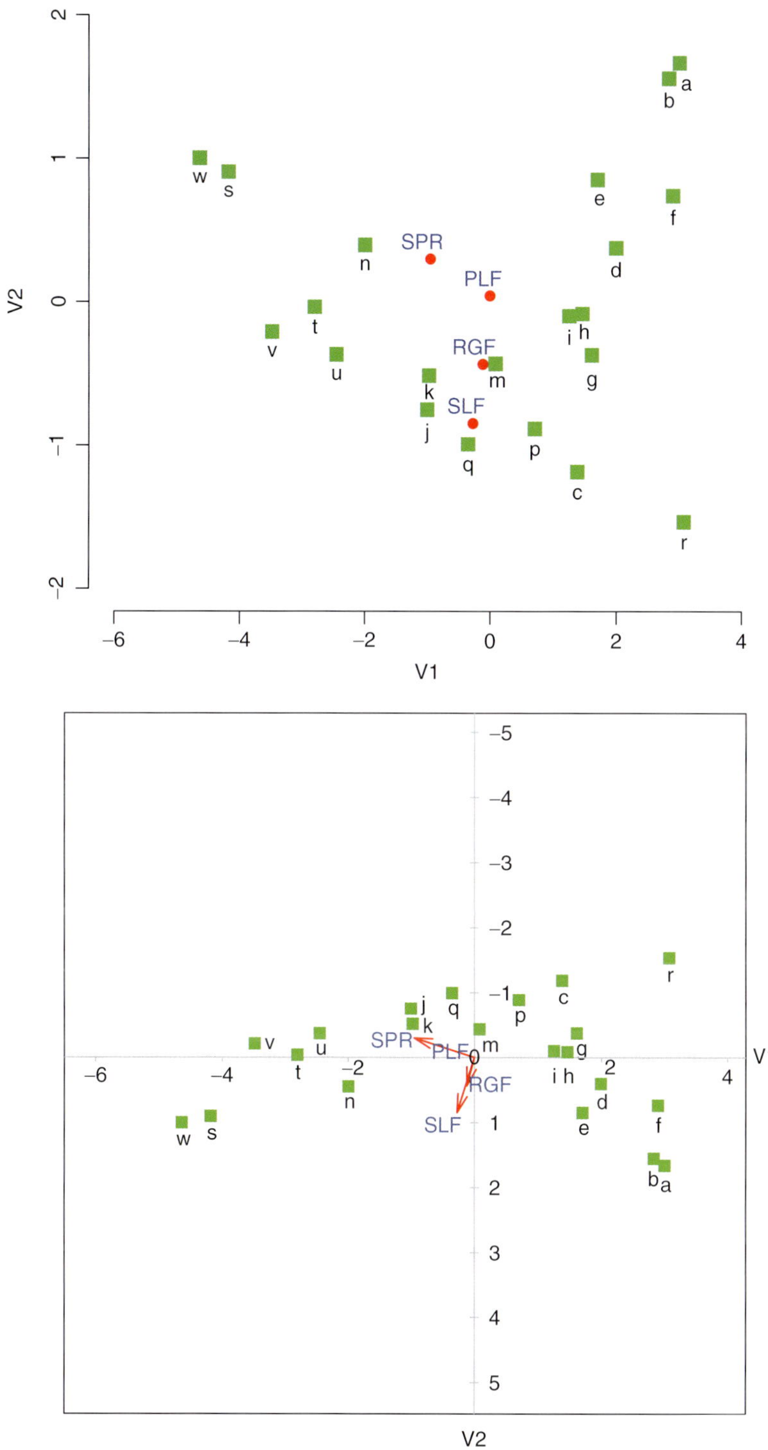

Figure 2.4 The Gabriel form of a biplot that is based upon the SVD, $\hat{\mathbf{X}}_2 = \mathbf{U\Sigma JV}'$.

Gabriel (1971). Figure 2.4 is obtained by adding the following R code to the code given above for Figure 2.3:

```
> plot(x = x, y = y, xlim = c(-6,4), ylim = c(-2,2), pch = 15,
  col = "green", cex = 1.2, xlab = "V1", ylab = "V2",
  frame.plot = FALSE)
> text(x = x, y = y, label = dimnames(aircraft.mat)[[1]], pos = 1)
> text(x = svd.X.centered$v[,1], y = svd.X.centered$v[,2], label
  = dimnames(aircraft.mat)[[2]], pos = 2, offset = 0.4, cex = 0.8)
> windows()
> PCAbipl(cbind(x,y), reflect = "y", colours = c("green",
  rep("black",8)), pch.samples = 15, pch.samples.size = 1.2,
  exp.factor = 1.4, n.int = c(5,3), offset = c(0, 0, 0.5, 0.5),
  pos.m = c(1,4), offset.m = c(-0.25, -0.25), pos = "Hor")
> arrows(0, 0, svd.X.centered$v[-3,1], svd.X.centered$v[-3,2],
  length = 0.15, angle = 15, lwd = 2, col = "red")
> text(x = -svd.X.centered$v[,1], y = svd.X.centered$v[,2],
  label = dimnames(aircraft.mat)[[2]], pos = 2, offset = 0.075,
  cex = 0.8)
```

We note from the approximation $\hat{\mathbf{X}}_{[r]} = \mathbf{UJ\Sigma JV}'$ that $\hat{\mathbf{X}}_{[r]}$ can be written as

$$\begin{aligned}\hat{\mathbf{X}}_{[r]} &= (\mathbf{UJ\Sigma})(\mathbf{VJ})' \\ &= (\mathbf{UJ\Sigma Q})(\mathbf{VJQ})' \\ &= \mathbf{A}_{[\mathrm{r}]}\mathbf{B}_{[\mathrm{r}]}.\end{aligned} \tag{2.8}$$

Since (2.8) is valid for any $p \times p$ orthogonal matrix $\mathbf{Q}$, it follows that the configurations in Figures 2.3 and 2.4 may be subjected to orthogonal rotations and/or reflections about the horizontal or vertical axes without violating the inner product representation above. The same code on different computers can thus result in apparently different representations, but one is just an orthogonal rotation and/or reflection of the other.

What are the practical implications of the biplot representation (2.8)? Instead of answering this question immediately we turn to our standpoint of understanding a biplot as an extension of an ordinary scatterplot. Although Figure 2.4 is a biplot, there are no calibrated axes representing the variables as in Figure 2.1. Therefore, in the next section we address the problem of converting the markers or arrows representing the variables in Figure 2.4 into calibrated axes analogous to ordinary scatterplots.

2.3 Calibrated biplot axes

We have seen in Section 2.2 that the biplot of Figure 2.4 uses an inner product representation. This inner product interpretation can be described as follows. The biplot axes are shown as vectors $\mathbf{v}_k$ whose end-points V_k have coordinates given by the first two elements of the kth row of $\mathbf{V}$. Then, the value $\hat{x}_{ik}$ associated with a point P_i and a vector $\mathbf{v}_k$ is the product of the lengths OP_i and OV_k and the cosine of the angle θ subtended at the origin. The matrix $\hat{\mathbf{X}}$ gives all np inner product values. Although a unit aspect ratio is essential (see Section 2.3.1), we have seen in (2.8) that it is legitimate to rotate and reflect diagrams based on inner products. Thus, at first glance, biplot representations of the same data matrix may seem to differ, but one is merely a rotation or reflection of the other: essentially the inter-sample distances and the projections of the samples onto the axes remain unchanged.

This inner product calculation is not easy to visualize except when comparing the relative values of two points P_i and P_j on the same variable V_k. Then, one only has to compare the lengths of the projections of P_i and P_j onto OV_k. This process does not work when comparing across variables $\boldsymbol{h}$ and $\boldsymbol{k}$, because then one has to take into account the different lengths of OV_k and OV_h.

All points P that project onto the same point on OV_k will have the same inner product. It follows that we may label that point with an appropriate unique value. This is the basis for the recommendation of Gower and Hand (1996) that the biplot axes be calibrated like ordinary coordinate axes. Figure 2.5 shows Figure 2.4 (reflected about the horizontal scaffolding axis) augmented in this manner. The four variables are now represented by four nonorthogonal axes, known as biplot axes, which extend throughout the diagram and are concurrent at, but not rooted in, the origin. The principal axes are of no further interest so have been removed. The biplot axes are used in precisely the same way as the Cartesian axes they approximate. That is, when a point representing an aircraft is projected orthogonally onto an axis, one may read off the value of the corresponding variable. This process will give approximate values that do not in general agree precisely with those given in Table 1.1 but reproduce the entries in the matrix $\hat{\mathbf{X}}_{[r]}$.

Figure 2.5 can be reproduced using the following function call (see Chapter 3 for a detailed discussion of the function `PCAbipl`):

```
PCAbipl(aircraft.data[,-1], colours = "green", pch.samples = 15,
   pch.samples.size = 1.2, n.int = c(5,3,5,3), reflect = "x",
   offset = c(1.2, 1.2, 0.3, 0), side.label = c(rep("right",3),
   "left"), pos.m = c(1,4,4,1), offset.m = rep(-0.15, 4))
```

In Figure 2.5, the scale markers are in the units of the variables of Table 1.1. Thus the biplot allows one to draw a scatter diagram and relate samples (here aircraft) to the values of associated variables. It gives a visualization of Table 1.1 that can be inspected for any interesting features. The salient feature of Figure 2.5 is the way that most of the aircraft are regularly placed from $\boldsymbol{a}$ to $\boldsymbol{w}$. Table 1.1 lists the aircraft in the temporal order of their development, and the ordering reflects increasing flight range coupled with increasing payloads. In this respect $\boldsymbol{r}$, the F-8A, is in an anomalous position because its specific power is very low, even lower than those of much earlier aircraft.

It should be apparent that this figure has all the characteristics of more familiar scatterplots:

- points, representing the 21 samples;
- labelled axes;
- calibrated axes.

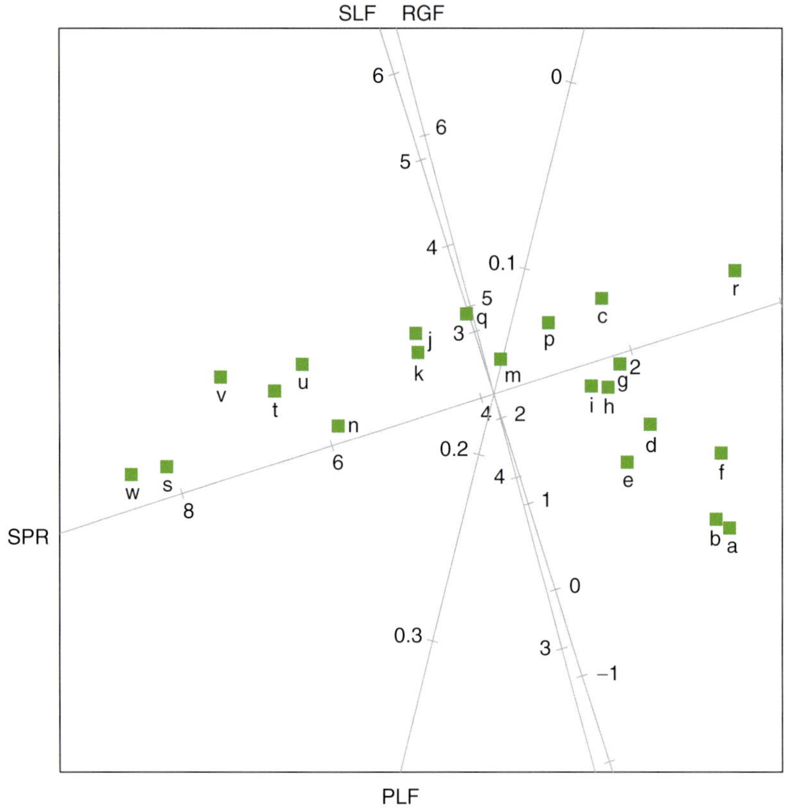

Figure 2.5 A two-dimensional biplot approximation of the aircraft data of Table 1.1 according to the Gower and Hand (1996) representation. Note the aspect ratio of unity.

Care has been taken with the construction of Figure 2.5 that the aspect ratio is equal to unity. This is not shown explicitly, but the square form of this figure (and others) is intended as an indication.

The main difference between the biplot in Figure 2.5 and an ordinary scatterplot is that there are more axes than dimensions and that the axes are not orthogonal. Indeed, it would not be possible to show four sets of mutually orthogonal axes in two dimensions. There is a corresponding exact figure in four dimensions and the biplot is an approximation to it. This biplot is read in the usual way by projecting from a sample point onto an axis and reading off the nearest marker, using a little visual interpolation if desired. If the approximation is good, the predictions too will be good.

Having shown a biplot with calibrated axes representing the original variables we now give details on how to calculate these calibrations: whenever a diagram depends on an inner product interpretation, the process of calibrating axes may be generalized as we now show.

Calibrated axes are used throughout this book for a variety of biplots associated with numerical variables. We point out that a simple methodology is common to all applications based on the use of an inner product **AB** where

$$\mathbf{A}: p \times 2 = \begin{bmatrix} \mathbf{a}_1' \\ \mathbf{a}_2' \\ \vdots \\ \mathbf{a}_p' \end{bmatrix} \quad \text{and} \quad \mathbf{B}': q \times 2 = \begin{bmatrix} \mathbf{b}_1' \\ \mathbf{b}_2' \\ \vdots \\ \mathbf{b}_q' \end{bmatrix}. \tag{2.9}$$

Thus, we may plot the rows of **A** as the coordinates of a set of points and the rows of $\mathbf{B}'$ give the directions of axes to be calibrated. Figure 2.6 shows the ith point $\mathbf{a}_i$ and the kth axis defined by $\mathbf{b}_k$.

The inner product $\mathbf{a}_i'\mathbf{b}_k$ is constant (μ, say) for all points on the line projecting $\mathbf{a}_i$ onto $\mathbf{b}_k$. Therefore, the point of projection may be calibrated by labelling this point with the value μ. This constant applies to the point of projection itself, $\lambda\mathbf{b}_k$. It follows that, for the point $\lambda\mathbf{b}_k$ to be calibrated μ, it must satisfy the inner product:

$$\lambda\mathbf{b}_k'\mathbf{b}_k = \mu \tag{2.10}$$

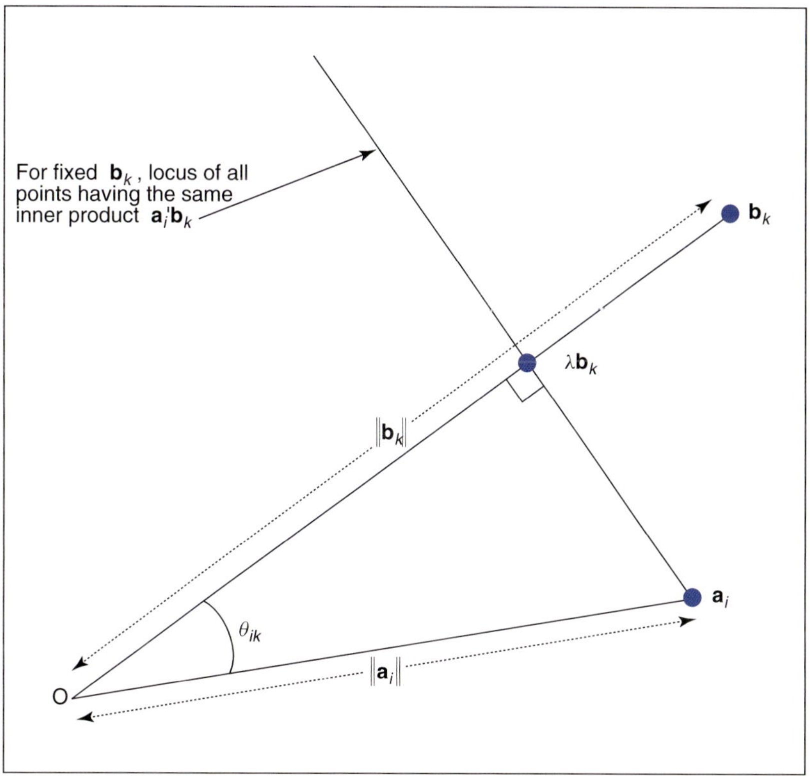

Figure 2.6 The projection of $\mathbf{a}_i$ onto $\mathbf{b}_k$ is $\lambda\mathbf{b}_k$. The inner product has the value $\mu = \|\mathbf{a}_i\|.\|\mathbf{b}_k\|.\cos\theta_{ik}$ which is constant for all points on the line of projection. The point $\lambda\mathbf{b}_k$ may be given the calibration marker μ.

so that $\lambda = \mu/\mathbf{b}_k'\mathbf{b}_k$ and $\mu\mathbf{b}_k/(\mathbf{b}_k'\mathbf{b}_k)$ gives the coordinates of the point on the $\mathbf{b}_k$-axis that is calibrated with a value of μ. Normally, μ will be set to values 1, 2, 3, ..., or other convenient steps for the calibration, to give the values required by the inner products.

Often, the inner product being approximated gives transformed values of some original variables $\mathbf{a}_i'\mathbf{b}_k = f(x_{ik})$ and one wants to calibrate in the original units of measurement. Suppose α represents a value to be calibrated in the original units; then we must set $\mu = f(\alpha)$, where the function will vary with different methods. For example, in PCA the data are centred, in correspondence analysis (CA) the original counts are replaced by row and/or column scaled deviations from an independence model, in metric scaling dissimilarities are defined by a variety of coefficients that are functions of the original variables, and in nonmetric scaling by monotonic transformations defined in terms of smooth spline functions or merely by step-functions. Another possibility is where the calibration steps are kept equal in the transformed units but labelled with the untransformed values; this is especially common with logarithmic transformations. Calibrated axes may be constructed for all such methods.

2.3.1 Lambda scaling

When plotting points given by the rows of $\mathbf{A}$ and $\mathbf{B}'$ one set will often be seen to have much greater dispersion than the other (see, for example, Figure 2.4 where the dispersion of the sample points overshadows that of the points representing the variables). This can be remedied as follows. First observe that

$$\mathbf{AB} = (\lambda\mathbf{A})(\mathbf{B}/\lambda), \tag{2.11}$$

so that the inner product is unchanged when $\mathbf{A}$ is scaled by λ, provided that $\mathbf{B}$ is inversely scaled. This simple fact may be used to choose λ in some optimal way to improve the look of the display. One way of choosing λ is to arrange that the average squared distance of the points in $\lambda\mathbf{A}$ and $\mathbf{B}/\lambda$ is the same. If $\mathbf{A}$ has p rows and $\mathbf{B}'$ has q rows and both are centred, this requires

$$\lambda^2\|\mathbf{A}\|^2\Big/p = \lambda^{-2}\|\mathbf{B}\|^2\Big/q. \tag{2.12}$$

giving the required scaling

$$\lambda^4 = \tfrac{p}{q}\|\mathbf{B}\|^2\Big/\|\mathbf{A}\|^2. \tag{2.13}$$

We term the above method lambda scaling.

Lambda scaling is not the only criterion available; one might prefer to work in terms of distances rather than squared distances or work in terms of maximum distances. Indeed, the inner product is invariant for quite general transformations $\mathbf{AB} = (\mathbf{AT}')(\mathbf{B}'\mathbf{T}^{-1})'$ but such general transformations are liable to induce conflicts such as changing Euclidean and centroid properties. However, whenever the inner product is maintained everything written above about the calibration of axes remains valid.

Lambda scaling has only a trivial proportionate effect on distances, but it is important to be aware that general scaling affects distance severely; this is especially relevant in PCA, canonical variate analysis (CVA), some forms of CA that approximate Pythagorean distance, Mahalanobis distance and chi-squared distance.

We illustrate the above procedure for calibrating a biplot axis with and without lambda scaling using the first four columns of the reaction-kinetic data set available as `ReactionKinetic.data`. For reference purposes we give this data set in Table 2.1. The following code shows how to implement the calibration procedure to equip a biplot with calibrated axes. Figure 2.7 shows the sample point ***11*** and biplot axis for variable ***y***.

```
function (X = ReactionKinetic.data[,1:4], add = c(2,2),
          shift = 0, lambda = 1, n.int = 5)
{ options(pty = "s")
  par(mar = c(3,3,3,3))
  # obtain biplot scaffolding
  X.svd <- svd(X)   # X not scaled or centered
  col.means <- apply(X,2,mean)
  sds  <- apply(X,2,function(x) sqrt(var(x)))
  UDelta <- X.svd$u %*% diag(X.svd$d)[,1:2]*lambda
  V <- X.svd$v[,1:2]/lambda
  # setup of plotting region
  limx <- c(min(V[,1], UDelta[,1]), max(V[,1], UDelta[,1])) + shift
  limy <- c(min(V[,2], UDelta[,2]), max(V[,2], UDelta[,2]))
  plot(rbind(UDelta,V), asp = 1, xlim = limx*1.1, ylim = limy*1.1,
   type = "n", xlab = "", ylab = "", xaxt = "n", yaxt = "n",
   xaxs = "i", yaxs = "i", main = "")
  points(x = 0, y = 0, pch = "O", cex = 2, col = "green")
  # plot row co-ordinate
  points(x = UDelta[11,1], y = UDelta[11,2], pch = 15)
  text(x = UDelta[11,1], y = UDelta[11,2], text = "11", pos = 1,
   cex = 0.70)
  # plot axis for column 'y'
  eq.line <- Draw.line2(x = c(0, V[1,1]), y = c(0,V[1,2]))
  # plot column marker
  arrows(x0 = 0, y0 = 0, x1 = V[1,1], y1 = V[1,2], col = "red",
   lwd = 2, length = 0.15)
  # plot row marker
  arrows(x0 = 0, y0 = 0, x1 = UDelta[11,1], y1 = UDelta[11,2],
   col = "red", lty = 2, lwd = 1.85, length = 0.1)
  # obtain 'nice' markers
  markers.x <- pretty(range(X[,1]), n = n.int)
  # ensure 0 towards middle of figure
  if (add[2]>0)
    for(i in 1:add[2])
      markers.x <- c(markers.x, markers.x[length(markers.x)]+
        (markers.x[length(markers.x)] -
       markers.x[length(markers.x)-1]))
  if (add[1]>0)
    for(i in 1:add[1])
      markers.x <- c(markers.x[1] - (markers.x[2] - markers.x[1]),
       markers.x)
  markers.v <- markers.x
  # apply eq (2.10)
  calibrations.x <- (markers.v / sum(V[1,1:2]^2)) * V[1,1]
  calibrations.y <- (markers.v / sum(V[1,1:2]^2)) * V[1,2]
```

```
  points(x = calibrations.x, y = calibrations.y, pch = 16,
   cex = 0.65)
  text(x = calibrations.x, y = calibrations.y, text = markers.x,
   pos = 4, cex = 0.70)
  # calculate 1sd marker
  marker.sd.v <-  sds[1]
  calibrations.sd.x <- (marker.sd.v / sum(V[1,1:2]^2)) * V[1,1]
  calibrations.sd.y <- (marker.sd.v / sum(V[1,1:2]^2)) * V[1,2]
  points(x = calibrations.sd.x, y = calibrations.sd.y, pch = 16,
   cex = 0.65)
  text(x = calibrations.sd.x, y = calibrations.sd.y, text = "1sd",
   pos = 4, cex = 0.70, col = "green")
  # orthogonal to projection of point 11 on axis y
  abline(a = UDelta[11,2] + 1/eq.line$gradient*UDelta[11,1],
   b = -1/eq.line$gradient, col = "blue", lty = 2)
}
```

Table 2.1 The first four columns of the reaction-kinetic data set. Variable ***y*** is a response variable; variables ***x1***, ***x2*** and ***x3*** are explanatory variables.

ID	y	x1	x2	x3
1	0.0000	0.0755	0.0758	0.0000
2	0.0430	0.0701	0.0672	0.0042
3	0.0687	0.0638	0.0552	0.0096
4	0.0756	0.0594	0.0463	0.0136
5	0.0789	0.0580	0.0430	0.0150
6	0.0795	0.0570	0.0408	0.0159
7	0.0794	0.0553	0.0369	0.0177
8	0.0761	0.0541	0.0332	0.0191
9	0.0679	0.0532	0.0310	0.0201
10	0.0585	0.0528	0.0296	0.0207
11	0.0490	0.0517	0.0278	0.0219
12	0.0369	0.0516	0.0274	0.0220
13	0.0247	0.0513	0.0272	0.0222
14	0.0150	0.0516	0.0271	0.0221
15	0.0134	0.0515	0.0272	0.0221
16	0.0124	0.0511	0.0270	0.0224
17	0.0120	0.0513	0.0271	0.0222
18	0.0105	0.0513	0.0271	0.0223
19	0.0090	0.0515	0.0271	0.0222
20	0.0074	0.0517	0.0273	0.0220
21	0.0060	0.0512	0.0271	0.0223
22	0.0045	0.0514	0.0270	0.0222
23	0.0030	0.0513	0.0271	0.0223
24	0.0015	0.0515	0.0271	0.0222

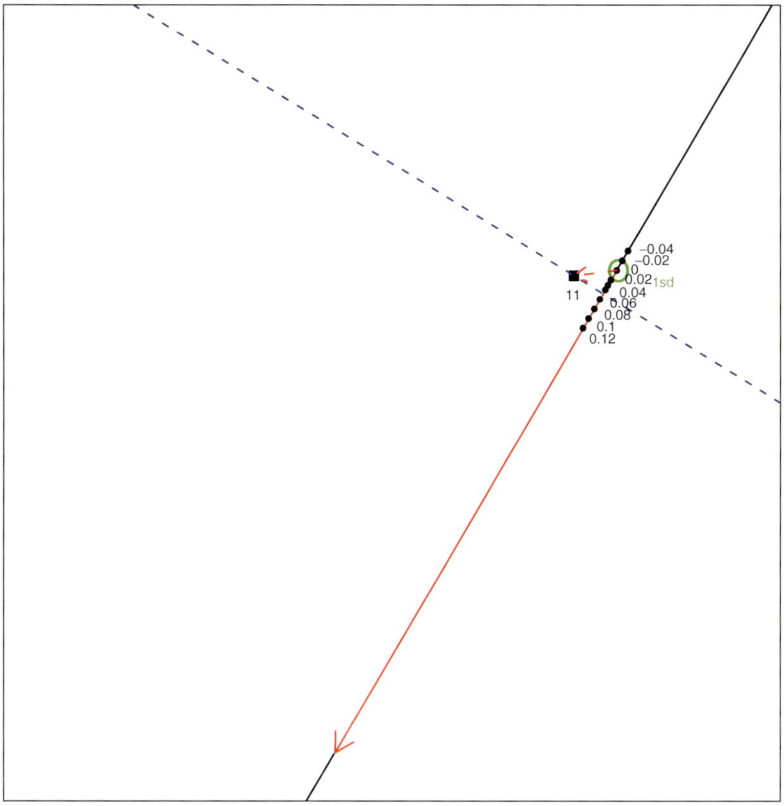

Figure 2.7 Calibrating a biplot axis. Shown are the sample point ***11*** and axis ***y*** for the data in Table 2.1. The origin is indicated by the green circle.

Using the above code with arguments `lambda = 3` and `shift = 0.1` results in the much better scaling shown in Figure 2.8. Although precisely the same y-value for ***11*** is read from the biplot axes in Figures 2.7 and 2.8, it is much easier to see in Figure 2.8. Note that the approximated y-value for ***11*** is very close to the actual value of 0.0490.

Now consider normalizing the data matrix to unit variances and zero means: let

$$\mathbf{X}_{\text{Norm}} : 24 \times 4 = \mathbf{U\Sigma V}' \approx \underset{24\times 2}{(\mathbf{U\Sigma})_2}\underset{2\times 4}{(\mathbf{V}_2)'} = \underset{24\times 2}{(\mathbf{U\Sigma}^{1/2})_2}\Big(\underset{2\times 4}{(\mathbf{V\Sigma}^{1/2})_2}\Big)'. \tag{2.14}$$

The right-hand side of (2.14) points to another form of scaling which we shall call *sigma scaling*. Sigma scaling refers to how the diagonal matrix $\mathbf{\Sigma}$ is divided between the coordinates of the row points of $\mathbf{X}$ and the coordinates for plotting its columns.

We illustrate the calibration procedure by displaying ***11*** and axis ***y*** calibrated in the original units although it is now not $\mathbf{X}$ that is approximated but the matrix $\mathbf{X}_{\text{Norm}}$. We first plot $(\mathbf{U\Sigma})_2 : 24 \times 2$ versus $\mathbf{V}_2 : 4 \times 2$ in Figure 2.9 and then $(\mathbf{U\Sigma}^{1/2})_2 : 24 \times 2$ versus $(\mathbf{V\Sigma}^{1/2})_2 : 4 \times 2$ in Figure 2.10. Due to all variables in $\mathbf{X}_{\text{Norm}}$ having zero means the origin in both Figures 2.9 and 2.10 approximates the mean. The unscaled mean for variable ***y*** is 0.0347. The y-value of ***11*** in $\mathbf{X}_{\text{Norm}}$ is 0.4676. It is this value that

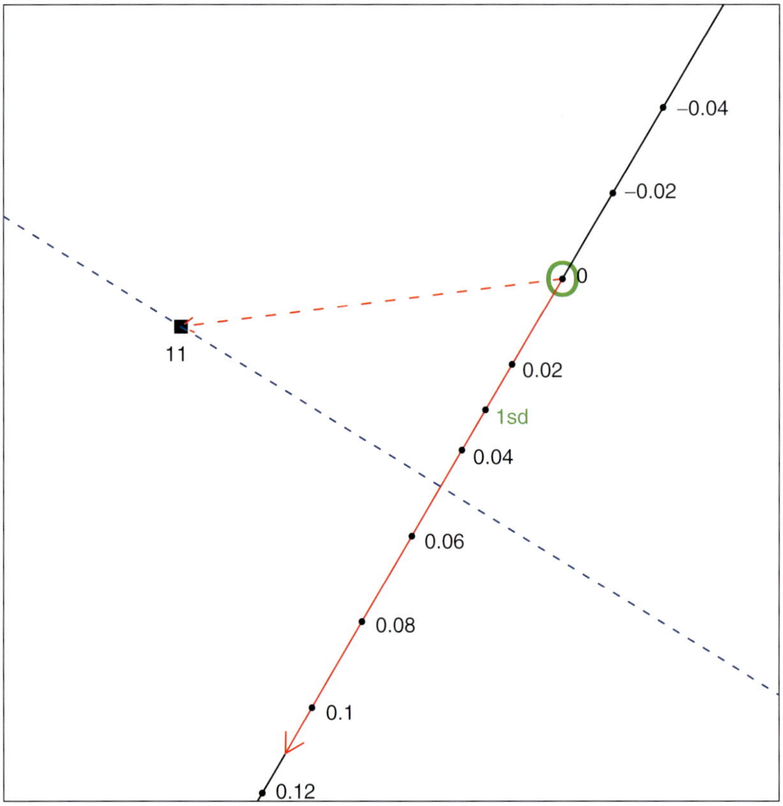

Figure 2.8 Calibrating a biplot axis. Similar to Figure 2.7 but with `lambda = 3` and `shift = 0.1`. The origin is indicated by the green circle. The calibration marker '1sd' approximates a distance of one standard deviation from the origin. The actual standard deviation is 0.0305.

is approximated by the intersection of the blue dotted line and the red arrow, but the calibrations are transformed into the original scale using the calibration procedure. The point '1sd' approximates the transformed mean of zero plus the transformed standard deviation of unity, but the calibration is in terms of the original mean of 0.0347 plus the original standard deviation of 0.0305. Since $\mathbf{V}_2\mathbf{V}_2'$ approximates $\mathbf{VV}' = \mathbf{I}$, which is the covariance matrix of the scaled data, the tip of the red arrow coincides in Figure 2.9 with the position of the point '1sd', but this is not so in Figure 2.10.

The calibration of the biplot axes in Figure 2.9 may seem a trivial operation, but let us take a closer look at the principles involved. Useful scale values should be in terms of the original data but the biplot scaffolding is in terms of the normalized data matrix. 'Nice' scale values of the first column of our original data matrix are thus needed and not those of the first column of $\mathbf{X}_{\text{Norm}}$. The values in the first column of $\mathbf{X}_{\text{Norm}}$ range over the interval $[-3.0304;3.2016]$. The R function `pretty` allows us to obtain the required nice values 0.00, 0.02, 0.04, 0.06, 0.08 easily using the instruction

```
> markers.x <- pretty(range(X[,1]), n = 5)
```

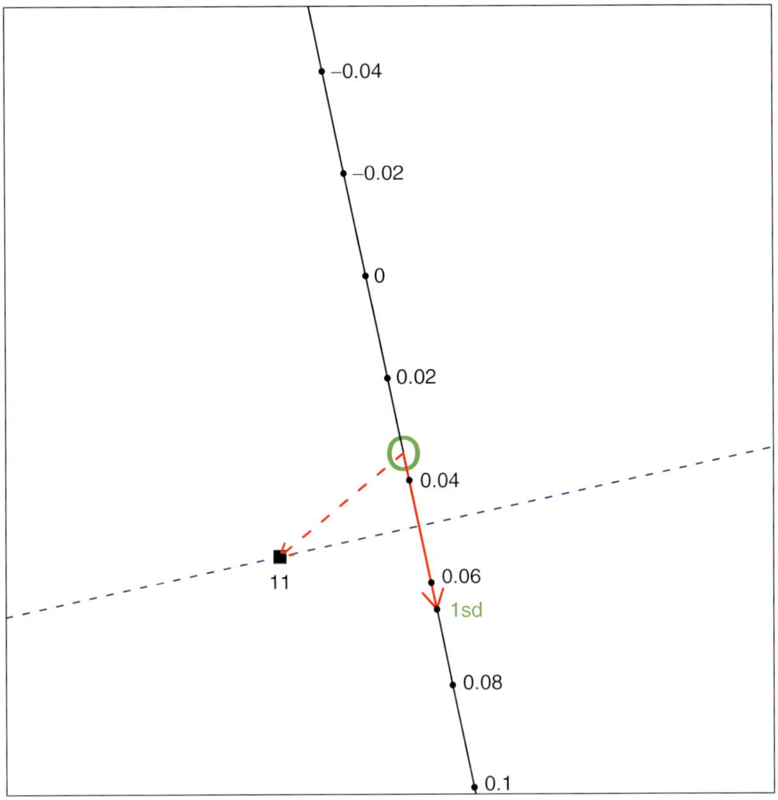

Figure 2.9 Calibrating a biplot axis in the case of normalized data. Biplot drawn by plotting $(\mathbf{U\Sigma})_2$ versus $\mathbf{V}_2$.

We now need the scaffolding coordinates of points on the biplot axis to be labelled with the above 'nice' values. Therefore we have to express `markers.x` in terms of the scaffolding using the transformation

```
> markers.v <- (markers.x - mean(X[,1]))/sqrt(var(X[,1]))
```

Next, (2.10) is applied to the transformed markers, `markers.v`, to obtain the required coordinates:

```
> coords.horizontal <- (markers.v / sum(V[1,1:2]^2)) * V[1,1]
> coords.vertical <- (markers.v / sum(V[1,1:2]^2)) * V[1,2]
```

The two vectors `coords.horizontal` and `coords.vertical` provide the coordinates of the markers on the biplot axis to be labelled with the 'nice' values 0.00, 0.02, 0.04, 0.06, 0.08.

As an exercise, the reader can construct a biplot axis similar to the one in Figure 2.9 but equipped with a scale in terms of $5 + \log(\mathbf{X}_{\text{Norm}})$. *Hints*: Firstly, calculate nice scale values using

```
> markers.x <- pretty(range(5+log(scale(X[,1]))), n = 5)
```

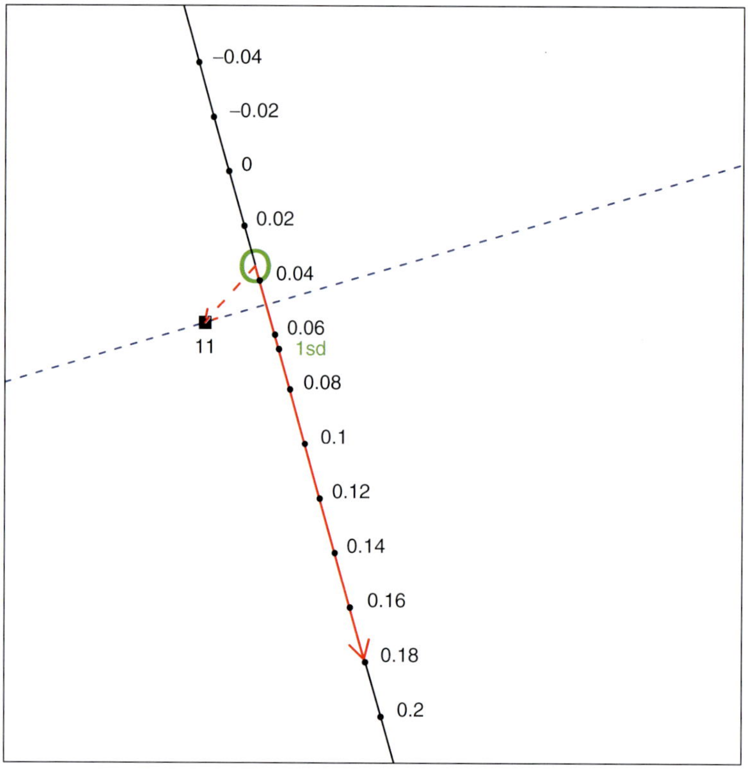

Figure 2.10 Calibrating a biplot axis in the case of normalized data. Biplot drawn by plotting $(\mathbf{U\Sigma}^{1/2})_2$ versus $(\mathbf{V\Sigma}^{1/2})_2$.

Secondly, obtain the corresponding representations on the biplot scaffolding using the inverse relationship

```
> markers.v <- exp(markers.x - 5)
```

Now apply (2.10) to find the coordinates of `markers.x` in terms of the biplot scaffolding:

```
> coords.horizontal <- (markers.v / sum(V[1,1:2]^2)) * V[1,1]
> coords.vertical <- (markers.v / sum(V[1,1:2]^2)) * V[1,2]
```

Plot `coords.vertical` versus `coords.horizontal` and label with `markers.x`.

The R code for constructing Figures 2.9 and 2.10 is as follows:

```
function (X = ReactionKinetic.data[,1:4], add = c(2,2),
          exp.factor = 0.8, shift = -2, lambda = 1, n.int = 5)
{ options(pty = "s")
  par(mar = c(3,3,3,3))
  # obtain biplot scaffolding
  X.scaled.svd <- svd(scale(X, scale = TRUE, center = TRUE))
  col.means <- apply(X,2,mean)
  sds  <- apply(X,2,function(x) sqrt(var(x)))
  USigma <- X.scaled.svd$u %*% diag(X.scaled.svd$d)[,1:2]*lambda
```

```
  V <- X.scaled.svd$v[,1:2]/lambda
  # setup of plotting region
  limx <- c(min(V[,1], USigma[,1]), max(V[,1], USigma[,1])) + shift
  limy <- c(min(V[,2], USigma[,2]), max(V[,2], USigma[,2]))
  plot(rbind(USigma, V), asp = 1, xlim = limx*exp.factor, ylim =
   limy*exp.factor, type = "n", xlab = "", ylab = "", xaxt = "n",
   yaxt = "n",  xaxs = "i", yaxs = "i", main = "")
  points(x = 0, y = 0, pch = "O", cex = 2, col = "green")
  # plot row co-ordinate
  points(x = USigma[11,1], y = USigma[11,2], pch = 15)
  text(x = USigma[11,1], y = USigma[11,2], label = "11", pos = 1,
   cex = 0.70)
  # plot axis for column 'y'
  eq.line <- Draw.line2(x = c(0,V[1,1]), y = c(0,V[1,2]))
  # plot column marker
  arrows(x0 = 0, y0 = 0, x1 = V[1,1], y1 = V[1,2], col = "red",
   lwd = 2, length = 0.15)
  # plot row marker
  arrows(x0 = 0, y0 = 0, x1 = USigma[11,1], y1 = USigma[11,2],
   col = "red", lty = 2, lwd = 1.85, length = 0.1)
  # obtain 'nice' markers
  markers.x <- pretty(range(X[,1]), n = n.int)
  # ensure 0 toward middle of figure
  if(add[2]>0)
    for(i in 1:add[2])
      markers.x <- c(markers.x, markers.x[length(markers.x)]+
      markers.x[length(markers.x)]-markers.x[length(markers.x)-1])
  if(add[1]>0)
    for(i in 1:add[1])
      markers.x <- c(markers.x[1] - (markers.x[2] - markers.x[1]),
      markers.x)
  markers.v <- (markers.x - col.means[1])/sds[1]
  # apply eq (2.10)
  calibrations.x <- (markers.v / sum(V[1,1:2]^2)) * V[1,1]
  calibrations.y <- (markers.v / sum(V[1,1:2]^2)) * V[1,2]
  points(x = calibrations.x, y = calibrations.y, pch = 16,
   cex = 0.65)
  text(x = calibrations.x, y = calibrations.y, label = markers.x,
   pos = 4, cex = 0.70)
  # calcuate 1sd marker
  marker.sd.x <- sds[1] + col.means[1]
  marker.sd.v <- (marker.sd.x - col.means[1])/sds[1]
  calibrations.sd.x <- (marker.sd.v / sum(V[1,1:2]^2)) * V[1,1]
  calibrations.sd.y <- (marker.sd.v / sum(V[1,1:2]^2)) * V[1,2]
  points(x = calibrations.sd.x, y = calibrations.sd.y, pch = 16,
   cex = 0.65)
  text(x = calibrations.sd.x, y = calibrations.sd.y, label = "1sd",
   pos = 4, cex = 0.70, col = "green")
  # orthogonal to projection of point 11 on axis y
  abline(a = UDelta[11,2] + 1/eq.line$gradient*UDelta[11,1],
   b = -1/eq.line$gradient, col = "blue", lty = 2)
}
```

The above code produces Figure 2.10 by changing `USigma` and `V` to `USigmaHalf` and `VSigmaHalf`, respectively, by setting

```
> USigmaHalf <- X.scaled.svd$u %*% diag(sqrt(X.scaled.svd$d))[,1:2]
  *lambda
> VSigmaHalf <- X.scaled.svd$v %*% diag(sqrt(X.scaled.svd$d))[,1:2]
  /lambda
```

together with the changes `add = c(2,6)`, `exp.factor = 1.2` and `shift = 0.5`. Note that in Figure 2.10 the same predicted value for ***11*** is obtained as in Figure 2.9 but the red arrow is lengthened to extend beyond the one standard deviation marker.

Scaling (lambda and sigma), rotation and axis shifts (Section 2.4) are all devices that can be used to enhance the visual quality of the display. These devices are best used interactively after the initial construction of a biplot, and we have taken advantage of them throughout the book to edit the final biplot.

The above calibration procedure and R code are incorporated in several functions in our library `UBbipl`.

2.4 Refining the biplot display

Figure 2.5 can be improved. In conventional scatterplots there is some freedom, which can be used to advantage, in positioning the axes. This freedom carries over to biplots.

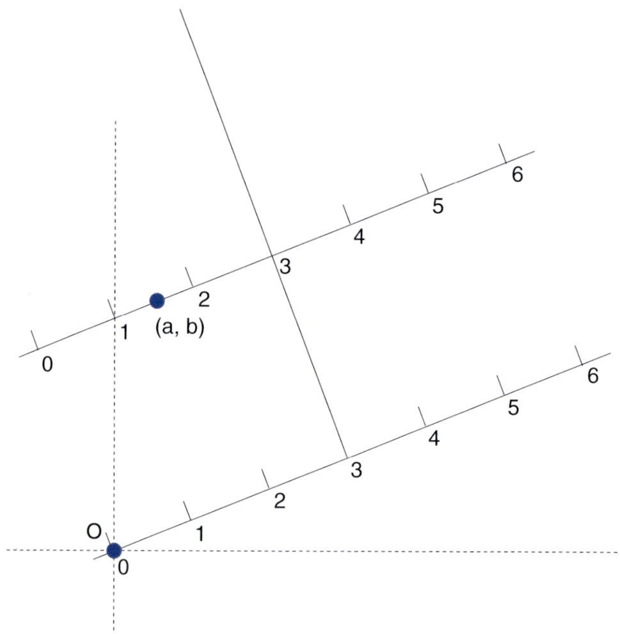

Figure 2.11 Orthogonal parallel translation. The upper axis is scaled in the same way as the lower parallel axis and equal-valued scale markers are on lines orthogonal to both axes. Projection onto either scale delivers the same result. Orthogonal axes through an origin O are shown as dotted lines; these are used solely for plotting purposes and would be deleted in any biplot. A point (a, b) relative to these plotting axes is shown.

We illustrate the principles for linear axes, but they may be readily extended to other cases. Figure 2.5 is plotted, as is usual, in such a way that the centroid of the points is at the origin. This is a natural choice of origin because it is known that the best fit of (2.1) passes through the centroid of all the points. However, it does tend to force the axes, with their scale markers, to intermingle with the points, which does not help legibility. As all we wish to do is to project orthogonally onto each axis, any axis may be moved parallel to itself – provided we ensure that the markers are moved consistently. By 'consistent' we mean that the line joining the same marker on two parallel axes is orthogonal to them both, as is shown in Figure 2.11. We call this process *orthogonal parallel translation*.

The axis may be made to pass through any chosen point (a, b) relative to the plotting axes and the same point (a, b) may be chosen for all axes, thus ensuring concurrency. By choosing judiciously, we may take advantage of this simple fact to separate the axes from the points, as is shown in Figure 2.12, resulting in an improved Figure 2.5. Complete separation is always possible but not always desirable when it induces a remote origin just to accommodate separation of a few points. Furthermore, the plotting area needs to be enlarged to facilitate reading off the values of samples $\boldsymbol{w}$ and $\boldsymbol{s}$.

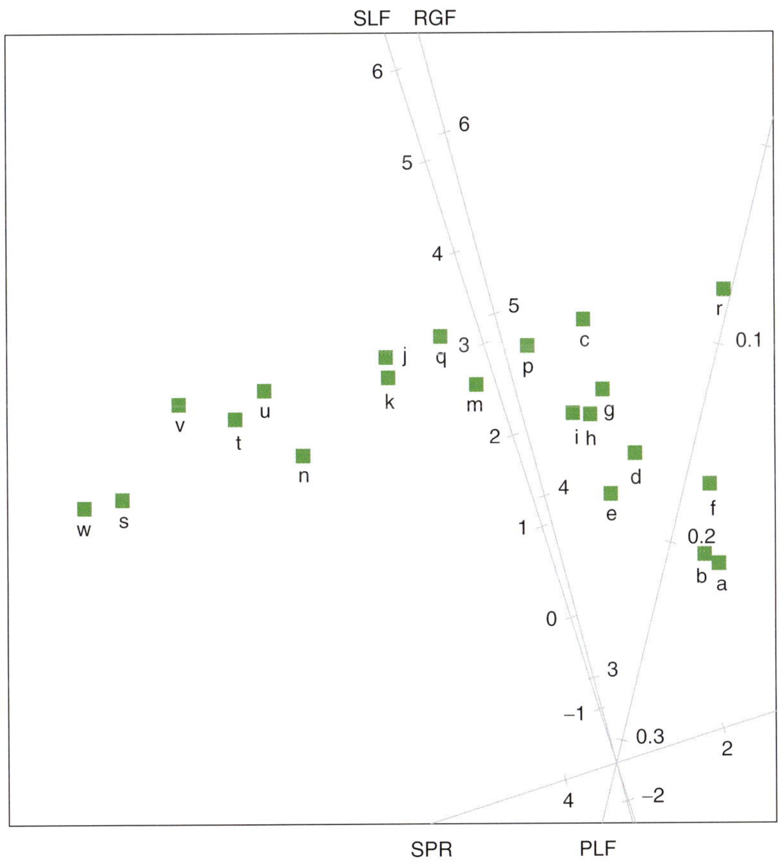

Figure 2.12 Biplot of the aircraft data with orthogonal parallel translation of the axes to separate the axes from the samples.

Orthogonal parallel translation is incorporated in several of the functions available in UBbipl. The two arguments orthog.transx and orthog.transy regulate the amount of orthogonal parallel translation in a horizontal and vertical direction, respectively. The argument reflect allows reflection about the x- and y-axes. Figure 2.12 is constructed by the following call to function PCAbipl.

```
PCAbipl(X = aircraft.data[,2:5], pch.samples = 15,
   pch.samples.size = 1.2, colours = "green", orthog.transx =
   rep(1.75, 4),  orthog.transy = rep(-4,4), rotate.degrees = 180,
   offset = c(1.3, 0.5, 0.3, 0.5), n.int = c(5,3,5,3),
   pos.m = c(1,4,4,4), offset.m = rep(-0.2, 4))
```

With two axes in two dimensions we would normally arrange that the origin corresponds to simple values on both variables. With p axes in $r < p$ dimensions, this cannot be done for all variables, but we may ensure that the new origin corresponds to simple values on r of the axes. This has been done in Figure 2.13, where the origin has been chosen

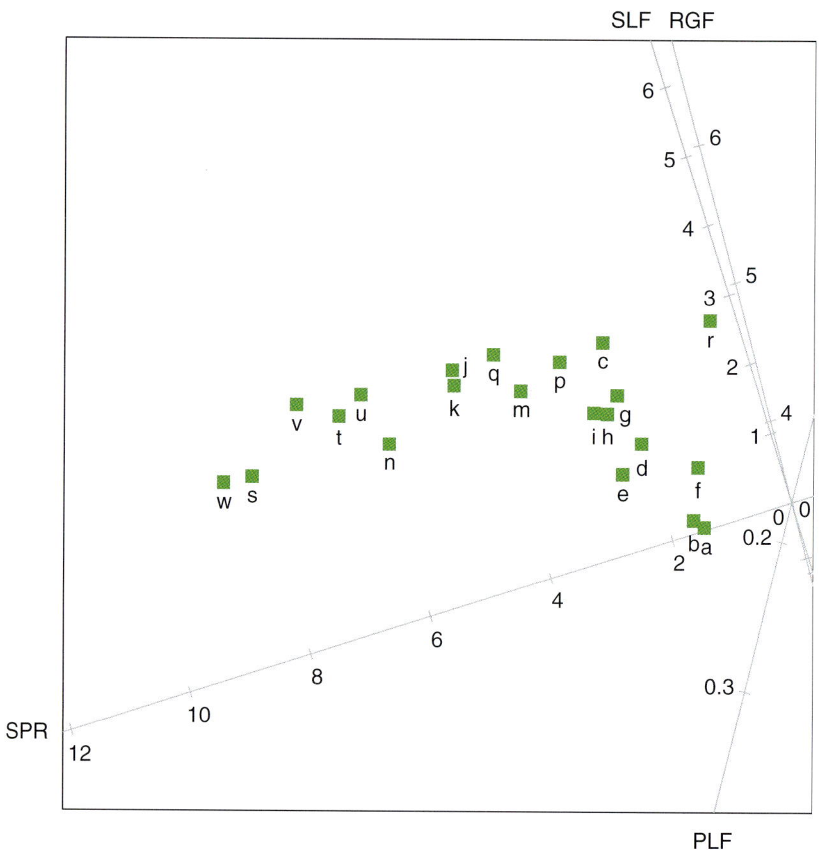

Figure 2.13 The biplot of the aircraft data with orthogonal parallel translation of the axes, where the origin has been chosen as $\boldsymbol{SPR} = 0$ and $\boldsymbol{SLF} = 0$.

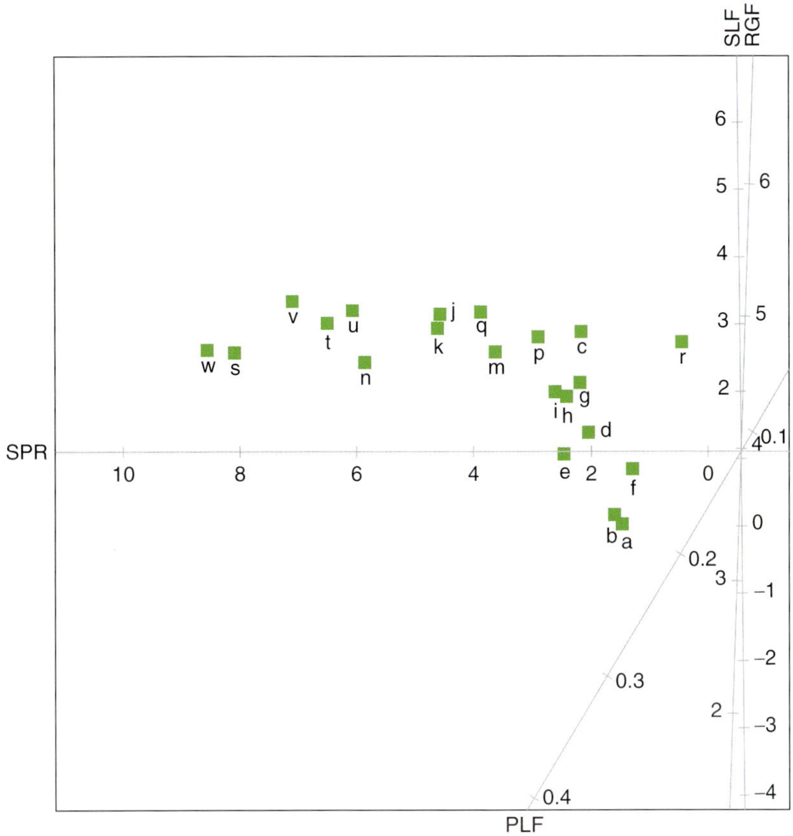

Figure 2.14 The biplot of Figure 2.5 with orthogonal parallel translation rotated such that the ***SPR***-axis is horizontal, similar to the first Cartesian axis of a scatterplot.

to give simple values of the variables ***SPR*** (0) and ***SLF*** (0) by making the following changes in the call to `PCAbipl`:

```
orthog.transx = rep(4.4, 4), orthog.transy = rep(-1.275, 4),
exp.factor = 1.55
```

We may go further, and note that the projections are unaffected if the whole figure is rotated through any angle, as has been done in Figure 2.14 by specifying `rotate.degrees = 162.7` in the function call to `PCAbipl`. We can also reflect the configuration about the horizontal scaffolding axis (by setting argument `reflect = "x"`) or the vertical scaffolding axis (by setting argument `reflect = "y"`). This has been done in Figure 2.15 by specifying

```
orthog.transx = rep(-3.84, 4), orthog.transy = rep(-2, 4),
reflect = "y"
```

so that the ***SPR***-axis attains its maximum on the right-hand side.

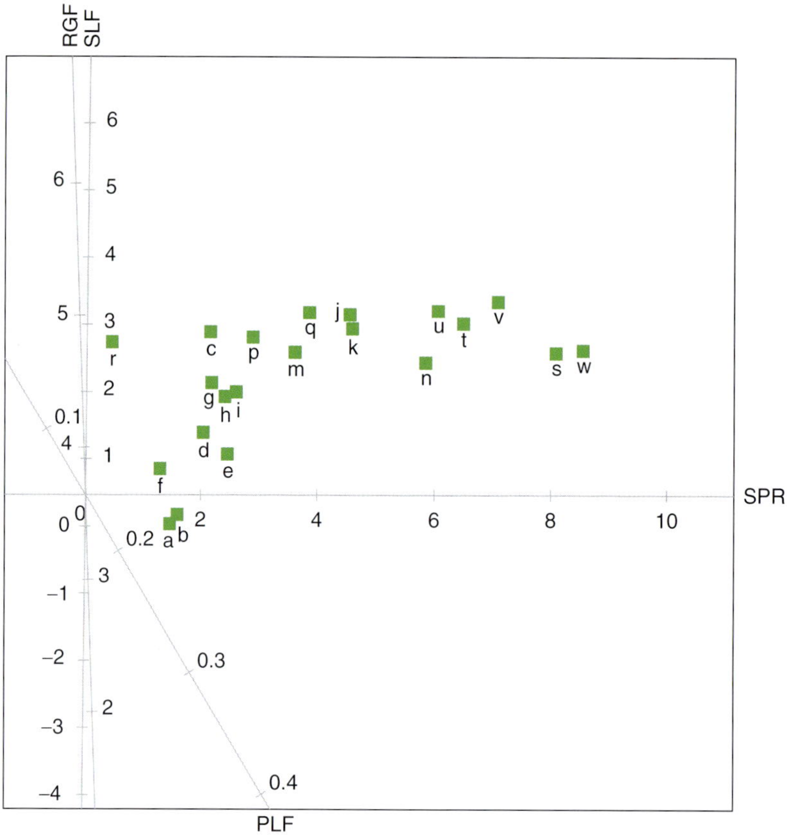

Figure 2.15 The biplot of Figure 2.5 with orthogonal parallel translation rotated and reflected such that the ***SPR***-axis is horizontal, with increasing calibrations from left to right similar to the first Cartesian axis of a scatterplot.

These operations are purely cosmetic but the results reinforce the close relationship that biplots have with conventional Cartesian plots.

2.5 Scaling the data

All the above biplots of the aircraft data have dealt with the raw data of Table 1.1. However, the variables are not on commensurable scales. In particular, ***PLF*** is measured on a much smaller scale than the other variables, a fact that manifests itself in the figures by the insignificant role played by this variable. This lack of commensurability is a serious problem because certain types of biplot are not scale-invariant. In such cases a common practice is to adjust for it by normalizing all variables to have the same sum of squares about their mean, as is done in Table 2.2. The chosen sum of squares is immaterial; typical values are unity, the sample size n or even $n - 1$, as in Table 2.2. We may proceed as with Table 1.1 using the data of Table 2.2 as input to our functions for constructing the biplot and, corresponding to Figure 2.5, we have Figure 2.16. The

Table 2.2 The data of Table 1.1 normalized to have zero means and equal sums of squares.

	Aircraft	SPR	RGF	PLF	SLF
a	FH-1	−1.000	−1.868	−0.016	−2.122
b	FJ-1	−0.942	−1.332	−0.151	−2.122
c	F-86A	−0.704	0.609	0.108	0.606
d	F9F-2	−0.752	0.373	1.213	−1.148
e	F-94A	−0.577	−0.590	1.473	−1.245
f	F3D-1	−1.073	−1.158	−0.196	−1.343
g	F-89A	−0.697	−0.811	−1.887	0.119
h	XF10F-1	−0.594	0.262	−0.568	−0.466
i	F9F-6	−0.518	−1.016	−0.140	0.021
j	F100-A	0.311	0.688	−0.331	0.898
k	F4D-1	0.320	−1.048	0.920	1.191
m	F11F-1	−0.090	−0.259	−0.275	0.509
n	F-101A	0.856	0.073	0.052	0.216
p	F3H-2	−0.395	−0.006	0.120	0.703
q	F102-A	0.021	1.430	−0.748	0.703
r	F-8A	−1.428	0.799	−1.797	0.353
s	F-104A	1.801	0.026	0.943	0.411
t	F-105B	1.130	1.130	2.239	0.606
u	YF-107A	0.952	1.698	−0.692	0.606
v	F-106A	1.385	1.446	−0.884	0.898
w	F-4B	1.995	−0.448	0.616	0.606

problem with this figure is that the scales on the biplot axes are in terms of the normalized values of the variables. It is easy to convert these to their original units of measurement, using as input to `PCAbipl` the original aircraft data as given in Table 1.1 setting argument `scaled.mat = TRUE`. The resulting biplot is given in Figure 2.17.

As before, we may improve legibility by moving the origin and rotating axes, as has been done in Figures 2.18 and 2.19. Arguments `orthog.transx = rep(3.43, 4)` and `orthog.transy = rep(-0.68, 4)` are used in producing Figure 2.18. Reflection, rotation and orthogonal translation are needed for the biplot in Figure 2.19. These operations are carried out with the following argument settings:

```
orthog.transx = rep(2.35, 4), orthog.transy = rep(2.5, 4),
exp.factor = 2.5, reflect = "y", rotate.degrees = 214.2
```

2.6 A closer look at biplot axes

So far, we have been concerned entirely with the visualization of Table 1.1. Since our aim is to use a biplot analogous to a scatterplot, we would like to use the biplot axes to read off the values of ***SPR***, ***RGF***, ***PLF*** and ***SLF*** for any of the aircraft. To be precise, we would like to use projection onto the biplot axes to read off approximate values of the variables in Table 1.1 for each of the 21 points marked with aircraft names. This process we term *prediction*. Our biplot functions contain an argument `predictions.sample` to visually aid the prediction process. As an illustration we set `predictions.sample`

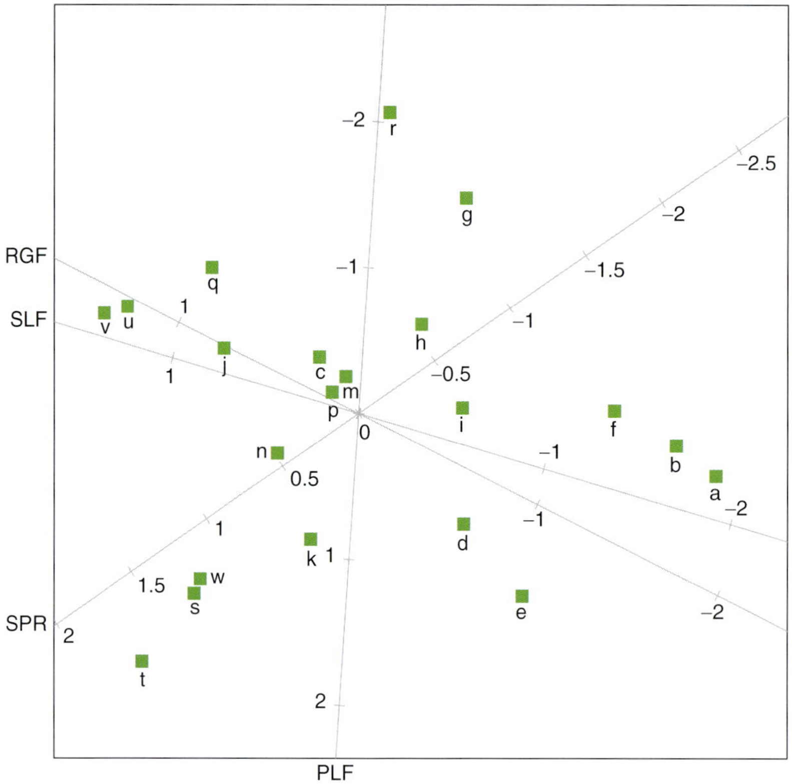

Figure 2.16 Two-dimensional biplot of the normalized aircraft data as given in Table 2.2 with scales in the normalized units.

`= c(8,9,13)` to predict the ***SPR***, ***RGF***, ***PLF*** and ***SLF*** for aircraft ***h***, ***i*** and ***n*** from the biplot axes in Figure 2.17. This is shown in Figure 2.20.

When argument `predictions.sample` is not set to `NULL` the biplot function not only shows explicitly the orthogonal projections onto the biplot axes as illustrated in Figure 2.20 but also returns the actual predictions made. The output resulting from setting `predictions.sample = c(8,9,13)` is shown in Table 2.3 together with the actual values. As an exercise, the reader may check that these predictions remain exactly the same regardless of applying orthogonal parallel translation, reflection or rotation.

Table 2.3 Predicted ***SPR***, ***RGF***, ***PLF*** and ***SLF*** values for aircraft ***h***, ***i*** and ***n*** together with the corresponding actual values in Table 1.1.

	Predictions			Actual values		
	s8	s9	s13	h	i	n
SPR	2.5266	2.6768	4.9705	2.426	2.607	5.855
RGF	4.4299	4.1956	4.6644	4.650	3.840	4.530
PLF	0.1108	0.1597	0.1947	0.117	0.155	0.172
SLF	2.0847	1.7540	2.6307	1.800	2.300	2.500

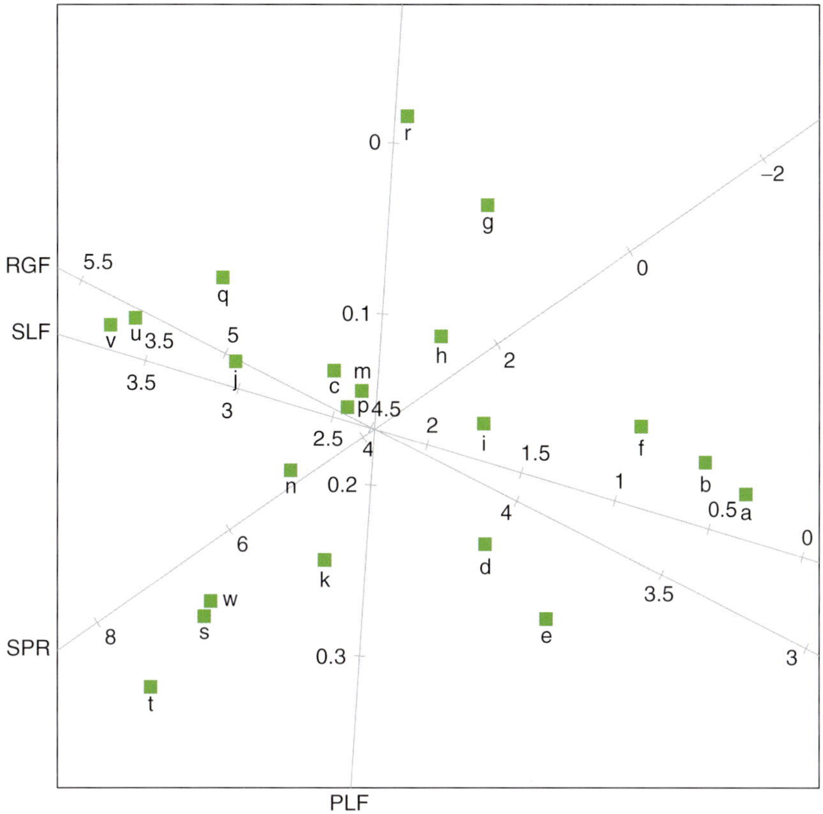

Figure 2.17 The biplot of Figure 2.16 but with the biplot axes calibrated in the original units of measurement.

Opposite to prediction, we have the process of *interpolation* where axes are utilized to place a (new) sample point by vector addition in the display. With an ordinary Cartesian scatterplot, we use the single set of calibrated orthogonal axes for both interpolation and prediction; however, this is invalid with the p nonorthogonal biplot axes.

In Figure 2.21, the process of prediction is illustrated in part (a). When interpolating the values obtained in (a) by completing the parallelogram, a different sample representation is obtained in (b). In parts (c) and (d), this process is repeated for the sample representation in (b). It is clear that a single set of nonorthogonal axes for both interpolation and prediction results in inconsistent representation of the same sample point. For this reason, generally biplots have to be equipped with different axes for prediction and for interpolation. It is important to remember to use the correct set of axes when performing predictions or interpolations. All the biplots considered thus far in this chapter have been equipped with prediction axes, resulting in valid predictions from the given biplot axes.

In order to interpolate a (new) sample whose values are given in a p-component row vector $\mathbf{x}'$ into one of the biplots considered in this chapter, its coordinates relative to the scaffolding of the biplot are given by $\mathbf{x}'\mathbf{V}_r = \mathbf{x}'(\mathbf{VJ})_r$. This is trivial to compute, so $\mathbf{x}'$ may be easily interpolated and shown on any computer screen or printout. However, when away from the computer, some visual method of interpolation may be useful. It is

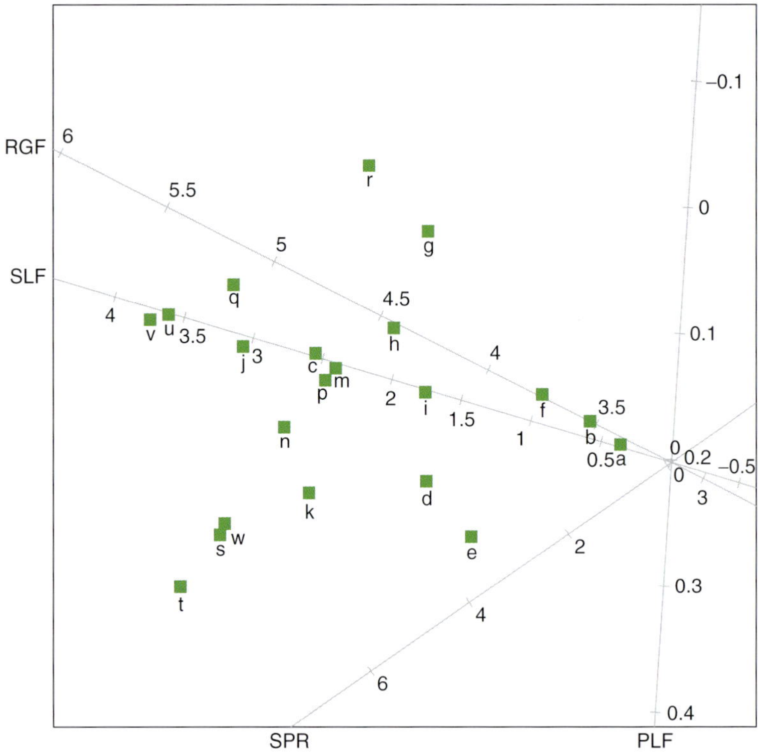

Figure 2.18 Two-dimensional biplot of the normalized aircraft data as given in Table 2.2 with the origin moved to ***SPR*** $= 0$ and ***SLF*** $= 0$. Calibrations on biplot axes in terms of the original (unnormalized) data.

not possible to use the above biplots by erecting normals at the values $x_1, x_2, x_3, \ldots, x_p$ on the corresponding biplot axes because only in exceptional circumstances will these normals be concurrent at a unique point, as they would be with Cartesian axes. Gower and Hand (1996) show that in the biplots described earlier in this chapter, biplot axes may be used for interpolation, provided they are calibrated with a scale that is inversely related to the one used for prediction. The coordinates for one unit for interpolation on the kth axis are given by $\mathbf{e}_k'\mathbf{V}_r$, the kth row of $\mathbf{V}_r$. These scales (converted back to their original units of measurement) are shown in Figure 2.22 for the normalized data of Table 2.2. Notice that the directions of the biplot axes are identical to those in Figure 2.17; it is only the calibrations that differ (inversely related). Our biplot functions included in `UBbipl` have an argument `ax.type` with default setting `ax.type = "predictive"`. Changing the setting to `ax.type = "interpolative"` results in a biplot with interpolative biplot axes like the one in Figure 2.22.

Not only do the scales for prediction and interpolation differ but also they are used differently. We have that

$$\mathbf{x}'\mathbf{V}_r = \sum_{k=1}^{p} x_k(\mathbf{e}_k'\mathbf{V}_r) = p \times \left(\frac{1}{p} \sum_{k=1}^{p} x_k(\mathbf{e}_k'\mathbf{V}_r) \right) = p \times centroid. \tag{2.15}$$

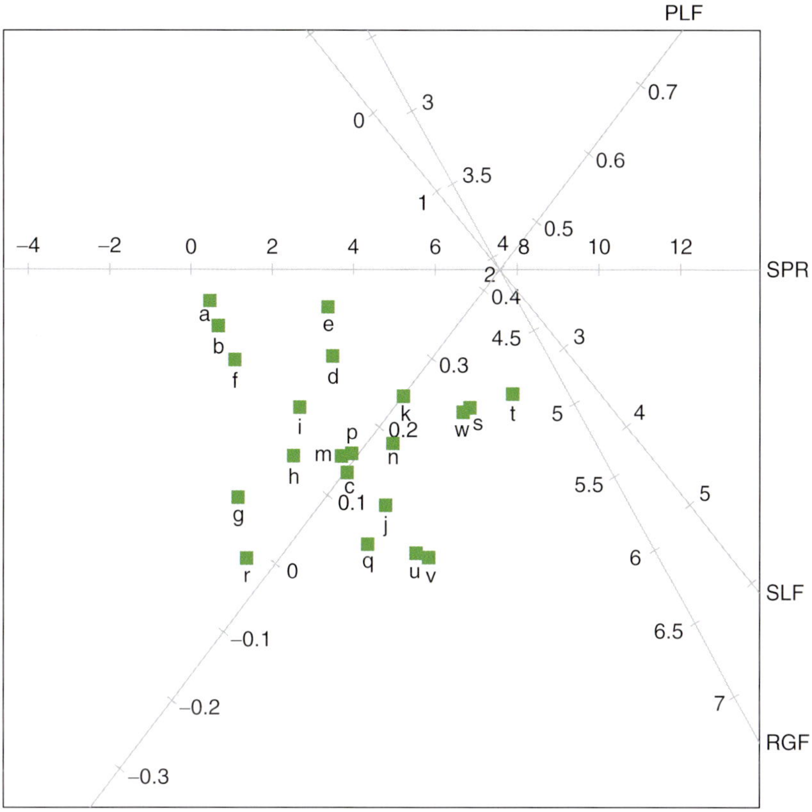

Figure 2.19 The two-dimensional biplot of Figure 2.18 with orthogonal parallel translation rotated such that the ***SPR***-axis is horizontal with calibrations increasing from left to right, similar to the first Cartesian axis of a scatterplot.

Thus, to interpolate, we need the vector-sum of the points corresponding to the markers $x_1, x_2, x_3, \ldots, x_p$. Geometrically, the sum of two vectors is usually thought of as being formed as the diagonal of the parallelogram that has the two vectors for its sides. Several vectors may be summed by constructing a series of such parallelograms. We may get the same result more simply by finding the centroid of the points given by the markers $x_1, x_2, x_3, \ldots, x_p$ and extending the result p times from the origin. How this is done is indicated in Figure 2.23 for the point with ***SPR*** = 8, ***RGF*** = 4, ***PLF*** = 0.3 and ***SLF*** = 3. Note that, if we interpolate one of the original samples, say ***p*** (F3H-2), its interpolated value will occupy the same position as it does in the analysis shown in Figure 2.17.

We have implemented the vector-sum method for interpolating a new sample point in our R function `vectorsum.interp`. First, the biplot in Figure 2.23 is constructed using `PCAbipl` with `scaled.mat=TRUE, ax.type="interpolative"` as described earlier. Then `vectorsum.interp` is called with `vertex.points = 4`, allowing the user to select with the mouse the four values on the respective axes of the point to be interpolated. A polygon is then drawn connecting these vertex points and the coordinates

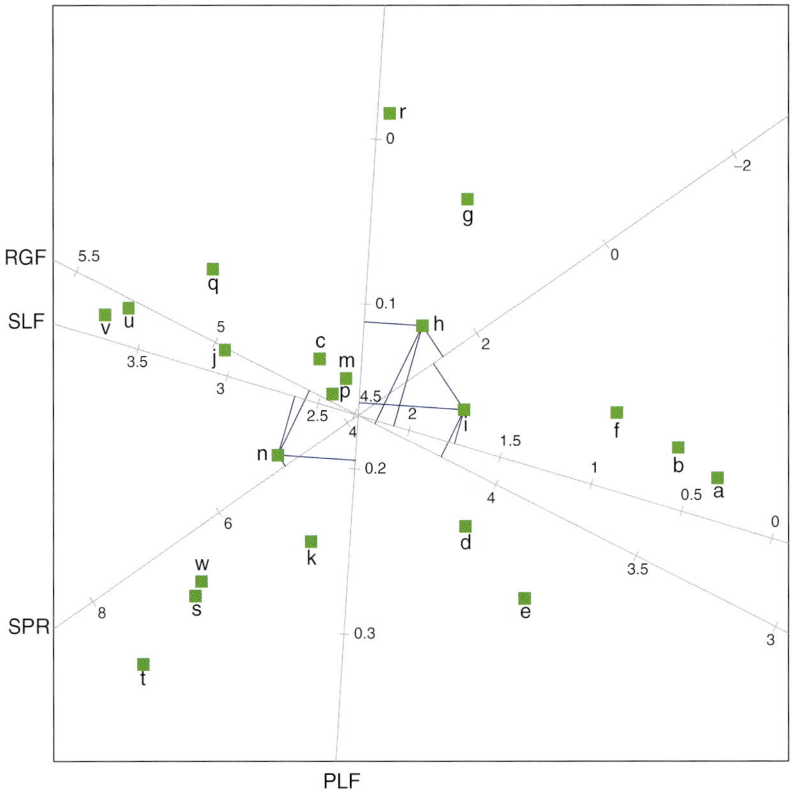

Figure 2.20 The two-dimensional biplot of Figure 2.18, illustrating the use of the biplot axes for prediction.

(cx, cy) of the centroid of the vertices of this polygon are calculated using the formula

$$(cx, cy) = \left(\frac{1}{n} \sum_{i=1}^{n} x_i, \frac{1}{n} \sum_{i=1}^{n} y_i \right). \tag{2.16}$$

The centroid of the polygon (quadrangle in Figure 2.23) is then drawn together with the arrows pointing to the interpolated position of the new sample point.

As with prediction, we may shift the interpolation biplot axes to positions that are more convenient. The process depends on the observation that for any set of markers $\mu_1, \mu_2, \mu_3, \ldots, \mu_p$,

$$\mathbf{x}'\mathbf{V}_r = \sum_{k=1}^{p} [(x_k - \mu_k)(\mathbf{e}_k'\mathbf{V}_r) + \mu_k(\mathbf{e}_k'\mathbf{V}_r)] = \sum_{k=1}^{p} [(x_k - \mu_k)(\mathbf{e}_k'\mathbf{V}_r)] + p\boldsymbol{\mu}', \tag{2.17}$$

where

$$\boldsymbol{\mu}' = \frac{1}{p} \sum_{k=1}^{p} \mu_k (\mathbf{e}_k'\mathbf{V}_r)$$

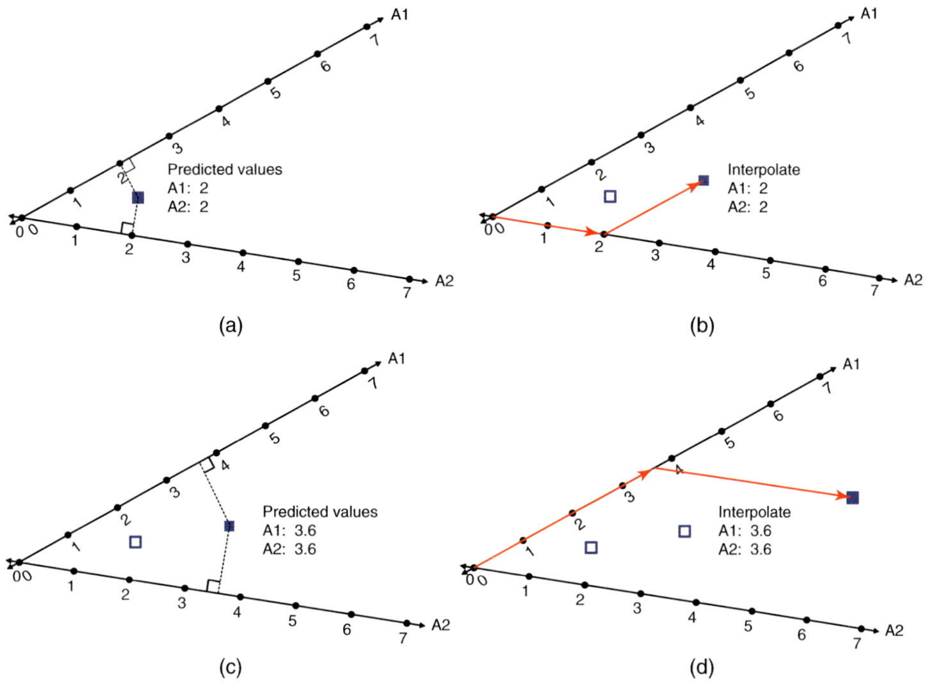

Figure 2.21 Why interpolation and prediction with a single set of nonorthogonal biplot axes are inconsistent.

is the centroid of the markers $\mu_1, \mu_2, \mu_3, \ldots, \mu_p$. The interpretation is that every point on the kth axis is translated obliquely by an amount $\mu_k(\mathbf{e}'_k\mathbf{V}_r)$, resulting in new axes, concurrent at $\boldsymbol{\mu}$. Contrast this process with the orthogonal translation method that is appropriate for the axes of predictive biplots. With interpolation, the position is better, because we may choose 'nice' values of μ_k on all the axes rather than just r of them. Figure 2.24 illustrates the process where μ_k has been chosen to be zero for each axis. To interpolate a point with the new axes, we proceed, as before, by finding the centroid of the markers but now we have to remember to add in the vector $\boldsymbol{\mu}$. All this means is that we continue to extend p times from the old origin O rather than from the new point of concurrency of the axes. This is why the point O is retained in Figure 2.24.

Oblique transformation is implemented in our `PCAbipl` function. It is only used with argument `ax.type = "interpolative"` and is invoked by assigning to argument `oblique.trans` a vector of p elements specifying the values of each variable at the point of concurrency of the biplot axes. Figure 2.25 shows the results of this process for the aircraft data, where we have chosen $\mu_1 = 0$, $\mu_2 = 0$, $\mu_3 = 0$ and, $\mu_4 = 0$.

In Figure 2.26 we show how to effect interpolating the new sample with $\boldsymbol{SPR} = 8$, $\boldsymbol{RGF} = 4$, $\boldsymbol{PLF} = 0.3$ and $\boldsymbol{SLF} = 3$ using the obliquely translated interpolative axes. Note the role of the original origin marked with the black cross. Note also that the interpolated position is exactly as in Figure 2.23. Interpolation performed using the graphical vector-sum method is implemented in the function `vectorsum.interp` described earlier in this section.

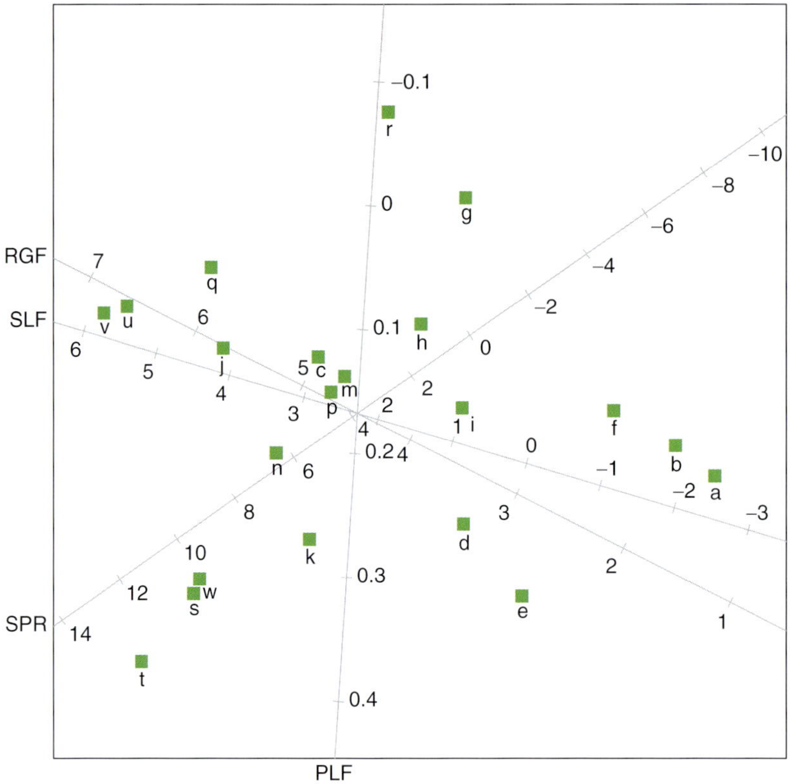

Figure 2.22 Interpolation biplot of the scaled aircraft data in Table 2.2.

The biplots constructed in this book will be fitted with prediction biplot axes unless stated otherwise. This is because it is intuitive for anyone analysing a plot to read off values for the variables on the axes. In general, the interpolation can be taken care of algebraically with a computer program, as is illustrated in the functions contained in `UBbipl`. As an example, we show in Figure 2.27 the result of interpolating the new point with values ***SPR*** $= 8$, ***RGF*** $= 4$, ***PLF*** $= 0.3$ and ***SLF*** $= 3$ using the algebraic formula for interpolation by specifying the argument `X.new.samples = matrix(c(8,4,0.3, 3), nrow = 1))`.

2.7 Adding new variables: the regression method

We have seen above how to add new samples to a PCA. Suppose now that we wish to add a new variable available in a centred column vector $\mathbf{x}^* : n \times 1$. We may add $\mathbf{x}^*$ to our PCA by using a regression method that assumes that $\mathbf{x}^*$ is approximately a linear function $\mathbf{x}^* = \mathbf{X}\mathbf{V}_r\mathbf{b}_r$ of the points $\mathbf{X}\mathbf{V}_r$. This is a multiple regression problem with solution

$$\hat{\mathbf{b}}_r : r \times 1 = (\mathbf{V}_r'\mathbf{X}'\mathbf{X}\mathbf{V}_r)^{-1}\mathbf{V}_r'\mathbf{X}'\mathbf{x}^*. \tag{2.18}$$

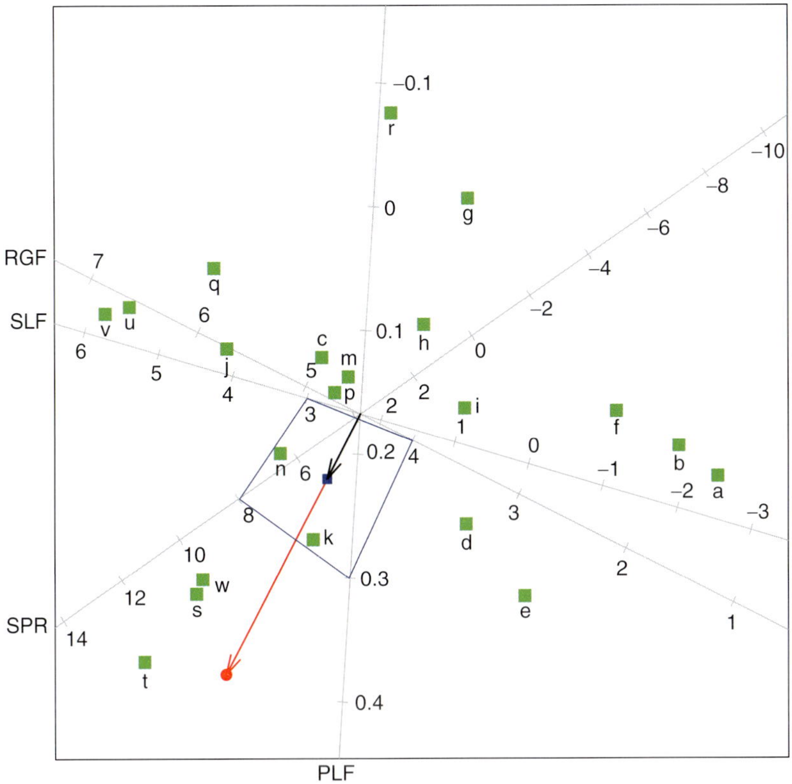

Figure 2.23 Illustration of vector-sum interpolation. The vertices of the blue quadrangle area give the values of the four variables to be interpolated. The end of the black arrow marks the centroid of the four vertices. The length of the red arrow is four (p) times the length of the black arrow and indicates the position of the interpolated point.

The r elements of the column vector $\hat{\mathbf{b}}_r$ may be taken as a point which, when joined to the origin, gives the direction of a new axis which can be calibrated and used in the usual way.

In (2.18) we have used the notation $\mathbf{V}_r$ introduced in Section 2.2 to denote the first r columns of the orthogonal matrix $\mathbf{V}$, thus avoiding complications with inverting singular matrices. We may simplify the expression for $\hat{\mathbf{b}}_r$ by using the notation for the rank-r approximation $\mathbf{U}_r\mathbf{\Sigma}_r\mathbf{V}_r'$ to $\mathbf{X} = \mathbf{U\Sigma V}'$. Then $\mathbf{V}_r'\mathbf{X}'\mathbf{X}\mathbf{V}_r = \mathbf{\Sigma}_r^2$ and $\mathbf{X}\mathbf{V}_r = \mathbf{U}_r\mathbf{\Sigma}_r$, giving

$$\hat{\mathbf{b}}_r = \mathbf{\Sigma}_r^{-1}\mathbf{U}_r'\mathbf{x}^*. \tag{2.19}$$

In terms of the $\mathbf{J}$-notation (2.19) may be rewritten as

$$\begin{bmatrix} \hat{\mathbf{b}}_r \\ \mathbf{0} \end{bmatrix} : p \times 1 = \mathbf{J\Sigma}^{-1}\mathbf{U}'\mathbf{x}^* = \mathbf{Jb}, \tag{2.20}$$

showing that different settings of r in $\mathbf{J}$ give the regression for any number of dimensions used in the PCA approximation, including the full p-dimensional solution, when $\mathbf{J} = \mathbf{I}$

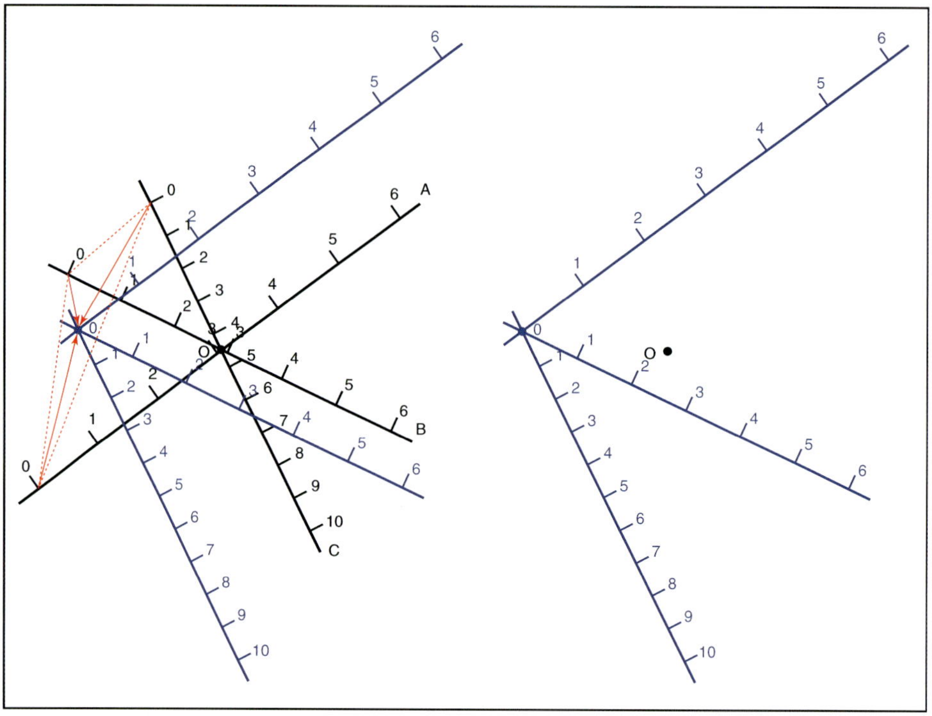

Figure 2.24 In the left-hand drawing, the black interpolative axes A, B and C intersect at an origin O that corresponds to ugly calibrated values. It is decided to translate the axes to intersect at the zero point on each axis. The intersection is at the centroid of the zero points on the original axes, shown as the vertices of a dashed triangle. Each axis is translated obliquely by the displacements shown by the arrow-headed vector to give new parallel axes A, B and C. This looks a complicated process but the result is the simple diagram shown in the right-hand drawing. The original origin O is retained for reasons discussed in the main text. Sample points are not shown; their positions relative to O are not changed by the oblique translation of the axes.

and $\mathbf{b} = \mathbf{\Sigma}^{-1}\mathbf{U}'\mathbf{x}^*$. Clearly, $\mathbf{x}^*$ may be replaced in (2.20) by any number of new columns and, in particular, replacing $\mathbf{x}^*$ by $\mathbf{X}$ gives

$$\mathbf{JB} : p \times p = \mathbf{J\Sigma}^{-1}\mathbf{U}'\mathbf{X} = \mathbf{J\Sigma}^{-1}\mathbf{U}'\mathbf{U\Sigma V}' = \mathbf{JV}'$$

showing that the regression method correctly derives the appropriate biplot axes for the primary data. Thus, the vector $\mathbf{Jb}$ acts similarly to a column of $\mathbf{VJ}$ and hopefully it will be acceptable when $\mathbf{x}^*$ refers to one or more entirely new variables. We will see during the course of this book that the regression method is available in other contexts (e.g. correspondence analysis and multidimensional scaling). We note that (2.19), equivalently (2.20), are examples of a *transition formula*, that relates samples to variables (see correspondence analysis). Examples of adding variables using the regression method are given in Chapter 3 after a discussion of the concept of axis predictivity for PCA biplots.

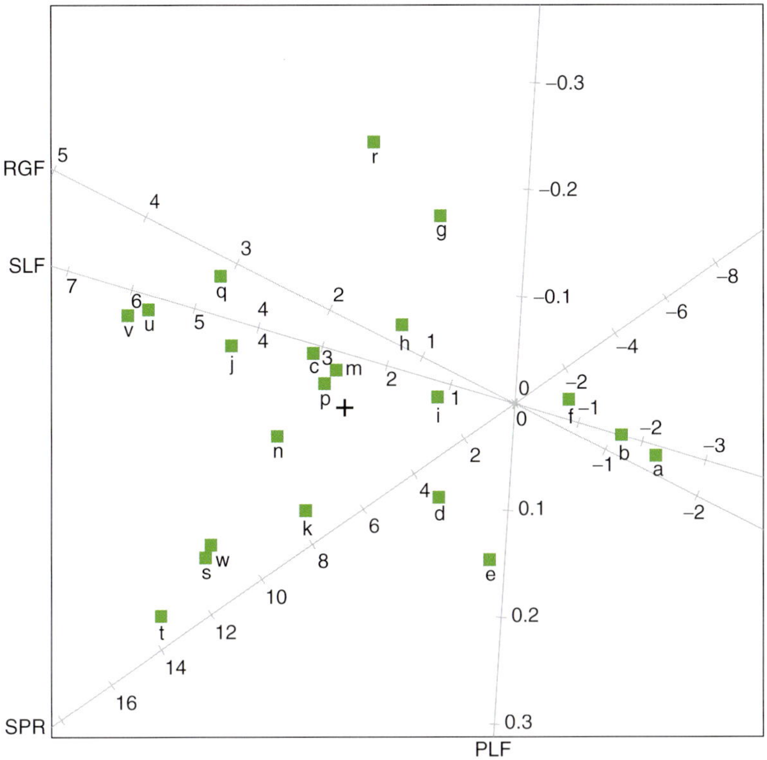

Figure 2.25 Interpolation biplot of the aircraft data with obliquely translated biplot axes, such that the point of concurrency is zero on each axis. The original origin is retained and marked with a black cross.

2.8 Biplots and large data sets

So far, we have used a very small data set to illustrate the essentials of constructing biplots. Can samples and variables of large data sets also be meaningfully represented in biplots? We first look at a moderately large data set consisting of a sample of 1135 responses to 90 questions (variables). This study was undertaken to investigate people's attitude to buying from a mail order catalogue. The investigator separated the questions into three components: ***Q1*** to ***Q33*** measured the reasons why people buy from mail order catalogues; ***Q34*** to ***Q57*** measured the perceived risks involved; and ***Q58*** to ***Q90*** measured risk relievers influencing their behaviour. Each item consists of a statement and the following six alternatives:

'agree completely', scored as 1;
'agree somewhat', scored as 2;
'agree a little', scored as 3;
'disagree a little', scored as 4;
'disagree somewhat', scored as 5;
'disagree completely', scored as 6.

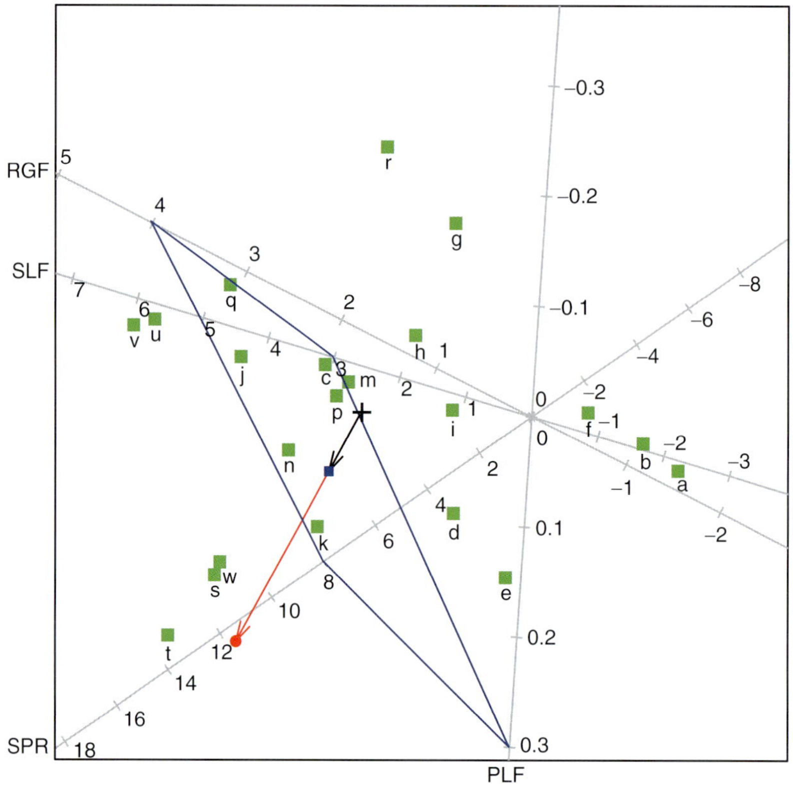

Figure 2.26 Interpolation biplot of the aircraft data with obliquely translated biplot axes, such that the point of concurrency is zero on each axis. The original origin is retained and marked with a black cross. A new point is interpolated graphically with values ***SPR*** = 8, ***RGF*** = 4, ***PLF*** = 0.3 and ***SLF*** = 3. The black arrow extends from the original origin to the centre of the polygon with vertices the values of the sample to be interpolated. The red arrow is $p = 4$ times the length of the black arrow and indicates the interpolated position of the new sample.

The data set is available as `mailcatbuyers.data` in `UBbipl`. Figure 2.28 is a biplot approximating the raw data matrix. In this biplot colour is used to distinguish the buying behaviour of different age groups. In Chapter 8, we consider other types of biplot more appropriate for questionnaire data of this kind.

The biplot clearly shows the main difficulty encountered with large data sets: overplotting and too much ink leave us with a graph so cluttered that it is hardly of practical use. Two factors contribute to the amount of ink in Figure 2.28: the number of variables and the number of samples. These two issues are addressed differently when considering biplots for use with large data sets.

First, we consider the number of variables – that is, the number of biplot axes. This issue can be addressed by having interactive computer software for turning on and off biplot axes. Using the `ax` argument of our function `PCAbipl` we can suppress plotting of any subset of axes. This facility was used to obtain the biplots in Figures 2.29 and 2.30.

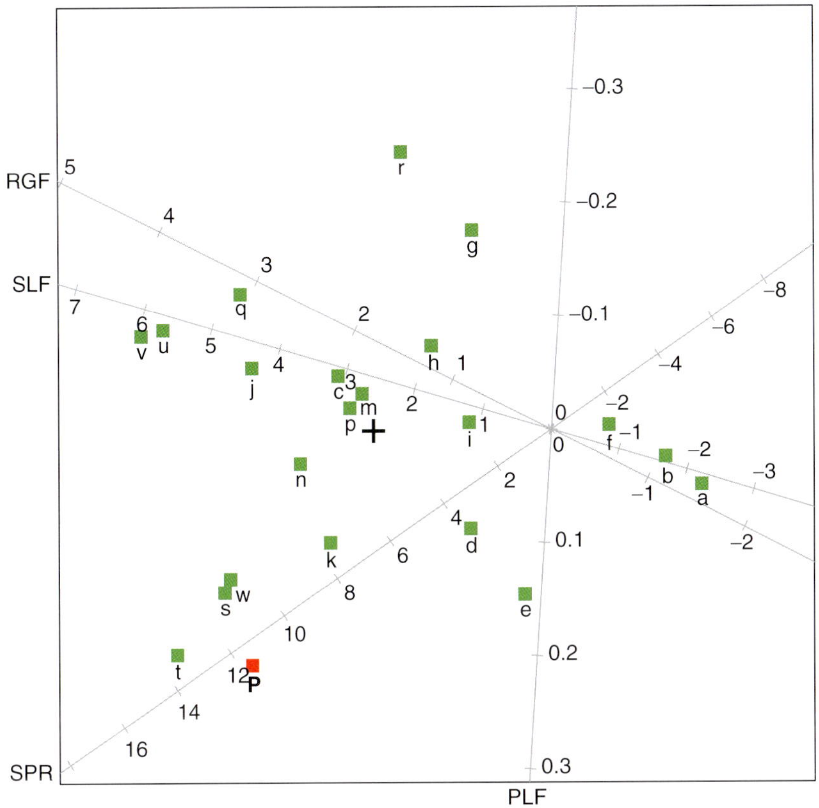

Figure 2.27 Interpolation biplot of the aircraft data with obliquely translated biplot axes, such that the point of concurrency is zero on each axis. The original origin is retained and marked with a black cross. A new point P is interpolated algebraically with values ***SPR*** = 8, ***RGF*** = 4, ***PLF*** = 0.3 and ***SLF*** = 3.

By interactively turning on and off the plotting of individual axes, very useful biplot displays of large data sets can be obtained. Figure 2.28 clearly shows three subsets in the variables relating to reasons for buying by mail catalogue, perceived risks in buying by mail catalogue and risk relievers when buying by mail catalogue. The top panel of Figure 2.30 suggests that ***Q49*** is out of its place or that perhaps the respondents wrongly interpret it. Furthermore, the axes relating to reasons for buying by mail catalogue are closer to one another than the other two sets of axes. This suggests a larger correlation among questions relating to the reasons than between questions relating to perceived risks and risk relievers when buying by mail catalogue.

This leaves us with the task of reducing the amount of ink in a biplot with many sample points. Of course, as in the case of axes, this goal can be accomplished by interactively turning on and off the sample points. A more fruitful idea is to enclose a specified subset of sample points and suppress plotting of the samples within the enclosure while showing only its boundary. We explore several possibilities for enclosing sample points in a two-dimensional biplot in the next section.

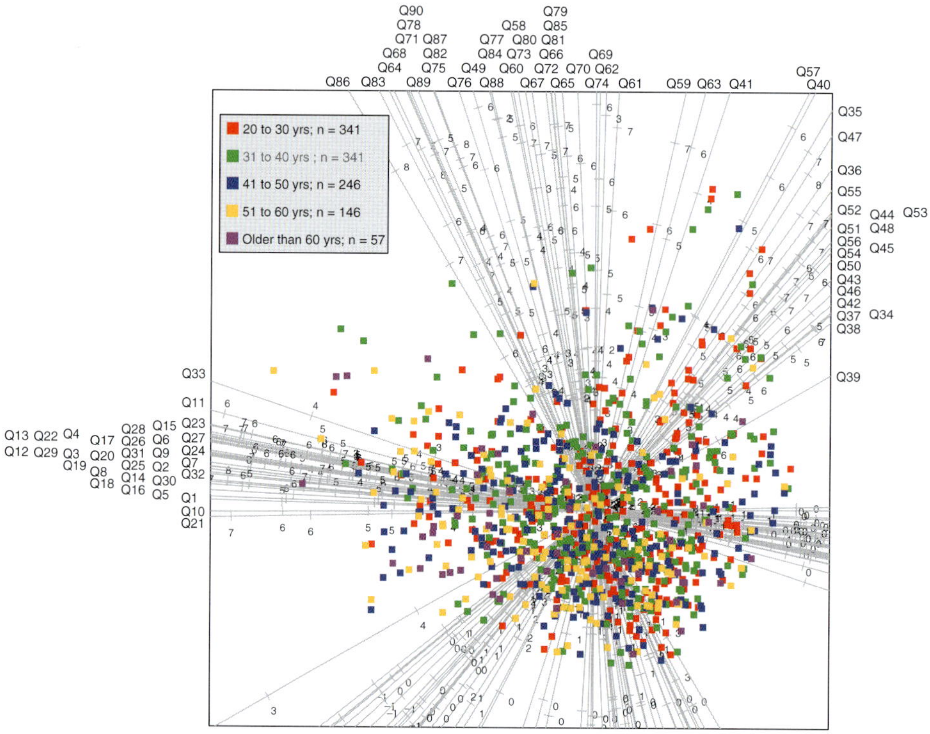

Figure 2.28 PCA biplot of 1135 respondents' answers to 90 questions addressing reasons, risks and risk relievers when buying from mail order catalogues. Respondents are colour-coded according to their age categories.

2.9 Enclosing a configuration of sample points

Flury (1988) described data giving head dimensions consisting of the following six variables measured for 200 young men and 59 young women: ***MFB*** (minimal frontal breadth), ***BAM*** (breadth of angulus mandibulae), ***TFH*** (true facial height), ***LGAN*** (length from glabella to apex nasi), ***LTN*** (length from tragion to nasion) and ***LTG*** (length from tragion to gnathion). These data are available in the R package `Flury` in the dataframes `swiss.heads` (the 200 males) and `f.swiss.heads` (the 59 females). We combined both these dataframes into the dataframe `Headdimensions.data` in `UBbipl`. Figure 2.31 is a scatterplot of the head dimensions data for variables ***LGAN*** and ***LTN*** resulting from the R instruction

```
> plot(Headdimensions.data[,c(5,6)], pch = 16, asp = 1)
```

Of course, Figure 2.31 is also a biplot.

We turn our attention now to several devices proposed for enclosing predefined sets of observations with the aim of finding a representative graphical summary of the shape of bivariate sample points belonging to a particular configuration.

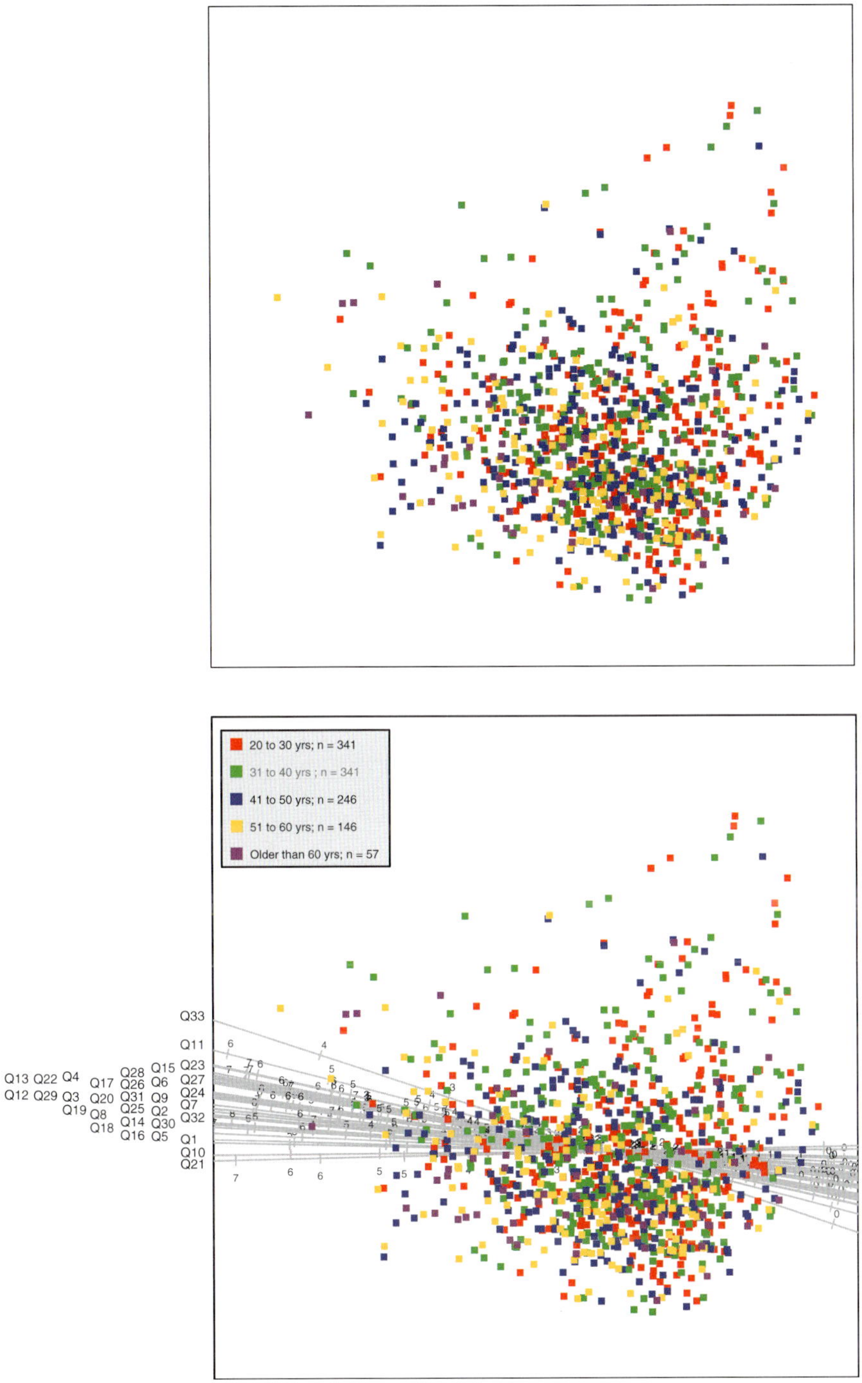

Figure 2.29 The biplot of Figure 2.28 but suppressing all or some of the axes: (top) suppressing all axes; (bottom) showing only axes relating to reasons for buying by mail order catalogue.

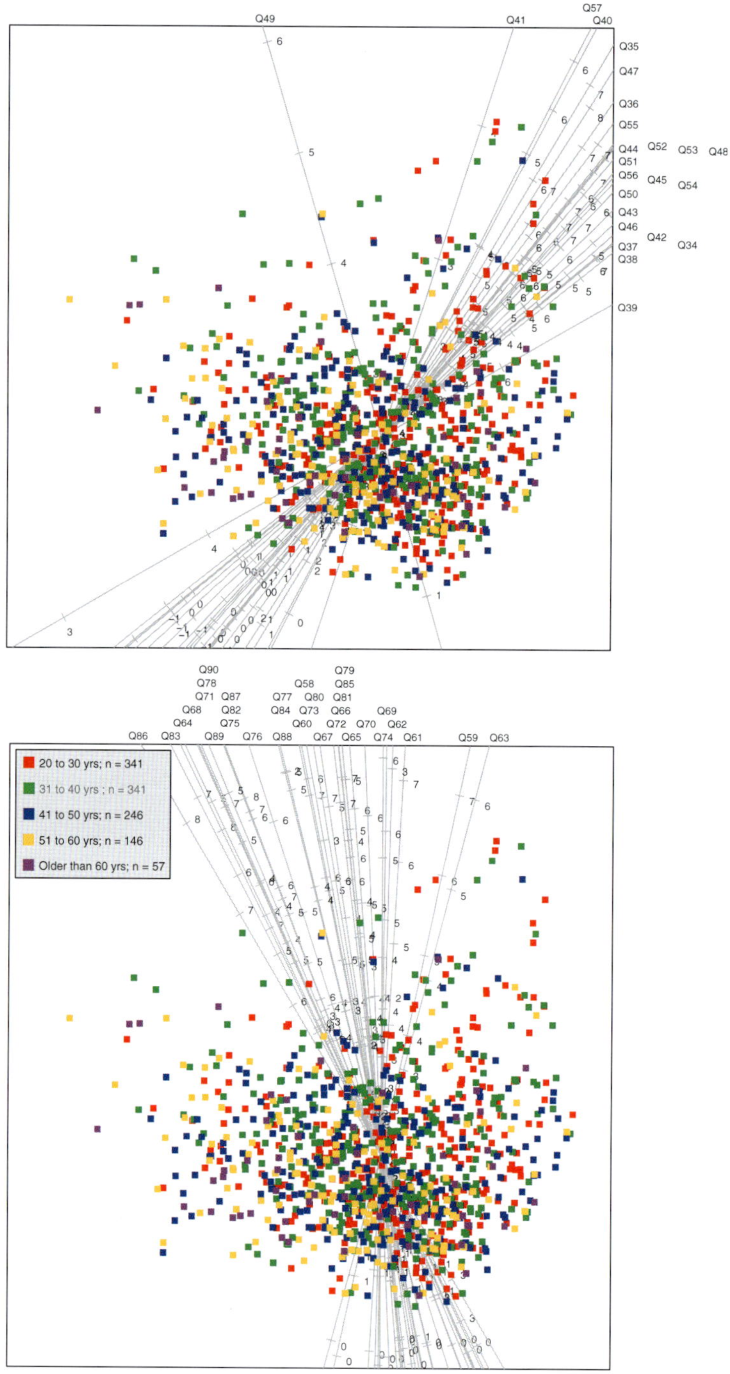

Figure 2.30 The biplot of Figure 2.28 but suppressing some of the axes: (top) showing only axes relating to perceived risks; (bottom) showing only axes relating to risk relievers.

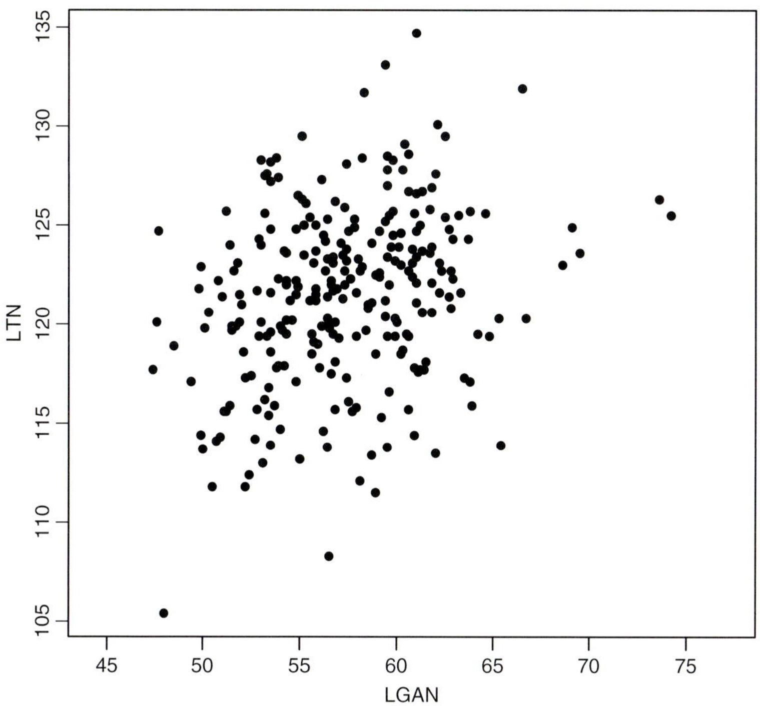

Figure 2.31 Scatterplot (biplot) of variables ***LTN*** and ***LGAN*** of the head dimensions data set described by Flury (1988).

2.9.1 Spanning ellipse

Constructing an ellipse surrounding chosen observations is a simple method for enclosing the data region. A spanning ellipse is defined as the smallest ellipse that covers all the objects. An algorithm for constructing such an ellipse is provided by Titterington (1976) and Silvey *et al.* (1978). The function `clusplot` of Pison *et al.* (1999) for drawing ellipses encircling clusters found in a cluster analysis incorporates the Titterington algorithm. This function is available in the R package `cluster`. The `clusplot` function is slightly modified into the function `MinSpanEllipse` included in `UBbipl` for enclosing the points in a two-dimensional scatterplot with a minimum spanning ellipse. Figure 2.32 shows a minimum spanning ellipse enclosing the scatterplot of the head dimensions data. This figure is a result of using the code

```
> library(cluster)
> MinSpanEllipse(Headdimensions.data[,5:6], asp = 1,
  clus = rep(1,259), col.p = "blue", col.clus = "red",
  pch.points = 16)
```

Perusal of Figure 2.32 shows the observations roughly to fall between 45 and 75 for ***LGAN*** and between 105 and 135 for ***LTN***. The shape of the spanning ellipse also suggests some degree of positive relationship between these variables.

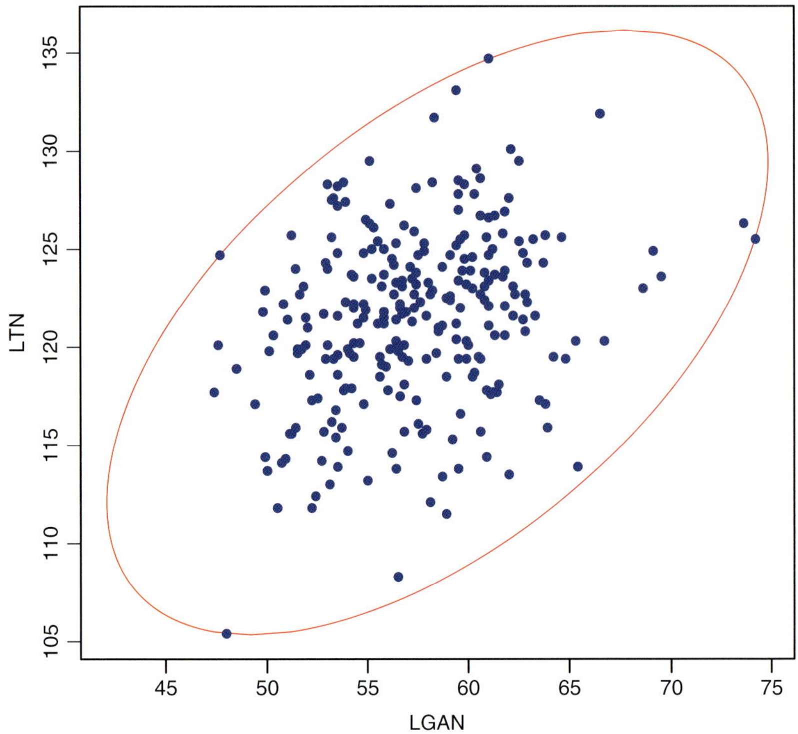

Figure 2.32 Spanning ellipse enclosing the configuration of two head dimensions variables.

2.9.2 Concentration ellipse

Consider a continuous random variable, say Y, with an unspecified distribution having a mean μ and finite variance σ^2. Let Y^* be another continuous random variable having a uniform distribution defined on the interval $(\mu - \sigma\sqrt{3}, \mu + \sigma\sqrt{3})$. It then follows from the expected value and variance of a uniform distribution that $\mathrm{E}(Y^*) = \mu$ and $\mathrm{var}(Y^*) = \sigma^2$. This led Cramér (1946) to suggest the interval

$$(\mu - \sigma\sqrt{3}, \mu + \sigma\sqrt{3}) \tag{2.21}$$

to be taken as a geometrical description of the concentration of the (unspecified) distribution of Y about its known mean μ. The interval (2.21) may be called a *concentration interval* and is not to be confused with a confidence interval for an unknown parameter μ. We note that if Y has a normal $(\mu; \sigma^2)$ distribution then $\mathrm{P}(\mu - \sigma\sqrt{3} < Y < \mu + \sigma\sqrt{3}) = 0.9167$.

In practice μ and σ^2 are usually unknown, so, if we have a random sample of size n then, replacing μ and σ^2 in (2.21) with their sample counterparts, we have the sample concentration interval

$$(\bar{x} - s\sqrt{3}, \bar{x} + s\sqrt{3}). \tag{2.22}$$

The interval (2.22) provides us with a geometrical description of the data points about their sample mean. If the data come from a normal distribution then approximately 91.67% of the data points will lie in the interval (2.22). If we have two samples and the concentration interval for sample 1 is contained within the concentration interval for sample 2 then we can say that the first sample is more concentrated about its mean than the second sample.

Cramér extended the idea leading to (2.21) to the random vector $\mathbf{Y}_p$ with specified $\boldsymbol{\mu}$ as expected vector and specified positive definite matrix $\boldsymbol{\Sigma} : p \times p$ as covariance matrix by considering the following question: what random vector $\mathbf{Y}_p^*$ has a density that is uniformly distributed over the interior of a p-dimensional ellipsoid centred at $\boldsymbol{\mu}$ such that $E(\mathbf{Y}_p^*) = \boldsymbol{\mu}$ and $\text{cov}(\mathbf{Y}_p^*) = \boldsymbol{\Sigma}$? Making use of the integrals (Cramér, 1946)

$$\int_{\mathbf{x}'\boldsymbol{\Sigma}\mathbf{x} < c^2} \cdots \int dx_1 \ldots dx_p = \frac{\pi^{p/2}}{\Gamma(\frac{p}{2}+1)} \cdot \frac{c^p}{\sqrt{|\boldsymbol{\Sigma}|}}, \tag{2.23}$$

$$\int_{\mathbf{x}'\boldsymbol{\Sigma}\mathbf{x} < c^2} \cdots \int x_i x_k dx_1 \ldots dx_p = \left(\frac{\pi^{p/2}}{\Gamma(\frac{p}{2}+1)} \cdot \frac{c^p}{\sqrt{|\boldsymbol{\Sigma}|}} \right) \cdot \frac{c^2}{p+2} \cdot \frac{\Sigma_{ki}}{|\boldsymbol{\Sigma}|}, \tag{2.24}$$

where Σ_{ki} denotes the cofactor of $\sigma_{ik} = \sigma_{ki}$ in $\boldsymbol{\Sigma}$ and $|\boldsymbol{\Sigma}|$ the determinant of $\boldsymbol{\Sigma}$, together with the properties of spherical and elliptical distributions (see, for example, Fang *et al.*, 1990), it can be shown that the random vector $\mathbf{Y}_p^*$ that is uniformly distributed over the interior of the ellipsoid defined by

$$(\mathbf{y} - \boldsymbol{\mu})'\boldsymbol{\Sigma}^{-1}(\mathbf{y} - \boldsymbol{\mu}) < p + 2, \tag{2.25}$$

has a mean of $\boldsymbol{\mu}$ and a covariance matrix of $\boldsymbol{\Sigma}$. The ellipsoid $(\mathbf{y} - \boldsymbol{\mu})'\boldsymbol{\Sigma}^{-1}(\mathbf{y} - \boldsymbol{\mu}) = p + 2$ is called a *concentration ellipsoid*. We observe that, for $p = 1$, equation (2.25) reduces to the interval (2.22). If we have a random sample of size n from the distribution of $\mathbf{Y}_p$ then we have the sample concentration ellipsoid defined by the locus of a point $\mathbf{y} : p \times 1$ satisfying

$$(\mathbf{y} - \bar{\mathbf{x}})'\mathbf{S}^{-1}(\mathbf{y} - \bar{\mathbf{x}}) = p + 2, \tag{2.26}$$

where $\bar{\mathbf{x}}$ and $\mathbf{S}$ are the usual (unbiased) estimates of $\boldsymbol{\mu}$ and $\boldsymbol{\Sigma}$, respectively. It is well known that if $\mathbf{Y}_p$ has a p-variate normal $(\boldsymbol{\mu}; \boldsymbol{\Sigma})$ distribution then

$$(\mathbf{Y}_p - \boldsymbol{\mu})'\boldsymbol{\Sigma}^{-1}(\mathbf{Y}_p - \boldsymbol{\mu}) \sim \chi_p^2. \tag{2.27}$$

In the case of (2.27) it follows that

$$P\{(\mathbf{Y}_p - \boldsymbol{\mu})'\boldsymbol{\Sigma}^{-1}(\mathbf{Y}_p - \boldsymbol{\mu}) < p + 2\} = \begin{cases} 0.9167 & \text{for } p = 1 \\ 0.8647 & \text{for } p = 2 \\ 0.8282 & \text{for } p = 3. \end{cases}$$

Therefore, in the case of a two-dimensional configuration of points we expect the sample *concentration ellipse* to enclose approximately 86.5% of the data points. Now, in the case of $p = 2$, write (2.26) as

$$(\mathbf{y} - \bar{\mathbf{x}})'\mathbf{S}^{-1}(\mathbf{y} - \bar{\mathbf{x}}) = \kappa^2. \tag{2.28}$$

The loci of the two-dimensional point $\mathbf{y}$ satisfying (2.28) for different values of κ define a set of ellipses. In the case of $\kappa = 2$ we have the sample concentration ellipse and when $\kappa = 1$ we have the *indicator ellipse* (Le Roux and Rouanet, 2004). Equation (2.28) then implies the following if the configuration of two-dimensional points can be considered to be a random sample from a bivariate normal distribution:

- approximately 39.35% of the points in the configuration lie within the boundary of the sample indicator ellipse;
- choosing $\kappa = (\chi^2_{2;1-\alpha})^{1/2}$, where $\chi^2_{2;1-\alpha}$ denotes the $(1-\alpha)100$th percentage point of the χ^2_2 distribution, results in an ellipse covering approximately $100\alpha\%$ of the configuration of two-dimensional points – that is, choosing $\kappa = 2.4477$ results in an ellipse covering approximately 95% of the points in the configuration; choosing $\kappa = 3.0349$ results in an ellipse covering approximately 99% of the points in the configuration.

We will use the term *κ-ellipse* to denote the choice of $\kappa = (\chi^2_{2;1-\alpha})^{1/2}$ resulting in the ellipse covering approximately $100\alpha\%$ of the configuration of two-dimensional points. In Figure 2.33 we show the results of a call to our function `ConCentrEllipse` with arguments `kappa = 2`, `sqrt(qchisq(0.95,2))` and `sqrt(qchisq(0.99,2)` respectively.

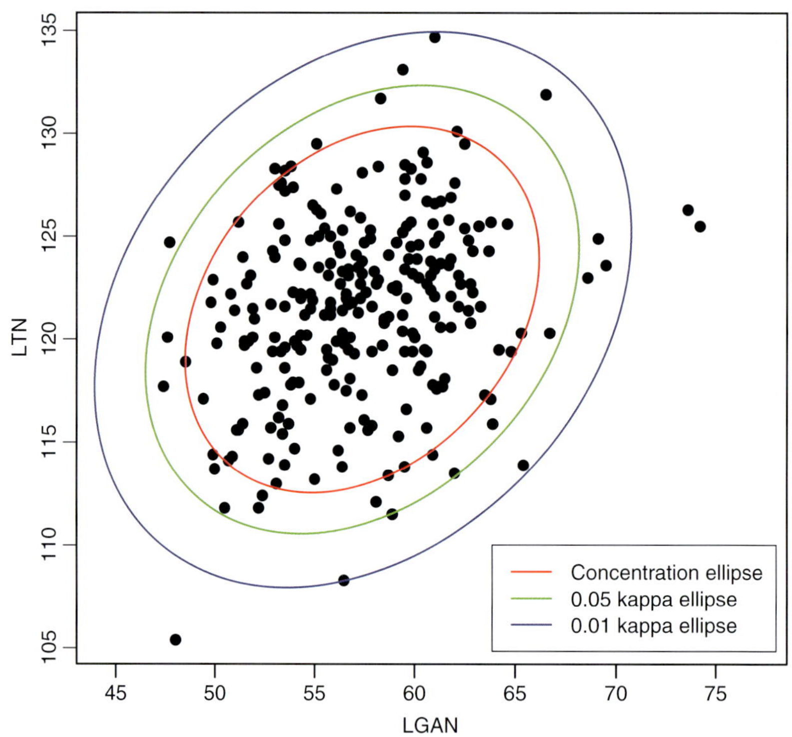

Figure 2.33 Concentration ellipse together with 0.05 and 0.01 kappa ellipses enclosing respectively approximately 86.5%, 95% and 99% of the configuration of two head dimensions variables.

2.9.3 Convex hull

Figure 2.33 provides a useful general idea of the area occupied by the observations, but it is evident that the spanning ellipse includes large areas without observations. This is also true for the κ-ellipse. Although there are no extreme outliers in the scatterplot, the restriction of the enclosure to be elliptical leads to an enclosed area that in general does not represent the shape of a specified cluster of points adequately.

A possible remedy is to consider the *convex hull* of a set, which is defined as the smallest convex set which contains all the points of the set. Figure 2.34 contains the convex hull of the scatterplot of the two variables of the head dimensions data set superimposed upon the spanning ellipse of Figure 2.32 and created by the following R code:

```
> MinSpanEllipse(Headdimensions.data[,5:6], asp = 1,
  clus = rep(1,259), col.p = "black", col.clus = "blue")
> points <- chull(Headdimensions.data[,5:6])
> lines(Headdimensions.data[c(points,points[1]), 5:6],
  col = "red")
```

In Figure 2.34 the area enclosed by the convex hull is much smaller than that inside the spanning ellipse. The shape of the convex hull is very different from that of the ellipse.

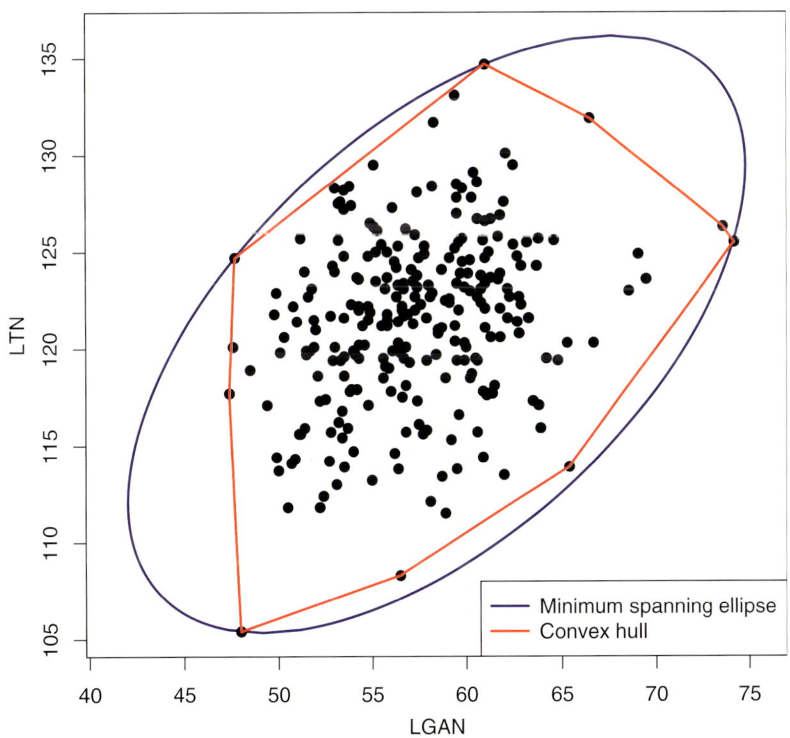

Figure 2.34 Convex hull enclosing the two-dimensional configuration of head dimensions variables ***LTN*** and ***LGAN***. The convex hull occupies a smaller area than the minimum spanning ellipse.

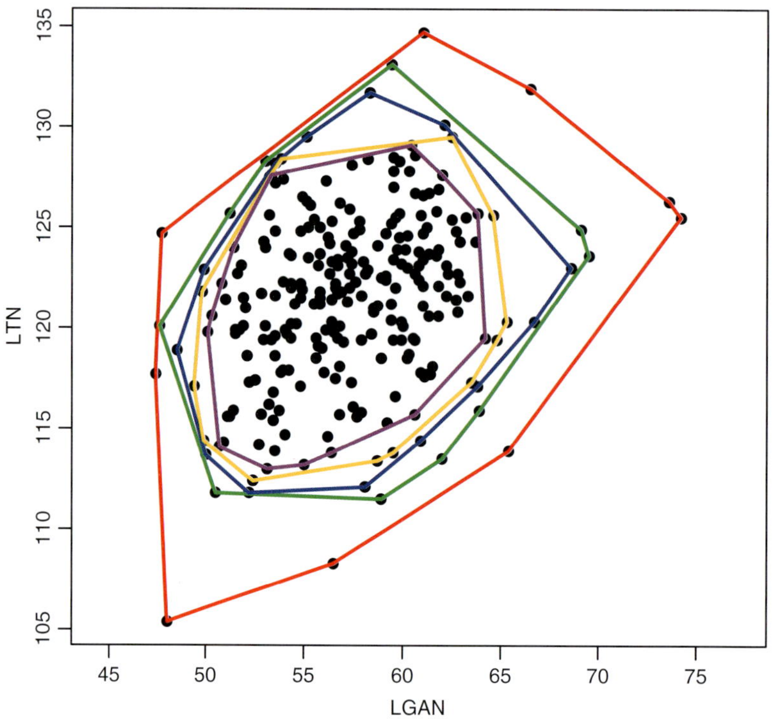

Figure 2.35 Convex hull peeling for enclosing the two-dimensional configuration of head dimensions variables ***LTN*** and ***LGAN***.

Outliers can still adversely affect the shape of the area enclosed by the convex hull – for example, the single observation with the smallest ***LTN*** value results in a convex hull with a point towards the bottom left-hand side and no other observations in the vicinity. To enclose only the area where the observations are concentrated, convex hull peeling could be used (see Green, 1985). Once the convex hull which encloses all the sample points is constructed, the sample points on the perimeter are designated the first convex layer and removed. The convex hull of the remaining points is constructed and the samples on the perimeter form the second convex layer. This process is repeated, removing the samples on the perimeter at each stage, to obtain a nested set of convex layers. The first five convex layers of the scatterplot of the head dimensions data set are shown in Figure 2.35. This figure can be reproduced using the R function `chull.peeling` provided in `UBbipl` with the call

```
> chull.peeling(Headdimensions.data[,5:6], asp = 1, k = 5, pch = 16)
```

2.9.4 Bagplot

The convex hull encloses the scatterplot of the observations, but the area of concentration of the sample points is completely ignored. Whether the sample points are evenly distributed or concentrated to one side with only a few scattered observations to the other side, the convex hull will be unchanged as long as the sample points on the perimeter remain the same. It is clear from Figure 2.35 that convex hull peeling could remove the influence of extreme data points, but the process is rather inflexible without utilizing fully the statistical properties of the data set.

In univariate data, the boxplot is a useful tool for summarizing the location, spread, skewness and outliers in data. Several suggestions have been made for generalizing this concept to bivariate data (see, for example, Goldberg and Iglewicz, 1992; Zani *et al.*, 1998; Liu *et al.*, 1999). Another such bivariate generalization by Rousseeuw *et al.* (1999), for which they coin the term *bagplot*, incorporates all the properties of the univariate boxplot.

A bag is constructed containing 50% of the data points. A fence is constructed by inflating the bag by a (default) factor of 3. Observations outside the fence are flagged as outliers. The whiskers can be represented in several forms: lines from the bag to the observations outside the bag but inside the fence; star-shaped whiskers connecting the observations outside the bag but inside the fence in such a way that it never cuts off a

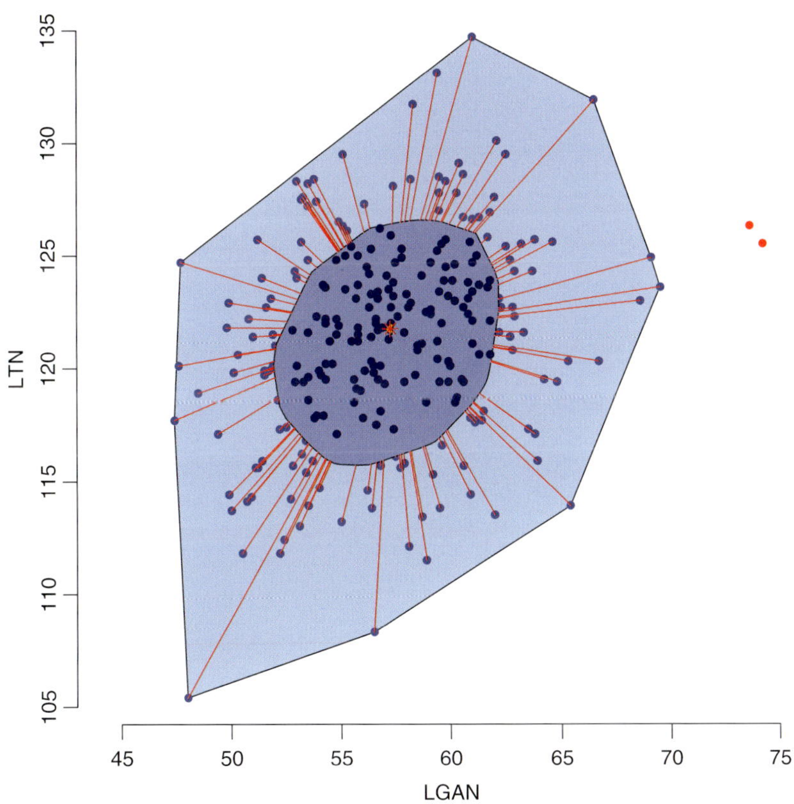

Figure 2.36 Bagplot of the two head dimension variables. The dark blue area is the bag. The fence, which is not shown, is formed by inflating the bag by a factor of 3. The loop is shown as the convex hull of the sample points inside the fence. Whiskers are denoted by the red lines. The two samples outside the fence can be considered as outliers.

part of the bag (Rousseeuw and Ruts, 1997); or a convex hull of the observations inside the fence. This convex hull is called the *loop*. An R function for constructing a bagplot is available in the package `aplpack`. A bagplot of the two head dimensions variables can be obtained from the following R code:

```
> library(aplpack)
> bagplot(Headdimensions.data[,5:6], factor = 3, cex = 0.8,
  asp = 1)
```

It is shown in Figure 2.36. Note that in order to get the correct aspect ratio argument `asp` must explicitly be assigned the value of unity. In Figure 2.37 the same bagplot is shown but with a fence formed by inflating the bag by a factor of 2. The reader can also experiment with functions `PCAbipl.bagplot` and `CVAbipl.bagplot` included in `UBbipl`.

The bagplot is a useful graphical summary of the scatterplot of bivariate observations. For the construction of the bagplot the univariate concept of rank is generalized to the concept of halfspace location depth of a point relative to a bivariate data set (see Rousseeuw *et al*., 1999). If less detail is required, especially for comparing several clusters in one plot, depth contours of the bivariate point data can be constructed. This

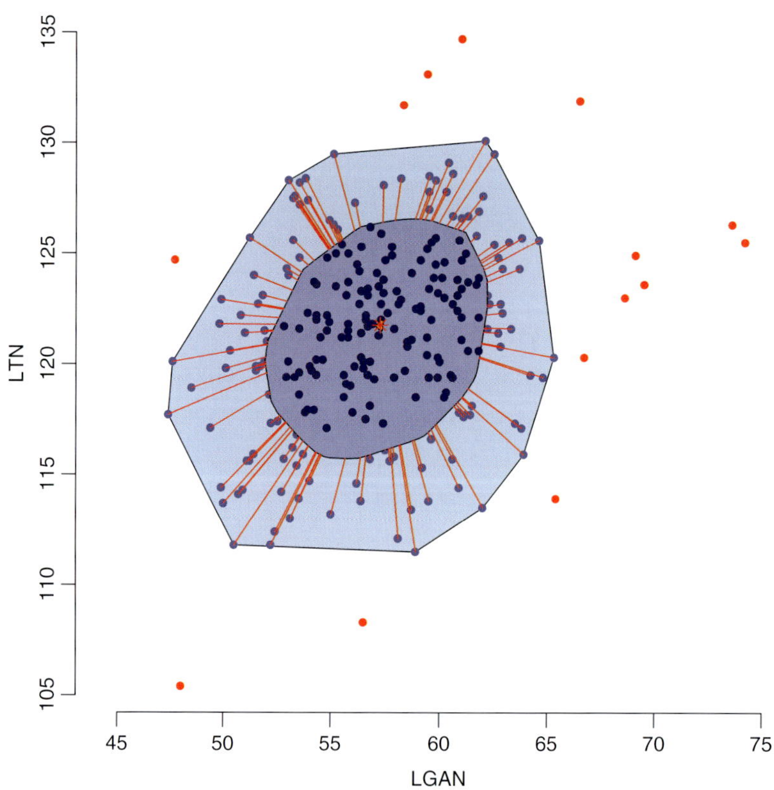

Figure 2.37 Bagplot of the two head dimensions variables. Similar to Figure 2.36 but with fence formed by inflating the bag by a factor of 2.

can be accomplished using the program `Isodepth` of Ruts and Rousseeuw (1996). The depth contours resulting from `Isodepth`, however, do not allow for specifying the exact proportion of bivariate data points enclosed by the contour. We propose the construction of depth contours using the interpolation algorithms provided in the original `bagplot` function of Rousseeuw and Ruts (1997) as graphical summaries of bivariate data points. The resulting contours will be called α-*bags*. The above functions have been modified and incorporated in the R library described in this book for constructing biplots. Instead of finding a bag containing $n/2$ observations, the 50% cut-off is replaced by a value α ranging between 0 and 1.

Typically, a value of 0.90 or 0.95 will be useful for enclosing a cluster of observations, excluding the 10% or 5% of the observations at the extremes of the cluster. Comparing the α-bag to depth contours or convex peeling, this method allows for controlling the number of observations outside the enclosure. Figure 2.38, constructed using the function `Scatterplotbags` available in `UBbipl`, contains the scatterplot of the head dimensions data with the 0.95-, 0.80-, 0.70-, 0.60- and 0.50-bags superimposed. The 0.50-bag corresponds with the bag of the bagplot. Note, however, that for large data sets the 0.50-bag and the bag of the bagplot are not identical due to the convention of using a random subsample of observations for computing the depth median and the bag in order to speed up the calculations.

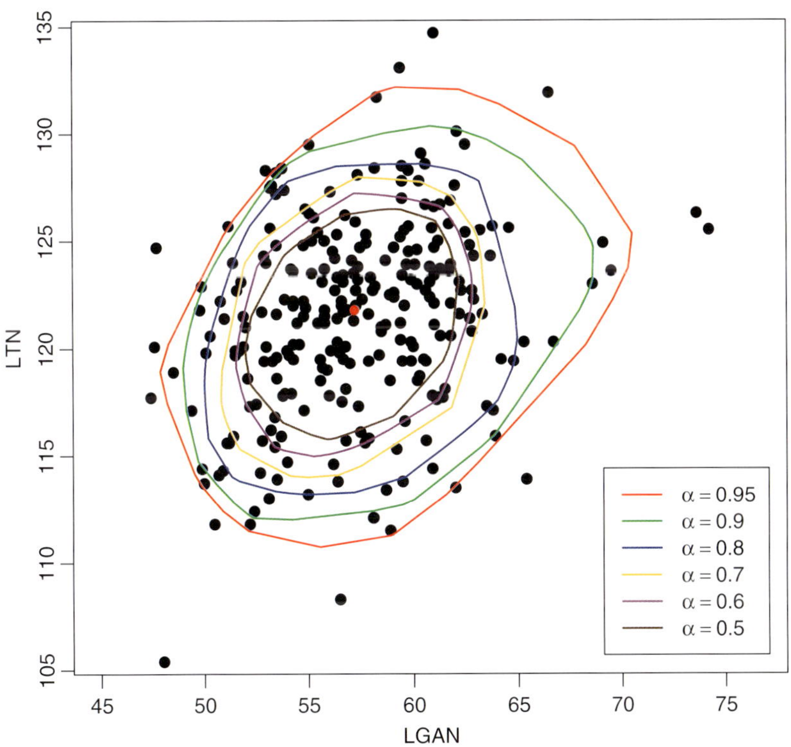

Figure 2.38 Scatterplot of two head dimensions variables with superimposed 0.95-, 0.90-, 0.80-, 0.70-, 0.60- and 0.50-bags. The Tukey median is shown as the red dot in the centre.

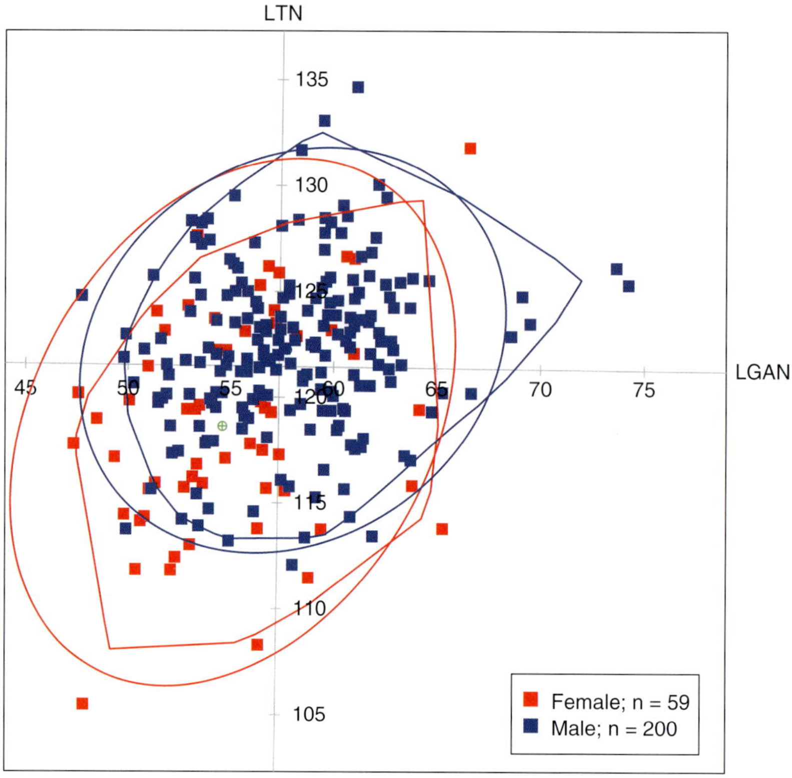

Figure 2.39 A comparison of the 0.95-bags and the 0.05 κ-ellipses for the females and males in the scatterplot of the two head dimensions variables ***LTN*** and ***LGAN***.

The biplot functions in `UBbipl` use a default value of 2500 sample points for the construction of the α-bags.

In Figure 2.39, we provide a comparison of 0.95-bags with 0.05 κ-ellipses. Separate bags and ellipses are drawn for the females and males for the two head dimensions variables ***LTN*** and ***LGAN***.

Typically, an α-bag will follow the bivariate scatter in the configuration of points more closely than the corresponding κ-ellipse because of the rigid form of the latter. This is illustrated in Figure 2.39. Note also, in the case of the females, the relatively large areas within the 0.05-ellipse not containing any data points at all. Since an α-bag is drawn about the depth median of the bivariate configuration, in contrast to the κ-ellipse that is drawn about the bivariate centroid, deviations from bivariate symmetry are reflected in differences in the general location of the two geometric structures.

2.9.5 Bivariate density plots

Other means for graphically summarizing the shape of a cloud of bivariate observations can also be computed. Hyndman (1996) proposes the use of highest density regions in

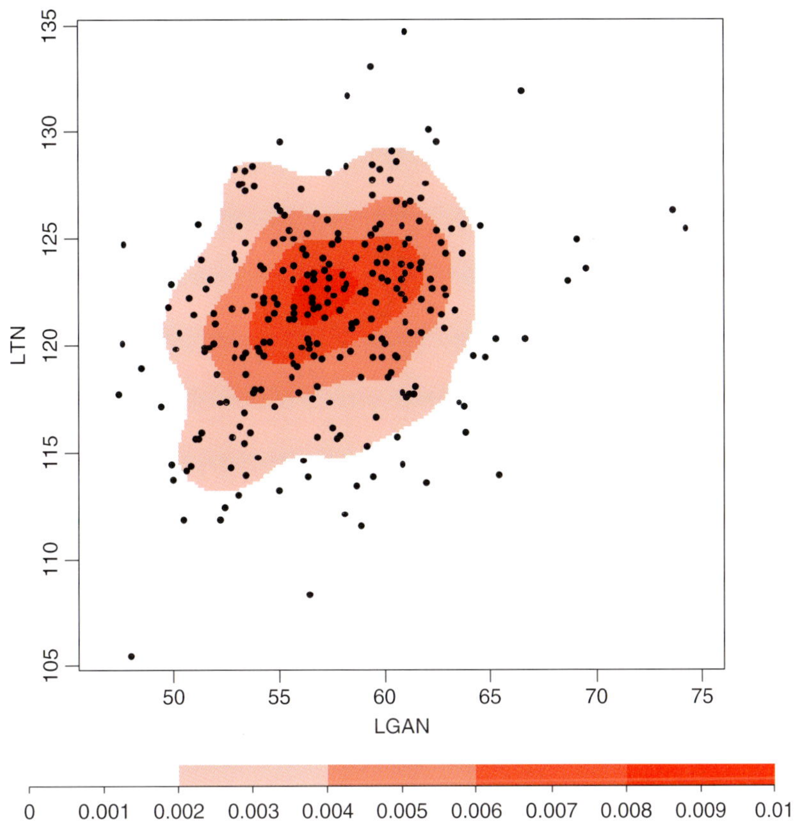

Figure 2.40 Scatterplot of two head dimensions variables data set superimposed upon a bivariate image plot of the data. Figure constructed using function `Scatterdensity` with five contour levels.

generalizing the boxplot. The highest density region is a probability region in contrast to the depth region defined in the construction of the bagplot. It differs from the other bivariate boxplot generalizations in its emphasis on density. For this reason the mode, rather than the median, is indicated. The α-bag provides a good summary of a unimodal cloud of sample points. However, instead of generalizing the boxplot, bivariate density estimation procedures (Scott, 1992; Venables and Ripley, 2002; Eilers and Goeman, 2004) can also be employed to represent multimodal clouds of points. We make use of the R function `kde2d` provided by Venables and Ripley in their package `MASS` to add a bivariate density plot in the form of an image plot to the scatterplot of the head dimensions data set. Figures 2.40 and 2.41 are constructed with our function `Scatterdensity` available in `UBbipl`.

All the above methods are invariant under rotation of the two-dimensional points and hence may be used with biplots. While α-bags are geared for unimodal distributions, density plots have the advantage that they are able to show multimodal characteristics of sample points in a scatterplot. However, when the sample points originate from different subgroups, the scatterplot (or biplot) may be overlaid with α-bags for each

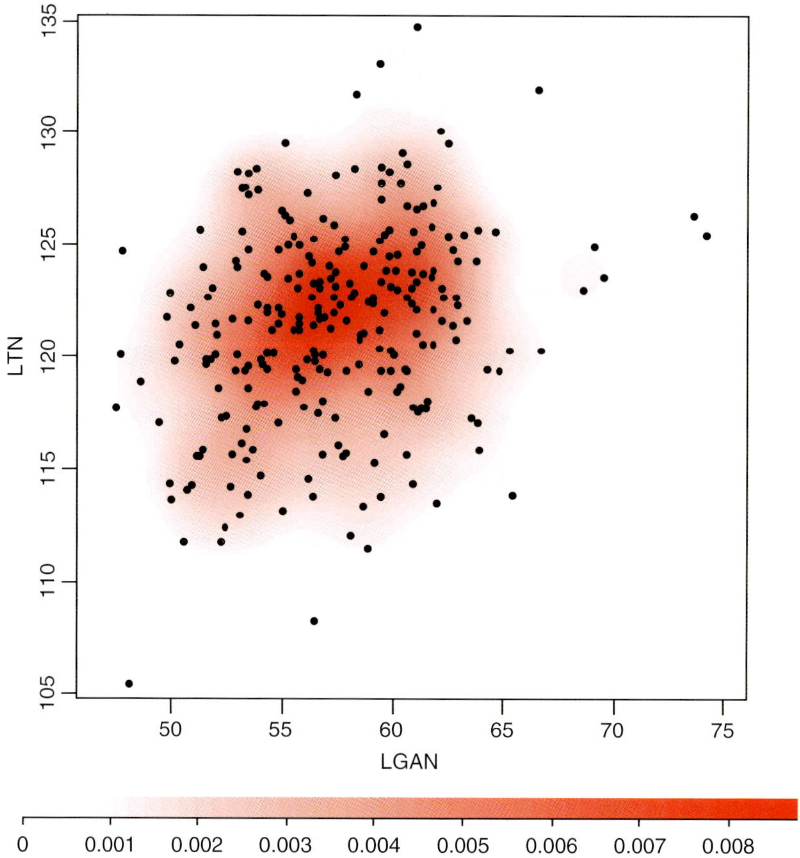

Figure 2.41 Scatterplot of two head dimensions variables data set superimposed upon a bivariate image plot of the data. Figure constructed using our function `Scatterdensity` such that a continuous background image plot is obtained.

group. In the next section, we provide an example of this. Overlaying a scatterplot with density estimates for several groups generally leads to a graph that cannot be easily comprehended. Therefore, we choose to use density estimates in the form of image plots as background for our biplots overlaid with α-bags and/or κ-ellipses. It follows from the above discussion that there are various methods for graphically summarizing the shape of a bivariate configuration. In this book, we concentrate on α-bags, κ-ellipses and image plots and will show by example that these are invaluable devices for providing bivariate descriptive summaries when used together with biplots of large data sets.

2.10 Buying by mail order catalogue data set revisited

The biplot given in Figure 2.28 is a two-dimensional approximation of the 90-dimensional sample points in the mail catalogue data set. The matrix consisting of the two-dimensional coordinates for plotting the biplot can be used as input to the R function `PCAbipl` to

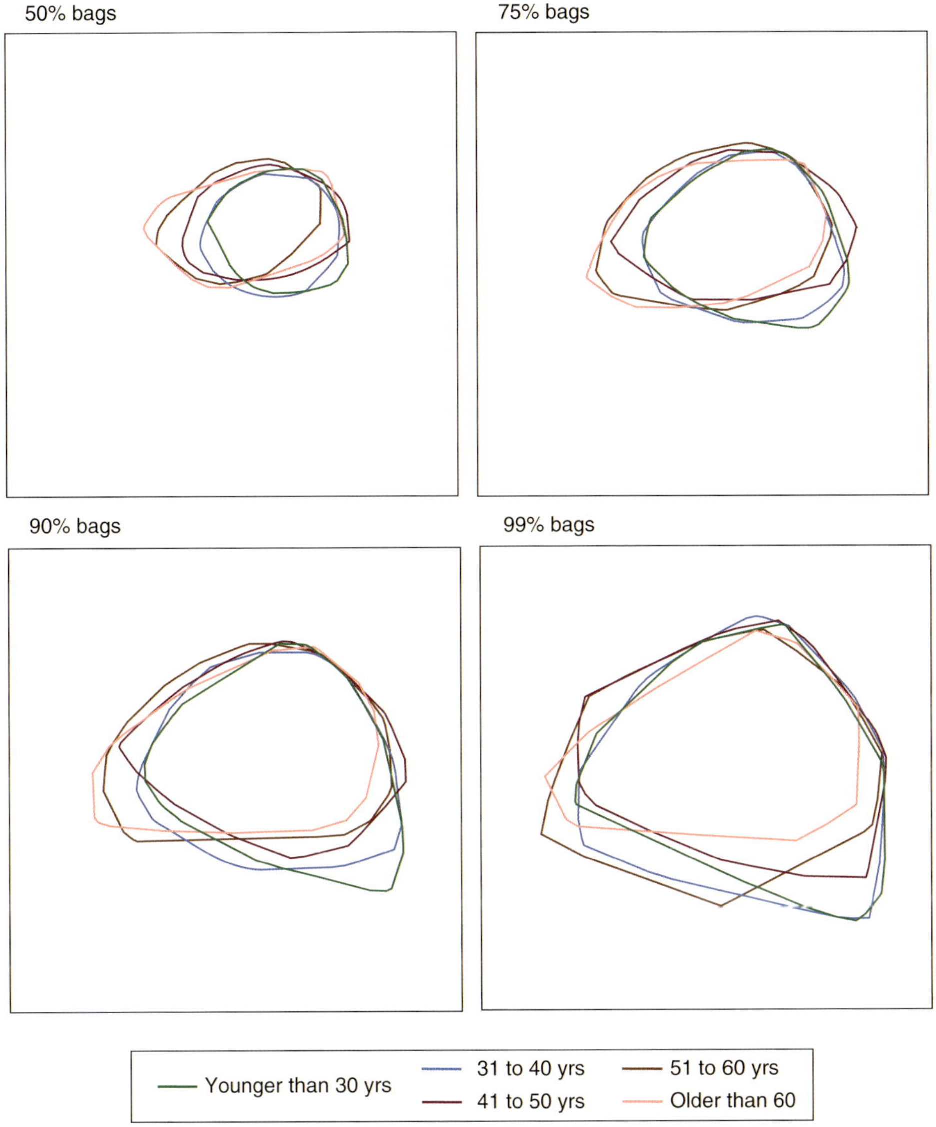

Figure 2.42 Biplots with selected α-bags of the mail order catalogue data set. The α-bags demarcate the deepest $100\alpha\%$ of the respondents in the different age groups.

equip the Figure 2.28 biplot with user-specified α-bags. Figure 2.42 shows this with 0.50-, 0.75-, 0.90- and 0.99-bags. Plotting of individual sample points as well as the axes has been suppressed but the user can interactively add sample points, means, or axes as required.

A detailed discussion with illustrations of the capabilities of `PCAbipl` and related biplot functions contained in `UBbipl` is given in Chapter 3 where a closer look is taken at PCA biplots.

2.11 Summary

In this chapter we have concentrated on assymmetric biplots as defined in Chapter 1. As with an ordinary scatterplot, we have seen that such a biplot consists of

- a set of points representing samples;
- labelled axes representing variables.

We have also studied how to calibrate the axes for prediction or interpolation. Our illustrations have used linear axes, but we have already discussed the possibility of using irregular calibrations, nonlinear axes and convex regions for categorical variables. We have pointed out the advantages of being able to shift axes and to rotate plots as an aid to making more readable biplots. The possibilities for scaling axes have been addressed. The size of a data set is a potential challenge to the usefulness of a biplot. We have discussed some measures for handling data sets containing many data points as well as data sets with many variables. Especially when there are a large number of points, there is a need to simplify the display by using various strategies for representing density and variability.

In this chapter, the focus has been on understanding the basics of biplot construction by analogy with properties of ordinary scatterplots for approximating a p-variable data matrix. We have also discussed in some detail R code for implementing all procedures. This we did for two reasons: firstly, we believe that programming a procedure leads to a better understanding of the procedure; and secondly, we would like to provide the reader with the necessary computer software to produce all the biplot material discussed in the book. An important issue has been avoided thus far: how good is the overall biplot approximation? How trustworthy are the biplot axes for predicting the values of the variables they represent? How accurate are the positions occupied by the individual samples in the biplot approximations? These and other topics have been deferred to chapters dealing with the specific types of biplots. In the next chapter, we give the algebraic and geometric basis of PCA biplots and discuss several interesting applications of PCA biplots.

3 Principal component analysis biplots

The general form of an asymmetric biplot was introduced in Chapter 2 as a multivariate scatterplot. In this chapter we turn our attention to the first and, arguably, the simplest and most popular form of asymmetric biplot, the principal component analysis (PCA) biplot. Although only referred to as multivariate scatterplots, many of the figures in Chapter 2 are indeed PCA biplots. The PCA biplot is asymmetric because it represents the samples and variables of $\mathbf{X}$; a symmetric form that mainly represents covariance or correlation is discussed in Chapter 10. In this chapter we first review algebraic and geometric properties of PCA before discussing some examples of its biplot.

Our main instrument for constructing PCA biplots is our R function `PCAbipl`. We discuss the capabilities of `PCAbipl` by introducing some interesting applications and enhancements of more conventional PCA biplots. Readers are urged to explore the potential of `PCAbipl` using their own data.

3.1 An example: risk management

Van Blerk (2000) discuss an example where biplot methodology is applied in risk management. Risk management is critical for institutions such as banks, pension funds, asset management and insurers to ensure that their liabilities can be met at all times. A popular quantity for measuring exposure to market risk is the so-called value at risk (VAR). This quantity is defined by Jorion (1997) as 'the expected maximum loss (or worst case) over a given time interval at a given confidence level'.

In practice the calculation of VAR is complicated. The institution considered here uses historical data and simulation to calculate the empirical probability distribution of the return on a financial instrument. The empirical πth percentile is then used as an estimate of the $(100 - \pi)\%$ VAR. Since risk management is concerned with expected losses the

Understanding Biplots John Gower, Sugnet Lubbe and Niël le Roux

Table 3.1 95% VAR of financial instruments constituting a portfolio.

Day	CM	IRD	MM	ALCO	SE	EDSA	EDM
1	−1.765	−0.248	−0.281	−0.296	−0.141	−0.226	−0.941
2	−0.818	−1.326	−0.281	−0.296	−0.142	−0.012	−3.384
3	−1.715	−1.140	−0.596	−0.296	−0.141	−0.182	−2.872
4	−1.771	−1.641	−0.596	−0.296	−0.145	−0.890	−1.946
5	−1.661	−1.302	−0.412	−0.476	−0.132	−0.215	−1.290
6	−0.022	−1.364	−0.608	−0.279	−0.216	−0.299	−1.377
7	−0.889	−1.137	−0.457	−0.453	−0.152	−0.255	−1.129
8	−0.914	−1.199	−0.457	−0.404	−0.147	−0.083	−1.137
9	−1.149	−1.182	−0.457	−0.404	−0.149	−0.357	−1.175
10	−1.273	−0.733	−0.457	−0.404	−0.156	−0.556	−0.894
11	−0.817	−0.851	−0.457	−0.404	−0.167	−0.379	−0.888
12	−1.207	−1.513	−0.457	−0.457	−0.161	−0.038	−0.804
13	−0.862	−1.819	−0.459	−0.457	−0.158	−0.139	−0.939
14	−2.552	−1.400	−0.459	−0.457	−0.165	−0.140	−0.914
15	−1.431	−1.320	−0.459	−0.457	−0.168	−0.137	−1.097
16	−2.838	−1.318	−0.459	−0.457	−0.158	−0.369	−0.162
17	−1.077	−1.273	−0.730	−0.457	−0.256	−0.089	−1.125
18	−1.026	−1.338	−0.730	−0.436	−0.277	−0.296	−1.024
19	−1.046	−1.307	−0.025	−0.031	−0.135	−0.365	−0.646
20	−0.627	−2.030	−0.025	−0.031	−0.132	−0.138	−0.613

choice of π invariably leads to a negative return and therefore $\pi\%$ is interpreted as the chance of observing a larger loss than this negative return. Plotting an instrument's VAR over time assists management in identifying trends. The portfolio considered here consists of seven financial instruments which are not performing independently of one another, calling for a multivariate approach for analysing the VAR figures calculated for each of these seven instruments on a daily basis.

Table 3.1 contains the daily 95% VAR for 20 consecutive working days for each instrument constituting the example portfolio. This data set is available as the dataframe `VAR95.data` in the package `UBbipl`, and Figure 3.1 is constructed with the following R command:

```
PCAbipl(X = VAR95.data[,-1], colours = "green", offset = c(0, 0,
   0.3, 0.3), offset.m = c(-0.25, -0.25, -0.25, -0.25, -0.25,
   -0.25, -0.25), pch.samples = 15, pch.sample.size = 1.1,
   pos = "Hor", pos.m = c(4,4,2,2,4,1,1), side.label = c("right",
   "right","left","left","right","right","left")
```

The call to `PCAbipl` resulting in Figure 3.1 accepts all the default settings except those specifying the input data matrix (`X`), the colour and size of the symbol for plotting the sample points (`colours` and `pch.samples.size`), the plotting symbol for the sample points (`pch.samples`), the placement of the labels for the axes (`pos`), the argument `offset` (the third and fourth values of `offset` are increased to add space between the side of the graph and the axis name ***SE*** and ***IRD*** on side 3; and ***EDM*** on side 4) and arguments for placement of the scales on the axes (`pos.m`, `offset.m` and `side.label`). In general, the construction of a PCA biplot is a two-stage procedure: first `PCAbipl` is

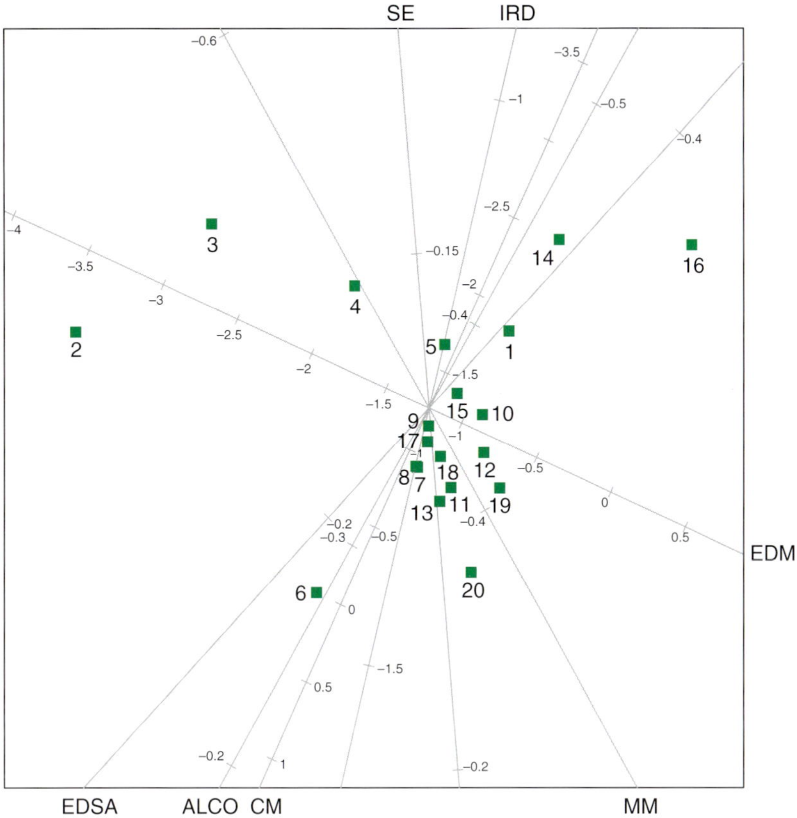

Figure 3.1 PCA biplot of 95% VAR of financial instruments data constituting a portfolio.

called with very few changes to the default settings; then the biplot is fine-tuned by utilizing the many arguments available for that purpose. As an example, consider the number of scale markers on each biplot axis in Figure 3.1. Some of the axes, like ***MM,*** have too few markers while others, like ***EDM,*** perhaps have too many. The argument `n.int` provides a mechanism for equipping each axis with just the right number of markers. This has been done in Figure 3.2.

The PCA biplot in Figure 3.1 is a useful multidimensional scatterplot representing the seven financial instruments over 20 days in a single display. Since the biplot axes shown here are prediction biplot axes, the approximate values for the individual financial instruments can be readily read off from the axes. Figure 3.2 illustrates how predictions for selected sample points can be obtained with `PCAbipl` using arguments `predictions.sample` and `ort.lty`. The following changes have been made in the call to `PCAbipl` to obtain Figure 3.2: `n.int = c(3,10,10,5,10,10,3), predictions.sample = c(2,16), ort.lty = 2.`

By default, the specified predictions are shown on the biplot as illustrated in Figure 3.2 and also returned as output on the R console – see Table 3.2. We note that on ***day2*** (and also on ***day3***) abs(VAR) of ***EDM*** was particularly large; this is in sharp contrast with

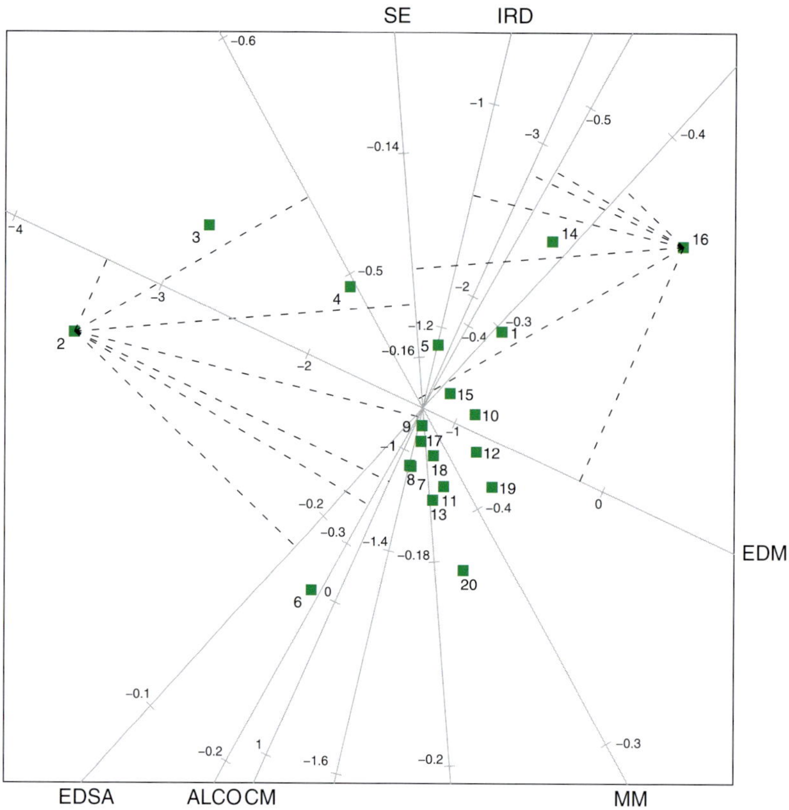

Figure 3.2 PCA biplot of 95% VAR of financial instruments constituting a portfolio. Predictions of samples ***2*** and ***16*** are shown on the biplot. Notice also how the number of markers on the axes has been changed by utilizing argument `n.int`.

Table 3.2 Predictions of VAR for sample points ***2*** and ***16*** for the different instruments.

	s2	s16
CM	−0.789	−2.803
IRD	−1.280	−1.082
MM	−0.533	−0.447
ALCO	−0.318	−0.471
SE	−0.155	−0.151
EDSA	−0.184	−0.372
EDM	−3.374	−0.156

day16 exhibiting a very large abs(VAR) of ***CM***. This is the type of useful information financial planners and managers need to notice. They can now research the circumstances surrounding ***EDM*** at the beginning of the period and those of ***CM*** towards the end of the period to suggest in future what circumstances lead to high risks on ***EDM*** and ***CM***.

One might argue that by a simple line plot over time of ***EDM,*** or of ***CM*** this feature would also be highlighted, but the advantage of the biplot is to give a global overview of the interlinked financial instruments. We return to this example later in the chapter.

3.2 Understanding PCA and constructing its biplot

According to Jolliffe (2002), PCA is essentially directed 'to reduce the dimensionality of a data set consisting of a large number of interrelated variables, while retaining as much as possible of the variation present in the data set. This is achieved by transforming to a new set of variables, the principal components (PCs), which are uncorrelated, and which are ordered so that the first *few* retain most of the variation present in *all* of the original variables.' While not disagreeing with Jolliffe, we take a rather different approach that emphasizes different aspects of the PCA transformation. This difference of emphasis will become clear in the following.

The fundamental problem of PCA is to approximate $\mathbf{X}$ by $\hat{\mathbf{X}}_{[r]}$ in r dimensions or, equivalently, of rank r. In PCA the columns of $\mathbf{X}$ refer to different variables, and before we can think about approximation we have to handle what might be incommensurabilities between variables. Thus, if we had variables measuring height in metres and weight in kilograms, we might have difficulties and certainly would not be content with any method of analysis that was sensitive to changes of scale, such as replacing metres and kilograms by feet and pounds. Some multivariate methods handle this problem routinely, but PCA is not one of them. The problem was avoided in our introductory example because all variables were on the same VAR scale. Other common methods of scaling are to normalize by dividing each variable by its standard error or, for positive variables, by taking logarithms.

Assuming that any necessary pre-scaling of the data has been attended to, PCA uses a least-squares criterion as the basis of approximation. To be precise, the sum of squares of the differences between corresponding members of $\mathbf{X}$ and $\hat{\mathbf{X}}_{[r]}$ is minimized. Algebraically this may be written:

$$\text{minimize tr}\{(\mathbf{X} - \hat{\mathbf{X}}_{[r]})'(\mathbf{X} - \hat{\mathbf{X}}_{[r]})\} \text{ or, equivalently, minimize } \|(\mathbf{X} - \hat{\mathbf{X}}_{[r]})\|^2. \quad (3.1)$$

Geometrically, we consider the rows of $\mathbf{X}$ as giving the coordinates of n points in p dimensions and are seeking the r-dimensional plane containing the points whose coordinates are given by the rows of $\hat{\mathbf{X}}_{[r]}$ that minimizes criterion (3.1). For a minimum, it is intuitive (and may be formally proved) that the best fit is obtained when $\hat{\mathbf{X}}_{[r]}$ is an orthogonal projection of $\mathbf{X}$. Furthermore, we know that the plane must pass through the centroid of the points given by $\mathbf{X}$. This is a simple consequence of Huygens' principle that the sum of squares about the mean is smaller than the sum of squares about any other point. Replacing $\mathbf{X}$ by $(\mathbf{I} - \frac{1}{n}\mathbf{11}')\mathbf{X}$ ensures that the centroid is at the origin. In the following, we assume that $\mathbf{X}$ has been centred in this way. Thus, it only remains to find the direction of the best-fitting plane. The solution to this least-squares problem is given by the Eckart-Young theorem (Eckart and Young, 1936).

As shown in (2.3), $\mathbf{X} : n \times p = \mathbf{U\Sigma V}'$ and the r-dimensional Eckart-Young approximation $\hat{\mathbf{X}}_{[r]} = \mathbf{U\Sigma JV}' = \mathbf{UJ\Sigma V}' = \mathbf{UJ\Sigma JV}' = \mathbf{XVJV}'$ minimizes the squared error loss (2.4). Then the coordinates of the r-dimensional approximation of the centred $\mathbf{X}$ are given by the first r columns of $\mathbf{U\Sigma J}$, $(\mathbf{U\Sigma})_r = \mathbf{XV}_r$ and the directions of the axes by $\mathbf{VJ}$. Thus we plot the coordinates given by the n rows of $\mathbf{U\Sigma J}$ to give the r-dimensional scatterplot and the p rows of $\mathbf{VJ}$ to give the directions of the axes. How to calibrate these axes is discussed below.

Note that $\mathbf{VJV}'$ is the required projection matrix. Thus, the rows of $\mathbf{X}$, $\mathbf{VJV}'$ give the projections of $\mathbf{X}$ onto the r-dimensional hyperplane relative to the original p orthogonal axes, while the rows of $\mathbf{XVJ}$ give the projections of the same points relative to r orthogonal axes in the hyperplane, the remaining $p - r$ dimensions being zero. In particular, the unit points on the original p axes project to $\mathbf{IVJV}'$ and $\mathbf{IVJ}$, respectively, showing that the first r columns of $\mathbf{VJ}$ give the direction cosines in r dimensions of the best-fitting plane. The columns of $\mathbf{V}$ are often termed the *principal components,* or the *principal component loadings,* and $\mathbf{XV}$ interpreted as new latent variables. Because $\mathbf{V}'\mathbf{X}'\mathbf{XV} = \mathbf{\Sigma}^2 = \mathbf{\Lambda}$, which is diagonal, these latent variables are uncorrelated and in the literature much is made of the possibilities of their interpretation. Here, we almost entirely ignore this aspect, which has its roots in factor analysis, and concentrate on representing and interpreting the original variables themselves. For us, the main interest in the uncorrelated property is that it gives us orthogonal axes in r dimensions that simplify the plotting of visualizations, but which need not themselves be shown. One important property is that

$$\|\mathbf{X}\|^2 = \|\hat{\mathbf{X}}_{[r]}\|^2 + \|\mathbf{X} - \hat{\mathbf{X}}_{[r]}\|^2, \tag{3.2}$$

showing that the total sum of squares is the sum of the fitted and residual sums of squares. This underpins the proper use of measures such as the 'variance accounted for' described more fully in Section 3.3. Also, it makes it clear that minimizing the sum of squares of the residuals is the same as maximizing the variation in the fitted plane, as stated in Jolliffe's definition of PCA.

3.2.1 Representation of sample points

Geometrically, we have seen that the Eckart-Young approximation orthogonally projects the points in $\mathbf{X}$ onto the best two (in general, r) dimensions for visualization.

Consider the three-dimensional data set given in Table 3.3 and represented graphically in Figure 3.3. To represent Figure 3.3 optimally in two dimensions, a plane, passing through the centroid, must be found that minimizes the sum of squares of the distances from the plane. We call this plane the biplot plane and denote it by $\mathscr{L}$. The plane $\mathscr{L}$ is shown in Figure 3.4.

Minimizing squared distances from original points amounts to orthogonal projections onto the plane. This is illustrated in Figure 3.5, where the first two principal components (the columns of $\mathbf{V}_r$) span the two-dimensional biplot space and the sum of the squares of the distances represented by the arrows is minimized.

In the PCA biplot, interpolation is achieved by orthogonal projection of each sample point onto the biplot space, and because $\mathbf{V}$ is orthogonal

$$\mathbf{x}'_{proj} = \mathbf{x}'\mathbf{V}_r\mathbf{V}'_r \tag{3.3}$$

is the representation of sample $\mathbf{x}$, projected onto in terms of the three Cartesian axes. The projection is illustrated in Figure 3.5 and the interpolated sample points are shown in Figure 3.6.

Table 3.3 Three-dimensional example data set.

Sample no.	X	Y	Z
1	13.5471	12.6356	21.7779
2	15.3202	5.4330	20.4339
3	17.7739	12.4688	19.3321
4	18.3664	14.7330	28.2576
5	18.0205	9.3600	25.1894
6	16.9528	11.8832	21.8291
7	17.5986	13.0341	21.5221
8	13.6643	11.2316	13.9913
9	19.3081	11.7703	28.3937
10	20.2114	8.0871	22.4164
11	15.7187	14.7718	18.7206
12	11.6222	17.9957	21.4343
13	14.4965	15.3002	20.6371
14	19.5328	11.8101	22.1391
15	16.1871	10.0019	16.2823
16	17.5598	10.0402	20.3788
17	17.2416	13.5807	25.4979
18	16.2913	14.9305	22.9105
19	17.7401	15.5304	22.6354
20	20.3395	10.9800	23.1621
21	8.5554	12.2874	12.9827
22	11.1397	12.4292	18.0374
23	11.9930	13.4574	21.6548
24	7.8248	4.4579	8.4648
25	10.7568	6.6412	17.6931

When representing the samples relative to orthogonal axes in $\mathscr{L}$, the coordinates of the projected points are given by

$$\mathbf{z}' = \mathbf{x}'\mathbf{V}_r, \tag{3.4}$$

which implies taking the biplot plane in Figure 3.6 and placing it in the x-y plane to produce a scatter diagram. The steps in creating this scatter diagram are: first, the data in Table 3.3 are column-centred to $\mathbf{X}$, say; then we obtain the SVD $\mathbf{X} = \mathbf{U\Sigma V}'$ with

$$\mathbf{V} = \begin{bmatrix} 0.5453 & -0.4502 & 0.7071 \\ 0.2784 & 0.8929 & 0.3540 \\ 0.7907 & -0.0038 & -0.6122 \end{bmatrix}. \tag{3.5}$$

Finally, using (3.4), the first two columns of $\mathbf{XV}$ are plotted, and the resulting scatter diagram is shown in Figure 3.7.

A figure similar to Figure 3.7 can also be constructed with the following call to `PCAbipl`:

```
PCAbipl(X = as.matrix(Table.3.3.data[,-1]), ax = NULL,
   pch.samples = 15, colours = "green")
```

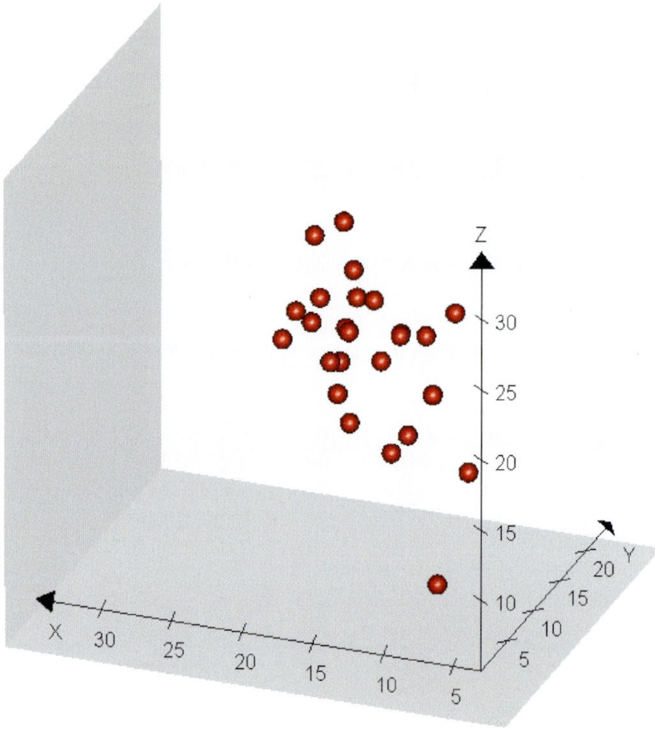

Figure 3.3 Three-dimensional representation of data from Table 3.3.

However, the figure resulting from the call to `PCAbipl` will differ in two important respects from Figure 3.7: there will be three biplot axes (unless their printing is suppressed by using the argument `ax = NULL`; printing of the scaffolding axes (V_1 and V_2 together with their calibrations) is always suppressed.

3.2.2 Interpolation biplot axes

However, the scatter diagram in Figure 3.7 is not a biplot since it only contains information on the sample points. No information is visible on the variables $\boldsymbol{X}$, $\boldsymbol{Y}$ and $\boldsymbol{Z}$ from Table 3.3. To add information on the variables in Figure 3.7, biplot axes are added to the display. Interpolation axes are used to interpolate (new) sample points. Any (new) sample $\mathbf{x}^* : p \times 1$ can be interpolated by orthogonal projection onto the biplot space $\mathscr{L}$ using (3.4) to give $\mathbf{z}^{*\prime} = \mathbf{x}^{*\prime}\mathbf{V}_r$. In order to construct interpolation biplot axes, note that

$$\mathbf{x} = \sum_{k=1}^{p} x_k \mathbf{e}_k \text{ and } \mathbf{z}^{*\prime} = \mathbf{x}^{*\prime}\mathbf{V}_r = \sum_{k=1}^{p} x_k^* \mathbf{e}_k' \mathbf{V}_r, \tag{3.6}$$

so that the interpolant $\mathbf{z}^*$ is obtained as the weighted sum of the orthogonal projections onto $\mathscr{L}$ of the unit points ($\mathbf{e}_k$) of the Cartesian axes.

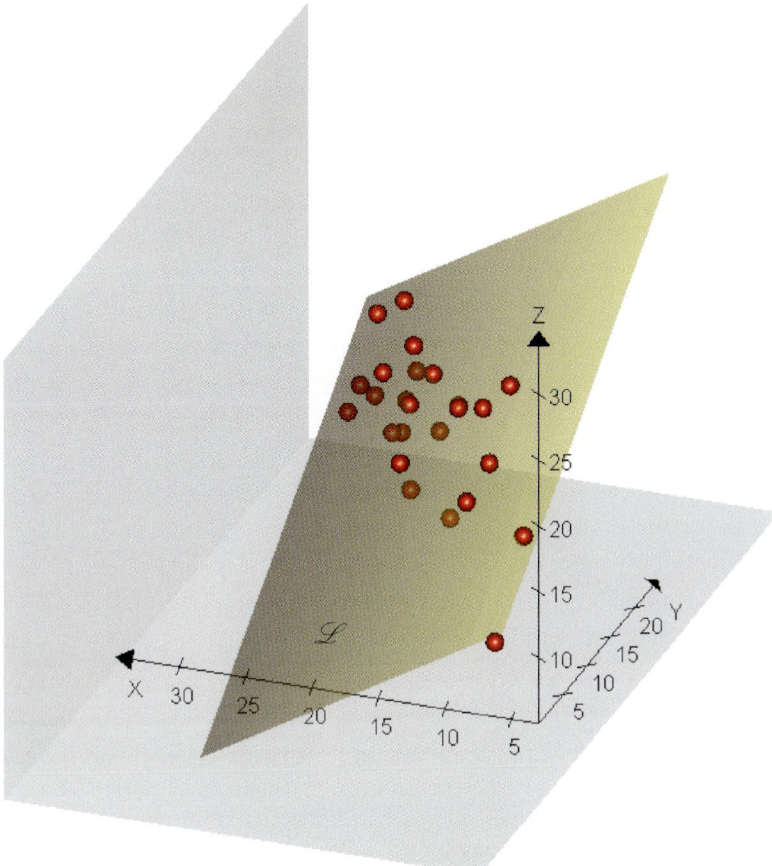

Figure 3.4 Two-dimensional PCA plane, denoted by $\mathscr{L}$, through the centroid, representing the three-dimensional data in Table 3.3. The brighter points are above the plane and the others below the plane.

From equation (3.6) it is clear that the kth Cartesian axis is represented in the biplot space by the orthogonal projection $\mathbf{e}_k'\mathbf{VJ}$ of the axis onto the biplot space. This is illustrated for the first Cartesian axis ($\boldsymbol{X}$) in Figure 3.8.

As mentioned earlier, the biplot space $\mathscr{L}$ necessarily passes through the centroid of the data points. In Figures 3.8 and 3.9 the Cartesian axes (in black) have been shifted to intersect at the centroid of the data configuration.

When placing the biplot space $\mathscr{L}$ in the two-dimensional x-y plane, the PCA biplot with all the interpolation axes is given in Figure 3.10. This biplot is obtained by the call

```
PCAbipl(X = as.matrix(Table.3.3.data[,-1]), ax.type =
   "interpolative", pch.samples = 15, colours = "green", pos =
   "Hor", offset = c(0.1,0.3,0.3,0.3), reflect = "x")
```

Notice that Figure 3.10 is just a reproduction of Figure 3.7 with interpolative biplot axes added but with the plotting of the scaffolding axes $\mathbf{v}_1$ and $\mathbf{v}_2$ suppressed. Also notice

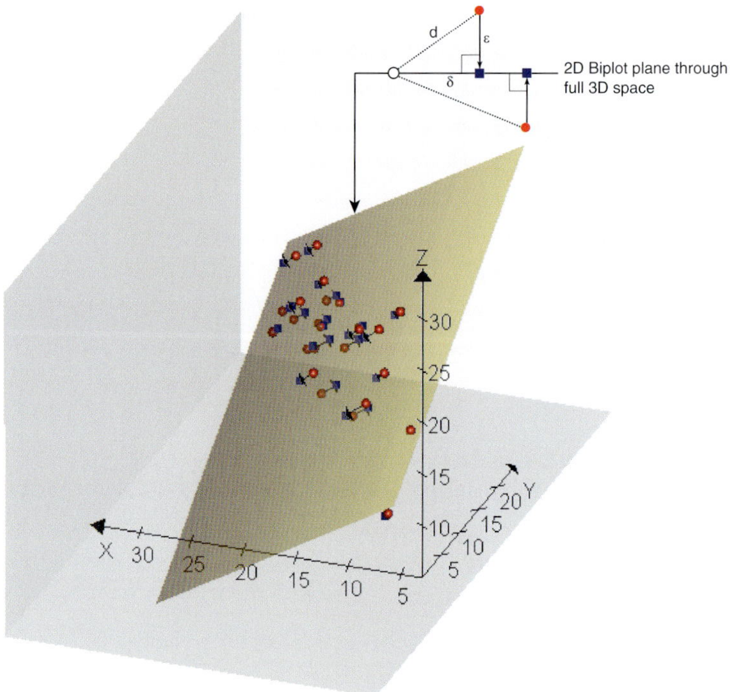

Figure 3.5 Projection of three-dimensional sample points onto the two-dimensional biplot space $\mathscr{L}$. Red points in full space are projected orthogonally onto the blue points in $\mathscr{L}$ such that the sum of the squared distances ε (indicated by the black arrows) is minimized. Actual distances in the full space are denoted by d; approximated distances in $\mathscr{L}$ denoted by δ. The plane of best fit passes through the centroid of the data.

that the calibrations of the axes are in terms of the original units used in Table 3.3. However, several features of Figure 3.10, for example the number of calibrations on each biplot axis, need to be fine-tuned. We will attend to fine-tuning of biplots in subsequent sections.

Recall from Section 2.6 that the biplot axes in Figure 3.10 cannot be used for reading off the $\boldsymbol{X}$, $\boldsymbol{Y}$ or $\boldsymbol{Z}$ value of any given sample point. However, as we have seen in Section 2.6, the interpolation calibrations can be used to determine the position of any new point in the biplot space using the vector-sum method. We illustrate this method step by step in Figure 3.11, introducing some functions for interactively enhancing biplots.

Figure 3.11 is the result of the following code:

```
> draw.arrow(col = "red", length = 0.15, angle = 15, lwd = 1.75)
```

The function `draw.arrow` is called three times, using the mouse to select the beginning and the end of the arrow.

```
> draw.polygon(vertex.points = 3, border = "blue", lwd = 1.75)
```

The function `draw.polygon` draws the polygon as selected by the mouse and returns the coordinates of the polygon's centroid as (−6.0009, 0.2896).

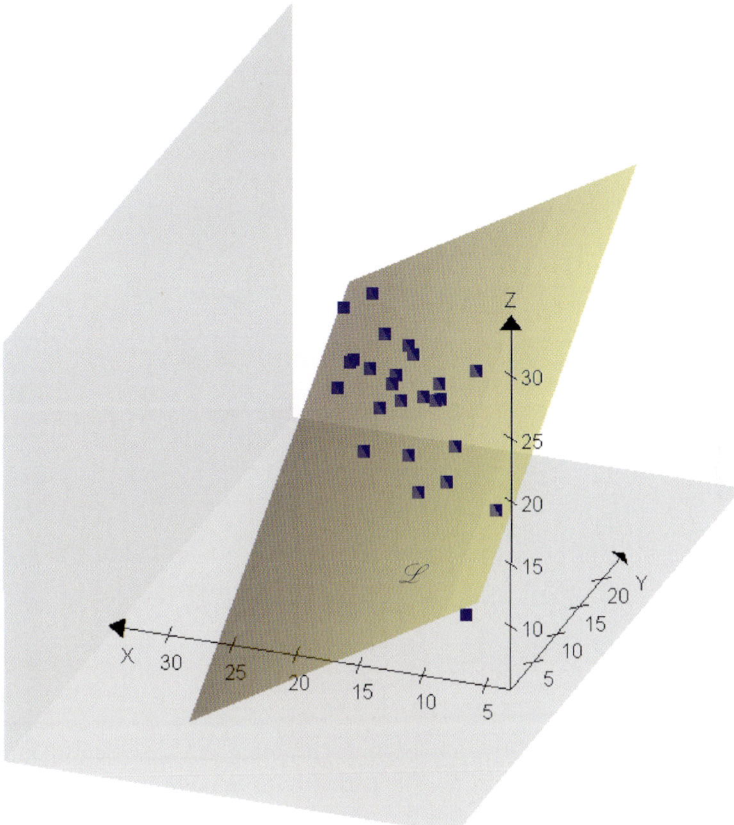

Figure 3.6 Interpolated sample points represented in terms of the three Cartesian axes.

```
> points(x = -6.0009, y = 0.2896, pch = 1, col = "blue",
  cex = 2, lwd = 2)
> draw.arrow(col = "black", length = 0.15, angle = 15, lwd = 1.75)
> arrows(-6.0009, 0.2896, -6.0009*3, 0.2896*3, col = "black",
  length = 0.15, angle = 15, lwd = 1.75, lty = 2)
> points(-6.0009*3, 0.2896*3, pch = 16, col = "red", cex = 0.8)
> draw.text(string = "outlier?", cex = 0.6)
```

The function `vectorsum.interp` is available for performing most of the above actions in a single function call.

In the next section we show in detail how to equip the PCA biplot with axes that can be used for prediction – that is, to allow reading off of variable values associated with the respective samples.

3.2.3 Prediction biplot axes

To illustrate the construction of a prediction biplot axis, the first Cartesian axis ($\boldsymbol{X}$) of the Table 3.3 data will be used.

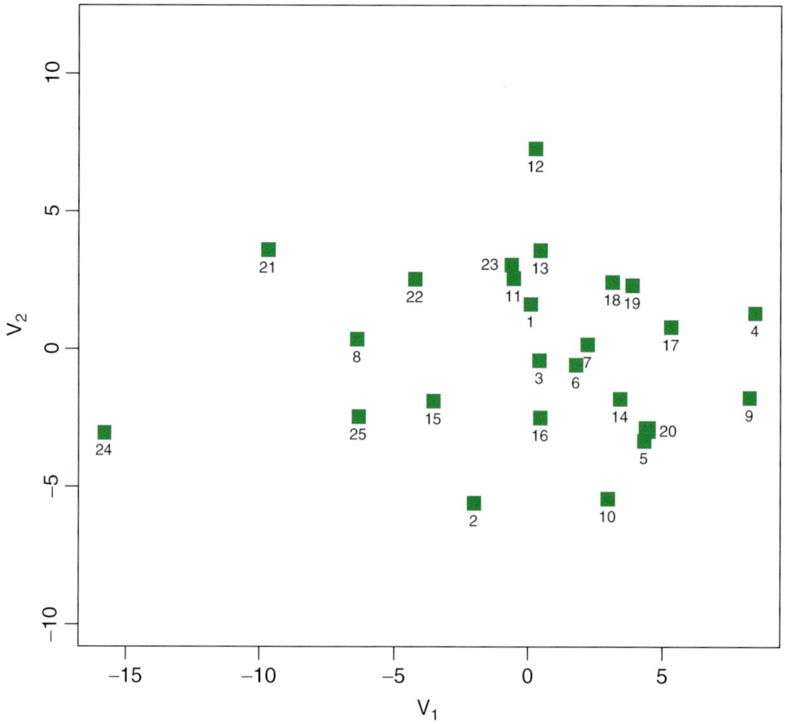

Figure 3.7 PCA display of the sample points of Table 3.3. The scaffolding axes are the first two columns of the matrix $\mathbf{V}$. The aspect ratio is unity.

In Figure 3.12 the biplot space $\mathscr{L}$ is the same subspace of the three-dimensional space $\mathscr{R}$ as above. Any point $\mathbf{z}^* : r \times 1$ in terms of the scaffolding (the first two columns of $\mathbf{V}$ in (3.5)) or basis for the biplot space, with $r = 2$ in Figure 3.12, is also a point $\mathbf{x}^* : p \times 1$ in terms of the basis for $\mathscr{R}$, where $p = 3$ in the figure. Such a point will project onto itself, giving

$$\mathbf{x}^{*\prime} = \mathbf{x}^{*\prime}\mathbf{V}_r(\mathbf{V}_r'\mathbf{V}_r)^{-1}\mathbf{V}_r'. \tag{3.7}$$

In Section 3.2.2 it was shown that the interpolation of a point $\mathbf{x}^*$, is given by $\mathbf{z}^{*\prime} = \mathbf{x}^{*\prime}\mathbf{V}_r$. Substituting into (3.7) above and since the columns of $\mathbf{V}_r$ are orthonormal, we have

$$\hat{\mathbf{x}}^{*\prime} = \mathbf{z}^{*\prime}\mathbf{V}_r' \tag{3.8}$$

as the prediction of $\mathbf{z}^*$. For constructing a prediction biplot axis, a $(p - 1)$-dimensional hyperplane $\mathscr{N}$ perpendicular to the Cartesian axis is needed. In Figure 3.13 a two-dimensional plane is constructed perpendicular to the first Cartesian axis through the marker '20'. The intersection of $\mathscr{L}$ and $\mathscr{N}$ is an $(r - 1)$-dimensional intersection space, indicated by a line in Figure 3.13. All points on this line will predict the value '20' for the $\boldsymbol{X}$-axis. In Figure 3.14 the plane $\mathscr{N}$ has been shifted orthogonal to the Cartesian axis $\boldsymbol{X}$ to go through marker '15'. All the points on this intersection space will predict the value '15' for the $\boldsymbol{X}$-axis.

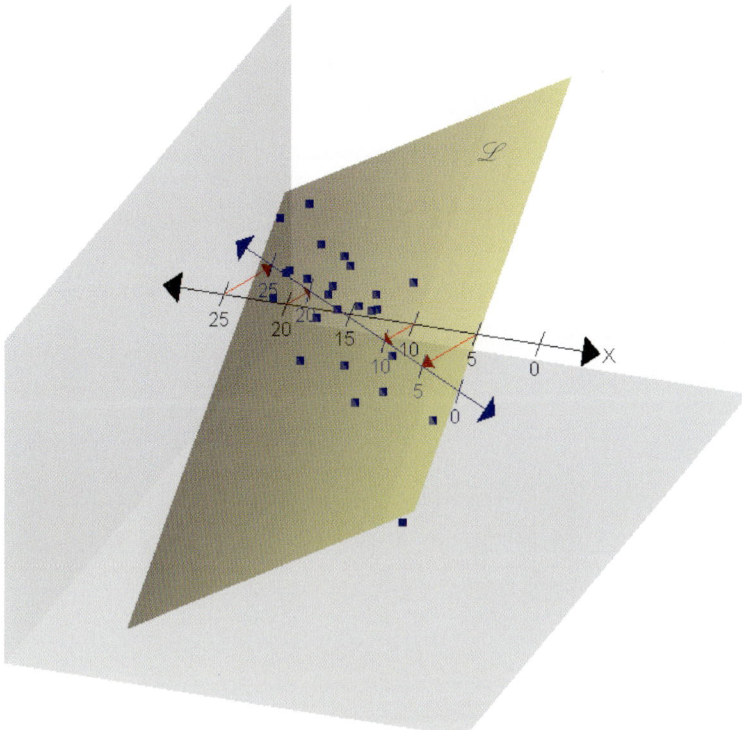

Figure 3.8 Constructing interpolation axis $\boldsymbol{X}$ in the biplot space.

As the plane $\mathscr{N}$ is shifted along the Cartesian axis $\boldsymbol{X}$, a series of parallel intersection spaces is obtained, as shown in Figure 3.15. Any line passing through the origin will pass through these intersection spaces and can be used as a prediction axis fitted with markers according to the value associated with the particular intersection space. To facilitate orthogonal projection onto the axis, similar to an ordinary scatterplot, the line orthogonal to these intersection spaces is chosen. For the data in Table 3.3 the prediction biplot axis is shown in Figure 3.16.

The call

```
PCAbipl(X = as.matrix(Table.3.3.data[,-1]), ax.type =
   "predictive", pch.samples = 15, pos = "Hor", offset =
   c(0.1, 0.3, 0.3, 0.3), reflect = "y", colours = "green",
   offset.m = rep(-0.25, 3), predictions.sample = c(9,24),
   ort.lty = 2)
```

results in the predictive PCA biplot in Figure 3.17 of the data in Table 3.3. Note that for `PCAbipl` the default type of axis is `ax.type = "predictive"`.

Now that the biplot basics have been illustrated by the specific construction of a PCA biplot, the next step is to evaluate how well the PCA biplot represents the original centred data matrix **X**.

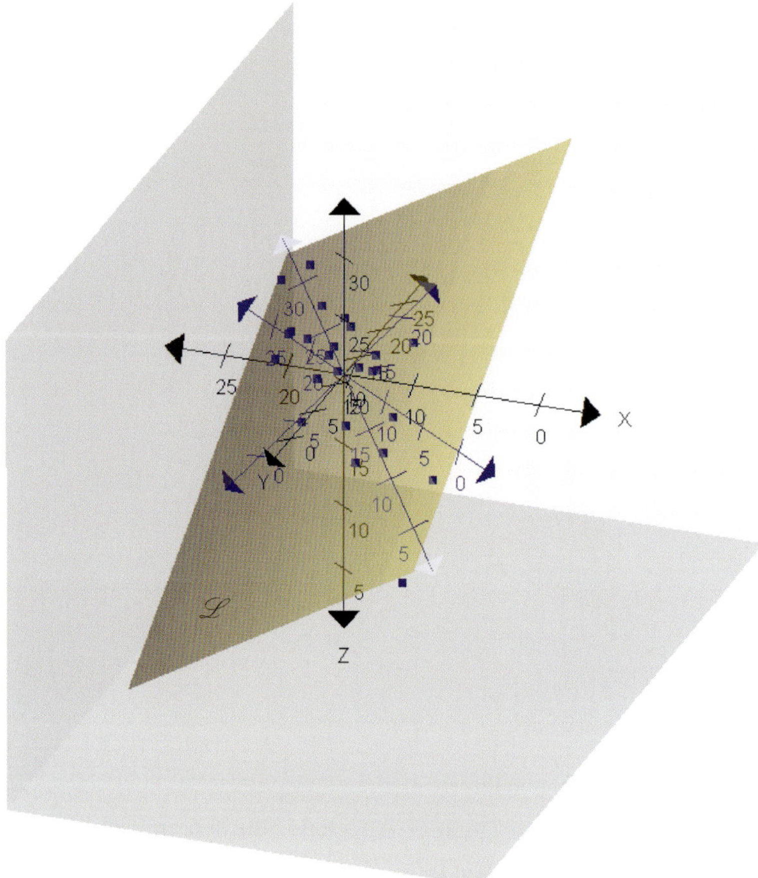

Figure 3.9 Interpolation biplot axes (in blue) fitted to the biplot display in the two-dimensional biplot plane.

3.3 Measures of fit for PCA biplots

For any observed vector $\mathbf{x}$ decomposed into a fitted part $\hat{\mathbf{x}}$ and residual $\mathbf{x} - \hat{\mathbf{x}}$, it follows that

$$\mathbf{x} = \hat{\mathbf{x}} + (\mathbf{x} - \hat{\mathbf{x}}). \tag{3.9}$$

The quality of the fit may be assessed by the *variance accounted for* ratio:

$$\frac{ss(\hat{\mathbf{x}})}{ss(\mathbf{x})} = \frac{\hat{\mathbf{x}}'\hat{\mathbf{x}}}{\mathbf{x}'\mathbf{x}}. \tag{3.10}$$

The expression 'variance accounted for' is used loosely and assumes, at least, that $\mathbf{x}$ is expressed in deviations from the mean. Much more important is that the ratio is meaningful only when the decomposition is orthogonal, that is,

$$ss(\mathbf{x}) = ss(\hat{\mathbf{x}}) + ss(\mathbf{x} - \hat{\mathbf{x}}). \tag{3.11}$$

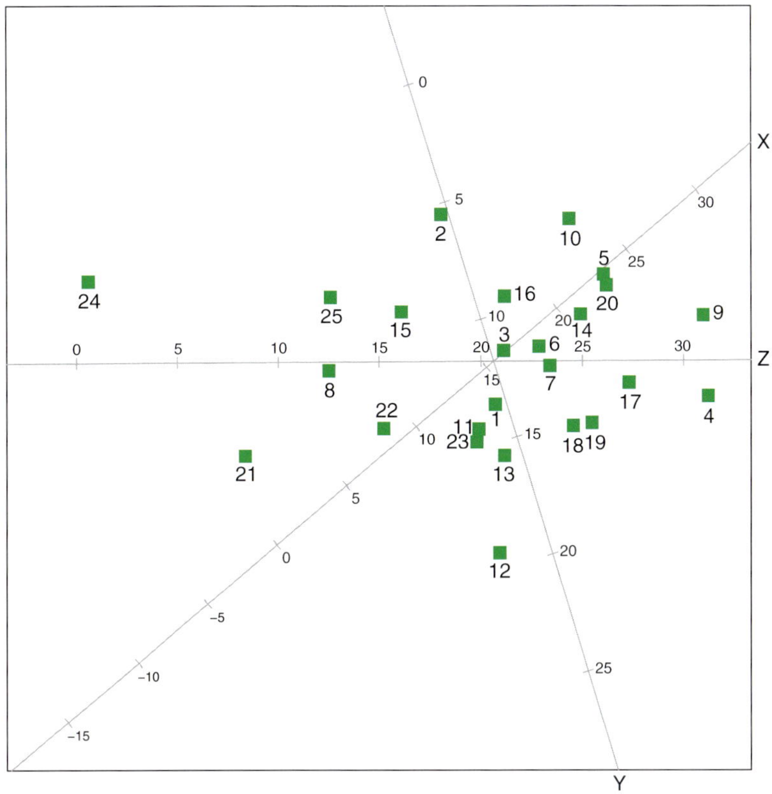

Figure 3.10 Interpolation PCA biplot of the data in Table 3.3.

Figure 3.18(a) shows an orthogonal breakdown where the sums of squares add up according to Pythagoras's theorem. Usually the residual sum of squares is minimized, so then any nonorthogonal analysis will have a larger residual sum of squares than with orthogonality. Figure 3.18(b) shows a nonorthogonal breakdown in which the fitted sum of squares is bigger than in Figure 3.18(a), giving a bigger 'variance accounted for' ratio, possibly even greater than one. Thus the ratio is useful only when orthogonality holds. Perhaps 'variance not accounted for' would be a safer measure.

The aim of the principal component method is to choose the r-dimensional subspace $\mathscr{L}$ which is best-fitting in the least-squares sense. Before evaluating this least-squares fit, we need to consider the orthogonality properties of the associated sums of squares decomposition.

We asserted above that

$$\mathbf{X} = \mathbf{XVJ} + (\mathbf{X}-\mathbf{XVJ}) = \hat{\mathbf{X}} + (\mathbf{X} - \hat{\mathbf{X}})$$

was an orthogonal decomposition in the sense that $\|\mathbf{X}\|^2 = \|\hat{\mathbf{X}}\|^2 + \|\mathbf{X} - \hat{\mathbf{X}}\|^2$. We now prove the much stronger result that

$$\mathbf{X}'\mathbf{X} = \hat{\mathbf{X}}'\hat{\mathbf{X}} + (\mathbf{X} - \hat{\mathbf{X}})'(\mathbf{X} - \hat{\mathbf{X}}). \tag{3.12}$$

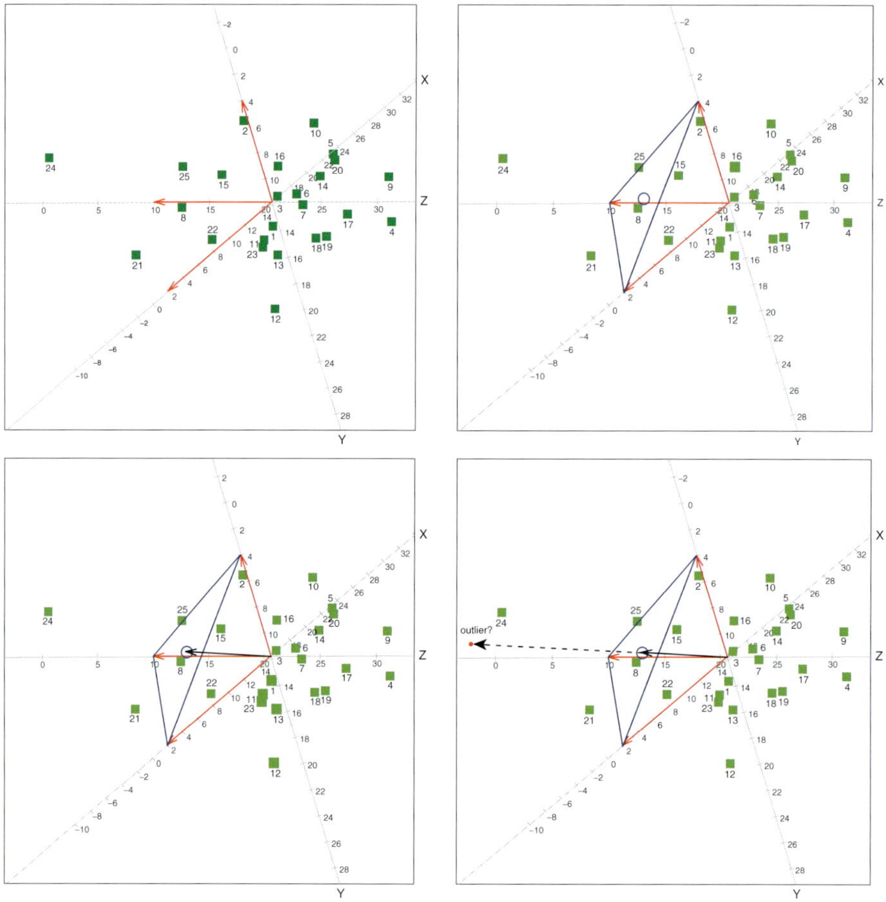

Figure 3.11 Interpolating a new point with $X = 2$, $Y = 4$, $Z = 10$. (Top left) The function `draw.arrow` is used three times to draw an arrow from the origin to the selected end point. (Top right) The function `draw.polygon` is called to construct the blue triangle by selecting with the mouse the three vertices and the centroid of the triangle is marked. (Bottom left) `draw.arrow` is again called upon to draw the black arrow from the origin to the centroid by selecting it with the mouse. (Bottom right) The black arrow is lengthed $p = 3$ times, indicated by a broken line; the position of the new point marked with a red solid circle and `draw.text` is used to interactively questions if this point is perhaps an outlier.

To show this, we need only establish that $(\mathbf{X} - \hat{\mathbf{X}})'\hat{\mathbf{X}} = \mathbf{0}$. From the SVD $\mathbf{X} = \mathbf{U\Sigma V}'$ we have that

$$(\mathbf{X} - \hat{\mathbf{X}})'\hat{\mathbf{X}} = \mathbf{V}(\mathbf{I} - \mathbf{J})\mathbf{\Sigma U}'(\mathbf{UJ\Sigma V}')$$

and the result follows immediately from the column orthonormality of $\mathbf{U}$ and the commutativity of diagonal matrices. We term the orthogonality result (3.12) *Type B orthogonality for a data matrix* $\mathbf{X}$.

Figure 3.12 Inverse relationship between interpolation and prediction. (Top) Interpolation axes in biplot plane. Interpolation is finding for the red sample point in the full space its representation at the tip of the arrow in the biplot space. (Bottom) The axes shown are the original axes passing through the biplot plane. Prediction is inferring the values of the original variables for a point in the biplot space.

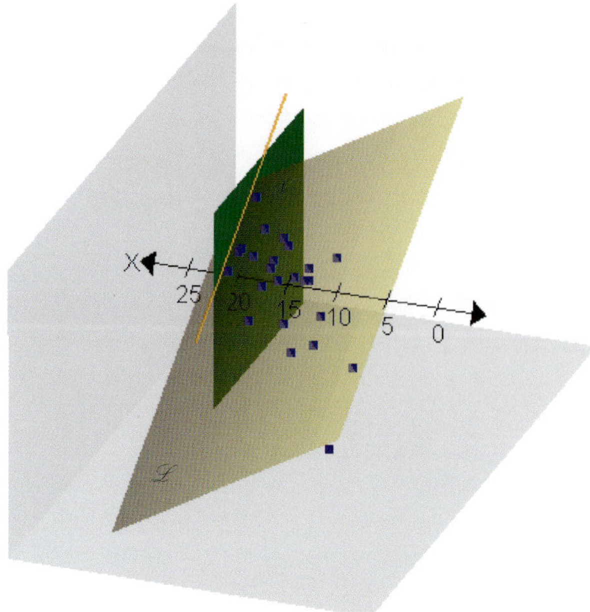

Figure 3.13 Plane $\mathscr{N}$ constructed orthogonal to Cartesian axis $\boldsymbol{X}$ through the marker '20'.

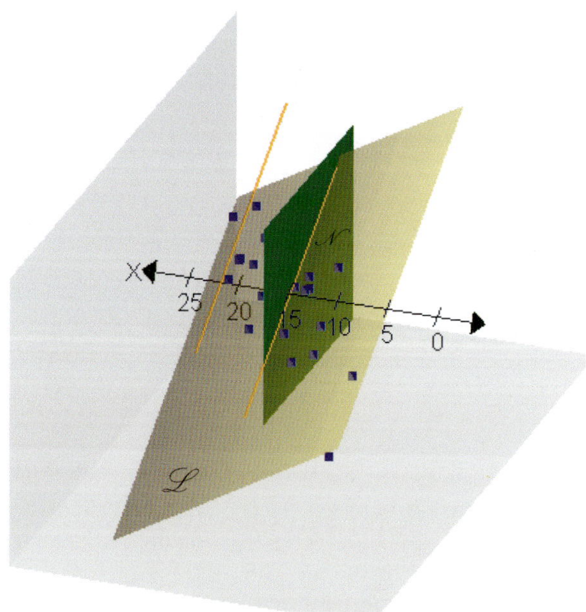

Figure 3.14 Plane $\mathscr{N}$ constructed orthogonal to Cartesian axis X through the marker '15'.

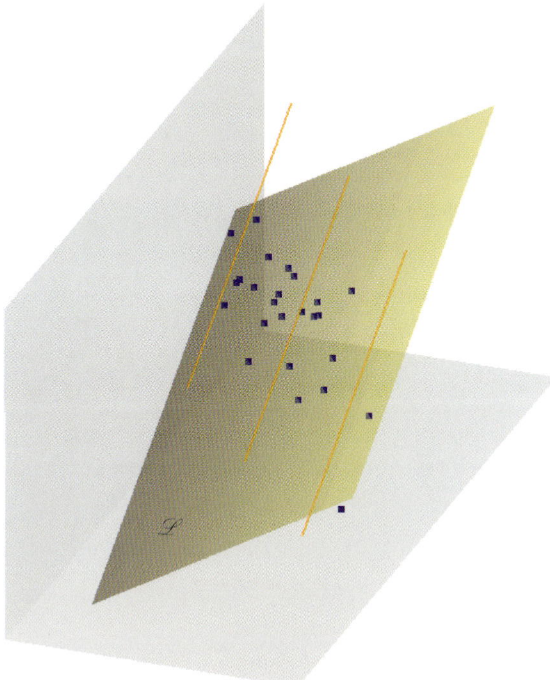

Figure 3.15 Parallel intersection spaces for predicting the values of the Cartesian axis $\boldsymbol{X}$.

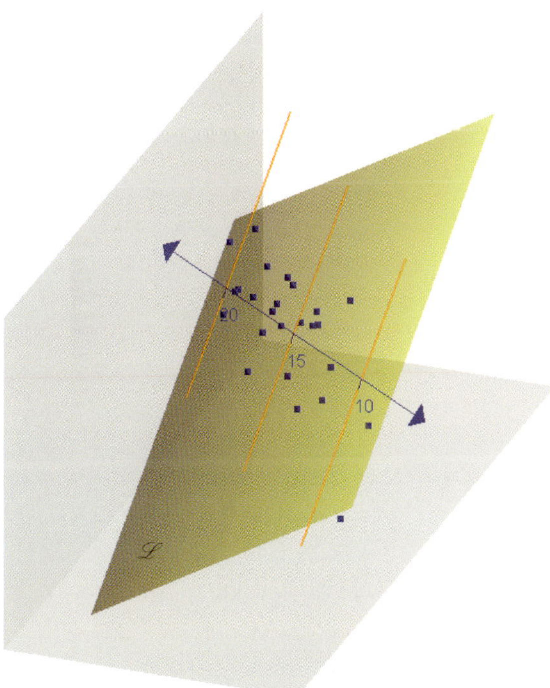

Figure 3.16 Prediction biplot axis for variable $\boldsymbol{X}$ of the Table 3.3 data.

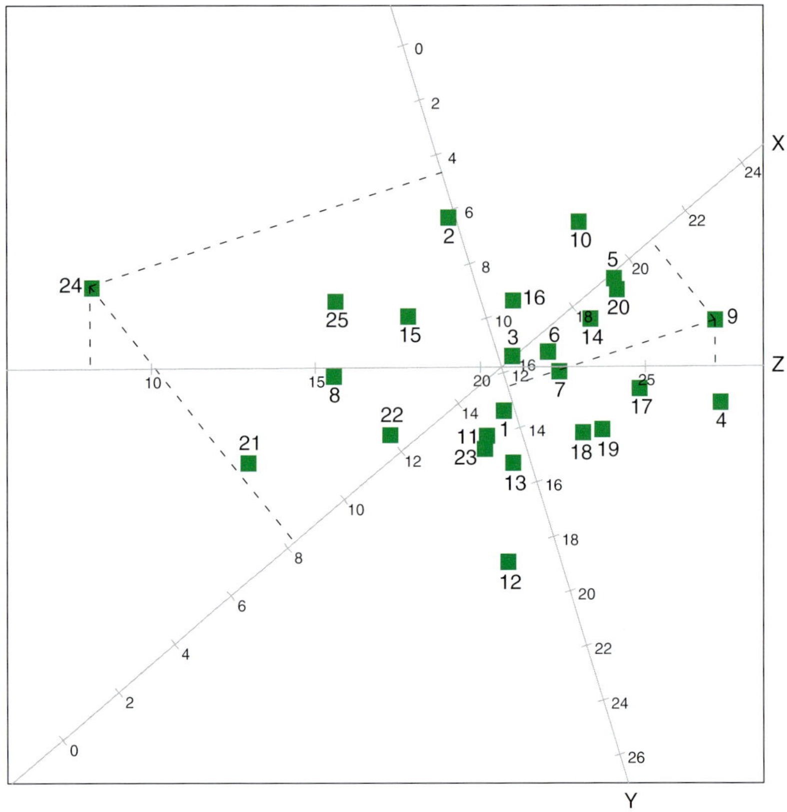

Figure 3.17 Prediction PCA biplot of the data in Table 3.3. Dotted lines show predictions for samples ***9*** and ***24*** as requested by the argument `predictions.sample = c(9,24)` and `ort.lty = 2` in the function call to `PCAbipl`.

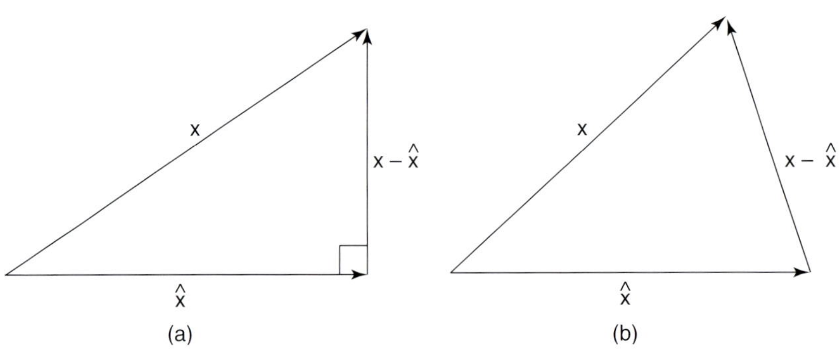

Figure 3.18 Breakdown of a sum of squares: (a) orthogonal; (b) nonorthogonal.

Moreover,

$$\mathbf{XX}' = \mathbf{XVJV}'\mathbf{X}' + \mathbf{XV}(\mathbf{I} - \mathbf{J})\mathbf{V}'\mathbf{X}'. \tag{3.13}$$

We immediately have that

$$\mathbf{XX}' = \hat{\mathbf{X}}\hat{\mathbf{X}}' + (\mathbf{X} - \hat{\mathbf{X}})(\mathbf{X} - \hat{\mathbf{X}})'. \tag{3.14}$$

The orthogonality result (3.14) is termed *Type A orthogonality for a data matrix* $\mathbf{X}$.

The standard result about the orthogonality of the fitted and residual sums of squares follows immediately from the traces of either $\mathbf{X}'\mathbf{X}$ or $\mathbf{XX}'$ but the above has shown that orthogonality applies to the *individual* terms of both matrices. In particular, it applies to the diagonal elements of $\mathbf{X}'\mathbf{X}$ and $\mathbf{XX}'$, equation (3.12) showing that we may apply it to each of the columns of $\mathbf{X}$ and (3.13) that we may apply it to the individual rows of $\mathbf{X}$. Similar results pertain to the off-diagonal elements but we do not need them.

The matrices $\mathbf{V}$ and $\mathbf{\Sigma}$ needed above can also be computed from the SVD of $\mathbf{X}'\mathbf{X}$, namely

$$\mathbf{X}'\mathbf{X} = \mathbf{V}\mathbf{\Sigma}^2\mathbf{V}' = \mathbf{V}\mathbf{\Lambda}\mathbf{V}'. \tag{3.15}$$

The residuals are given by $\mathbf{X}(\mathbf{I} - \mathbf{VJ})$ with sum of squares

$$\|\mathbf{X}(\mathbf{I} - \mathbf{VJ})\|^2 = \text{tr}\{(\mathbf{X} - \hat{\mathbf{X}})(\mathbf{X} - \hat{\mathbf{X}})'\} = \text{tr}\{\mathbf{XV}(\mathbf{I} - \mathbf{J})\mathbf{V}'\mathbf{X}'\}$$

from (3.13). Therefore the sum of squares for the residuals is

$$\text{tr}(\mathbf{X}'\mathbf{X}) - \text{tr}(\mathbf{X}'\mathbf{XVJV}') = \text{tr}(\mathbf{\Lambda} - \mathbf{\Lambda J}) = \sum_{j=r+1}^{p} \lambda_j. \tag{3.16}$$

The orthogonal analysis of variance ('Total sum of squares' = 'Fitted sum of squares' + 'Residual sum of squares') justifies a measure of overall *quality* of the display given by the ratio

$$\text{quality} = \frac{\text{tr}(\mathbf{\Lambda J})}{\text{tr}(\mathbf{\Lambda})} = \frac{\sum_{j=1}^{r} \lambda_j}{\sum_{j=1}^{p} \lambda_j} = \frac{\sum_{j=1}^{r} \sigma_j^2}{\sum_{j=1}^{r} \sigma_j^2}. \tag{3.17}$$

Overall quality is only part of the story, and we may also want to know about the quality of the variables as represented in r dimensions. Just as $\mathbf{XVJ}$ projects the samples, so $\mathbf{IVJ}$ projects the unit points on every coordinate axis, to give p points that may be regarded as representing the variables. Because $\mathbf{V}$ is an orthogonal matrix, its rows have unit sums of squares and it follows that the sums of squares of the rows of $\mathbf{VJ}$ measure the adequacy of representation for each variable. Algebraically this is given by the $p \times p$ diagonal matrix diag($\mathbf{VJV}'$). We define the measure *adequacy* as

$$\text{adequacy} = \sum_{j=1}^{r} v_{ij}^2 = i\text{th diagonal element of diag}(\mathbf{VJV}'). \tag{3.18}$$

This is a measure of the adequacy of the representation for the ith variable. Adequacy is used under different terminology in factor analysis, but happens also to be used, we think inadvisably, when examining the approximation of **X**. Because **V** is orthogonal, we attain the maximum adequacy of one when $r = p$ when the 'approximation' is exact. This is the basis of methods that plot a unit circle, with the rows of **VJ** represented as vectors from its centre (see Chapter 10). In Figure 2.9 this vector is shown with the '1sd' marker (the marker denoting the mean plus one standard deviation) on the biplot axis. For high adequacy a unit point must lie in or near the r-dimensional subspace and for low adequacy the corresponding variable must be nearly orthogonal to the subspace. The unit circle indicates 'perfect adequacy' and the further away the '1sd' marker is from the unit circle, the poorer the adequacy.

Clearly, the graphical representation in Figure 3.19 described by three variables cannot represent all variables adequately in two dimensions, whereas the sample points themselves may be shown perfectly in two dimensions. Therefore, one may be more interested in how close to the plane are the points represented by the rows of **X** rather than how close are the axes. A measure termed *predictivity,* described below, has been designed to address this kind of phenomenon.

For true two-dimensional data as in Figure 2.1, PCA gives an exact two-dimensional fit (i.e. unit overall quality) and the predictions obtained as projections $\hat{\mathbf{X}}$ onto the calibrated biplot axes will reproduce the data **X** exactly. Yet, the variables will have varying

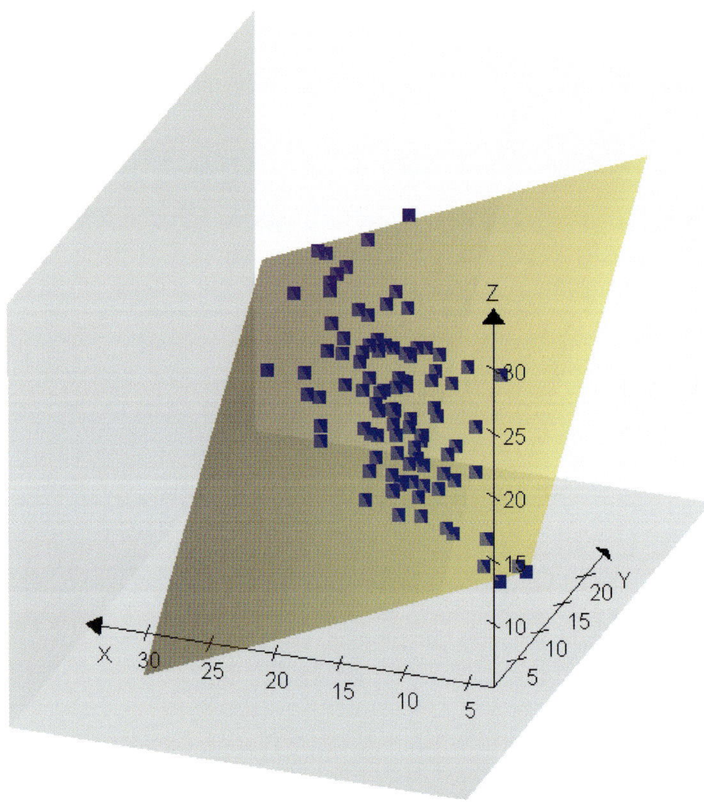

Figure 3.19 An artificial three-dimensional data set where all sample points lie on a plane in the three-dimensional space.

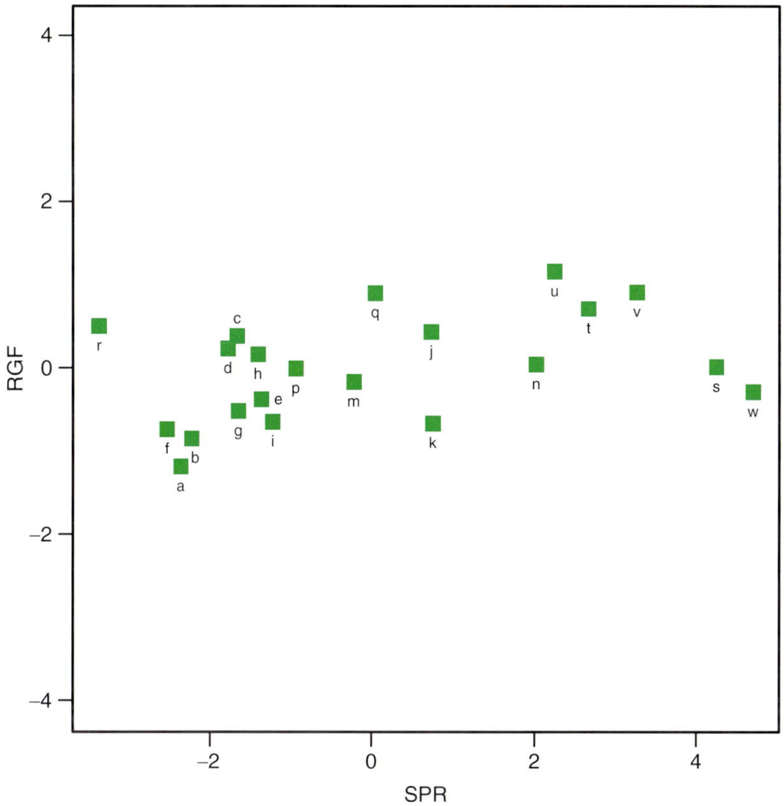

Figure 3.20 Scatter diagram of the centred variables ***RGF*** vs ***SPR*** of the aircraft data set.

adequacies. Clearly adequacy does not measure the fit to the data. We elaborate on this point by considering the following example.

Figure 3.20 shows a scatter diagram of the first two centred variables (***SPR*** and ***RGF***) of Table 1.1. We took these two centred variables, added four zero dummy variables and gave the resulting table a random orthogonal rotation in six dimensions to give the values shown in Table 3.4. A PCA biplot of the Table 3.4 data is shown in Figure 3.21.

By the method of construction, we know that the data points contained in Table 3.4 all lie on a plane in a six-dimensional space, a fact that is verified by the PCA biplot in Figure 3.21 which has two-dimensional overall quality equal to one. Note that the scatter diagram for the first two variables shown in Figure 3.20 is reproduced exactly by the biplot in Figure 3.21. The adequacies are given in Table 3.5.

The difference in the adequacy values and their nonunit values might seem to suggest that the variables contribute differently to prediction. However, calculating the fitted values using $\hat{\mathbf{X}} = \mathbf{XVJV}'$ confirms that the fit for the six variables is perfect. Indeed, these are the values found by projecting the sample points onto the calibrated axes and reading off the predicted values. The biplot axes are behaving perfectly notwithstanding their different adequacies. We have ended up with six axes in two dimensions so we know that there must be linear relationships among them, including the two that will recover the

Table 3.4 Six artificial variables constructed from the first two variables of Table 1.1 by first centring them, then adding four zero dummy variables and finally applying a random orthogonal rotation to the six-variable data set, rounding the final values to four decimal places.

	Var 1	Var 2	Var 3	Var 4	Var 5	Var 6
a	1.2547	0.9841	−0.4140	0.8834	−1.3810	1.2630
b	1.0845	0.9866	−0.4420	0.8032	−1.3329	0.9533
c	0.4454	0.9602	−0.5251	0.4914	−1.1157	−0.1733
d	0.5392	0.9874	−0.5273	0.5439	−1.1713	−0.0316
e	0.6137	0.6357	−0.2981	0.4770	−0.8335	0.4607
f	1.1539	1.1747	−0.5476	0.8909	−1.5461	0.8873
g	0.7635	0.7546	−0.3483	0.5829	−0.9998	0.6101
h	0.4340	0.7760	−0.4128	0.4324	−0.9236	−0.0063
i	0.6617	0.5026	−0.2082	0.4611	−0.7116	0.6827
j	−0.4129	−0.2927	0.1170	−0.2817	0.4226	−0.4473
k	−0.0250	−0.5455	0.3337	−0.1707	0.5659	0.5100
m	0.1351	0.0754	−0.0257	0.0862	−0.1175	0.1671
n	−0.7239	−1.0570	0.5417	−0.6521	1.2976	−0.2308
p	0.3298	0.4899	−0.2520	0.2995	−0.5998	0.0967
q	−0.3411	0.1725	−0.1598	−0.1121	−0.0751	−0.7912
r	1.0023	1.8891	−1.0134	1.0268	−2.2324	−0.1135
s	−1.4953	−2.2401	1.1540	−1.3635	2.7385	−0.4191
t	−1.1912	−1.2507	0.5889	−0.9308	1.6351	−0.8772
u	−1.2055	−0.9300	0.3882	−0.8443	1.3110	−1.2292
v	−1.4743	−1.5245	0.7143	−1.1452	1.9996	−1.1095
w	−1.5486	−2.5484	1.3365	−1.4786	3.0700	−0.2016

original variables ***SPR*** and ***RGF***. Attempts to find these involve factor rotation methods and are not discussed here (see Lawley and Maxwell, 1971).

We remark that readers who try to verify the above results by calling `PCAbipl` with the data given in Table 3.4 will find that, due to the rounding, the quality will not exactly equal unity nor will the adequacies be exactly as given in Table 3.5. It is better to use the true input matrix we obtained with the following function:

```
function (X = aircraft.data[,2:3], new.orthog.rot = FALSE)
{ X.new <- cbind(scale(X, scale = FALSE), matrix(0,
           nrow = nrow(X), ncol = 4))
  if (new.orthog.rot == TRUE)
    random.ortmat <- svd(matrix(rnorm(36), nrow = 6))$v
  else set.seed(3015)
  random.ortmat <- svd(matrix(rnorm(36), nrow = 6))$v
  X.rot <- X.new %*% random.ortmat
  dimnames(X.rot)[[2]] <- paste("V", 1:6, sep = "")
  X.rot
}
```

Thus, although adequacy is a popular measure, it has its limitations. Also, the overall quality, depending on the trace operator, is a little crude. Can we do better? In the definition of overall quality of fit we used the trace of the squared residuals. However, calculating the 'variance accounted for' ratio individually for each diagonal element

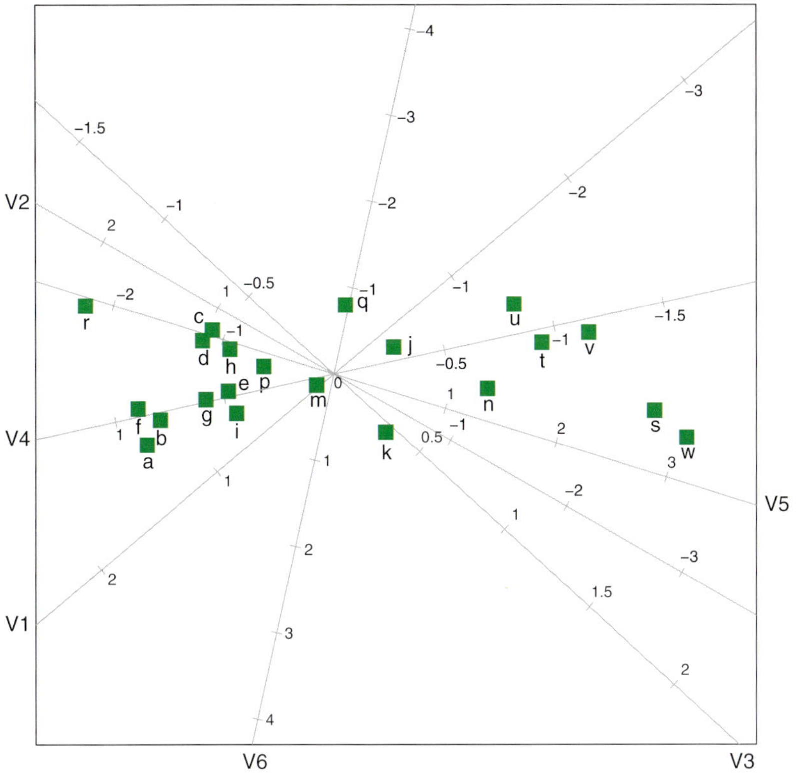

Figure 3.21 PCA biplot of data given in Table 3.4.

Table 3.5 Two-dimensional adequacies associated with the six variables of the artificially constructed two-dimensional data set in Table 3.4.

Variable	Var 1	Var 2	Var 3	Var 4	Var 5	Var 6
Adequacy	0.252	0.326	0.111	0.114	0.428	0.770

measures the degree to which the corresponding approximation agrees with the corresponding true element in $\mathbf{X}$. Since both $\mathbf{X}'\mathbf{X}$ and $\mathbf{X}\mathbf{X}'$ yield orthogonal decompositions of the sums of squares, we can define *axis predictivity* as the diagonal elements of the matrix $\mathbf{\Pi} : p \times p$ given by

$$\mathbf{\Pi} = \text{diag}(\hat{\mathbf{X}}'\hat{\mathbf{X}})[\text{diag}(\mathbf{X}'\mathbf{X})]^{-1} = \text{diag}(\mathbf{V\Lambda JV}')[\text{diag}(\mathbf{V\Lambda V}')]^{-1}, \tag{3.19}$$

and similarly *sample predictivity* as the diagonal elements of the matrix $\mathbf{\Psi} : n \times n$ given by

$$\mathbf{\Psi} = \text{diag}(\hat{\mathbf{X}}\hat{\mathbf{X}}')[\text{diag}(\mathbf{X}\mathbf{X}')]^{-1} = \text{diag}(\mathbf{U\Lambda JU}')[\text{diag}(\mathbf{U\Lambda U}')]^{-1}. \tag{3.20}$$

The diagonal elements of $\mathbf{\Pi}$ are our measures of the predictive powers of each variable and clearly cannot exceed unity. We term the ith element the axis predictivity for the ith variable and denote it by Π_i. Axis predictivity is 'variance accounted for' per variable, and its interpretation depends critically on the orthogonal ANOVA and the independence result for the diagonals of $\mathbf{X}'\mathbf{X}$. It is this property that justifies expressing the overall quality as a weighted sum of the Π_i with weights

$$w_i = \frac{(\mathbf{V\Lambda V}')_{ii}}{\mathrm{tr}(\mathbf{\Lambda})}$$

given by the variances of the ith variable relative to total variance. Since

$$\Pi_i = \frac{(\mathbf{V\Lambda JV}')_{ii}}{(\mathbf{V\Lambda V}')_{ii}}$$

it follows that $\Pi_i(\mathbf{V\Lambda V}')_{ii} = (\mathbf{V\Lambda JV}')_{ii}$ and summing gives

$$\sum_{i=1}^{p} \Pi_i(\mathbf{V\Lambda V}')_{ii} = \mathrm{tr}(\mathbf{V\Lambda JV}') = \mathrm{tr}(\mathbf{\Lambda J})$$

which, after division by $\mathrm{tr}(\mathbf{\Lambda})$ gives

$$\text{quality} = \frac{\mathrm{tr}(\mathbf{\Lambda J})}{\mathrm{tr}(\mathbf{\Lambda})} = \sum_{i=1}^{p} \frac{(\mathbf{V\Lambda V}')_{ii}}{\mathrm{tr}(\mathbf{\Lambda})}\Pi_i = \sum_{i=1}^{p} w_i \Pi_i. \tag{3.21}$$

While axis predictivity is concerned with the quality of the representation of each variable, sample predictivity provides the quality of representation of samples. This measures how far is each sample from its r-dimensional approximation; a value of unity implies that the sample is in the plane of approximation and a value of zero that it is orthogonal to the plane. A word of warning: sample predictivity combines measures on variables that may not be commensurable, thus highlighting the problems discussed in Section 2.5.

Revisiting our example earlier, both the axis predictivities and sample predictivities of the artificial two-dimensional data set in six dimensions all equal unity. Turning to the complete set of aircraft data, a PCA achieved an overall two-dimensional quality measure of 0.9677. Table 3.6 gives the two-dimensional adequacies and predictivities. Variable ***SPR*** is the most adequately approximated and ***PLF*** the least adequate. In this case, the predictivities show a similar ranking. However, as we saw above, the predictivities give an immediate estimate of the success in predicting the values taken by $\mathbf{X}$ for its associated variables, while adequacy is more concerned with the dispositions of coordinate axes and any related distortions of the scale of each axis.

The axis predictivities are summarized in Figure 3.22, which plots the predictivities for each variable against the number of dimensions fitted. We see that ***SPR*** is nearly perfectly predicted in one dimension and that all variables except ***PLF*** require only

Table 3.6 Adequacies and axis predictivities for the two-dimensional PCA biplot of the aircraft data in Figure 2.5.

	SPR	RGF	PLF	SLF
Adequacy	0.999	0.205	0.002	0.794
Predictivity	1.000	0.571	0.245	0.957

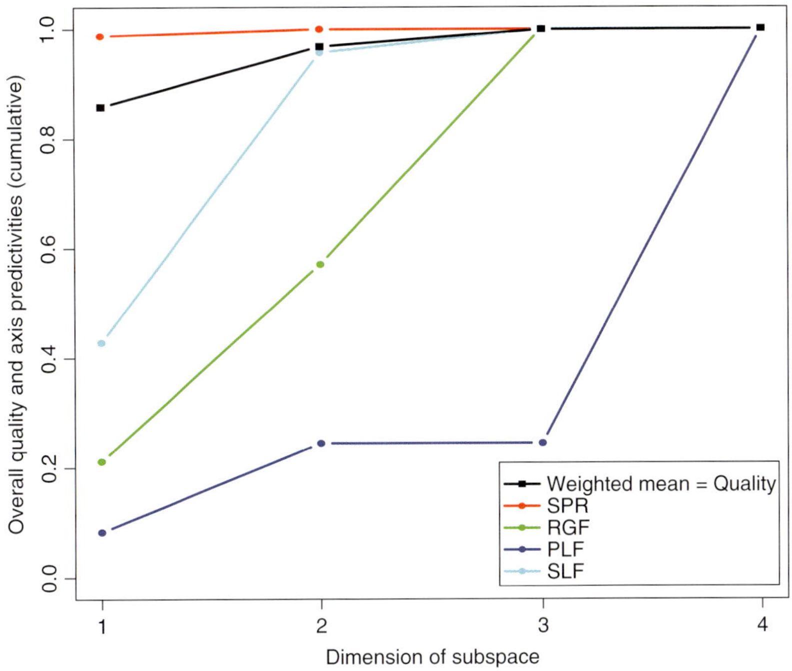

Figure 3.22 Plot of axis predictivity against dimensionality. Note that the dimensions are cumulative; for example, the three-dimensional fits include the first and second dimensions.

three dimensions. Of course the incommensurability of the variables is having a major effect. To obtain the overall quality, the predictivities have to be weighted proportionately to the variances of each variable. The variances are (5.5881, 0.4116, 0.0079 and 1.0533), giving the weights (0.7914, 0.0583, 0.0011, 0.1492) required to reproduce the overall cumulative qualities (0.7915 0.9677 0.9992 1.0000).

Perusal of Table 3.6 shows axis predictivity to be larger than axis adequacy. This can be proved to be true in general (see Gardner-Lubbe *et al.*, 2008), but is of academic interest as adequacy and predictivity measure different things and are not comparable. Adequacy may be of interest for comparing the approximation to **V**, as in factor analysis, but not when concerned with approximating **X**.

We provide a function `PCA.predictivities` for calculating the above measures of fit. `PCA.predictivities` takes a data matrix as first argument, centres the data and optionally scales it to unit variances. It returns a list with components `Quality`, `Weights`, `Adequacies`, `Sample.predictivities.original`, `Axis.predictivities.original`, `Sample.predictivities.new`, `Axis.predictivities.new.1`, and `Axis.predictivities.new.2`. Except for `Weights`, all other components are given for dimensions 1, 2,..., p. Tables 3.7 and 3.8 are examples of the output of `PCA.predictivities`. Because of the very high predictivity of ***SPR*** in the first dimension, together with relatively small predictivities of the other variables in that dimension, the warning encountered in Section 2.5 regarding pre-scaling of the data should be seriously considered here. We note from Table 3.8 that all samples except

Table 3.7 Adequacies and axis predictivities returned by `PCA.predictivities` for the centred (but unscaled) aircraft data.

	Adequacy				Axis predictivity			
Dimension	SPR	RGF	PLF	SLF	SPR	RGF	PLF	SLF
1	0.9110	0.0144	0.0001	0.0745	0.9878	0.2119	0.0828	0.4283
2	0.9992	0.2054	0.0018	0.7936	1.0000	0.5710	0.2446	0.9566
3	0.9995	0.9997	0.0018	0.9990	1.0000	1.0000	0.2446	1.0000
4	1.0000	1.0000	1.0000	1.0000	1.0000	1.0000	1.0000	1.0000

k and ***m*** have satisfactory predictivities in two dimensions but again the cautionary remark about pre-scaling the data should be kept in mind.

3.4 Predictivities of newly interpolated samples

Apart from calculating the predictivities for a sample point, the predictivity for newly interpolated samples can also be calculated, although such samples do not play any role in constructing the scaffolding underlying the biplot. We have seen that a new sample, say $\mathbf{x}^*$, can be interpolated into the biplot by the relation $\mathbf{z}^{*\prime} = \mathbf{x}^{*\prime}\mathbf{V}_r$. Therefore, its sample predictivity can be calculated by applying (3.20). The predictivity of the new sample $\mathbf{z}^{*\prime} = \mathbf{x}^{*\prime}\mathbf{V}_r$ follows then as

$$\Psi_{\mathbf{x}^*} = \hat{\mathbf{x}}^{*\prime}\hat{\mathbf{x}}^*(\mathbf{x}^{*\prime}\mathbf{x}^*)^{-1} = \mathbf{x}^{*\prime}\mathbf{VJV}'\mathbf{x}^*(\mathbf{x}^{*\prime}\mathbf{x}^*)^{-1}. \tag{3.22}$$

When interpolating a new sample algebraically into a PCA biplot and calculating its sample predictivity the meaning of $\mathbf{x}^*$ must be correctly understood. Remember that in this chapter we follow the convention that $\mathbf{X}$ denotes the column-centred data, therefore $\mathbf{x}^*$ denotes the observed new vector of values centred by the mean vector ($\bar{\mathbf{x}}$) used in the column centring of $\mathbf{X}$, that is,

$$\mathbf{x}^* = \mathbf{x}_{\text{New}} - \bar{\mathbf{x}}. \tag{3.23}$$

If in addition to centring, the columns of $\mathbf{X}$ are scaled, then the same scaling must be applied to the columns (elements) of (3.23) – for example, dividing the ith element of (3.23) by the standard deviation of the ith column of $\mathbf{X}$.

`PCAbipl` and `PCA.predictivities` have optional arguments `X.new.samples` and `X.new.vars` expecting raw data in matrix format. The necessary centring and scaling are then taken care of by the respective functions.

As an example of the sample predictivity associated with a new sample consider the following data. Anatomical characteristics of 37 wood samples were determined by microscopic methods. The following measurements were made: vessel diameter in micrometres (***VesD***), vessel element length in micrometres (***VesL***), fibre length in micrometres (***FibL***), ray height in micrometres (***RayH***), ray width in micrometres (***RayW***) and the number of vessels per square millimetre (***NumVes***). The 37 samples consisted of three known species: *Ocotea bullata, O. porosa* and *O. kenyensis*. The data are presented in Table 3.9 and a PCA biplot disregarding the group structure of this data set is given in Figure 3.23. Since the standard deviations of the six variables range from

Table 3.8 Sample predictivities returned by `PCA.predictivities` for the centred (but unscaled) aircraft data.

	Sample predictivities																				
Dim	a	b	c	d	e	f	g	h	i	j	k	m	n	p	q	r	s	t	u	v	w
1	0.76	0.76	0.57	0.86	0.79	0.94	0.85	0.95	0.80	0.64	0.38	0.02	0.96	0.35	0.09	0.80	0.95	0.98	0.89	0.98	0.95
2	1.00	1.00	1.00	0.89	0.98	1.00	0.90	0.95	0.81	1.00	0.48	0.56	1.00	0.91	0.83	1.00	1.00	0.98	0.91	0.98	0.99
3	1.00	1.00	1.00	1.00	1.00	1.00	0.99	1.00	1.00	1.00	1.00	1.00	1.00	1.00	1.00	1.00	1.00	1.00	1.00	1.00	1.00
4	1.00	1.00	1.00	1.00	1.00	1.00	1.00	1.00	1.00	1.00	1.00	1.00	1.00	1.00	1.00	1.00	1.00	1.00	1.00	1.00	1.00

Table 3.9 The *Ocotea* data set.

Sample	Species	VesD	VesL	FibL	RayH	RayW	NumVes
1	Obul	79	383	941	333	30	17
2	Obul	78	346	961	223	24	31
3	Obul	82	361	1039	316	27	25
4	Obul	79	324	1048	369	29	26
5	Obul	85	418	1051	347	34	14
6	Obul	111	448	1096	379	40	13
7	Obul	76	320	1130	347	29	13
8	Obul	103	371	1165	326	26	10
9	Obul	129	406	1165	428	44	11
10	Obul	74	281	1175	324	26	11
11	Obul	102	567	1221	395	40	11
12	Obul	95	415	1225	416	38	10
13	Obul	91	372	1234	375	26	11
14	Obul	113	314	1253	466	23	10
15	Obul	93	541	1267	347	34	14
16	Obul	94	437	1271	336	36	10
17	Obul	119	359	1280	412	32	11
18	Obul	104	387	1290	381	22	12
19	Obul	114	569	1369	568	52	11
20	Obul	141	621	1527	419	34	15
21	Oken	147	402	1391	440	32	9
22	Oken	142	393	1468	443	35	6
23	Oken	125	322	1530	459	34	11
24	Oken	156	401	1588	512	42	11
25	Oken	162	502	1591	369	42	8
26	Oken	103	378	1655	441	34	11
27	Oken	126	414	1759	459	42	8
28	Opor	122	346	981	393	40	14
29	Opor	139	133	993	342	33	14
30	Opor	130	368	1005	356	39	16
31	Opor	127	331	1027	473	38	20
32	Opor	112	309	1044	358	47	8
33	Opor	115	352	1048	300	36	14
34	Opor	130	471	1072	409	39	15
35	Opor	153	419	1077	392	48	20
36	Opor	135	370	1104	531	38	15
37	Opor	130	325	1166	428	36	12

5.23 to 214.05, the data were pre-centred and pre-scaled to unit variances. The biplot axes are calibrated in terms of the original measurement units. The usefulness of an initial PCA biplot could be improved by augmenting it with the interpolated centroids of the three *Ocotea* species. The mean vectors of these three species are given in Table 3.10.

To obtain the biplot in Figure 3.23 we first represent Table 3.10 as an R object `newsamples` in the form of a 3×6 matrix. The following function call creates the biplot in Figure 3.23:

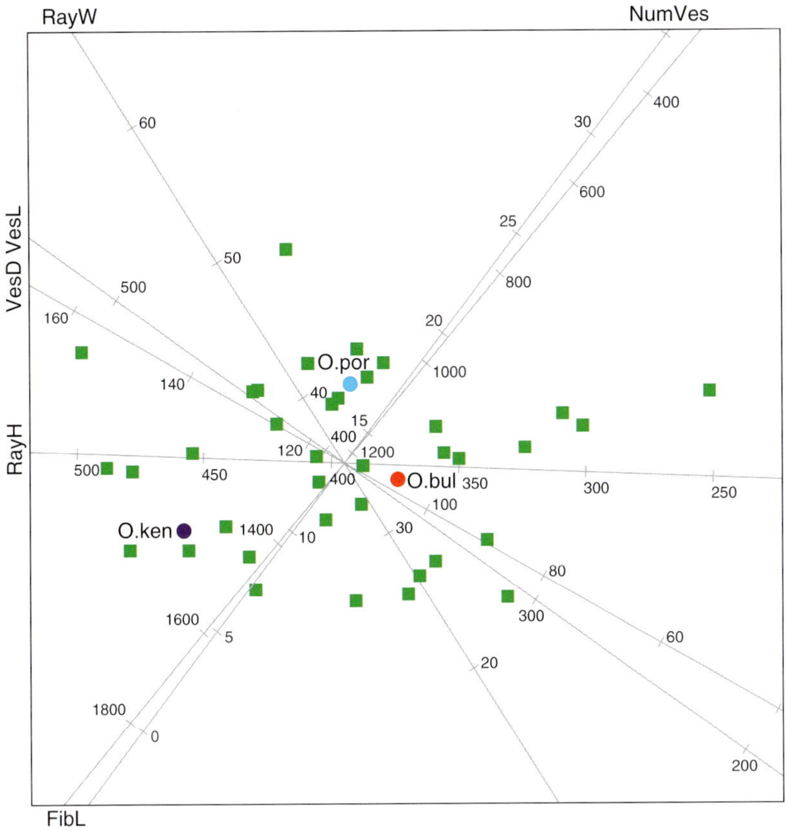

Figure 3.23 PCA biplot of the centred and scaled *Ocotea* data set disregarding the species group structure. The mean vectors of the three species *O. bullata, O. porosa* and *O. kenyensis* have been interpolated into the biplot and are shown by the coloured and appropriately labelled circles. If required, the individual sample points can be labelled by setting `label = TRUE` in the function call described in the text.

```
PCAbipl(Ocotea.data[,3:8], scaled.mat = TRUE, X.new.samples =
   newsamples, colours = "green", pch.samples = 15,
   pch.samples.size = 1.2, label = FALSE, pos = "Paral", offset =
   c(-0.2, 0.1, 0.1, 0), n.int = c(5,5,5,5,3,5), side.label =
   c(rep("right",5),"left"), pos.m = c(4,4,4,4,4,2), offset.m =
   c(-0.1, -0.1, 0.1, -0.1, -0.1, 0.1), pch.new = 16, pch.new.cols
   = c("red","blue","cyan"), pch.new.labels = c("O.bul","O.ken",
   "O.por"))
```

We draw attention to arguments `scaled.mat = TRUE` requesting the scaling of the input matrix, `X.new.samples = newsamples` providing the data for the new samples, `side.label = c(rep("right",5), "left")`, `pos.m = c(4,4,4,4,4,2)` and `offset.m = c(-0.1, -0.1, 0.1, -0.1, -0.1, 0.1)` for fine-tuning the calibrations on the respective axes.

Table 3.10 Means of species *O. bullata, O. porosa* and *O. kenyensis* for each of six variables.

	VesD	VesL	FibL	RayH	RayW	NumVes
O.bul	98.10	412.00	1185.40	375.35	32.30	14.30
O.ken	137.29	401.71	1568.86	446.14	37.29	9.14
O.por	129.30	342.40	1051.70	398.20	39.40	14.80

Table 3.11 Overall quality (%) of the Figure 3.23 biplot in each of the six dimensions.

	Dim 1	Dim 2	Dim 3	Dim 4	Dim 5	Dim 6
Quality	47.38	64.37	79.93	88.87	96.14	100.00

The call

```
PCA.predictivities(Ocotea.data[,3:8], scaled.mat = TRUE,
   X.new.samples = new.samples)
```

returns the measures of fit given in Tables 3.11–3.14.

From Table 3.11 we see that the Figure 3.23 biplot has an overall quality of 64%. Adding another dimension would increase the quality to 80% and constructing a three-dimensional biplot can thus be considered. Inspecting the axis predictivities in Table 3.12 shows that ***VesL*** has a poor predictivity of 0.29, while ***FibL*** and ***RayW*** have predictivities exceeding 0.79. Judging from the three-dimensional predictivities it seems that all variables have satisfactory predictivities in three dimensions. Turning attention to the individual samples, it is clear from Table 3.13 that some of the predictions for some sample points could be made very accurately from the Figure 2.23 biplot (e.g. ***S1***, ***S2***, ***S3***, ***S35***), while others have a very low sample predictivity (e.g. ***S37***). It is clear from Table 3.14 that interpolating the sample mean of *O. kenyensis* into the biplot results in a point whose predictions for the six variables can be very accurately inferred from the two-dimensional PCA biplot. For the two other interpolated new points adding a third dimension is desirable. In the next chapter we will see how a canonical variate analysis biplot allows us to represent the means for all three species exactly in two dimensions.

3.5 Adding new axes to a PCA biplot and defining their predictivities

We saw in Section 2.7 how to use the regression method for obtaining coordinates for constructing a biplot axis for a new variable. Let $\mathbf{x}^*$: $n \times 1$ denote a centred (and scaled, if required) vector of sample values for a new variable. Note that this new variable is centred and scaled using its own values analogous to the centring and scaling of the original matrix $\mathbf{X}$. It follows from (2.20) that the coordinates for adding a new biplot axis representing $\mathbf{x}^*$ are given by the first r nonzero elements (which we denote by $\hat{\mathbf{b}}_r$) of $\mathbf{J}\boldsymbol{\Sigma}^{-1}\mathbf{U}'\mathbf{x}^* = \mathbf{J}\mathbf{b}$ with $\mathbf{J}$ and $\boldsymbol{\Sigma}$ as defined in Section 3.2. This added variable does not play any role in providing the scaffolding axes for constructing the PCA biplot but its axis predictivity can be defined as follows. For predictivity, we have to predict the values of $\mathbf{x}^*$ given the r-dimensional PCA coordinates $\mathbf{XVJ}$. The usual orthogonal multiple

Table 3.12 Adequacies and axis predictivities of the Figure 3.23 biplot in each of the six dimensions.

	Adequacy						Axis predictivity					
	VesD	VesL	FibL	RayH	RayW	NumVes	VesD	VesL	FibL	RayH	RayW	NumVes
Dim 1	0.1906	0.0865	0.1882	0.2222	0.1574	0.1550	0.5418	0.2459	0.5350	0.6317	0.4473	0.4407
Dim 2	0.2495	0.1291	0.4479	0.2226	0.5394	0.4115	0.6018	0.2894	0.7998	0.6321	0.8369	0.7022
Dim 3	0.4278	0.8519	0.4923	0.2590	0.5413	0.4277	0.7683	0.9641	0.8412	0.6661	0.8386	0.7173
Dim 4	0.4713	0.8519	0.5892	0.4608	0.7108	0.9160	0.7916	0.9641	0.8932	0.7744	0.9296	0.9794
Dim 5	0.8917	0.8596	0.6457	0.9538	0.7275	0.9217	0.9749	0.9674	0.9179	0.9893	0.9368	0.9818
Dim 6	1.0000	1.0000	1.0000	1.0000	1.0000	1.0000	1.0000	1.0000	1.0000	1.0000	1.0000	1.0000

Table 3.13 Sample predictivities for the first five and last five of the original samples in the Figure 3.23 biplot.

	S1	S2	S3	S4	S5	...	S33	S34	S35	S36	S37
Dim 1	0.8694	0.8258	0.8397	0.7154	0.5523	...	0.4401	0.0903	0.0610	0.2064	0.0841
Dim 2	0.8754	0.8558	0.8588	0.7502	0.5579	...	0.5070	0.7492	0.8784	0.3497	0.0887
Dim 3	0.9272	0.8902	0.8987	0.7545	0.7637	...	0.5369	0.7940	0.9054	0.5535	0.9616
Dim 4	0.9482	0.9478	0.9899	0.9281	0.9331	...	0.7910	0.7940	0.9124	0.7214	0.9621
Dim 5	0.9868	0.9845	0.9901	0.9673	0.9959	...	0.9989	0.8038	0.9917	0.9182	0.9666
Dim 6	1.0000	1.0000	1.0000	1.0000	1.0000	...	1.0000	1.0000	1.0000	1.0000	1.0000

Table 3.14 Sample predictivities for the three new interpolated points in the Figure 3.23 biplot.

	O.bul	O.ken	O.por
Dim 1	0.5411	0.7790	0.0032
Dim 2	0.5901	0.9074	0.5127
Dim 3	0.9506	0.9175	0.9754
Dim 4	0.9525	0.9412	0.9904
Dim 5	0.9920	0.9717	0.9939
Dim 6	1.0000	1.0000	1.0000

regression decomposition using all p variables (i.e. when $\mathbf{J} = \mathbf{I}$) is (see Section 2.7)

$$\mathbf{x}^* = \mathbf{XVb} + (\mathbf{x}^* - \mathbf{XVb}). \tag{3.24}$$

Now, $\mathbf{XVb}$ may be partitioned into orthogonal components (i) $\mathbf{XVJb}$ and (ii) $\mathbf{XV}(\mathbf{I} - \mathbf{J})\mathbf{b}$. In (i) $\mathbf{XVJ}$ represents the r-dimensional data, while $\mathbf{Jb}$ selects the r coefficients of $\hat{\mathbf{b}}_r$. The term (ii) $\mathbf{XV}(\mathbf{I} - \mathbf{J})\mathbf{b}$ represents an equivalent decomposition in the remaining $p - r$ dimensions. Thus we have a full orthogonal breakdown,

$$\mathbf{x}^* = \mathbf{XVJb} + \mathbf{XV}(\mathbf{I} - \mathbf{J})\mathbf{b} + (\mathbf{x}^* - \mathbf{XVb}), \tag{3.25}$$

with an equivalent analysis of variance with three terms, the first of which represents the sum of squares in the r-dimensional regression approximations, the second the remaining sum of squares in the full p-dimensional space, while the final term is a residual sum of squares not accounted for by regression.

Axis predictivity may be evaluated, as usual, as the ratio of the fitted sum of squares to the original sum of squares:

$$\Pi_x = \frac{(\mathbf{b}'\mathbf{JV}'\mathbf{X}')(\mathbf{XVJb})}{\mathbf{x}^{*\prime}\mathbf{x}^*}. \tag{3.26}$$

Alternatively, we may replace $\mathbf{x}^{*\prime}\mathbf{x}^*$ by its fitted regression sum of squares,

$$\Pi_2 = \frac{(\mathbf{b}'\mathbf{JV}'\mathbf{X}')(\mathbf{XVJb})}{\mathbf{b}'\mathbf{V}'\mathbf{X}'\mathbf{XVb}}. \tag{3.27}$$

The measures (3.26) and (3.27) will coincide only when the regression fit is exact and the final residual in (3.25) vanishes. Normally (3.26), unlike (3.27), will not attain unity even when all p dimensions of the PCA fit are used. In the expressions for (3.26) and (3.27), $\mathbf{X}$ may be replaced by its SVD, in which case we have that

$$\mathbf{b'JV'X'XVJb} = \mathbf{b'JV'V\Sigma^2V'VJb} = \mathbf{x^{*\prime}U\Sigma^{-1}\Sigma JJ\Sigma\Sigma^{-1}U'x^*} = \mathbf{x^{*\prime}UJU'x^*},$$

giving

$$\Pi_{\mathrm{x}} = \frac{\mathbf{x^{*\prime}UJU'x^*}}{\mathbf{x^{*\prime}x^*}}. \tag{3.28}$$

Furthermore,

$$\mathbf{b'V'X'XVb} = \mathbf{b'V'V\Sigma^2V'Vb} = \mathbf{x^{*\prime}U\Sigma^{-1}\Sigma^2\Sigma^{-1}U'x^*} = \mathbf{x^{*\prime}UU'x^*}.$$

so that

$$\Pi_2 = \frac{\mathbf{x^{*\prime}UJU'x^*}}{\mathbf{x^{*\prime}UU'x^*}}. \tag{3.29}$$

The above expressions generalize in a straightforward way to add several new variables. It is clear from formulae (2.20), (3.28) and (3.29) that in order to add new axes with their associated predictivities all we need, in addition to the original SVD, are the values of all the samples on each of the new variables. There is no need to perform the actual regression.

As an example of adding new variables we again consider the *Ocotea* data. Knowledge of the ratios ***VesL*** to ***VesD*** and ***RayH*** to ***RayW*** is of practical importance. These two ratios (***VLDratio*** and ***RHWRatio***) have been added in the form of calibrated biplot axes to the Figure 3.23 PCA biplot. The augmented biplot is given in Figure 3.24.

The function call for obtaining the biplot in the bottom panel of Figure 3.24 is

```
> VLDratio <- Ocotea.data[,4]/Ocotea.data[,3]
> RHWratio <-Ocotea.data[,6]/Ocotea.data[,7]
> Ocotea.data.newvars <- data.frame(Ocotea.data, VLDratio =
   VLDratio, RHWratio =  RHWratio)
> PCAbipl(Ocotea.data[,3:8], scaled.mat = TRUE,
   X.new.vars = as.matrix(Ocotea.data.newvars[,9:10]),
   colours = "green", pch.samples = 15, pch.samples.size = 1.25,
   label = FALSE, pos = "Hor", offset = c(-0.2, 0.1, 0.1, 0.2),
   n.int = c(5,5,5,5,3,5,10,5),
   ax.col = list(ax.col = c(rep("grey",6),"red","red"),
   tickmarker.col = c(rep("grey",6),"red","red"),
   marker.col = c(rep("grey",6),"red","red")),
   ax.name.col = c(rep("black",6), "red","red"),
   pch.new = 16, pch.new.cols = c("red","blue","cyan"),
   pch.new.labels = c("O.bul","O.ken","O.por"),
   predictions.sample = c(10,35))
```

It follows from Table 3.16 that neither of the two added variables has high axis predictivity in two dimensions. Although this is particularly true of ***VLDratio***, there is a dramatic increase in its axis predictivity when a third dimension is added.

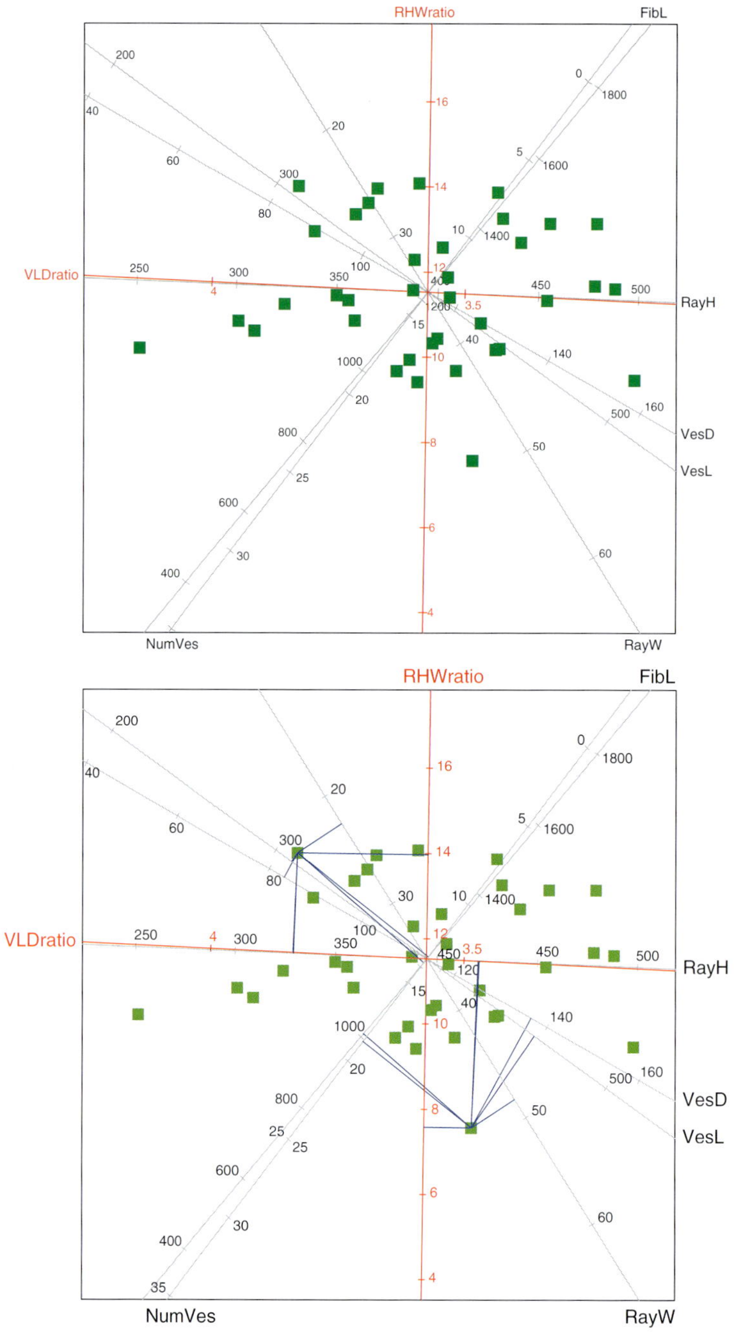

Figure 3.24 PCA biplot of ***Ocotea*** data with two new variables, ***VLDratio*** and ***RHWratio***, added using the regression method. The bottom panel illustrates that added variables in the form of prediction axes can be used exactly as the original biplot axes to provide predictions for sample points ***S10*** and ***S35***. These graphical predictions can be verified with the algebraically determined predictions given in Table 3.15.

Table 3.15 Predictions for samples S10 and S35 of both the original and the two added variables.

	S10 predictions	S35 predictions
VesD	83.27	136.50
VesL	309.74	456.43
FibL	1211.16	1009.14
RayH	329.05	420.48
RayW	22.55	48.33
NumVes	13.33	18.76
VLDratio	3.84	3.47
RHWratio	13.97	7.58

Table 3.16 Axis predictivities for the newly added variables, VLDratio and RHWratio.

	Axis predictivities for new variables according to formula			
	(3.26)		(3.27)	
	VLDratio	RHWratio	VLDratio	RHWratio
Dim 1	0.05	0.00	0.05	0.00
Dim 2	0.05	0.40	0.05	0.42
Dim 3	0.90	0.42	0.92	0.45
Dim 4	0.91	0.69	0.93	0.74
Dim 5	0.98	0.78	1.00	0.83
Dim 6	0.98	0.94	1.00	1.00

3.6 Scaling the data in a PCA biplot

In Section 2.5 we saw that different biplot representations of the aircraft data are obtained, depending on the initial scaling. With the data in Table 1.1 it was essential to scale the data since variable ***PLF*** is measured on a scale that is totally incommensurable with those of the other variables. Comparing this situation to Table 3.5, it is clear why the incommensurable variation in ***PLF*** did not contribute much to the PCA biplot display of the unscaled data, resulting in very poor adequacy of representation of ***PLF***. That PCA is not scale-invariant is something that has to be addressed, usually as in Chapter 2, by normalizing the columns of **X**; naturally, normalization affects measures of quality. When the columns of **X** are normalized, all variables have unit weights and then the overall quality is just the average of the p values of Π_i. Thus, relative qualities of the contributions given by the individual variables can be easily assessed.

Figure 2.16 gives the PCA biplot for the aircraft data normalized to unit sum of squares. For the normalized data set we have, from the output of `PCAbipl`,

$$\mathbf{\Lambda} = \begin{bmatrix} 41.0083 & 0 & 0 & 0 \\ 0 & 24.0087 & 0 & 0 \\ 0 & 0 & 9.2851 & 0 \\ 0 & 0 & 0 & 5.6979 \end{bmatrix} \text{ and } \mathbf{V}_2 = \begin{bmatrix} -0.5587 & 0.3789 \\ -0.5543 & -0.2740 \\ -0.0557 & 0.8657 \\ -0.6142 & -0.1785 \end{bmatrix}.$$

Table 3.17 Adequacies and predicitivities for the two-dimensional biplot of the normalized aircraft data.

	SPR	RGF	PLF	SLF
Adequacy	0.4557	0.3824	0.7528	0.4091
Predictivity	0.8124	0.7202	0.9065	0.8118

The overall quality of the two-dimensional PCA biplot is

$$0.8127 = (41.0083 + 24.0087)/(41.0083 + 24.0087 + 9.2851 + 5.6979).$$

Comparing Table 3.17 with the corresponding entries in Table 3.6, ***PLF*** now becomes the most adequately approximated variable, with ***RGF*** the least adequate. It may be verified from Table 3.17 that the average predictivity agrees with the overall quality of 0.8127. The fit measures for all four dimensions are obtained with the call

```
PCA.predictivities(aircraft.data[,-1], scaled.mat = TRUE)
```

The axis predictivities and quality are summarized in Figure 3.25. Even in one dimension, all variables start at a gratifyingly high level, except for ***PLF*** which from a bad start catches up immediately in the two-dimensional fit. A possible explanation is that most variables involved with aircraft development improve, which means that they increase

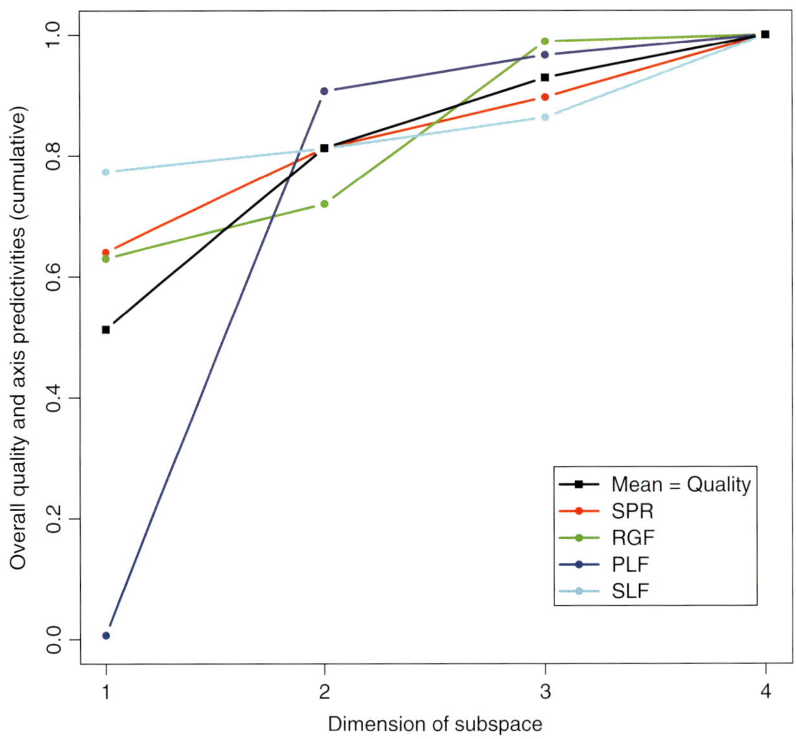

Figure 3.25 Plot of axis predictivity against dimensionality for the normalized aircraft data.

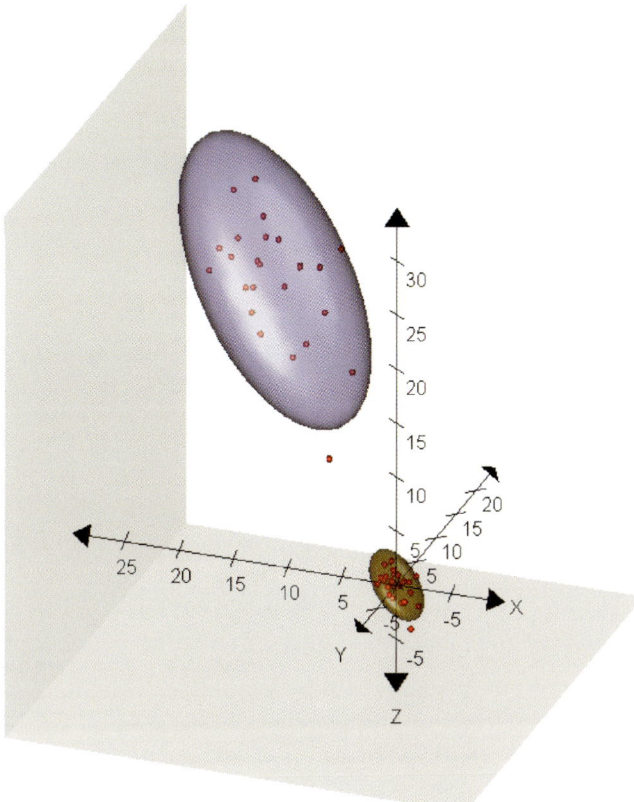

Figure 3.26 Three-dimensional display of the unscaled data of Table 3.3 (blue ellipse) and the scaled data given in Table 3.18 (brown ellipse).

with time, so generating a strong unidimensional trend. Pay load factor is an exception, with some very small values in later years, possibly to be identified with aircraft designed for surveillance rather than carrying capacity.

What happens when the data are scaled? To illustrate the effect of scaling, the data of Table 3.3 have been scaled to have mean zero and unit standard deviation (see Table 3.18). Compare Figure 3.3 with Figure 3.26. There was no change in the correlations in the data, resulting in an ellipsoid with the same orientation as that of Figure 3.3. However, since the variance for each of the three Cartesian axes is equal, the ellipsoid has a more spherical shape. A new plane of best fit is obtained for this data set (minimizing the distances between the samples and their projections onto the plane) and a new projection matrix results in new biplot axes. Many other scalings can be performed on the data, each resulting in a different PCA biplot representation. The benefit of scaling the data before constructing the biplot depends on the practical interpretation resulting from that specific scaling.

The effect of scaling the Table 3.3 data on the biplot is minimal as can be judged by comparing Figures 3.17 and 3.27. Only slight changes in the relative position of samples, for example ***9*** and ***11*** (just above ***23*** in Figure 3.17), can be observed.

Table 3.18 Scaled three-dimensional data from Table 3.3.

Sample no	X	Y	Z
1	−0.5551	0.2612	0.2566
2	−0.0538	−1.9743	−0.0441
3	0.6399	0.2094	−0.2906
4	0.8074	0.9122	1.7062
5	0.7096	−0.7555	1.0198
6	0.4078	0.0277	0.2680
7	0.5903	0.3849	0.1994
8	−0.5219	−0.1746	−1.4854
9	1.0736	−0.0074	1.7367
10	1.3290	−1.1505	0.3994
11	0.0589	0.9242	−0.4274
12	−1.0993	1.9248	0.1797
13	−0.2867	1.0882	0.0014
14	1.1371	0.0050	0.3374
15	0.1913	−0.5562	−0.9729
16	0.5794	−0.5444	−0.0564
17	0.4894	0.5545	1.0888
18	0.2207	0.9735	0.5100
19	0.6303	1.1597	0.4484
20	1.3652	−0.2527	0.5663
21	−1.9663	0.1531	−1.7111
22	−1.2357	0.1971	−0.5802
23	−0.9944	0.5163	0.2290
24	−2.1728	−2.2770	−2.7218
25	−1.3439	−1.5993	−0.6573

This apparent ineffectiveness of scaling is not true in general. Consider a data set from a normal distribution with mean vector $\boldsymbol{\mu}$ and covariance matrix $\boldsymbol{\Sigma}$, where

$$\boldsymbol{\mu} = \begin{bmatrix} 7 \\ 7 \\ 6 \end{bmatrix}, \quad \boldsymbol{\Sigma} = \begin{bmatrix} 1 & 0.2 & 0.6 \\ 0.2 & 1 & 0.36 \\ 0.6 & 0.36 & 4 \end{bmatrix}. \tag{3.30}$$

The resulting clouds of three-dimensonal data points before and after scaling each column to unit variance are expected to be like Figure 3.28. Again the correlation structure does not change with scaling but the ellipsoids have quite different shapes. This is in contrast with what happens in Figure 3.26 and can be explained by considering the correlation matrix associated with (3.30),

$$\begin{bmatrix} 1 & 0.2 & 0.3 \\ 0.2 & 1 & 0.18 \\ 0.3 & 0.18 & 1 \end{bmatrix},$$

which points to very little correlation between the three variables, therefore the cloud for the scaled data is close to a sphere in three dimensions. The cloud for the unscaled data is stretched in the z-direction since this variable has a much larger variance than for

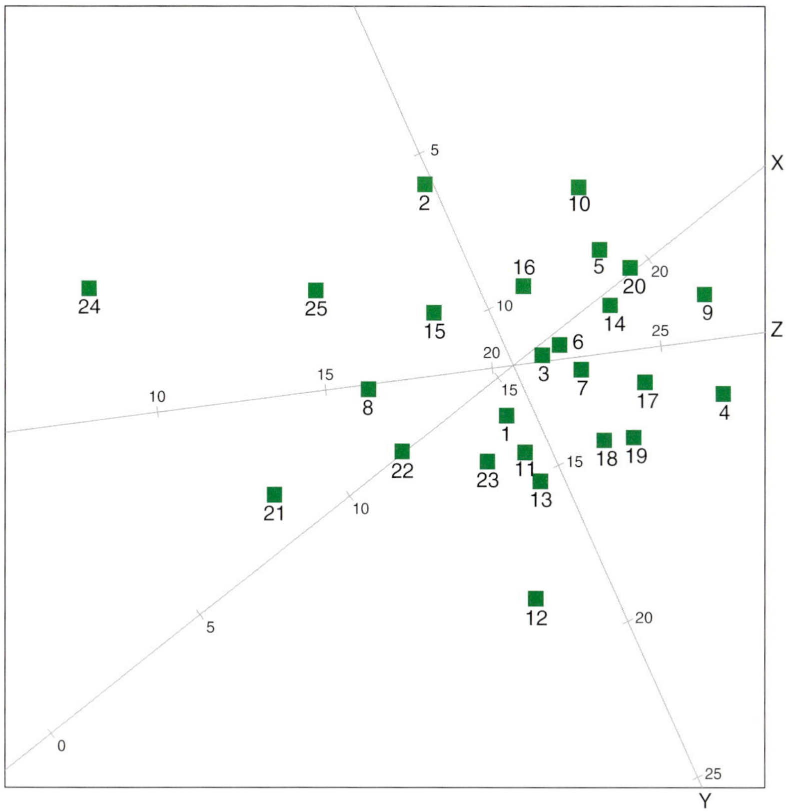

Figure 3.27 PCA biplot of the scaled data given in Table 3.18.

x and y. The plane of best fit, for PCA, will lie along the z-axis for the unscaled data, while virtually any plane could be chosen for the scaled data.

3.7 Functions for constructing a PCA biplot

3.7.1 Function `PCAbipl`

The main function for constructing PCA biplots, described in detail in this section, is `PCAbipl`. The reader is encouraged to study and experiment with the arguments of `PCAbipl` to appreciate all its capabilities. Several functions are called by `PCAbipl` for drawing the biplot, adding features to it and changing its appearance; provision is made for one-, two- and three-dimensional biplots. These functions are generally not directly called by the user and will not be discussed in detail in this book, but a complete set of help files is available on the *Understanding Biplots* website. There are also several functions closely related to `PCAbipl` that users may like to call to construct specific types of PCA biplots or adding enhancements to an existing biplot. These functions are briefly introduced in sections 3.7.2–3.7.9.

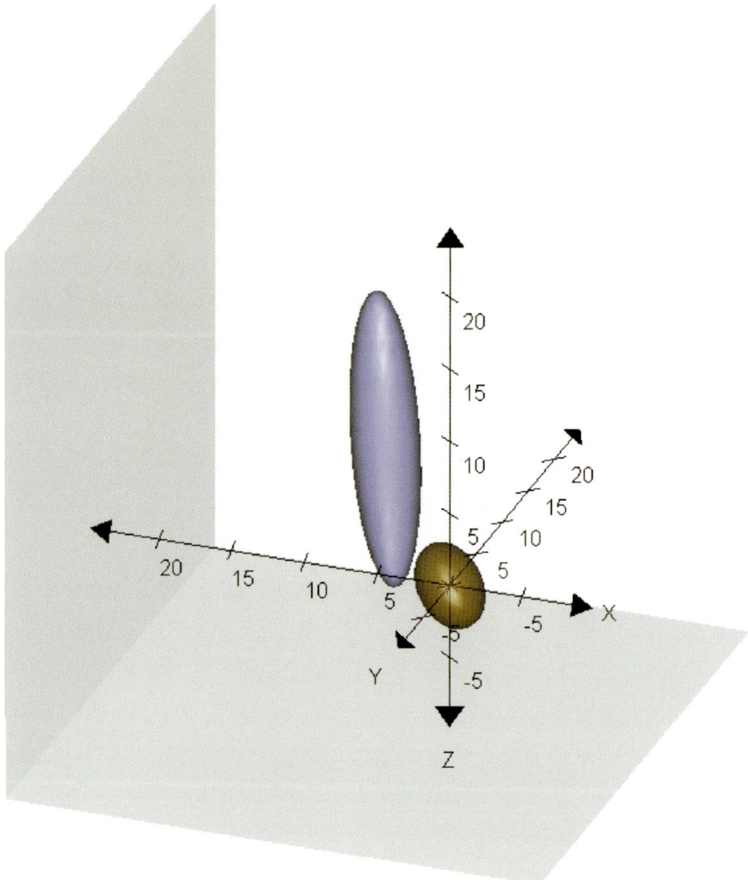

Figure 3.28 Clouds of points for the unscaled and scaled covariance matrices in (3.30).

Usage

```
PCAbipl (X = Ocotea.data[,3:8], G = NULL, X.new.samples = NULL,
   X.new.vars = NULL, scaled.mat = FALSE, e.vects = 1:ncol(X),
   dim.biplot = c(2,1,3), adequacies.print = FALSE,
   alpha = 0.95, alpha.3d = 0.1, aspect.3d = "iso",
   ax.col.3d = "black", ax = 1:sum(c(ncol(X),ncol(X.new.vars))),
   ax.name.col = rep("black",sum(c(ncol(X),ncol(X.new.vars)))),
   ax.name.size = 0.75, ax.type = c("predictive","interpolative"),
   ax.col = list(ax.col = rep("grey",sum(c(ncol(X),
   ncol(X.new.vars)))), tickmarker.col = rep("grey",
   sum(c(ncol(X),ncol(X.new.vars)))), marker.col = rep("black",
   sum(c(ncol(X),ncol(X.new.vars))))),
   between = c(1,-1,0,1), between.columns = -1,
   cex.3d = 0.6, char.legend.size = c(1.2, 0.7), c.hull.n = 10,
   colour.scheme = NULL, colours = c(1:8,3:1),
   colours.means = NULL, col.plane.3d = "lightgrey",
   col.text.3d = "black", columns = 1, constant = 0.1,
```

```
correlation.biplot = FALSE, density.plot = FALSE,
ellipse.kappa = NULL, ellipse.alpha = NULL, exp.factor = 1.2,
factor.x = 2, factor.y = 2, font.3d = 2, ID.labs = FALSE,
ID.3d = 1:nrow(X), label = TRUE, label.size = 0.75,
large.scale = FALSE, legend.type = c(means = FALSE,
samples = FALSE, bags = FALSE), line.length = c(1,1),
line.size = 2.5, line.type = 1:ncol(G), line.width = 1,
markers = TRUE, marker.size = 0.6, max.num = 2500,
means.plot = FALSE, n.int = rep(5,sum(c(ncol(X),
ncol(X.new.vars)))), oblique.trans = NULL, offset = rep(0,4),
offset.m = rep(0.5,sum(c(ncol(X),ncol(X.new.vars)))),
ort.lty = 1, orthog.transx = rep(0,sum(c(ncol(X),
ncol(X.new.vars)))), orthog.transy = rep(0,sum(c(ncol(X),
ncol(X.new.vars)))), output = 1:10, parlegendmar = c(3,1,3,1),
parplotmar = rep(3,4), pch.means = 0:10, pch.means.size = 1,
pch.new = 1, pch.new.cols = "black", pch.new.labels = NULL,
pch.new.size = 0.75, pch.new.labels.size = 0.75,
pch.samples = 0:10, pch.samples.size = 1,
pos = c("Orthog","Hor","Paral"),
pos.m = rep(1,sum(c(ncol(X),ncol(X.new.vars)))),
predictions.3D = TRUE, predictions.mean = NULL,
predictions.sample = NULL, predictivity.print = FALSE,
quality.print = FALSE, reflect = c(FALSE,"x","y"),
rotate.degrees = 0, select.origin = FALSE,
side.label = rep("right",sum(c(ncol(X),ncol(X.new.vars)))),
size.ax.3d = 0.5, size.means.3d = 10, size.points.3d = 5,
specify.bags = NULL, specify.classes = dimnames(G)[[2]],
specify.ellipses = dimnames(G)[[2]], specify.xaxis = NULL,
text.width.mult = 1, Title = "",
Titles.3d = c("","","x","y","z"), Tukey.median = TRUE, ...)
```

Arguments

`X`	Required argument. Data matrix of size $n \times p$ representing observations on p variables of n samples.
`G`	Optional argument. Indicator matrix defining g groups of samples in the data.
`X.new.samples`	Matrix of size $s \times p$ representing s new samples of observations on the original p variables.
`X.new.vars`	Matrix of size $n \times t$ representing observations of the original n samples on t new variables.
`scaled.mat`	Logical argument indicating whether `X` is to be standardized to unit column variances prior to performing the PCA. Defaults to FALSE.
`e.vects`	Vector specifying the eigenvectors to be used for constructing the scaffolding for the PCA biplot. Defaults to those associated with the largest eigenvalues.
`dim.biplot`	Integer value of 1, 2 or 3 specifying the dimension of the PCA biplot. Defaults to 2.

`adequacies.print`	Logical value requesting or suppressing the adequacies of the individual axes to be printed in brackets after the titles of the respective biplot axes. Defaults to FALSE.
`alpha`	Numerical value between zero and one indicating the size of α-bags. Defaults to 0.95.
`alpha.3d`	The alpha parameter associated with a plot made by the R package `rgl`. Regulates transparency of plane added to a three-dimensional biplot. Defaults to 0.1.
`aspect.3d`	Aspect ratio for three-dimensional plot as required by the R package `rgl`. Defaults to `"iso"`.
`ax.col.3d`	Specify colour of biplot axes in three-dimensional biplots. Defaults to `"black"`.
`ax`	Integer-valued vector specifying which biplot axes are shown in a biplot. By default all p biplot axes are shown.
`ax.name.col`	Vector of size p specifying colours of labels of biplot axes. Defaults to `"black"` for all axis labels.
`ax.name.size`	Size of label of biplot axis. Defaults to 0.75.
`ax.type`	One of `"interpolative"` or `"predictive"` specifying the type of biplot axes. Defaults to `"predictive"`.
`ax.col`	A list with three components specifying the colour for each of the biplot axes; the colours of the tick marks on each biplot axis and the colour of the markers associated with the tick marks. Default is `"grey"` for all axes, `"grey"` for all tick marks and `"black"` for all the markers.
`between`	Formatting legend: distance between legend components. Defaults to `c(1, -1, 0, 1)`.
`between.columns`	Formatting legend: distance between columns when `"columns"` argument is 2 or more. Default is −1.
`cex.3d`	Three-dimensional biplots: specifies size of labels of three-dimensional biplot axes. Defaults to 0.6.
`char.legend.size`	Formatting legend: numeric vector of size 2 for controlling the size of plotting symbol (the first element) and the size of its legend (the second element) for individual samples in the legend to the biplot. Default is `c(1.2, 0.7)`.
`c.hull.n`	Integer value. If the number of samples in a group is less than or equal to `c.hull.n`, a convex hull is constructed for that group instead of an α-bag. Default value is 10.
`colour.scheme`	User-defined colour scheme to use. Default is NULL. If a vector of specified colour (character) names is specified, the argument colours must be an integer specifying the number of colours to be interpolated between the names provided in `colour.scheme`.
`colours`	If `colour.scheme` is not NULL then `colours` is an integer value. If `colour.scheme` is NULL then `colours` is a vector of colour names or integer values specifying the `colours` to be used in the biplot. Default is `c(1:8, 3:1)`.
`colours.means`	Vector-valued colour specification to be used in the legend for the means.

`col.plane.3d`	Three-dimensional biplots: colour of optional plane added to a three-dimensional biplot. Defaults to `"lightgrey"`.
`col.text.3d`	Three-dimensional biplots: specifies colour of labels of three-dimensional biplot axes. Defaults to `"black"`.
`columns`	Formatting legend: integer value specifying number of column blocks in a legend. Defaults to a single column block.
`constant`	Numeric value for regulating the vertical translation of horizontal biplot axes in a one-dimensional biplot. A negative value results in biplot axes to be shifted downwards and a positive value in an upwards displacement. Defaults to 0.1.
`correlation.biplot`	PCA biplot that optimizes the approximation of the correlations among the variables. Defaults to FALSE. Note that this is not the same as the 'true' correlation biplot discussed in Chapter 10 as an example of a monoplot.
`density.plot`	One-dimensional biplot: logical value specifying if density estimates are added to a one-dimensional biplot. Defaults to FALSE.
`ellipse.alpha`	The value of α resulting in a κ-ellipse covering approximately $100\alpha\%$ of the configuration of two-dimensional points. See Section 2.9.2.
`ellipse.kappa`	The value of κ to construct a κ-ellipse. See Section 2.9.2.
`exp.factor`	Default is 1.20. Positive numeric value regulating scaffolding axes of the biplot. Larger values for zooming out and smaller values for zooming in with respect to sample points in the biplot display.
`factor.x`	Three-dimensional biplot: numeric value regulating diameter in the x direction of a plane added to a three-dimensional biplot. Default is 2.
`factor.y`	Three-dimensional biplot: numeric value regulating diameter in the y direction of a plane added to a three-dimensional biplot. Default is 2.
`font.3d`	Three-dimensional biplot: specifies font to be used for text in a three-dimensional biplot. Default is 2.
`ID.labs`	Three-dimensional biplot: logical value specifying if samples in a three-dimensional biplot are labelled in the three-dimensional biplot. Defaults to FALSE.
`ID.3d`	A vector specifying the labels of samples in a three-dimensional biplot. Defaults to 1, 2, 3 and so on.
`label`	A vector specifying if samples in a one- or two-dimensional biplot are to be labelled. Defaults to TRUE.
`label.size`	Scalar value regulating size of labels of samples in a one- or two-dimensional biplot. Defaults to 0.75.
`large.scale`	Logical value specifying if a large-scale version of the biplot is required. Mainly used in the case of several groups to focus on the area occupied by the group means. Defaults to FALSE.

`legend.type`	Three-component logical vector specifying the type of legend to construct: group mean plotting character; sample plotting character; line type of α-bag or convex hull. Legend appears in a separate graph window. If no legend is needed, `legend.type = c(means = FALSE, samples = FALSE, bags = FALSE)` is used. This is the default for PCA biplots.
`line.length`	A numeric vector consisting of two components: the first component determines the length of the tick marks on all biplot axes; the second component determines the distance of the scale marker to the tick mark. The default is `c(1,1)`.
`line.size`	Numeric value determining the length of line used in legend for α-bag or convex hull. Default is 2.6.
`line.type`	Integer-valued vector specifying line type for constructing α-bags. Default is 1, requesting a solid line.
`line.width`	Numeric value specifying line width for constructing α-bags. Default is 1.
`markers`	Logical value regulating display of the markers associated with the tick marks on the biplot axes. If TRUE, all markers are displayed; if FALSE, only the markers of the two extreme tick marks on each biplot axis are displayed. Default is TRUE.
`marker.size`	Numeric value regulating the size of the markers associated with the tick marks on the biplot axes. Default is 0.6.
`max.num`	Integer value. If the number of samples in a group is more than `max.num` a random sample of size `max.num` is selected for constructing an α-bag. Default value is 2500.
`means.plot`	Logical value regulating if group means are displayed. Defaults to FALSE for a PCA biplot.
`n.int`	Integer-valued vector of size equal to the number of biplot axes to control the number of tickmarks on each individual biplot axis. Default is 5 for each axis.
`oblique.trans`	Only used with `ax.type = "interpolative"`. Default is NULL. Otherwise a vector of p elements specifying the values of each variable at the point of concurrency of the biplot axes.
`offset`	A four-component numeric vector controlling the distance a biplot axis title is printed from the side of the figure. Sides are numbered 1 to 4 according to R conventions clockwise starting from the bottom horizontal side. Default is `rep(0,4)`.
`offset.m`	Used together with arguments `pos.m` and `side.label` to optimize placement of labels of calibrations on biplot axes. A p-component numeric vector controlling the offset from the position specified by `pos.m` and `side.label` for each axis. Default is an offset of 0.5 for each axis.
`ort.lty`	Two-component vector specifying the type and colour of the line used for predicting values for samples or means from biplot prediction axes. Default is a solid black line.

`orthog.transx` Only used when `dim.biplot = 2` and `ax.type = "predictive"`. Numeric vector of size p specifying the x-coordinate of the parallel transformation of each axis. Defaults to zero for each axis.

`orthog.transy` Only used when `dim.biplot = 2` and `ax.type = "predictive"`. Numeric vector of size p specifying the y-coordinate of the parallel transformation of each axis. Defaults to zero for each axis.

`output` Integer-valued vector specifying components of `VALUE` to be printed. Defaults to 1:10, and NULL suppresses printing of all output. The output list has the following components: 1, `Z`; 2, `Z.axes`; 3, `V`; 4, `Eigenvectors`; 5, `e.vals`; 6, `PCA.quality`; 7, `adequacy`; 8, `Predictivity`; 9, `Predictions`; 10, `origin.pos`.

`parlegendmar` Four-component numeric vector for regulating the legend margins. Defaults to `c(3,1,3,1)`.

`parplotmar` Four-component numeric vector for regulating the figure margins. Defaults to `rep(3,4)`.

`pch.means` Vector of size equal to the number of groups specifying plotting characters for group means. Defaults to 0:10.

`pch.means.size` Numeric value regulating the size of the plotting character for displaying group means. Defaults to 1.

`pch.new` Defaults to `"1"` for open circles. If FALSE, new points are indicated by `"N1"`, `"N2"`, and so on.

`pch.new.cols` Colour specification for new points. Defaults to `"black"`.

`pch.new.labels` Label specifications for plotting character of new points. Defaults to NULL.

`pch.new.size` Numeric value specifying size of plotting character for new sample points. Defaults to 0.75.

`pch.samples` Vector specifying plotting symbols for individual samples. Defaults to 0:10.

`pch.samples.size` Size of plotting symbol for individual samples. Defaults to 1.

`pos` One of `"Orthog"`, `"Hor"` or `"Paral"` specifying titles of axes to appear orthogonal to the side of the figure; always horizontally or always parallel to the side of the figure. Default is `"Orthog"`.

`pos.m` Used together with arguments `offset.m` and `side.label` to optimize placement of labels (markers) of calibrations on biplot axes. A p-component integer-valued vector controlling placement of markers on each biplot axis: 1 = below, 2 = to left, 3 = above and 4 = to the right of specified coordinates. Defaults to 1 for each axis.

`predictions.3D` Logical value specifying whether a matrix with the predicted values accompanying a three-dimensional PCA biplot is to be returned. Default is TRUE.

`predictions.mean` Integer-valued vector specifying which group means are to be predicted from the prediction axes on a one- or two-dimensional biplot. Default is NULL.

`predictions.sample` Integer-valued vector specifying which samples are to be predicted from the prediction axes on a one- or two-dimensional biplot. Default is NULL.

`predictivity.print` Logical value requesting or suppressing the individual axis predictivities to be added in brackets to the title of the biplot axis. Default is FALSE.

`quality.print` Logical value requesting or suppressing overall PCA quality to be printed in the legend to the biplot. Default is FALSE.

`reflect` Defaults to FALSE. Possible values are `"x"` for reflection about x-axis or `"y"` for reflection about y-axis.

`rotate.degrees` Default is 0. Positive value results in anti-clockwise rotation and negative value in clockwise rotation.

`select.origin` Allows interactive selection of the position of the origin using the mouse. Defaults to FALSE.

`side.label` Used together with arguments `offset.m` and `pos.m` to optimize placement of labels (markers) of calibrations on biplot axes. A p-component vector with possible values `"left"` or `"right"` (the default) controlling on which side of a biplot axis the markers are placed.

`size.ax.3d` Numeric value specifying width of three-dimensional biplot axes. Defaults to 0.5.

`size.means.3d` Numeric value specifying size of plotting character for group means in a three-dimensional biplot. Defaults to 10.

`size.points.3d` Numeric value specifying size of plotting character for sample points in a three-dimensional biplot. Defaults to 5.

`specify.bags` Integer-valued or character vector with the index numbers or names of those groups for which α-bags are to be drawn on the biplot. Default of NULL specifies no α-bags to be drawn.

`specify.classes` Integer-valued or character vector with the index numbers or names of those groups whose samples are to be drawn on the biplot. Default specifies all groups to be shown.

`specify.ellipses` Integer-valued or character vector with the index numbers or names of those groups for which κ-ellipses are to be drawn on the biplot. Default of NULL specifies no κ-ellipses to be drawn.

`specify.xaxis` Allows the user to specify a rotation resulting in the specified axis appearing horizontal in a two-dimensional biplot. Default is NULL. Other admissible specifications are integer value, character name of axis (variable) or "`maxpred`" requesting the biplot to be rotated such that the biplot axis with the highest axis predictivity is in a horizontal position.

`text.width.mult` Numeric value specifying distance between block elements in a legend to a biplot. Default is 1.

`Title`	Character string to provide a title for a one- or two-dimensional biplot. Default is `" "`.
`Titles.3d`	Character string to provide a title for a three-dimensional biplot. Defaults to `c(" ", " ", "x", "y", "z")`.
`Tukey.median`	Logical value requesting or suppressing Tukey medians to be added to a biplot. Default = FALSE.
`...`	Optional arguments passed to the plot function controlling the appearance of the biplot.

Value

`Z`	Matrix with each row containing the details of the point to be plotted.
`Z.axes`	A list with each component a matrix containing the details of a biplot axis to be plotted.
`V`	Matrix consisting of the eigenvectors as columns.
`Eigenvectors`	Vectors used for the biplot scaffolding.
`e.vals`	All the eigenvalues.
`PCA.quality`	The overall quality of the display.
`Adequacy`	Adequacies associated with the variables.
`Predictivity`	The axis predictivities.
`Predictions`	Predictions of the means or of the samples as specified.
`origin.pos`	Scaffolding coordinates of the origin.

3.7.2 Function `PCAbipl.zoom`

This is a function for zooming into an interactively selected area of a one- or two dimensional PCA biplot. In addition to the arguments of `PCAbipl`, this function has the argument `zoomval` for controlling the amount of zooming. Specify `zoomval = NULL` for no zooming, a value less than unity for zooming in and a value larger than unity for zooming out. When `PCAbipl.zoom` is called, a graph window with the original PCA biplot is opened. The cursor, when moved over the biplot, changes to a cross. Position the cross at the bottom left-hand corner of that part of the biplot in need of zooming and select. A second graph window is opened with the zoomed biplot.

3.7.3 Function `PCAbipl.density`

This is a function for constructing two-dimensional PCA biplots on top of a density plot of a two-dimensional PCA approximation of the input matrix. The R function `kde2d` described by Venables and Ripley (2002) and available in package `MASS` is used to perform two-dimensional kernel density estimation with an axis-aligned bivariate normal kernel, evaluated on a square grid. `PCAbipl.density` has the following arguments not found in `PCAbipl`:

`bandwidth`	A two-component vector specifying the bandwith to be used by `kde2d` in the x and y directions. The default of NULL indicates that the default bandwith of `kde2d` is used.
`colours.density`	A vector of at least two components specifying the colours of the density response surface. There are `cuts.density-1` colours interpolated between the components of `colours.density`. Default is `c("green", "yellow", "red")`.
`cuts.density`	Number of interpolated colours to be used in the density response surface. Default is 50.
`dens.ax.cex`	Arguments `dens.ax.cex`, `dens.ax.tcl` and `dens.mgp` are used for fine-tuning the colour key. Default of `dens.ax.cex` is 0.6.
`dens.ax.tcl`	Arguments `dens.ax.cex`, `dens.ax.tcl` and `dens.mgp` are used for fine-tuning the colour key. Default of `dens.ax.tcl` is – 0.2.
`dens.mgp`	Arguments `dens.ax.cex`, `dens.ax.tcl` and `dens.mgp` are used for fine-tuning the colour key. Default of `dens.mgp` is `c(0, -0.25, 0)`.
`draw.densitycontours`	Specify if density contours are added to the density response surface. Default is FALSE.
`layout.heights`	A two-component vector specifying the top and bottom panels in the graph window. The top panel is where the biplot is constructed and the bottom panel provides for the colour key. Default is `c(100, 10)`.
`parlegendmar.dens`	Control the margins surrounding the colour key. Default is `c(2,5,0,5)`.
`smooth.n`	Integer value to be used as the argument `n` in function `kde2d` to specify the number of grid points to be used in each direction. Defaults to 100.
`specify.density.class`	Single integer or string specifying the sample points to be used for the density plot. Specifying `"allsamples"` results in all the rows of the input matrix to be used. When argument `G` is used as an indicator matrix specifying a group structure in the data, any one of the groups can be specified by selecting one of the column names of `G`.

3.7.4 Function `PCAbipl.density.zoom`

Analogous to `PCAbipl.zoom`, the function `PCAbipl.density.zoom` provides zooming functionality for `PCAbipl.density` through its argument `zoomval`.

3.7.5 Function `PCA.predictivities`

A function for calculating the various measures of fit discussed in this chapter for PCA biplots. It has the four arguments, discussed in Section 3.7.1 for `PCAbipl`: `X`, `X.new.samples = NULL`, `X.new.vars = NULL`, and `scaled.mat = FALSE`. It is assumed that `X` is an $n \times p$ data matrix with p not larger than n. By default `X` is centred but not scaled.

The various measures of fit are provided for dimensions $1, 2, \ldots, p$ and the output (value) of `PCA.predictivities` is a list with the following components.

`Quality`	Overall quality of the display calculated according to (3.17).
`Weights`	The weights calculated according to (3.21).
`Sample.predictivities.original`	Sample predictivities for original n samples as defined in (3.20).
`Adequacies`	Adequacies calculated according to (3.18).
`Axis.predictivities.original`	Axes predictivities for original p variables as defined in (3.19).
`Sample.predictivities.new`	Sample predictivities for new samples as defined in (3.22).
`Axis.predictivities.new.1`	Axes predictivities for new variables as defined in (3.28).
`Axis.predictivities.new.2`	Axes predictivities for new variables as defined in (3.29).

3.7.6 Function `PCA.predictions.mat`

The function `PCA.predictions.mat` takes the three arguments: `X`, `scaled.mat = FALSE` and `e.vects = 1:2` as described in Section 3.7.1. Input `scale(X)` to obtain predictions for the `X`-values centred to zero means and unit variances; `X` and `scaled.mat = FALSE` to obtained predictions for the unscaled `X`-values; or `X` and `scaled.mat = TRUE` to obtain predictions for the scaled values that are transformed into the original scales of measurement. Predictions can be obtained in any dimension by a proper specification of argument `e.vects`.

3.7.7 Function `vector.sum.interp`

A function call to `vector.sum.interp` results in interpolating a new sample into a PCA biplot as described in Section 3.2.2. The following arguments are available: `vertex.points = 3`, `p = vertex.points`, `pch.centroid = 15`, `col.centroid =`

"black", size.centroid = 1, pch.interp = 16, col.interp = "red", size .interp = 1.2, length = 0.15, angle = 15, lty = 1, lwd = 1.75, ... where the argument ... refers to any additional argument(s) passed to the function polygon called by vector.sum.interp.

3.7.8 Function circle.projection.interactive

This function has the following arguments:

cent.line	Specifies if a centre line is to be drawn. Default is FALSE.
Colr	Specifies colour of circle. Default is "black".
Origin	Specifies position of origin of biplot. Default is c(0,0).
...	Optional arguments passed to the lines function controlling appearance of circle.

The function circle.projection.interactive is called after a PCA biplot with prediction axes has been created on a graph window. On calling this function the cursor changes to a cross when moved over the biplot. Position the cursor at any point on the biplot (usually one of the sample points). A circle is drawn with its centre the midpoint of the line from the origin to the selected point. The intersections of the circle with the prediction axes give the predictions on all variables for the chosen sample. This method of prediction is called circular prediction and plays an important role in nonlinear biplots (see Chapter 5).

3.7.9 Utility functions

The four functions draw.arrow, draw.rect, draw.text and draw.polygon are useful for annotating any biplot and are called after a biplot has been drawn. They all contain a ... argument for passing arguments to change the properties of the arrow, rectangle, text or polygon to be created. After calling draw.arrow the two end positions for the arrow must be selected on the biplot; similarly, the bottom left and top right position for the rectangle must be selected after calling draw.rect. The function draw.text has a required argument string specifying the string value to be printed at the selected point on the biplot. Printing in the figure margins is allowed and control over the exact printing position is managed through the ... argument. In several of the biplots shown in this book we have made use of draw.text to manually adjust some of the axes labels for improved legibility. The draw.polygon function has a required integer-valued argument vertex.points specifying the number of vertex points needed. The default vertex.points = 3 results in a triangle being drawn after selecting the three positions for the vertices on the biplot. The function draw.polygon returns the coordinates of the vertices as well as the coordinates of the centroid of the drawn polygon.

3.8 Some novel applications and enhancements of PCA biplots

3.8.1 Risk management example revisited

In the biplot in Figure 3.1 the point (***CM*** = 0, ***IRD*** = 0, ***MM*** = 0, ***ALCO*** = 0, ***SE*** = 0, ***EDSA*** = 0, ***EDM*** = 0) is interpolated by adding the argument `X.new.samples = matrix(rep(0,7), nrow = 1))` in the call to `PCAbipl`. This results in the top panel of Figure 3.29. If the interpolated point is wanted for the scaled data then the argument `scaled.mat = TRUE` is also needed. The reader can verify that the result is a biplot where the interpolated point does not appear. The reason is that the interpolated point lies outside the original plotting area. To make this point visible, increase the default setting of `exp.factor`. The bottom panel of Figure 3.29 was obtained with the settings `scaled.mat = TRUE, exp.factor = 2`. Note the annotation using functions `draw.arrow` and `draw.text`. The interpolated point can be viewed as an ideal point where there is zero loss for all seven instruments. Clearly the 'best' situation using the unscaled data was achieved on ***day20***.

One problem, especially with the biplot in the bottom panel of Figure 3.29, is that several data points are so bundled together that it is impossible to identify a particular day. With even moderately large data sets this problem is bound to change for the worse, therefore we provide an R function, `PCAbipl.zoom`, to interactively zoom into any required part of a biplot. When this function is called with the argument `zoomval = x`, the window with the drawn PCA biplot is activated, the mouse pointer changes to a cross and the user can move the cross to select the bottom left-hand corner to zoom into. The value x controls the amount of zooming such that the aspect ratio is kept constant at unity. Figure 3.30 gives an example of the zooming function.

The Figure 3.29 biplot can be further enhanced by adding a trend line showing the seven-dimensional movement over time. A simple solution would be to connect the sample points in the PCA biplot in temporal order. Since `PCAbipl` returns the coordinates of the sample points in the biplot, this is easy to accomplish. The connecting lines are shown in Figure 3.31 for both the unscaled and the scaled data, but it is obvious that the trend is difficult to follow, with too many interconnecting lines. Had the data set consisted of 100 days' VAR values, such a connecting line would render the biplot useless. It is easy to correct this: we need some form of smoothing. In Figure 3.32 a nonparametric regression smoother was fitted to each of the two dimensions separately. The R commands

```
> z1 <- X.cent %*% Eigenvectors[,1]
# Eigenvectors returned by PCAbipl
> z2 <- X.cent %*% Eigenvectors[,2]
> zfit1 <- fitted(loess(z1 ~ I(1:20)))
> zfit2 <- fitted(loess(z2 ~ I(1:20)))
```

yield the values to be connected to form this trend line. In this example, the default span for the `loess` function was used. However, the amount of smoothing can be controlled by this parameter and any other smoothing technique can be applied similarly.

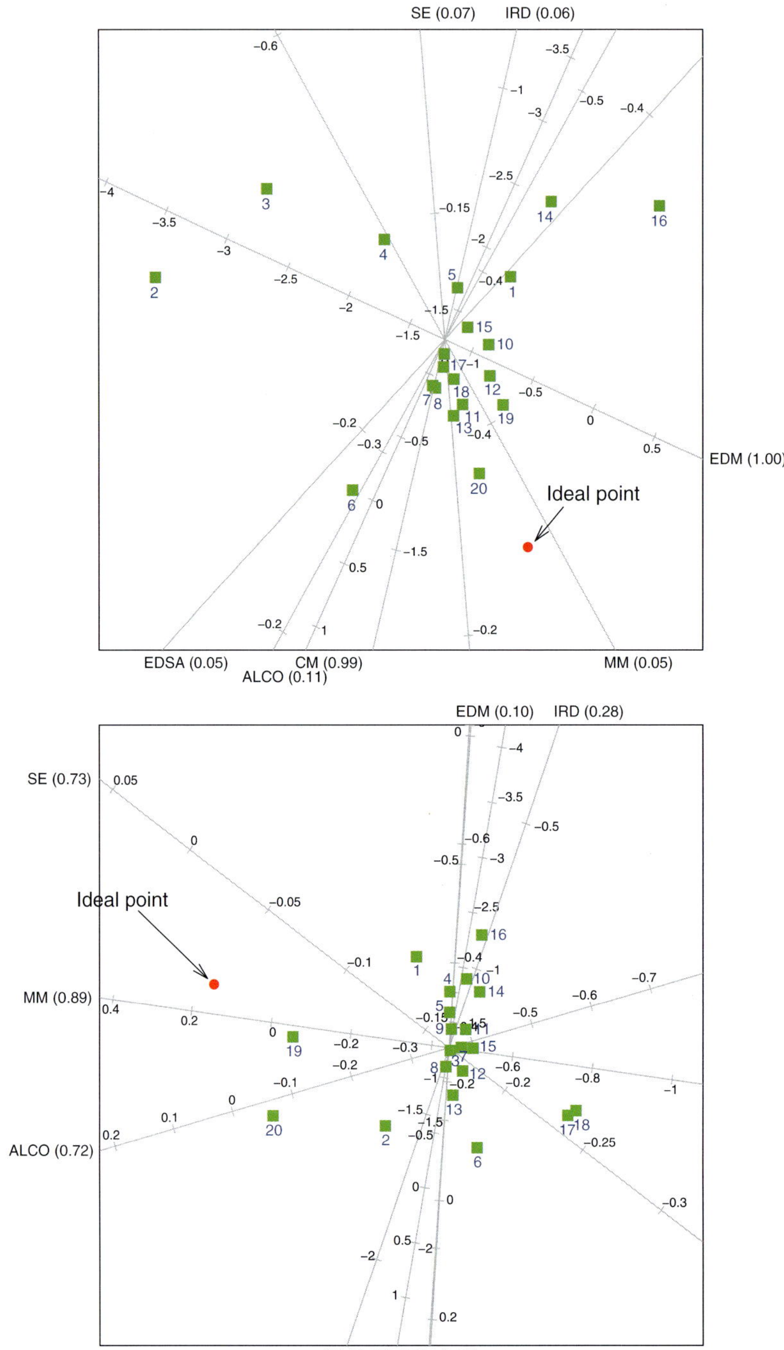

Figure 3.29 PCA biplot of the risk management data with an ideal point interpolated. (Top) Unscaled data as in Figure 3.1. (Bottom) Data normalized to unit variances. Annotation obtained with function calls `draw.arrow(angle = 15, length = 0.15)` and `draw.text(1,"Ideal point", xpd = NA)`. Axis predictivities are added in brackets after the axis labels.

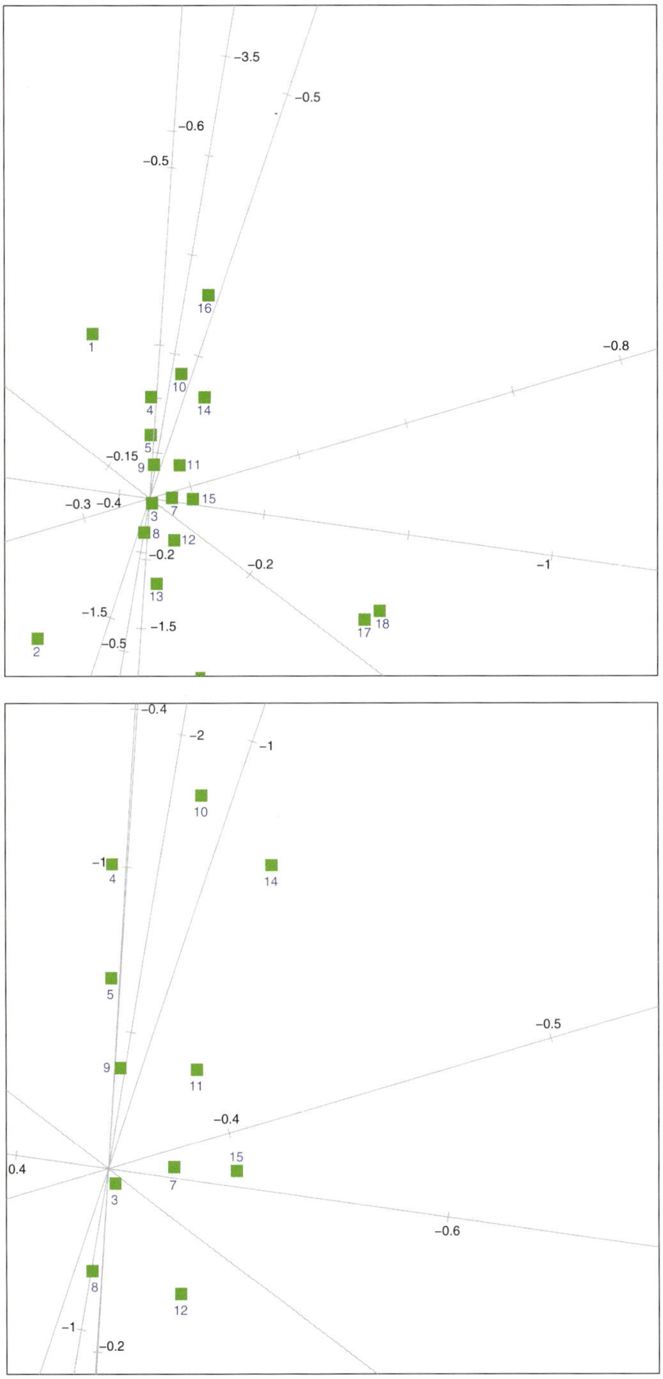

Figure 3.30 Zooming into the biplot given in the bottom panel of Figure 3.29. A function call is made to `PCAbipl.zoom` with argument `zoomval` specified as 0.6 in the top panel and as 0.2 in the bottom panel.

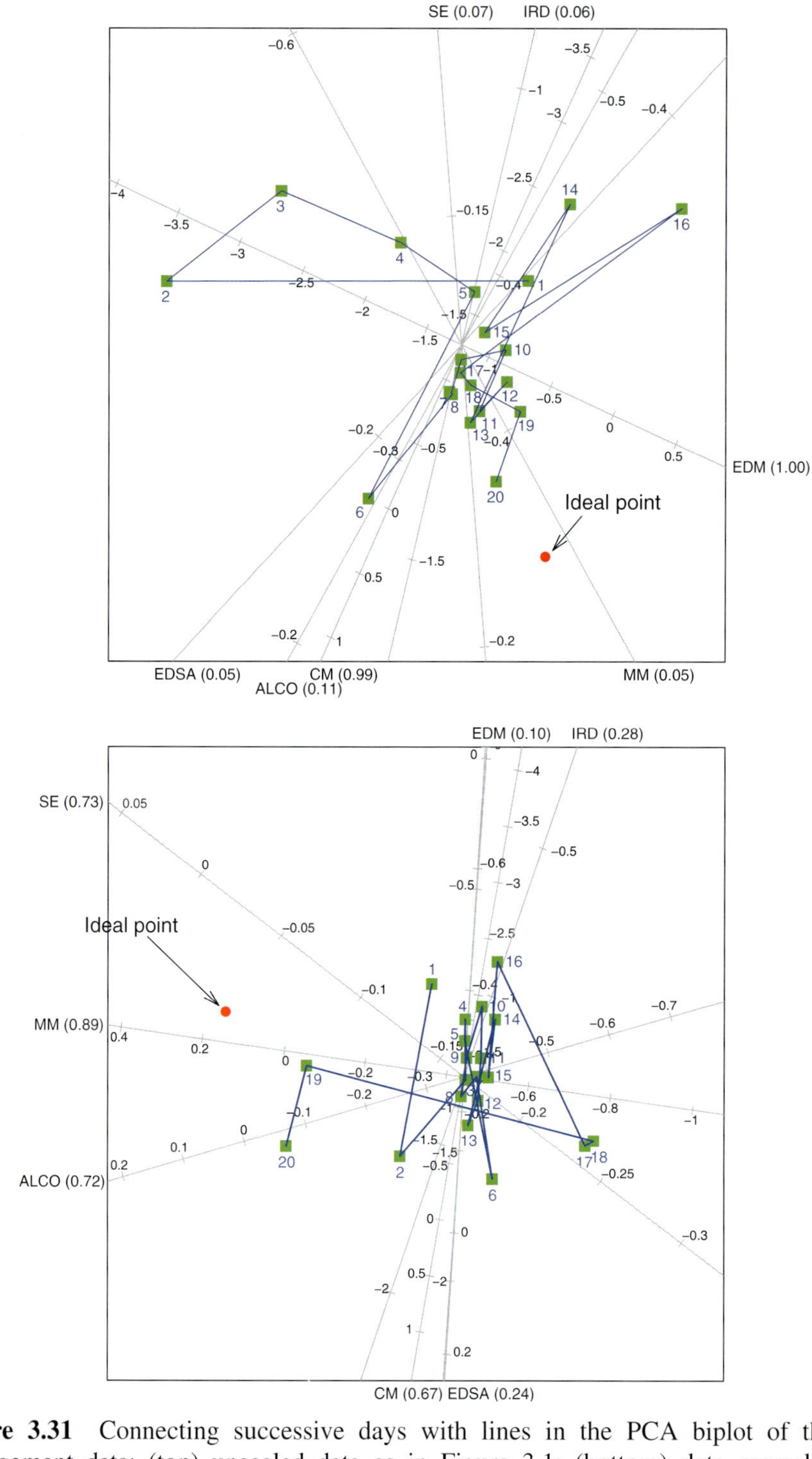

Figure 3.31 Connecting successive days with lines in the PCA biplot of the risk management data: (top) unscaled data as in Figure 3.1; (bottom) data normalized to unit variances.

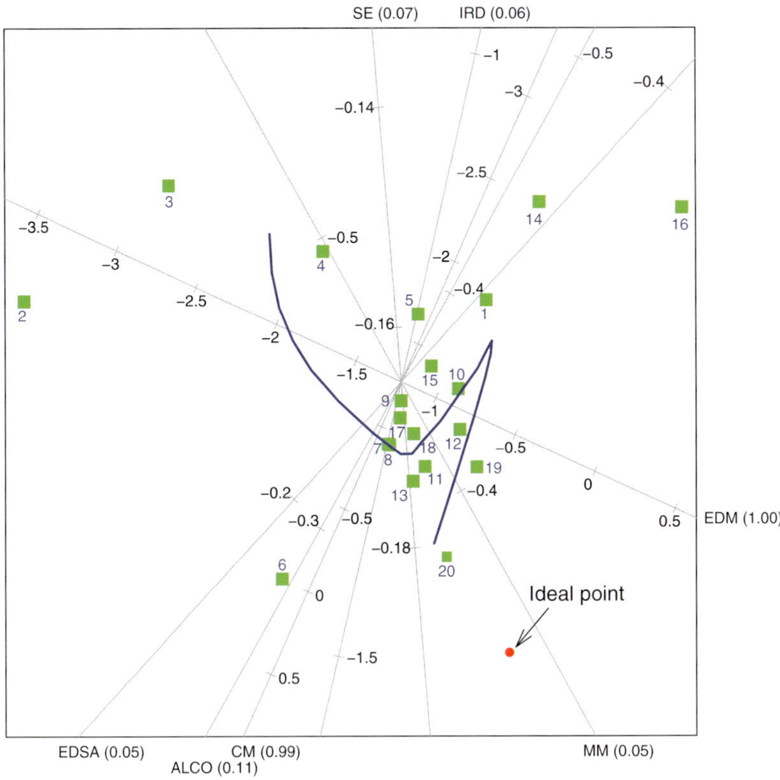

Figure 3.32 PCA biplot of the unscaled risk management data with smooth trend line.

In Figure 3.32 the biplot provides the risk manager with a single visual representation showing the performance of the seven financial instruments. While ***EDM*** improved and stabilized over the period, other instruments such as ***CM*** exhibited up and down behaviour with a sudden change on about ***day15***.

3.8.2 Quality as a multidimensional process

A manufacturing company is monitoring 15 different variables in a production process. In an effort to quantify the overall product quality, this particular manufacturing company devised a quality index value. Throughout the period of a calendar month the 15 variables were measured during the production process. At the end of the month, the means and standard deviations of the 15 selected variables were transformed into a single quality index value in the interval [0,100]. We shall treat this transformation process as a company-specific black box but emphasize that the biplot procedure described below is independent of what is inside this black box. Although the index values coming out of the black box are useful for top-level management reporting, they give no indication of what the causes of a poor index value could be. The monthly mean values for January 2000 to March 2001 are given in Table 3.19. The variable names indicate eight different

Table 3.19 Monthly mean values of process quality data.

	A1	A2	A3	A4	A5	B5	C6	C7	C4	D6	D7	D4	C5	E5	C8
Target	50.0000	79.0000	1.0000	1.5000	21.5000	21.5000	13.5000	60.0000	4.0000	26.0000	44.0000	4.5000	20.5000	14.4000	31.0000
Jan00	52.1429	79.6921	0.9838	1.3261	21.7913	21.7913	13.6533	57.2769	5.6172	26.6848	44.1449	4.4253	21.1304	14.4043	29.7404
Feb00	51.8194	79.6480	0.9689	1.4501	21.7026	21.7026	12.8269	55.5428	5.3329	26.5088	43.3836	4.7137	21.0632	14.4053	30.0839
Mar00	51.2883	79.3036	0.9645	1.4301	21.9850	21.9850	14.5462	55.6089	5.1844	26.8377	43.4200	4.4371	21.4700	14.5750	30.0242
Apr00	51.1715	78.4583	1.0340	1.6277	21.8286	21.8286	15.8880	57.0383	4.6620	26.4532	42.6851	4.7684	21.3643	14.4429	31.1838
May00	47.9080	78.0719	0.9539	1.6117	22.1742	22.1742	13.1614	54.9745	4.1805	25.3487	41.6083	4.6769	21.1355	14.6567	30.0964
Jun00	48.9479	78.7573	1.0617	1.5024	21.9667	21.9667	11.9847	58.7851	3.0873	25.5832	42.9429	4.6417	20.7000	14.3714	29.8877
Jul00	49.7307	79.2724	1.0704	1.9038	21.3190	21.3190	14.6813	62.3389	3.2039	26.1395	43.1948	5.8265	20.2561	14.1762	31.8072
Aug00	50.9614	80.4318	0.9771	1.3479	21.0732	21.0732	13.3213	61.4085	2.8640	28.4579	45.2571	4.6279	21.4268	14.2415	30.6000
Sep00	49.0064	79.7186	0.9100	1.3799	20.8026	20.8026	9.9144	57.7668	3.2018	28.3067	44.5784	4.9123	21.1923	14.3282	29.5640
Oct00	48.9670	79.9079	0.7634	1.3619	20.2195	20.2195	9.5232	56.3606	3.5400	28.9799	45.0961	4.8704	20.1390	14.4683	29.6432
Nov00	49.0967	78.4098	0.7746	1.5025	20.6745	20.6745	10.3263	56.9594	3.6787	27.4899	43.7873	4.8270	20.6059	14.3824	29.2240
Dec00	48.1956	78.1670	0.8458	1.3891	20.8000	20.8000	11.0603	59.3170	3.6811	31.2640	45.5633	4.7664	20.9500	14.1000	29.2000
Jan01	50.5007	79.2910	0.7746	1.2869	21.1300	21.1300	12.8229	57.6100	3.9680	29.5918	45.1305	5.0812	19.8700	14.2500	29.1319
Feb01	48.5487	78.8882	0.8191	1.4455	20.8852	20.8852	12.7963	56.8522	4.5222	28.7501	44.6700	5.4576	20.2111	14.3519	30.1912
Mar01	49.1918	78.9735	1.0433	1.2782	21.0381	21.0381	16.8893	64.5517	3.1575	30.2973	46.1898	4.8632	19.7667	14.3952	31.5034

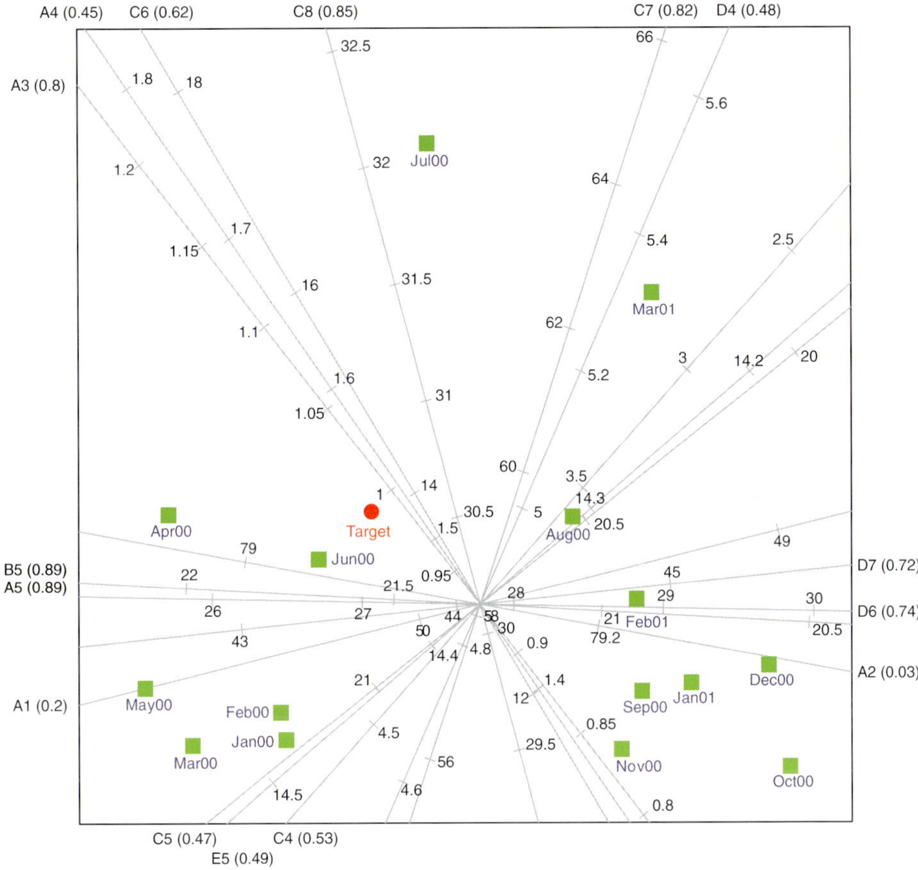

Figure 3.33 PCA biplot of the scaled process quality data with a multidimensional target interpolated.

characteristics measured at five stages, A, B, C, D and E, in the manufacturing process. A target value specified by company management for each variable is included in the table.

The PCA biplot of the process quality data (available as R object `ProcessQuality.data`) in Figure 3.33 improves the summary and interpretation of the process. The target values are interpolated into the biplot as $\mathbf{z}'_{\text{target}} = \mathbf{t}'\mathbf{V}_r$ using formula (3.4) with $\mathbf{t}$ the scaled target values, to put each month's performance into perspective. The function call for constructing the Figure 3.33 biplot is

```
PCAbipl(ProcessQuality.data[-1,], scaled.mat = TRUE, X.new =
   matrix(ProcessQuality.data[1,], nrow = 1), colours = "green",
   pch.samples = 15, pch.new = 16, pch.new.cols = "red",
   pch.new.labels = "Target")
```

After obtaining the initial biplot several arguments are available (see Section 3.7.1) for refining the general appearance of the biplot.

The biplot representation of the process quality data shows that on average ***Jun00*** produced a product much closer to target than did ***Jul00***. Since the biplot axes are concurrent at the centroid, it is clear that over the 15-month period the process was not centred about the target. The smoothed trend line (see Figure 3.34) shows how the process started in early 2000 with high levels for ***C4***, ***E5***, ***C5***, ***A1***, ***A5*** and ***B5*** and low levels of ***A3***, ***A4***, ***C6***, ***C8***, ***C7***, ***D4***, ***D7***, ***D6*** and ***A2***. There was some movement towards the target, especially on ***C5***, ***E5***, ***C4***, ***C7*** and ***D4***. However, these variables changed too much and this, together with lower values for ***A1***, ***A5***, ***B5***, ***A3***, ***A4***, ***C6***, ***C8*** and higher values for ***D6***, ***D7*** and ***A2***, resulted in poor quality at the end of 2000. At this stage the senior management requested a quality improvement plan which resulted in a sharp turnaround, but mainly on ***A3***, ***A4***, ***C6*** and ***C8***.

To assess what quality product is associated with each position in the biplot, *quality regions* can be constructed according to company management's ruling: index values above 80 are considered to be good quality, while index values below 50 are considered poor quality. These quality regions are constructed as follows:

Start: Overlay the biplot space, $\mathscr{L}$, with a two-dimensional $m \times m$ grid, coded as $\mathbf{E}$: $m^2 \times 2$

Initialize: Set $i = 1$

Repeat:

- Let $\mathbf{z}^{*\prime}$ be the ith row of $\mathbf{E}$
- Find the predicted values of $\mathbf{z}^*$ for all the variables using the prediction formula (3.8) as $\hat{\mathbf{x}}^{*\prime} = \mathbf{z}^{*\prime}\mathbf{V}_r' : 1 \times p$
- $\hat{\mathbf{x}}^{*\prime} \Longrightarrow$ Black box $\Longrightarrow$ Index value
- Colour grid point $\begin{cases} \text{Red} & \text{if } 0 \le \text{Index value} < 50 \\ \text{Blue} & \text{if } 50 \le \text{Index value} < 80 \\ \text{Green} & \text{if } 80 \le \text{Index value} < 100 \end{cases}$
- Set $i = i + 1$

Figure 3.34 shows the PCA biplot of the scaled process quality data with quality regions added. Also shown is a smoothed trend line. This line is constructed by smoothing each dimension separately. The R function `loess` is applied to the sample pairs (z_{1j}, t_1), $(z_{2j}, t_2), \ldots, (z_{nj}, t_n)$ for $j = 1, 2, \ldots, r$ (in Figure 3.34, $r = 2$) and $t_i = i$. The resulting `loess` fitted values provide the coordinates $(\hat{z}_{11}, \hat{z}_{12}), (\hat{z}_{21}, \hat{z}_{22}), \ldots, (\hat{z}_{n1}, \hat{z}_{n2})$ which are then connected to trace a smooth path over time leading to the smooth yellow line shown in Figure 3.34.

We see that ***C6*** for ***Mar01*** is quite high, 16.9, compared to the target of 13.5. This sample is also located in the satisfactory (rather than good) quality region, indicating that there is a problem of some kind for this month. We can see that some leeway around the target is allowed, but the very narrow satisfactory quality area shows how steep the penalty is when moving further away from the target. The current variability in ***A2***, ***A5***, ***B5***, ***D6***, ***D7***, ***A1*** is such that these variables stay in the good quality regions while the current variability in especially ***C5***, ***E5*** and ***C4*** can easily lead to an overall poor quality index value.

The overall quality for the PCA biplot in Figures 3.33 and 3.34 is 59.82%. This might seem very low, 40% of the variability is not represented, but keeping in mind that the number of dimensions has been reduced from 14 to 2, this is not too bad. The overall

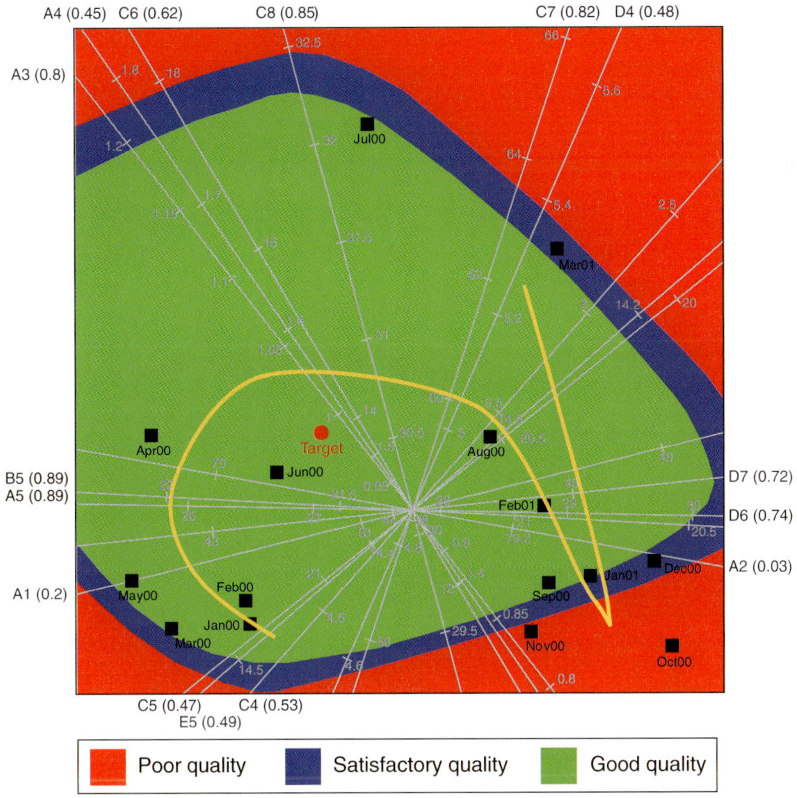

Figure 3.34 PCA biplot of process quality data with a target, smooth trend line and quality regions added. Axis predictivities are added in brackets after the axis labels.

quality, adequacies, axis predictivities, sample predictivities as well as the predictivity for the target are given in Tables 3.20–3.24. These values are returned by the following call to function `PCA.predictivities`:

```
PCA.predictivities(ProcessQuality.data[-1,],
    X.new.samples = matrix(ProcessQuality.data[1,],nrow = 1),
    scaled.mat = TRUE)
```

We see that ***A2*** has a particularly low axis predictivity value. When interpreting the trend line, we mentioned that ***A2*** starts at a low level and reaches a high level at the end of 2000. Comparing the observed data in Table 3.19 with this statement shows that this is not the case. Axis predictivity is therefore an important measure of the weight that should be given to the interpretation of each variable. Similarly, ***A3***, ***A5***, ***B5*** and ***C8*** have axis predictivities above 0.8. These variables clearly closely follow the pattern suggested by the biplot and trend line.

We also see that the predictivities for ***Mar00*** and ***Apr00*** are the highest, 0.85, and that for ***Aug00*** the lowest, 0.14. Comparing the predictive values for these three samples to the observed values, we notice that ***Aug00*** has quite large differences especially on ***A1***, ***A2***, ***A4***, ***C7*** and ***D4*** (see Table 3.25).

Table 3.20 Overall quality attained associated with Figure 3.34 biplot in all dimensions.

Dim 1	Dim 2	Dim 3	Dim 4	Dim 5	Dim 6	Dim 7	Dim 8	Dim 9	Dim 10	Dim 11	Dim 12	Dim 13	Dim 14
37.8	59.8	74.9	82.7	89.7	94.6	97.5	98.6	99.4	99.7	99.9	100.0	100.0	100.0

The sample predictivity of the target is calculated as 0.46. This moderate value indicates that the distance from the biplot space to the target in the full space is relatively large. Since the biplot space is 'as close as possible' to the sample points, this suggests that overall the sample points are not very close to the target.

3.8.3 Using axis predictivities in biplots

Axis predictivities can be used as a criterion for selecting axes to be shown in a predictive PCA biplot. In the mail order catalogue data shown as a PCA biplot in Figure 2.8 there are several axes with very low axis predictivity. In Figure 3.35 we again show the PCA biplot of the scaled mail order catalogue data but we use argument `ax` of `PCAbipl` to suppress plotting of all axes with predictivity less than 0.45. In addition, we specify `predictivity.print = TRUE` for including in the label of each axis its predictivity.

A PCA biplot can be rotated so that the axis with maximum predictivity is drawn in a horizontal position by calling `PCAbipl` with argument `specify.xaxis = "maxpred"`. If in this call the argument `select.origin` is set to TRUE then the origin of this biplot can be interactively changed to any desired position. This is shown in Figure 3.36 for the *Ocotea* data.

In Figure 3.37 we show how to carry out individual orthogonal parallel translation of the biplot axes so that the biplot axes are moved to peripheral positions that do not interfere with the sample points. This leads to a biplot with approximately orthogonal predictive axes similar to conventional scatterplots. The function call to `PCAbipl` resulting in Figure 3.37 needs the settings `orthog.transy = c(-4.9, -4.2, -5.5, -4, 5.3, 5.5)` and `select.origin = FALSE`.

3.8.4 One-dimensional PCA biplots

As an example of a one-dimensional biplot we consider mining data from a copper flotation plant. The complete data set of nearly 500 samples is described by Aldrich *et al.* (2004) and is available as `CopperFroth.data` in library `UBbipl`. In this section we ignore the group structure but in Section 4.10 we will revisit `CopperFroth.data` by considering biplots that take into account predefined groups in the data.

One problem with one-dimensional biplots is that all sample points and all the biplot axes appear on a single straight line. `PCAbipl` shifts all the biplot axes parallel to the line containing the samples, as shown in Figure 3.38. The axes are calibrated in the usual way to allow for predictions to be made. We illustrate this in Figure 3.38 for samples ***s20***, ***s50*** and ***s250***. Compare these graphically determined values with the algebraically computed

Table 3.21 Adequacies associated with Figure 3.34 biplot in all dimensions.

	A1	A2	A3	A4	A5	B5	C6	C7	C4	D6	D7	D4	C5	E5	C8
Dim 1	0.034	0.006	0.074	0.036	0.157	0.157	0.042	0.023	0.055	0.130	0.126	0.021	0.061	0.062	0.017
Dim 2	0.036	0.006	0.191	0.110	0.157	0.157	0.158	0.230	0.120	0.130	0.127	0.128	0.098	0.105	0.246
Dim 3	0.273	0.201	0.219	0.274	0.159	0.159	0.220	0.267	0.142	0.160	0.235	0.215	0.114	0.110	0.253
Dim 4	0.386	0.520	0.222	0.380	0.184	0.184	0.308	0.288	0.143	0.248	0.257	0.326	0.137	0.162	0.254
Dim 5	0.433	0.524	0.297	0.380	0.186	0.186	0.407	0.335	0.441	0.256	0.259	0.466	0.375	0.191	0.265
Dim 6	0.465	0.675	0.298	0.399	0.202	0.202	0.408	0.335	0.511	0.301	0.260	0.476	0.440	0.707	0.321
Dim 7	0.467	0.700	0.304	0.445	0.316	0.316	0.432	0.344	0.542	0.360	0.263	0.480	0.795	0.752	0.485
Dim 8	0.744	0.822	0.320	0.490	0.378	0.378	0.432	0.384	0.565	0.418	0.283	0.680	0.866	0.752	0.486
Dim 9	0.780	0.828	0.621	0.490	0.406	0.406	0.604	0.402	0.881	0.443	0.301	0.698	0.897	0.754	0.490
Dim 10	0.814	0.831	0.739	0.491	0.406	0.406	0.635	0.786	0.943	0.609	0.320	0.712	0.953	0.774	0.582
Dim 11	0.833	0.933	0.760	0.903	0.432	0.432	0.638	0.804	0.960	0.785	0.324	0.843	0.985	0.779	0.590
Dim 12	0.922	0.945	0.932	0.928	0.435	0.435	0.638	0.812	0.975	0.867	0.329	0.968	0.995	0.958	0.858
Dim 13	0.928	0.947	0.942	1.000	0.435	0.435	0.694	0.977	0.996	1.000	0.841	0.975	0.996	0.960	0.873
Dim 14	1.000	1.000	1.000	1.000	0.500	0.500	1.000	1.000	1.000	1.000	1.000	1.000	1.000	1.000	1.000

Table 3.22 Axis predictivities associated with Figure 3.34 biplot in all dimensions.

	A1	A2	A3	A4	A5	B5	C6	C7	C4	D6	D7	D4	C5	E5	C8
Dim 1	0.195	0.034	0.418	0.206	0.888	0.888	0.237	0.132	0.310	0.736	0.713	0.122	0.348	0.349	0.095
Dim 2	0.201	0.034	0.805	0.450	0.889	0.889	0.620	0.815	0.527	0.736	0.717	0.475	0.470	0.492	0.852
Dim 3	0.739	0.476	0.868	0.821	0.893	0.893	0.761	0.898	0.577	0.805	0.962	0.671	0.506	0.504	0.868
Dim 4	0.870	0.847	0.872	0.944	0.923	0.923	0.864	0.924	0.578	0.907	0.988	0.800	0.533	0.564	0.869
Dim 5	0.919	0.852	0.950	0.944	0.924	0.924	0.967	0.972	0.889	0.915	0.990	0.948	0.782	0.594	0.881
Dim 6	0.943	0.964	0.951	0.959	0.937	0.937	0.967	0.972	0.942	0.949	0.991	0.954	0.830	0.977	0.923
Dim 7	0.944	0.974	0.953	0.978	0.985	0.985	0.978	0.976	0.955	0.974	0.992	0.956	0.981	0.997	0.992
Dim 8	0.993	0.996	0.956	0.986	0.996	0.996	0.978	0.983	0.959	0.984	0.996	0.991	0.993	0.997	0.992
Dim 9	0.997	0.996	0.993	0.986	0.999	0.999	0.999	0.985	0.997	0.987	0.998	0.994	0.997	0.997	0.993
Dim 10	0.998	0.997	0.997	0.986	0.999	0.999	1.000	0.999	0.999	0.993	0.998	0.994	0.999	0.997	0.996
Dim 11	0.999	1.000	0.998	1.000	1.000	1.000	1.000	0.999	1.000	0.999	0.999	0.998	1.000	0.998	0.996
Dim 12	1.000	1.000	1.000	1.000	1.000	1.000	1.000	1.000	1.000	1.000	0.999	1.000	1.000	1.000	1.000
Dim 13	1.000	1.000	1.000	1.000	1.000	1.000	1.000	1.000	1.000	1.000	1.000	1.000	1.000	1.000	1.000
Dim 14	1.000	1.000	1.000	1.000	1.000	1.000	1.000	1.000	1.000	1.000	1.000	1.000	1.000	1.000	1.000

Table 3.23 Sample predictivities associated with Figure 3.34 biplot in all dimensions.

	Jan00	Feb00	Mar00	Apr00	May00	Jun00	Jul00	Aug00	Sep00	Oct00	Nov00	Dec00	Jan01	Feb01	Mar01
Dim 1	0.338	0.509	0.692	0.787	0.554	0.307	0.010	0.077	0.411	0.599	0.243	0.535	0.478	0.348	0.138
Dim 2	0.495	0.651	0.849	0.850	0.587	0.330	0.783	0.142	0.523	0.755	0.486	0.558	0.540	0.348	0.564
Dim 3	0.912	0.782	0.953	0.850	0.888	0.436	0.862	0.638	0.526	0.766	0.857	0.575	0.563	0.526	0.693
Dim 4	0.918	0.911	0.960	0.852	0.956	0.467	0.988	0.703	0.640	0.806	0.858	0.662	0.564	0.526	0.942
Dim 5	0.936	0.964	0.969	0.864	0.961	0.754	0.992	0.955	0.912	0.815	0.859	0.713	0.741	0.897	0.952
Dim 6	0.952	0.973	0.972	0.885	0.988	0.759	0.997	0.959	0.930	0.983	0.860	0.977	0.779	0.898	0.997
Dim 7	0.966	0.977	0.984	0.970	0.988	0.988	0.998	0.969	0.936	0.988	0.867	0.992	0.944	0.908	0.997
Dim 8	0.967	0.977	0.987	0.985	0.995	0.989	0.998	0.970	0.980	0.991	0.988	0.993	0.944	0.985	0.998
Dim 9	0.993	0.991	0.992	0.989	0.998	0.995	0.998	0.996	0.981	0.993	0.988	0.996	0.999	0.986	1.000
Dim 10	1.000	0.999	0.993	0.997	0.999	0.998	0.998	0.997	0.981	0.996	0.997	0.997	1.000	0.988	1.000
Dim 11	1.000	0.999	0.996	1.000	0.999	0.999	1.000	0.997	0.990	1.000	1.000	1.000	1.000	0.997	1.000
Dim 12	1.000	0.999	0.999	1.000	1.000	0.999	1.000	1.000	1.000	1.000	1.000	1.000	1.000	1.000	1.000
Dim 13	1.000	1.000	1.000	1.000	1.000	1.000	1.000	1.000	1.000	1.000	1.000	1.000	1.000	1.000	1.000

Table 3.24 Sample predictivities associated with the target in Figure 3.34 biplot in all dimensions.

Dim 1	Dim 2	Dim 3	Dim 4	Dim 5	Dim 6	Dim 7	Dim 8	Dim 9	Dim 10	Dim 11	Dim 12	Dim 13	Dim 14
0.269	0.459	0.479	0.517	0.537	0.612	0.621	0.769	0.852	0.854	0.857	0.981	0.987	1.000

Table 3.25 Comparison of predicted and observed values for low and high sample predictivity.

	A1	A2	A3	A4	A5	B5	C6	C7	C4	D6	D7	D4	C5	E5	C8
Predicted															
Mar00	50.7	79.0	0.97	1.49	22.0	22.0	13.2	55.2	5.0	25.9	42.7	4.53	21.8	14.5	29.9
Apr00	50.6	79.0	1.06	1.60	22.1	22.1	15.0	58.0	4.5	25.7	42.7	4.80	21.1	14.5	30.8
Aug00	49.5	79.2	0.93	1.47	21.1	21.1	13.1	59.7	3.6	28.4	44.6	5.02	20.5	14.3	30.4
Observed															
Mar00	51.3	79.3	0.96	1.43	22.0	22.0	14.5	55.6	5.2	26.8	43.4	4.44	21.5	14.6	30.0
Apr00	51.2	78.5	1.03	1.63	21.8	21.8	15.9	57.0	4.7	26.5	42.7	4.77	21.4	14.4	31.2
Aug00	51.0	80.4	0.98	1.35	21.1	21.1	13.3	61.4	2.9	28.5	45.3	4.63	21.4	14.2	30.6

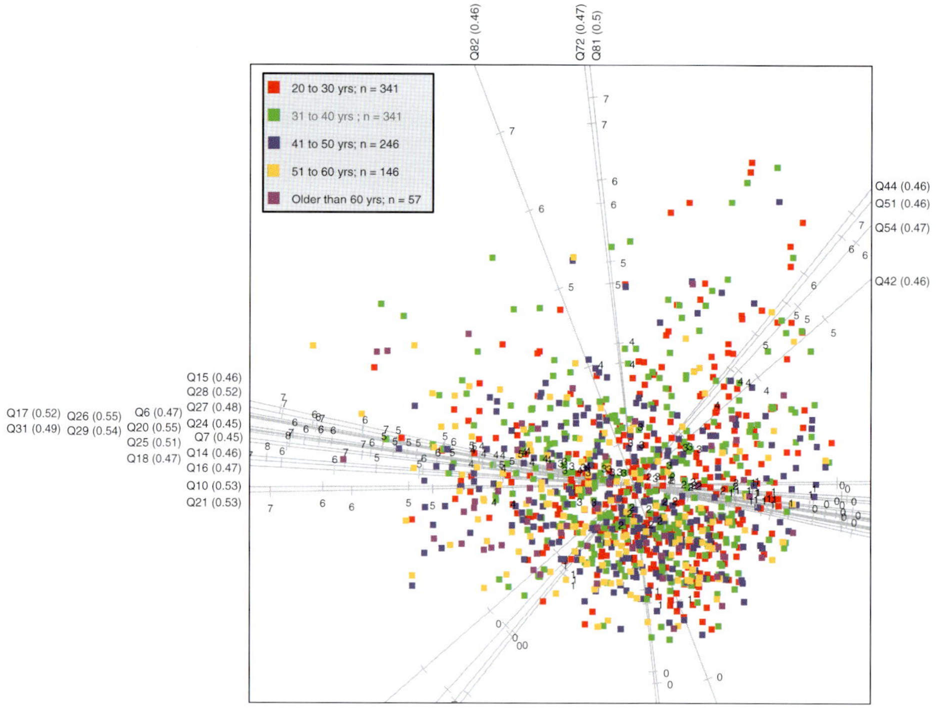

Figure 3.35 Revisiting the mail order catalogue data of Section 2.8 by constructing its PCA biplot with only those axes shown with predictivities larger than 0.45 and adding the predictivity value to the axis label.

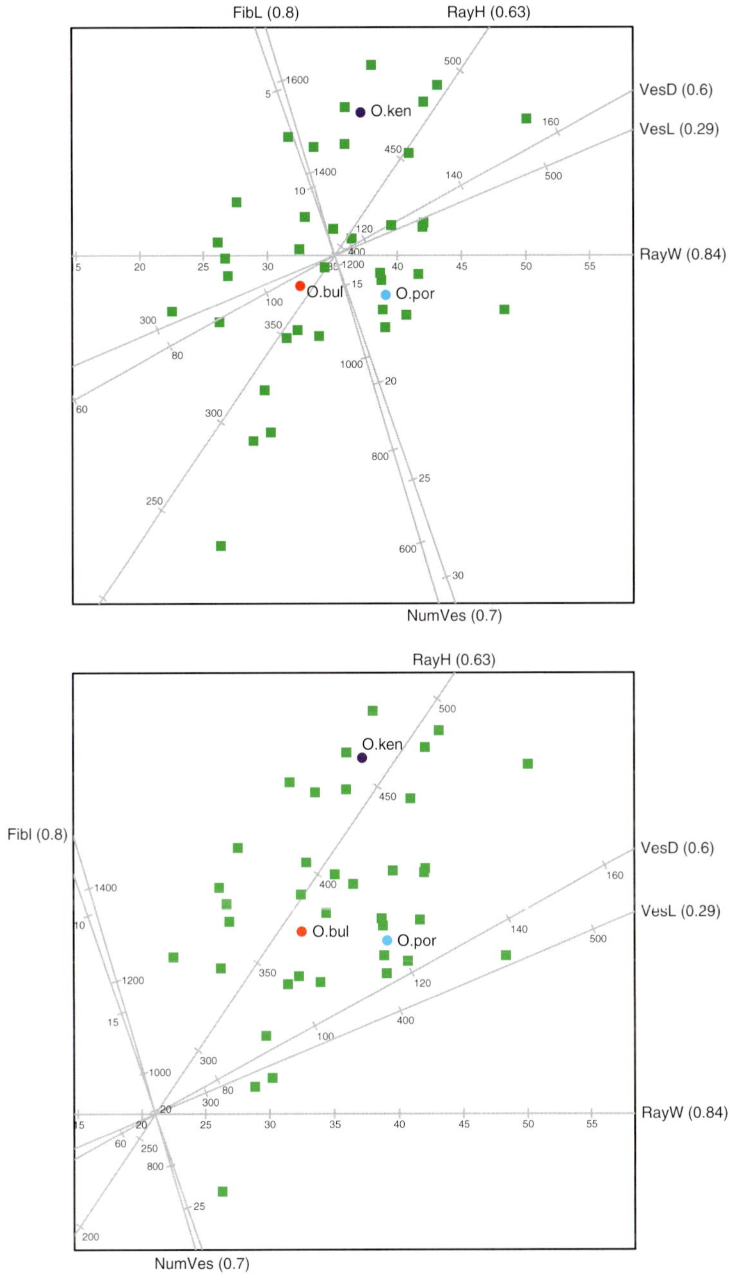

Figure 3.36 Some enhancements to the Figure 3.23 PCA biplot of the scaled *Ocotea* data. In the top panel the biplot is rotated so that the axis with the highest predictivity, ***RayW***, is in the horizontal position. In the bottom panel the orthogonal parallel translation procedure (see Section 2.8) has been called upon by the argument `select.origin = TRUE` to translate the origin to allow a less obscured configuration of sample points. In both panels the axis predictivities have been added to the labels of the respective axes.

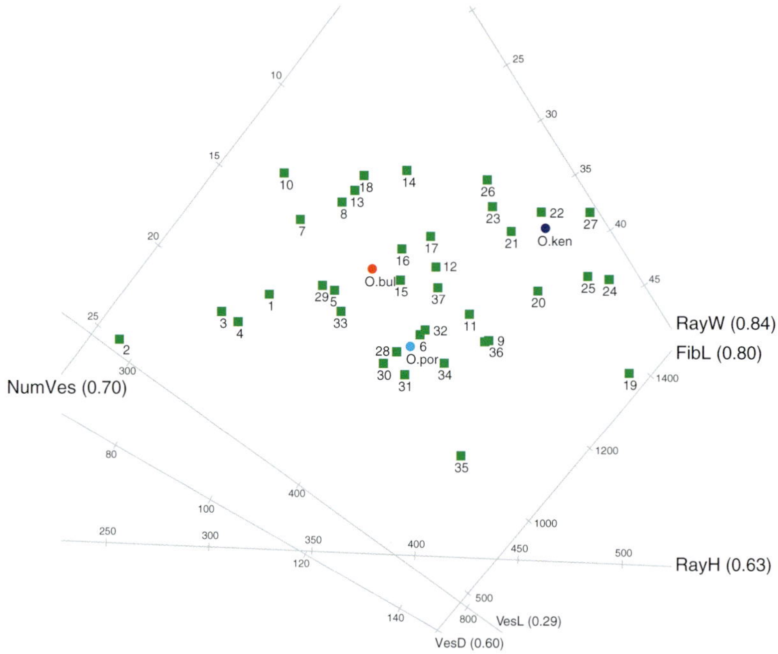

Figure 3.37 Orthogonal parallel translation of the individual biplot axes of the PCA biplot in Figure 3.23. This leads to a biplot where all biplot axes are moved to approximately orthogonal peripheral positions not interfering with the sample points, similar to ordinary scatterplots.

Table 3.26 Quality, axis predictivities and predictions for three samples associated with the one-dimensional PCA biplot in Figure 3.38.

		Quality of 1D display: 61.62%		
Var	Axis predictivity	s20	s50	s250
X1	0.47	32.54	32.62	32.15
X2	0.03	6476.79	6492.19	6398.15
X3	0.82	330.58	344.25	260.75
X4	0.82	−5.17	−5.19	−5.02
X5	0.93	9.16	10.02	4.78
X6	0.94	−8.30	−9.00	−4.71
X7	0.82	22617.39	20929.64	31237.43
X8	0.11	1.04	1.03	1.07

values as returned by `PCAbipl` in Table 3.26. Since there are so many sample points we suppress labelling them in the biplot shown, but `PCAbipl` optionally allows for adding a density plot to the one-dimensional PCA biplot. This is also shown in Figure 3.38 and suggests some structure in the data even in one dimension. We have also requested the axes predictivities to be added to the labels of the biplot axes.

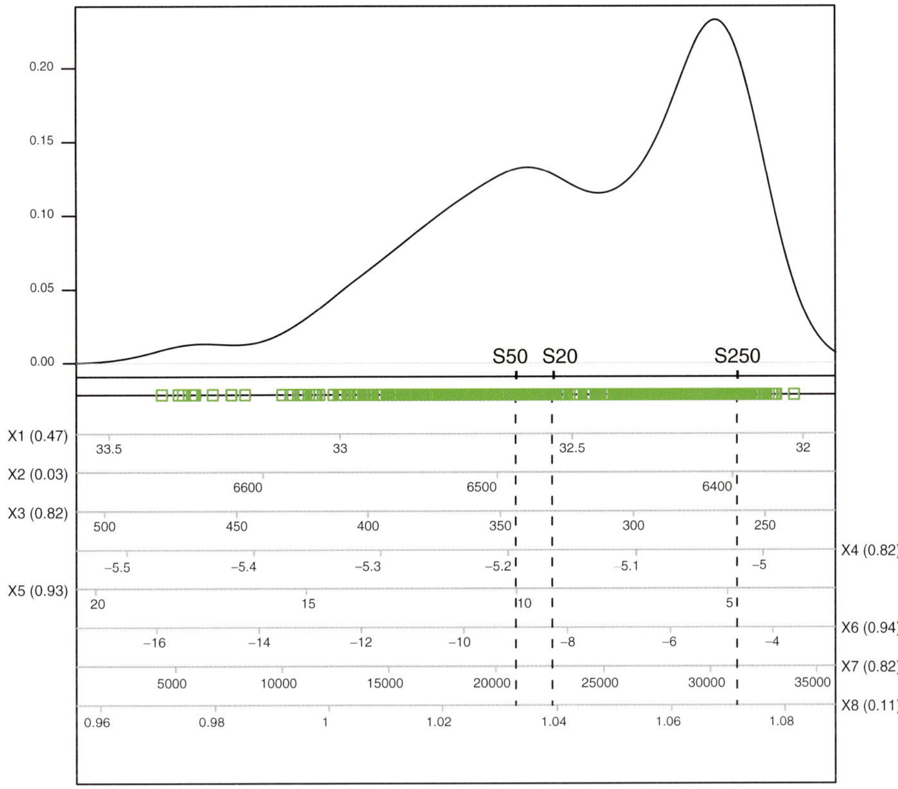

Figure 3.38 One-dimensional PCA biplot of the copper froth data with density estimate added.

The full function call resulting in Figure 3.38 is

```
PCAbipl(X = CopperFroth.data[,1:8], scaled.mat = TRUE,
   dim.biplot = 1, density.plot = TRUE, colours = "green",
   constant = -0.6, label = FALSE, means.plot = FALSE,
   n.int = c(5,15,8,5,5,5,5,25), offset = c(0, 2.4, 0, 0.2),
   ort.lty = 2, predictivity.print = TRUE,
   predictions.sample = c(20,50,250))
```

The biplot representation of the sample points, together with their density estimate in Figure 3.38, has also a specific statistical interpretation: it is a graphical display of the principal component scores with respect to the first principal component. Similar plots can be made for the other principal components by specifying them in the argument `e.vects` argument of `PCAbipl`. This is shown for principal components 2 to 5 in Figure 3.39.

3.8.5 Three-dimensional PCA biplots

Three-dimensional PCA biplots are obtained by specifying `dim.biplot = 3` in calls to `PCAbipl`. Package `rgl` is required and on calling `PCAbipl` with `dim.biplot = 3` the

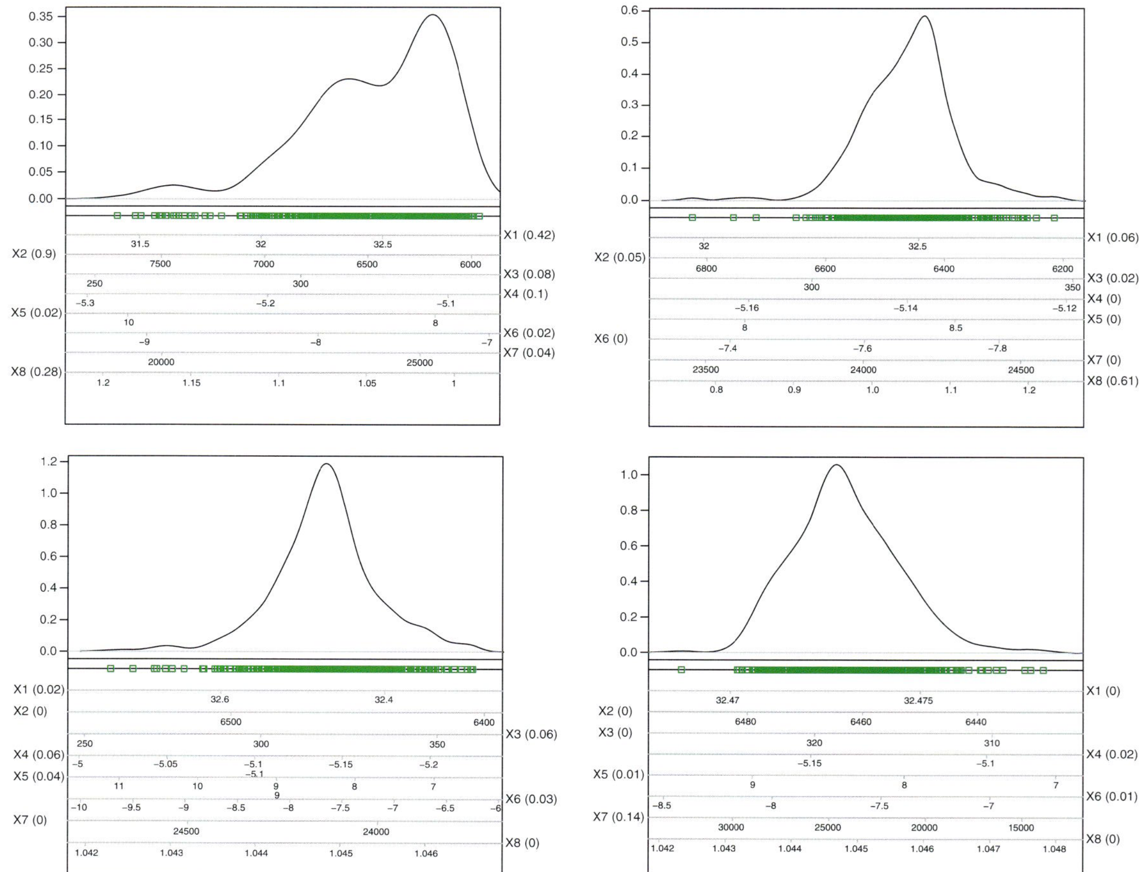

Figure 3.39 One-dimensional PCA biplots of copper froth data: (top left) second principal component scores; (top right) third principal component scores; (bottom left) fourth principal component scores; (bottom right) fifth principal component scores. These four biplots are constructed by assigning in the call to `PCAbipl` argument `e.vects` successively the single integer values 2, 3, 4 and 5.

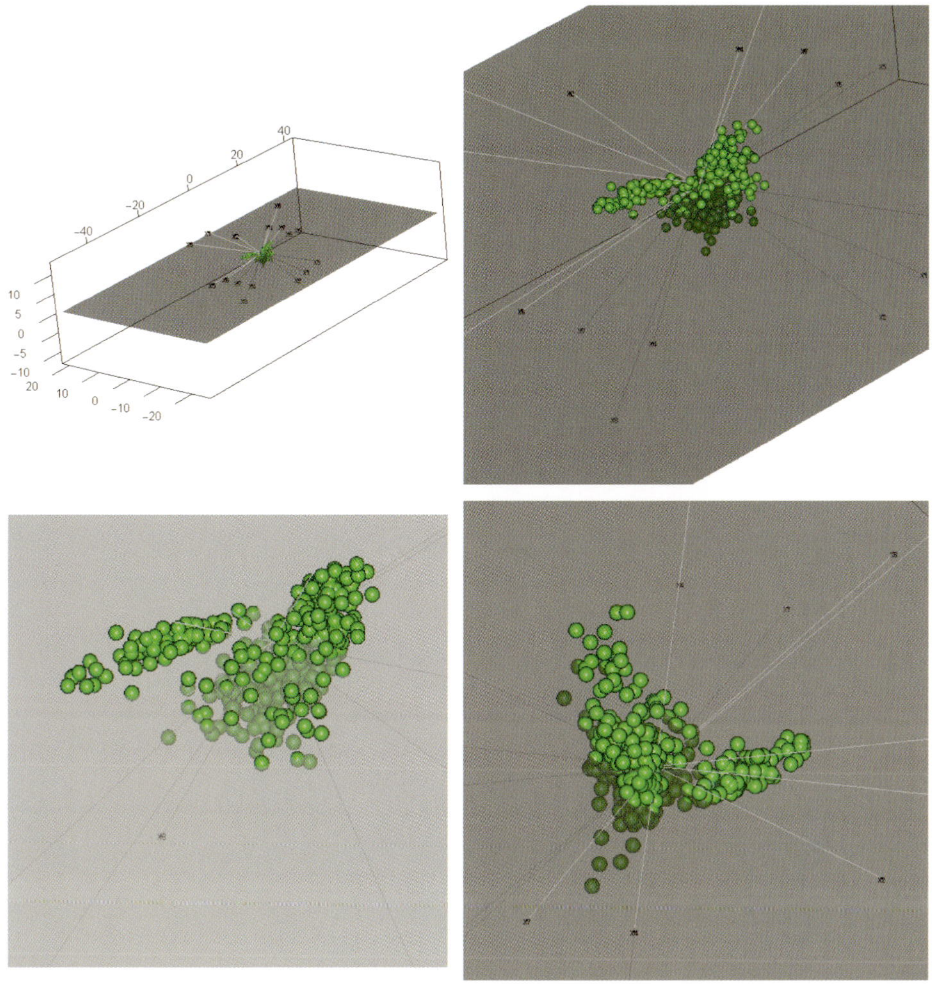

Figure 3.40 Snapshots of a three-dimensional PCA biplot of the copper froth data. The quality of the three-dimensional PCA biplot is 94.23%. The top right snapshot shows a zoomed view of the biplot in the top left panel. The snapshots in the bottom panels are also zoomed views of the top left biplot but have been rotated as well. The three-dimensional biplots clearly show some group structure in the data.

biplot is drawn in an `rgl` window. The plot can then be interactively rotated and zoomed using the mouse buttons. Figure 3.40 contains snapshots of the results of the call.

```
PCAbipl(CopperFroth.data[,1:8], scaled.mat = TRUE,
   dim.biplot = 3, colours = "green", predictions.3D = TRUE,
   size.points.3d = 0.25)
```

Note that by default the three-dimensional biplot contains a grey coloured plane through the origin separating points above the plane from those below the plane. We suggest that readers try the above call and then interactively explore the three-dimensional PCA biplot. In addition to the biplot, the call returns a matrix with the three-dimensional predictions of all samples on all the variables. As with all other biplots, the `e.vects` argument allows for any three principal axes to be chosen for the biplot scaffolding.

3.8.6 Changing the scaffolding axes in conventional two-dimensional PCA biplots

In Figure 3.41 we show for the copper froth data two-dimensional PCA biplots using various principal components to form the scaffolding for the biplot. The two scaffolding axes are selected according to the first two components of the vector assigned to argument `e.vects` of `PCAbipl`. Some structure is seen in these biplots. Note also that the predictivities of all

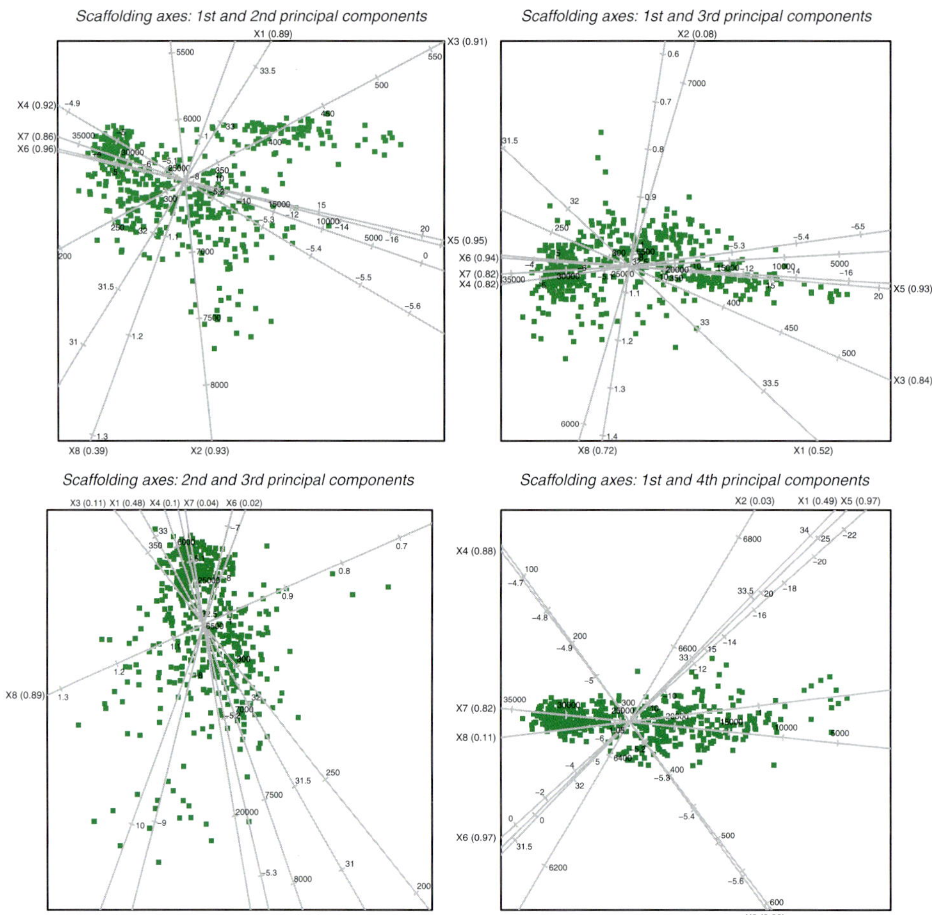

Figure 3.41 Two-dimensional PCA biplots of the copper froth data obtained for various combinations of the two scaffolding axes. Quality of the displays: (top left) 85% (maximum for a two-dimensional biplot); (top right) 70.85%; (bottom left) 32.61%; (bottom right) 64.27%.

variables with the exception of ***X8*** are high when the first two principal components are used as scaffolding, ***X2*** is very well represented by the first principal component, while ***X8*** catches up when the third principal component is one of the scaffolding axes.

3.8.7 Alpha-bags, kappa-ellipses, density surfaces and zooming

In Figures 3.42 and 3.43 we show some enhancements of the PCA biplot of the copper froth data shown in the top left panel of Figure 3.41. In the top panel of Figure 3.42 we add a 0.95-bag enclosing the inner 95% of the sample points as well as a concentration ellipse. These enhancements were added by including the following arguments in the function call to `PCAbipl: alpha = 0.95, ellipse.kappa = 2, specify.bags = 1, specify.ellipses = 1`. Instead of calling `PCAbipl` we have called `PCAbipl.density` to construct the biplot in the bottom panel of Figure 3.42. As can be seen from the latter biplot, the sample points obscure the surface of highest density. Therefore we suppress plotting the sample points in Figure 3.43 while specifying `draw.densitycontours = TRUE, cuts.density = 20` in the call to `PCAbipl.density` for adding contour lines to the density surface. In the bottom panel we have translated the origin interactively, as explained earlier, to allow a much clearer view of the higher density surface.

The function `PCAbipl.zoom` can be used to interactively zoom into any part of interest in a PCA biplot. The results of calling `PCAbipl.zoom` with `zoomval = 0.5` (`zoomval = 0.2` in bottom panel) followed by selecting the bottom left-hand corner of the area to be zoomed are shown in Figure 3.44, respectively.

3.8.8 Predictions by circle projection

Finally, we demonstrate in Figure 3.45 how to obtain predictions on all variables simultaneously in a PCA biplot for any chosen sample point using the function `circle.projection.interactive`. In the top panel of Figure 3.45 the PCA biplot is constructed in the usual way. Then `circle.projection.interactive` is called with argument `colr = "blue"`. Now we select a sample point in the biplot. The blue circle is then drawn as shown in the top panel of Figure 3.45. The intersections of the circle with the biplot axes give the respective predictions. The function call to `circle.projection.interactive` can be repeated for as many samples as needed. Note that the vertical lines to the respective axes are only drawn here to demonstrate the orthogonality of the projections. With a little more effort `circle.projection.interactive` can also be used after an interactive translation of the origin. This is demonstrated in the bottom panel of Figure 3.45. When `PCAbipl` is called with `select.origin = TRUE` it returns also the `usr` coordinates of the newly selected point of intersection of the axes. These coordinates are then assigned to argument `origin` of `circle.projection.interactive`, resulting in the circles drawn in the bottom panel of Figure 3.45. The reader can verify that the predictions remain the same. As a reference, Table 3.27 gives the predictions for the sample points determined algebraically by `PCAbipl`. An argument `cent.line` is available for adding a centre line to the circle as shown for the red circle in the bottom panel. Adding this line helps to identify the point whose predictions are shown.

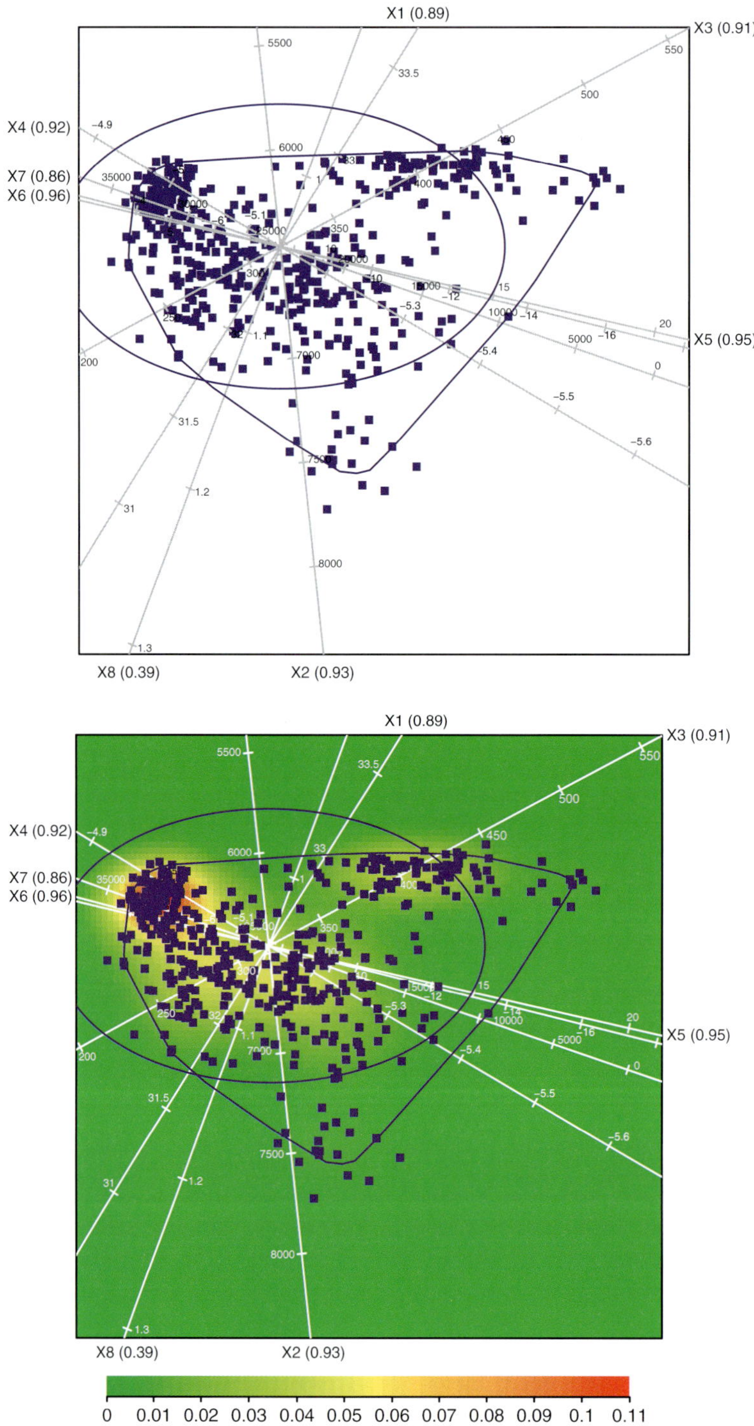

Figure 3.42 PCA biplot with 0.95-bag, concentration ellipse and density surface added.

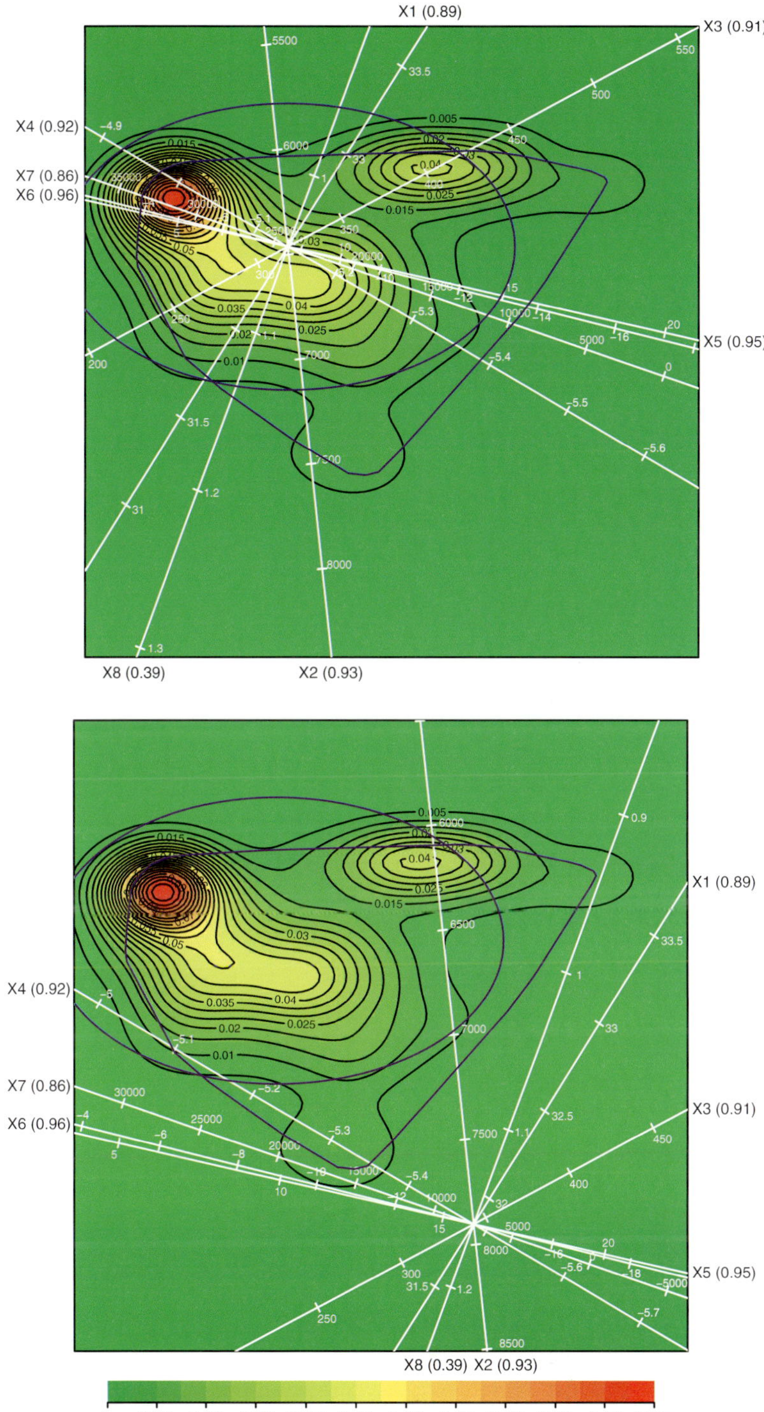

Figure 3.43 Improving the biplot in Figure 3.42 by suppressing the plotting of sample points, adding density contours and translating the axes in the bottom panel.

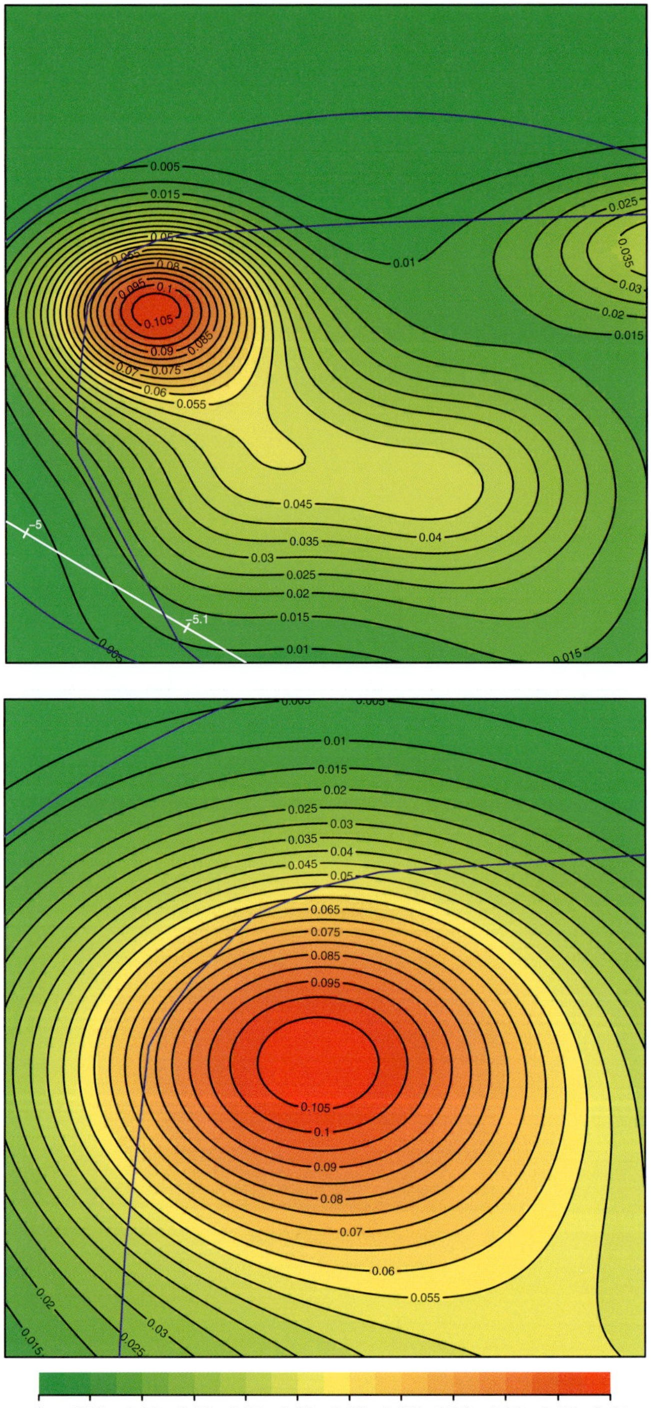

Figure 3.44 Zooming the biplot of Figure 3.43, with zoomval chosen as: (top) 0.5; (bottom) 0.2.

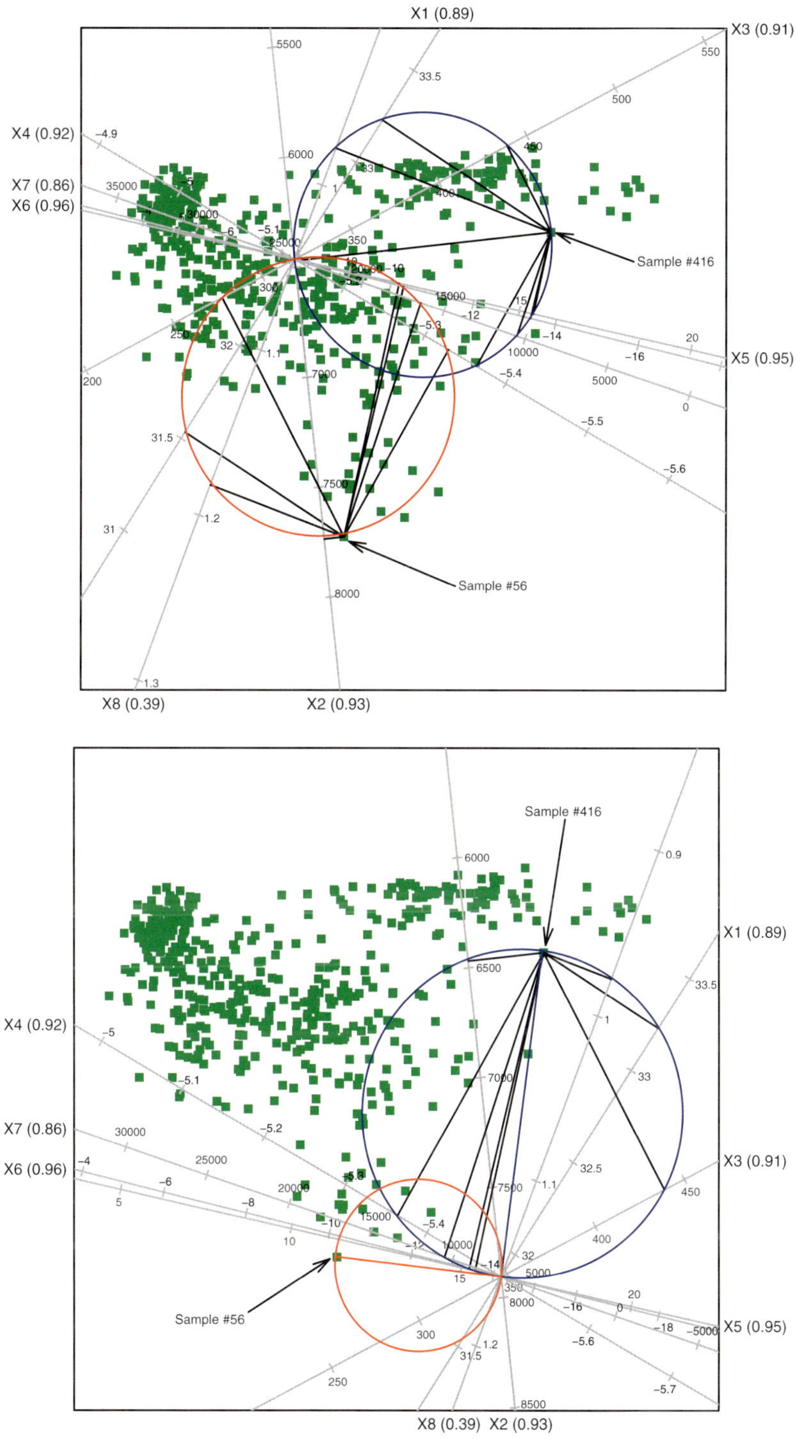

Figure 3.45 Finding predictions by circular projection where the intersection of the biplot axes: (top) coincides with the origin; (bottom) has been interactively translated.

Table 3.27 Predictions for samples ***s416*** and ***s56***.

Var	s416	s56
X1	33.23	31.53
X2	6469.34	7735.64
X3	438.91	277.49
X4	−5.36	−5.33
X5	15.41	11.45
X6	−13.40	−10.30
X7	10494.21	16423.32
X8	0.98	1.18

The annotations in the biplot in the bottom panel of Figure 3.45 were added using the function calls

```
> draw.arrow(length = 0.15, angle = 15)    (twice)
> draw.text(string = "Sample #56", cex = 0.7)
> draw.text(string = "Sample #416", cex = 0.7)
```

and making the appropriate selections on the biplot.

3.9 Conclusion

In this chapter we have discussed PCA as a method for approximating a centred (and scaled if needed) data matrix. This allowed us to find the scaffolding upon which to construct a PCA biplot in one, two or three dimensions as an extension of an ordinary scatterplot. Understanding the basics of a PCA biplot is essential for its interpretation. However, understanding the basics and constructing the initial PCA biplot are but the first steps in a biplot analysis of a data matrix. The final stage is not only a matter of understanding but is also an art: fine-tuning and adding enhancements to come up with a final biplot highlighting the secrets of a data matrix. Therefore, we urge the reader to experiment with the various functions provided in `UBbipl` and use their creativity to enable novel extensions.

Several of the examples encountered in Chapter 3 contain information regarding some predefined grouping or class structure. Although this information can be included in a PCA biplot by, for example, interpolating group means into the biplot, the procedure of finding the scaffolding underpinning the PCA biplot does not utilize this information. This deficiency is remedied in the next chapter.

4 Canonical variate analysis biplots

In contrast to PCA, canonical variate analysis (CVA) focuses on observations grouped into K classes. We shall be interested in both between- and within-class variation, particularly as to how these may be exhibited in graphical form. The main tool of CVA is to transform the observed variables into what are termed canonical variables that have the property that the squared distances between the means of the groups are given by Mahalanobis's D^2, defined formally in Section 4.2. Mahalanobis distance is monotonically related to the probability of misclassification when assigning a sample to one of two groups each with a multinormal distribution with the same covariance matrix but with different means. This probability of misclassification is given by

$$\frac{1}{\sqrt{2\pi}}\int_{D/2}^{-\infty} e^{-1/2y^2}\,dy$$

(see Rao, 1952). This result establishes a close relationship with discriminant analysis which, as we shall see, extends to biplot applications. However, the Mahalanobis distances are of interest in their own right and we shall approximate them in graphical displays with associated axes calibrated to give the values of the original variables.

Having transformed to canonical variables, the table of the group means may be treated for biplot purposes in a similar way to the PCA analysis of $\mathbf{X}$, described in Chapter 3. Unlike PCA, CVA is invariant to the measurement scales used, thus avoiding scaling problems.

The algebra may be developed in several ways, but first we establish the context by discussing a simple example.

4.1 An example: revisiting the *Ocotea* data

Stinkwood (*Ocotea bullata*) is a large tree, indigenous to South Africa, belonging to the family *Lauraceae*. When the wood is freshly cut it has an unpleasant smell from which

Understanding Biplots John Gower, Sugnet Lubbe and Niël le Roux

its name derives. 'Old Cape furniture' refers to furniture produced in the Cape region in South Africa during 1652–1900 according to the prevailing Western styles. Stinkwood was one of the principal woods used in the Cape during the 18th and 19th centuries for the manufacture of chairs, cupboard panel surrounds, fronts of drawers, framing and styles of doors. The kind of wood used plays an important role in the buying and selling of old Cape furniture. Furniture made from stinkwood alone, or from a combination of stinkwood and yellow wood, has a high prestige value. It is therefore important that the private owner, as well as the trader, must be able to accurately identify the kind of wood. Many people working with antiques claim that they can identify the wood by visual inspection. However, it is clear that in most publications on old Cape furniture, even where in-depth research was conducted, information on the kind of wood used is very scant. The identification of the kind of wood is usually based on face value and according to tradition. The importance of the scientific identification of the kind of wood becomes critical when authors contradict each other; for example, Botha (1977) states that the 'tolletjiestoel' was usually made of stinkwood but Baraitser and Obholzer (1981) declare that it was rarely made of stinkwood, yellow wood or teak. South Africa is a country with few indigenous timbers. It was thus imperative that wood was imported. Imbuia (*O. porosa*) was one of the woods imported from South America and eventually became an accepted substitute for stinkwood. Swart and Van der Merwe (1980) describe the identification problems concerning *O. bullata* (indigenous stinkwood) and *O. porosa* (exotic imbuia). Both are species from the genus *Ocotea* from the family *Lauraceae*. The substantial price difference between furniture made from these two species from one genus necessitates accurate discrimination.

Swart (1980) gathered research material with nondestructive measures from two indigenous species of the *Lauraceae* family: *O. bullata* and *O. kenyensis*. Both qualitative and quantitative wood anatomical features were measured. In the first phase the genus *Ocotea* was distinguished from the other species in the family, using mainly microscopic qualitative anatomical features. Based on these features, the indigenous *Ocotea* species, as well as the exotic imbuia, could be positively identified as belonging to the genus *Ocotea*. In a second phase, distinctions were required between the species within the genus. The two indigenous species *O. bullata* and *O. kenyensis*, and also the two furniture woods *O. bullata* and the exotic *O. porosa* (imbuia) must be differentiated. Up to that time these distinctions were based on physical macroscopic features such as colour and smell, but the large variability in these features leads to very subjective identifications.

A controversy about a 19th-century Cape chair was the primary incentive for this example, originally discussed by Burden *et al.* (2001) and Le Roux and Gardner (2005). The data given in Table 3.9 show the wood anatomical features of 20 *O.bullata*, 10 *O.porosa* and 7 *O.kenyensis* samples investigated by Swart (1985). For each of the 37 samples, 50 microscopic measurements were made on each of the six variables discussed in Section 3.4. The mean values of the 50 measurements for each of the samples were calculated for each of the six wood anatomical features. These mean values were used as the observed values for each of the samples, thus forming the data set given in Table 3.9. Recall that a predictive PCA biplot of the scaled *Ocotea* data set is given in Figure 3.23, showing the mean vectors of *O.bullata, O.porosa* and *O.kenyensis*. The sample predictivities of the mean vectors are given in Table 3.14. Before considering a CVA biplot of the *Ocotea* data, we reflect on some aspects of the PCA biplot in Figure 3.23 and the function `PCAbipl` for creating it. First of all, we remember that

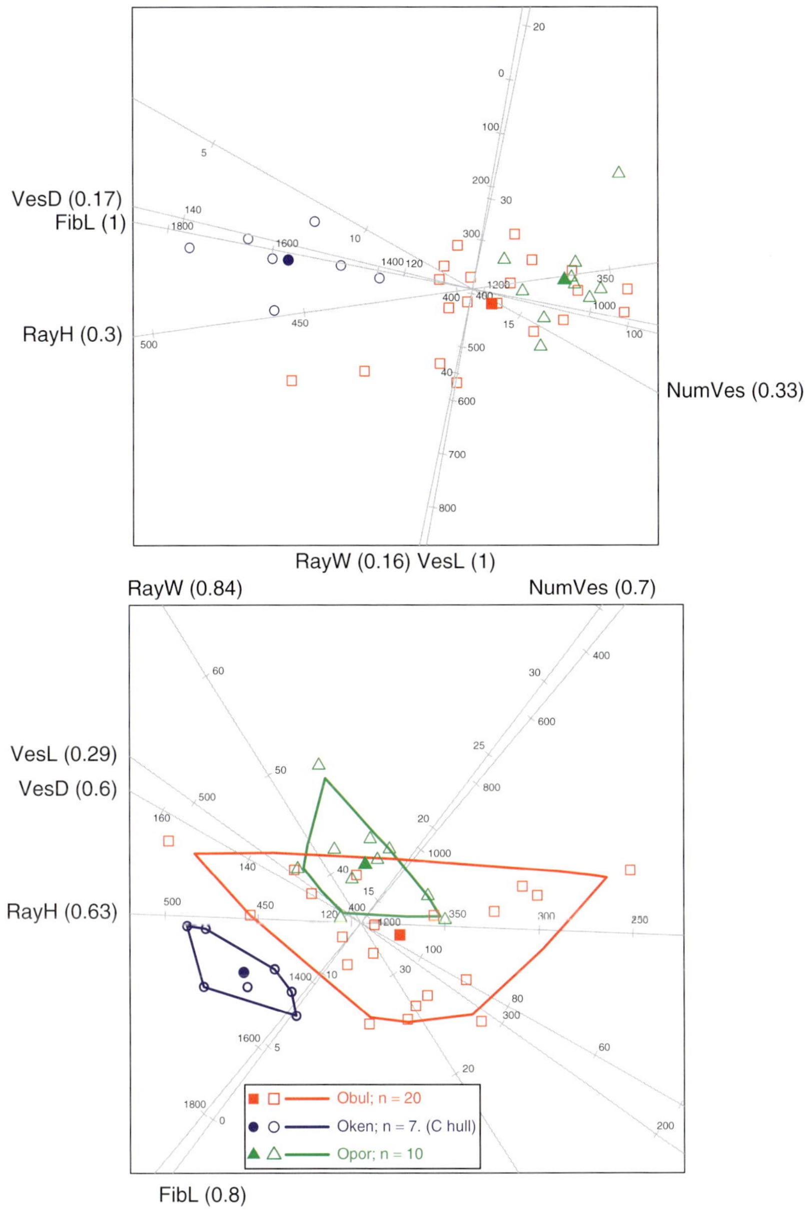

Figure 4.1 PCA biplot of *Ocotea* data: (top) unscaled variables; (bottom) variables scaled to unit variances.

a PCA biplot is not scale-invariant; secondly, `PCAbipl` allows through its argument `G` information of a predetermined group structure to be incorporated in the biplot. How this is done is illustrated in Figure 4.1. The top panel of Figure 4.1 shows a PCA biplot of the unscaled (but centred) `Ocotea.data`; by contrast, the biplot in the bottom panel is a PCA biplot of the scaled `Ocotea.data`. It is clear that the two biplots are entirely different – not only the positions of the samples and the axes but also the axis predictivities are different. The biplot in the bottom panel is obtained with the call

```
PCAbipl (Ocotea.data[,3:8], scaled.mat = TRUE,
   G = indmat(Ocotea.data[,2]), means.plot = TRUE,
   colours = c("red","blue","green"), alpha = 0.95, pch.samples =
   0:2, pch.samples.size = 1.25, pch.means = c(15,16,17),
   pch.means.size = 1.5, label = FALSE, pos = "Hor",
   line.type = rep(1,3), line.width = rep(2,3), specify.bags = 1:3,
   legend.type = c(T,T,T), Tukey.median = FALSE,
   n.int = c(5,5,5,5,3,5), offset = c(-0.2, 0.05, 0.1, 0),
   side.label = c(rep("right",5),"left"), pos.m = c(4,4,4,4,4,2),
   offset.m = c(-0.1, -0.1, 0.1, -0.1, -0.1, 0.1),
   rotate.degrees = 180, predictivity.print = TRUE,
   parplotmar = c(3, 3.5, 3, 2.5))
```

This call involves several arguments we have not yet encountered. The argument `G` expects an $n \times K$ indicator matrix specifying the class membership of each sample, with a one in position (i,k) if the ith sample belongs to the class associated with the kth column and a zero otherwise. We provide the utility function `indmat` for converting a vector (in the above call the second column of our R dataframe `Ocotea.data`) of group labels into an indicator matrix. When `PCAbipl` is called with a nonnull `G` argument the group means are calculated and interpolated into the biplot if argument `means.plot` is set to TRUE. The arguments of `colours`, `pch.samples`, `pch.means` all must now be K-component vectors allowing the various groups to be displayed in different colours and plotting characters if required. Bagplots can be drawn for any group selected by argument `specify.bags`. Finally, the legend describing the groups displayed in the biplot is controlled by the argument `legend.type`. We also draw attention to the argument `rotate.degrees = 180` in the call above which ensures compatibility with the configuration in Figure 3.23.

The bags in the bottom panel of Figure 4.1 are 0.95-bags. These bags show an overlap between ***Obul*** and ***Opor***, with both these groups well separated from ***Oken***. The reader can experiment by varying alpha to quantify the amount of overlap or by adding kappa-ellipses. Furthermore, by using the argument `specify.density.class` of `PCAbipl.density` the reader can experiment with adding density surfaces to the PCA biplot using all the data or only the data for a specified group.

From the discussion and illustrations above, it follows that information about group structure can be incorporated into a PCA biplot, thereby adding value to its usefulness in practice; but including the group structure as explained above does not have any influence on the biplot scaffolding. Axis predictivities and sample predictivities of the original samples remain exactly the same whether group means are shown or not. Despite its usefulness, it remains a passive way of incorporating group structure in a biplot. Is there a better way of displaying multidimensional group differences graphically – especially when we would like to classify a new item? Leaving the details of its construction for Section 4.2 and subsequent sections, we give a CVA biplot of the *Ocotea* data in Figure 4.2. This biplot is a result of the call

```
CVAbipl(Ocotea.data[,3:8], G = indmat(Ocotea.data[,2]),
   colours = c("red","blue","green"), pch.samples = 0:2,
   pch.samples.size = 1.25, predictivity.print = TRUE)
```

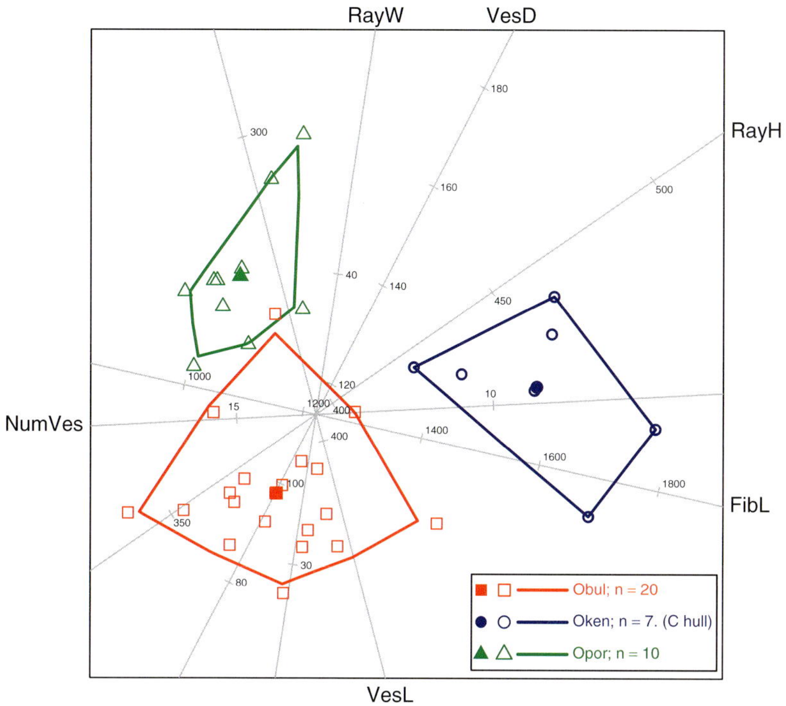

Figure 4.2 CVA biplot of the *Ocotea* data set with six variables. Group means are indicated by solid symbols according to the key included. 0.95-bags are shown.

For reference purposes and to be used when comparing different methods later on, we give the mean values and standard deviations of each of the variables by species in Table 4.1. This table shows considerable differences in the mean values of the three species for each of the wood anatomical features, but the standard deviations show that a considerable amount of overlap between the samples from the three species can also be expected. A formal test of equality of covariance matrices rejects the null hypothesis at the 5% significance level. Thus any statistical method for investigating group differences depending on the assumption of equal covariance matrix should be regarded with caution.

Table 4.1 Mean values and standard deviations of wood anatomical features for each of the three species of the genus *Ocotea*.

Species	Statistic	VesD (μm)	VesL (μm)	FibL (μm)	RayH (μm)	RayW (μm)	NumVes
O. bullata	Mean	98.10	412.00	1185.40	375.40	32.30	14.30
	SD	18.76	94.17	140.40	69.22	7.730	6.01
O. kenyensis	Mean	137.30	401.70	1568.90	446.10	37.30	9.10
	SD	20.54	53.54	120.93	42.22	4.50	1.95
O. porosa	Mean	129.30	342.40	1051.70	398.20	39.40	14.80
	SD	11.80	87.76	55.94	67.08	4.72	3.52

One of the assumptions of CVA is that the within-class covariance matrices are equal among all classes.

On comparing the CVA biplot in Figure 4.2 with the corresponding PCA biplot in Figure 4.1 we draw attention to the following:

- The PCA biplot is scale-dependent; the CVA biplot is scale-invariant.
- Axis predictivities and predictivities for the group means in the PCA biplot are not equal to unity, but in the CVA biplot they all attain the maximum value of unity (see Section 4.2)
- There is a higher degree of separation between the group means in the CVA biplot than in the PCA biplot.
- In the PCA biplot the interpolated group means do not contribute to the scaffolding axes; in the CVA biplot the scaffolding axes are determined by the group means.

Before discussing the theoretical basis of these differences for a complete understanding of CVA biplots, let us study our example in a little more detail. It is clear from Figure 4.2 that the species are well separated by the CVA, although there is some overlap between stinkwood (***Obul***) and imbuia (***Opor***). The leave-one-out cross validation error rate was calculated and the proportion of incorrect classifications found to be 0.081. Since it is well known that the error rate can sometimes be improved by using only a subset of the variables (see Flury, 1997), the possibility of using fewer variables was investigated. In Table 4.2 only the smallest leave-one-out cross validation error rate for each subset size is given, together with the associated subset of variables.

Table 4.2 shows that a better classification rate is obtained with the optimal subset of three, four or five variables than with the complete set of six variables. Based on the principle of parsimony (Occam's razor) the variables ***FibL***, ***VesL*** and ***VesD*** were selected for use in the final analysis. Testing for significant differences between the class covariance matrices based upon the three variables selected, a p-value of 0.065 is obtained, leading to the nonrejection of the hypothesis at the 5% significance level.

After selecting these three variables on a statistical basis, it transpired that there is the added practical advantage that smaller (thin strip) wood samples are needed to make these measurements compared to the other three variables. Therefore methods based on only these features can be viewed as nondestructive, which is an important aspect when taking wood samples for microscopic analysis from precious old Cape furniture. Furthermore, very small articles can be identified. The CVA biplot based on variables ***FibL***, ***VesL*** and ***VesD*** is given in Figure 4.3.

Table 4.2 Smallest CVA cross validation error rates for different sized subsets of the variables of the *Ocotea* data set.

Subset size	Cross validation error rate	Associated variables
1	0.243	FibL
2	0.081	FibL, VesL
3	0.054	FibL, VesL, VesD
4	0.054	FibL, VesL, VesD, RayW
5	0.054	FibL, VesL, VesD, RayW, RayH
6	0.081	complete set

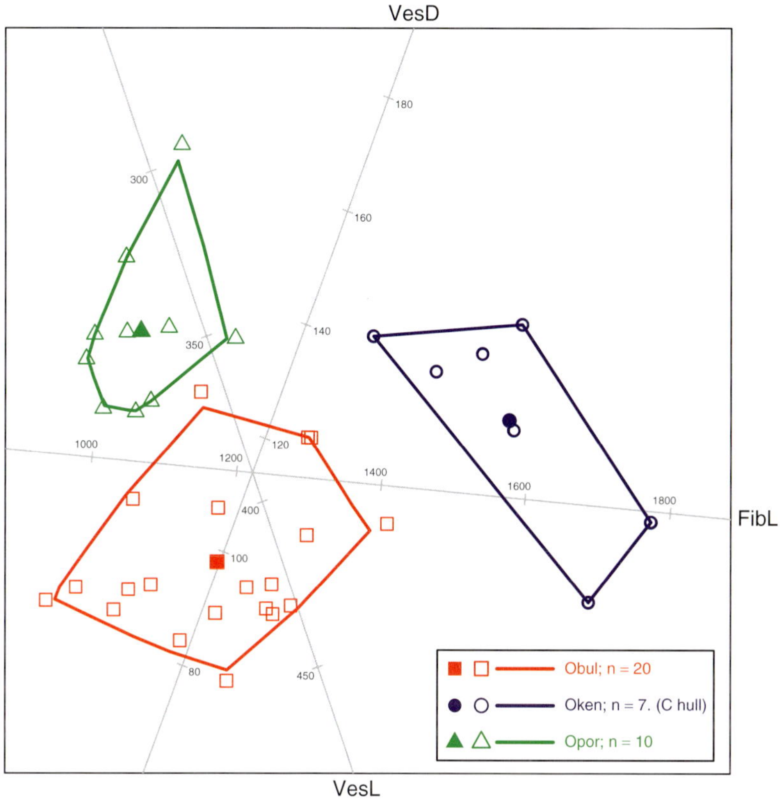

Figure 4.3 CVA biplot of the *Ocotea* data set using only variables ***FibL***, ***VesL*** and ***VesD***. Group means are indicated by the solid symbols in the key. 0.95-bags are shown.

Comparison of Figure 4.3 with Figure 4.2 shows that the distribution of the samples and the orientation of the three biplot axes for ***FibL***, ***VesL*** and ***VesD*** remain very much the same as when using all six variables. Referring to the distribution of the sample points around their respective class means shows that the indigenous ***Oken*** are well separated from the other two species, while the indigenous ***Obul*** and the exotic ***Opor*** are closer to each other. Note, however, that the two indigenous species ***Obul*** and ***Oken*** overlap on each of the individual axes (variables) and that separation of the species is only obtained by the simultaneous analysis of all three variables. Furthermore, there is overlap between ***Oken*** and ***Opor*** with respect to ***VesL*** and ***VesD*** when each variable is considered individually. The two species are, however, well separated when orthogonally projected onto the ***FibL***-axis. On the other hand, there is much more overlap between the projections of ***Obul*** and ***Opor*** on the ***FibL***-axis, than with respect to the remaining two variables.

Comparing the class mean values in Table 4.1 with the predicted values obtained by orthogonal projection onto the biplot axes confirms that the mean values can be accurately read off from the graphical display. It will be shown in Section 4.2 that these CVA biplots are in fact exact representations of the class means in two dimensions.

Table 4.3 Wood anatomical measurements obtained for a specimen of unknown origin.

Wood anatomical feature	Mean values (μm)
VesD	134
VesL	375
FibL	1170

We now return to the problem of identifying the kind of wood used in the manufacture of the controversial chair. Using mainly qualitative, microscopic wood anatomical features, the sample was initially identified as belonging to the family *Lauraceae* and then as a representative of the genus *Ocotea*. In the second step of the classification, a small inconspicuous sample of wood from the chair was obtained and the microscopic measurement values of ***FibL***, ***VesL*** and ***VesD*** were used to calculate a mean value for each variable similar to the methods used for collecting the data set in Table 3.9. These values are given in Table 4.3.

In Figure 4.4 classification regions are added to the CVA biplot of Figure 4.3. Each region contains the points nearest the relevant mean. The interpolated position of the

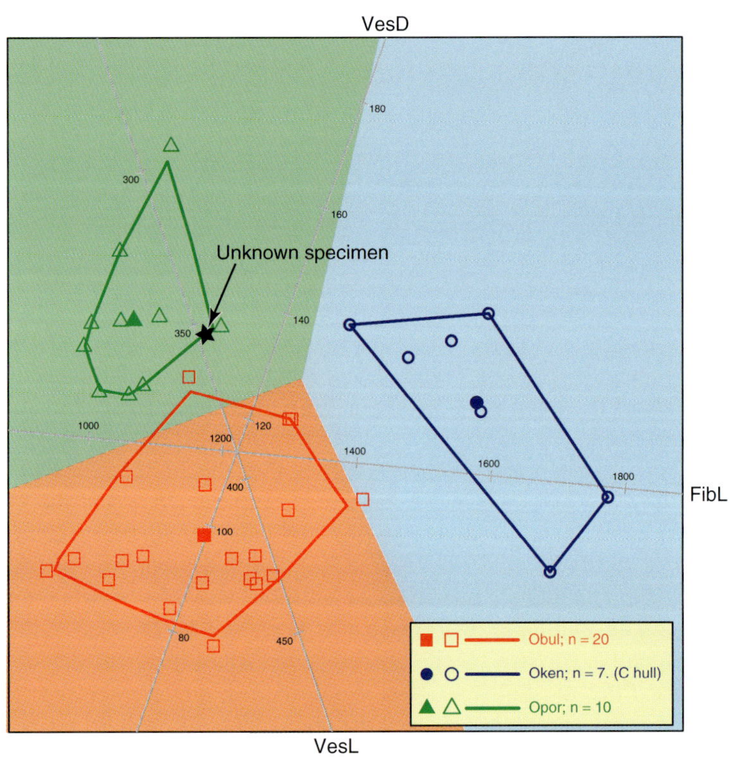

Figure 4.4 CVA biplot using variables ***FibL***, ***VesL*** and ***VesD*** of the *Ocotea* data set. Classification regions have been added. Group means are indicated by solid symbols in the legend. The black star is the interpolated position of the specimen of unknown origin. 0.95-bags are shown.

sample with values given in Table 4.3 is also added. Judging by Figure 4.4, it is clear that the wood sample of unknown origin should be classified as ***Opor***. The piece of furniture is therefore not made of stinkwood, but most probably of imported imbuia. Using the biplot as a graphical display in the classification process showed the amount of separation obtained between the species. Furthermore, the Pythagorean distances from the new sample to the class means are Mahalanobis distances that can be used as a measure of certainty associated with the classification. If the new sample point fell just inside the ***Opor*** classification region, but almost on the border with ***Obul***, it could be argued that it is not clear whether the piece of furniture is from expensive stinkwood or the less valuable imbuia. Since the observation is well inside the ***Opor*** classification region, it can be assumed with some confidence that the wood type is indeed imbuia and not stinkwood.

In this example, we have mentioned that the means are positioned exactly, together with exact classification regions; the individual samples are approximations obtained by projection from their exact three-dimensional positions. The accuracy of the means is valid only because three points can always be set in two dimensions. If we had had more than three species then a two-dimensional representation of the means would also be a projected approximation and the classification regions too would be approximations. This is treated more fully below.

4.2 Understanding CVA and constructing its biplot

Consider the data matrix $\mathbf{X} : n \times p$ centred such that $\mathbf{1}'\mathbf{X} = \mathbf{0}'$. The data contained in $\mathbf{X}$ consist of p measurements made for each of K classes. The class sizes are $n_1, n_2, \ldots, n_K$, respectively, such that $\sum_{k=1}^{K} n_k = n$. *Let* $\mathbf{N} = \text{diag}(n_1, n_2, \ldots, n_K)$. Then the matrix of group means can be calculated as

$$\overline{\mathbf{X}} : K \times p = \mathbf{N}^{-1}\mathbf{G}'\mathbf{X} = (\mathbf{G}'\mathbf{G})^{-1}\mathbf{G}'\mathbf{X},$$

where $\mathbf{G}$: $n \times K$ denotes an indicator matrix defining membership of the K classes.

The sums-of-squares-and-products (SSP) matrix of $\mathbf{X}$ can be partitioned into a within-class SSP matrix and between-class SSP matrix such that $\mathbf{T} = \mathbf{W} + \mathbf{B}$ (Total = Within + Between), where

$$\mathbf{W} = \mathbf{X}'\mathbf{X} - \overline{\mathbf{X}}'\mathbf{N}\overline{\mathbf{X}} = \mathbf{X}'[\mathbf{I} - \mathbf{G}(\mathbf{G}'\mathbf{G})^{-1}\mathbf{G}']\mathbf{X} \tag{4.1}$$

and

$$\mathbf{B} = \overline{\mathbf{X}}'\mathbf{N}\overline{\mathbf{X}} = \mathbf{X}'\mathbf{G}(\mathbf{G}'\mathbf{G})^{-1}\mathbf{G}'\mathbf{X}. \tag{4.2}$$

The crucial thing is to find a transformation of the variables such that the Pythagorean distances between the group means of the transformed variables are Mahalanobis distances. Writing $\overline{\mathbf{x}}_k$ and $\overline{\mathbf{x}}_h$ for the means of the kth and hth groups, respectively, then the Mahalanobis distance δ_{kh} between the two group means is given by

$$\delta_{kh} = \sqrt{(\overline{\mathbf{x}}_k - \overline{\mathbf{x}}_h)'\mathbf{W}^{-1}(\overline{\mathbf{x}}_k - \overline{\mathbf{x}}_h)},$$

so we are looking for a nonsingular transformation matrix $\mathbf{L}$: $p \times p$ such that $\mathbf{x}'\mathbf{L}\mathbf{L}'\mathbf{x} = \mathbf{x}'\mathbf{W}^{-1}\mathbf{x}$ or $\mathbf{L}\mathbf{L}' = \mathbf{W}^{-1}$. Consider the eigenvector equation

$$\mathbf{W}\mathbf{L} = \mathbf{L}\mathbf{\Lambda}, \tag{4.3}$$

where we scale the eigenvectors to give $\mathbf{L}'\mathbf{W}\mathbf{L} = \mathbf{I}$, noting that although the eigenvectors are always orthogonal, their scaling is arbitrary. Our choice of scaling gives $\mathbf{L}'\mathbf{L} = \mathbf{\Lambda}^{-1}$ and $\mathbf{L}\mathbf{L}'\mathbf{W}\mathbf{L}\mathbf{L}' = \mathbf{L}\mathbf{I}\mathbf{L}'$ or $\mathbf{L}\mathbf{L}' = \mathbf{W}^{-1}$, as required.

Given a sample $\mathbf{x}$ $(x_1, x_2, \ldots, x_p)$, the transformed values $\mathbf{y}' = \mathbf{x}'\mathbf{L}$ of the p variables are said to be *canonical variables* and the p-dimensional space generated by the rows of $\mathbf{L}$ is called the *canonical space*. In particular, the data matrix $\mathbf{X}$ transforms to $\mathbf{XL}$ and the group means $\overline{\mathbf{X}}$ transform to $\overline{\mathbf{X}}\mathbf{L}$ in the canonical space. The transformed group means $\overline{\mathbf{X}}\mathbf{L}$ are the means of the canonical variables and are called the *canonical means* for short. Furthermore, the inner product associated with the canonical means is $\overline{\mathbf{X}}\mathbf{L}\mathbf{L}'\overline{\mathbf{X}}' = \overline{\mathbf{X}}\mathbf{W}^{-1}\overline{\mathbf{X}}'$, confirming the Mahalanobis distances between the means $\overline{\mathbf{X}}$. Thus we have a property that is of practical consequence: Mahalanobis distances between the means $\overline{\mathbf{X}}$ are represented in the canonical space as ordinary Pythagorean distances between the canonical means.

Since $\mathbf{L}$ is nonsingular, the rank of the matrix $\overline{\mathbf{X}}\mathbf{L}$ is the same as the rank of $\overline{\mathbf{X}} : K \times p$. However, the weighted sum of the rows of $\overline{\mathbf{X}}$ vanishes so rank$(\overline{\mathbf{X}})$ $\leq \min(K-1, p)$. Therefore, the points given by the K rows of $\overline{\mathbf{X}}\mathbf{L}$ will occupy at most $m = \min(K-1, p)$ dimensions of the canonical space and may be approximated in fewer dimensions by PCA (Chapter 3) using the SVD of $\overline{\mathbf{X}}\mathbf{L}$. The whole process of approximating $\overline{\mathbf{X}}$ in the canonical space may thus be regarded as a two-step process of first finding $\mathbf{L}$ as the scaled solution to (4.3) and then in the second step of finding the SVD of $\overline{\mathbf{X}}\mathbf{L}$.

We may proceed directly to the eigenvalue decomposition:

$$(\mathbf{L}'\overline{\mathbf{X}}'\mathbf{C}\overline{\mathbf{X}}\mathbf{L})\mathbf{V} = \mathbf{V}\mathbf{\Lambda}, \tag{4.4}$$

where $\mathbf{C}$ represents a centring operation (discussed below) and the columns of $\mathbf{V}$ are orthogonal eigenvectors. For the reasons stated above, this equation can have only m non-zero eigenvalues, implying zero coordinates for the means in the remaining dimensions. We note that equation (4.4) may be written

$$\mathbf{L}\mathbf{L}'(\overline{\mathbf{X}}'\mathbf{C}\overline{\mathbf{X}})\mathbf{L}\mathbf{V} = \mathbf{L}\mathbf{V}\mathbf{\Lambda}$$

or

$$\overline{\mathbf{X}}'\mathbf{C}\overline{\mathbf{X}}(\mathbf{L}\mathbf{V}) = \mathbf{W}(\mathbf{L}\mathbf{V})\mathbf{\Lambda}, \tag{4.5}$$

which is a two-sided eigenvalue problem with eigenvectors $\mathbf{M} = \mathbf{L}\mathbf{V}$. The eigenvectors should be normalized so that $\mathbf{M}'\mathbf{W}\mathbf{M} = \mathbf{V}'\mathbf{L}'\mathbf{W}\mathbf{L}\mathbf{V} = \mathbf{V}'\mathbf{I}\mathbf{V} = \mathbf{I}$. The solution $\mathbf{L}\mathbf{V}$ incorporates both the transformation to canonical variables and the PCA orthogonal transformation $\mathbf{V}$, thus subsuming both steps into one calculation. Therefore it is common practice to use the two-sided eigenvalue form for computation, so avoiding the two separate steps discussed above. However, we think that the two-step form is the more informative. Once we have transformed to canonical variables, we are concerned essentially with PCA, and everything said in Chapter 3 remains valid. In particular, the PCA approximation $\hat{\overline{\mathbf{X}}\mathbf{L}}$ to the canonical means $\overline{\mathbf{X}}\mathbf{L}$ is $(\overline{\mathbf{X}}\mathbf{L})\mathbf{V}\mathbf{J}\mathbf{V}'$, giving

$$\hat{\overline{\mathbf{X}}} = \overline{\mathbf{X}}(\mathbf{L}\mathbf{V})\mathbf{J}(\mathbf{V}'\mathbf{L}^{-1}) = \overline{\mathbf{X}}(\mathbf{M}\mathbf{J}\mathbf{M}^{-1}), \tag{4.6}$$

indicating, as usual, the dimensionality of the approximation by the diagonal matrix $\mathbf{J}$.

After the PCA of the canonical means $\bar{\mathbf{X}}\mathbf{L}$, we plot in as many dimensions as we require the PCA approximation $\mathbf{XLVJ}$, with the group means $\bar{\mathbf{X}}\mathbf{LVJ}$. It is interesting to note that we might have proceeded immediately to the PCA of the canonical variables $\mathbf{XL}$, which would require the eigenvectors of

$$(\mathbf{XL})'(\mathbf{XL}) = \mathbf{L}'\mathbf{TL} = \mathbf{L}'(\mathbf{B} + \mathbf{W})\mathbf{L} = \mathbf{V\Lambda V}' + \mathbf{I} = \mathbf{V}(\mathbf{\Lambda} + \mathbf{I})\mathbf{V}'.$$

Thus, we obtain the same eigenvectors as (4.4) with $\mathbf{C} = \mathbf{N}$ but the between-group eigenvalues are increased by unity, indicating the inclusion of the between-group variation in the m dominant dimensions. Of course, working with the group means achieves the same ends by using a smaller calculation, but this result emphasizes that we are essentially concerned with a simple PCA of the canonical variables.

We note that the two-sided equation also arises from finding the linear combination $\mathbf{x}'\boldsymbol{\beta}$ of the variables that maximizes the between-class to the within-class variance ratio

$$\frac{\boldsymbol{\beta}'(\bar{\mathbf{X}}'\mathbf{C}\bar{\mathbf{X}})\boldsymbol{\beta}}{\boldsymbol{\beta}'\mathbf{W}\boldsymbol{\beta}}. \tag{4.7}$$

This is a useful property but only gives a one-dimensional solution for $\boldsymbol{\beta}$, though the above two-step approach fully justifies the retention of the remaining eigenvectors, albeit without the variance ratio justification, unless one accepts the often quoted property that the rth vector maximizes the ratio conditional on being orthogonal (in the metric $\mathbf{W}$) to the previous $r - 1$ vectors. In our approach, this property is satisfied globally as a natural consequence of the least-squares property of the SVD that, in turn, generates the two-sided eigenvector equation. The one-dimensional solution, given by the first eigenvector $\mathbf{Lv}_1$ of the two-sided eigenvalue equation, is prominent in the statistical literature as the linear discriminant function (LDF), especially when $K = 2$ and therefore $m = 1$. It offers a linear combination of the variables with optimal discriminatory properties (i.e. maximizing the variance ratio (4.7)). We have seen that Mahalanobis distance discriminates by using all dimensions and that CVA, with its classification regions, supports discrimination using one or more dimensions to approximate Mahalanobis distance. The LDF is merely the one-dimensional version.

Finally, we discuss the matrix $\mathbf{C}$. In PCA we require the data to be expressed in deviations from its column means, thus ensuring the requirement that the best-fitting plane contains the centroid. In CVA we have the choice of whether to work in deviations from the group means, weighted or not by the group sizes. In the unweighted case, $\mathbf{C} = \mathbf{I} - K^{-1}\mathbf{11}'$; in the weighted case, $\mathbf{C} = \mathbf{N}$. The choice makes no difference to Mahalanobis distance but ensures that the best-fitting plane goes through either the unweighted centroid of the group means or their weighted centroid. Lower-dimensional approximations are affected, as might be expected. In the unweighted form, all group means are treated equally, irrespective of sample sizes. In the weighted form the groups with the most samples will be better represented than those with few samples. Circumstances will dictate which is the better. If one wants a general appreciation of the Mahalanobis distances between the groups then the unweighted form seems better. If one were classifying a new sample, as with the stinkwood example, then the weighted form would seem to be the better. However, with the stinkwood example, it makes no difference which form is used, as both give an exact representation of the three group means in two dimensions; only orientation changes. In general, classifying using the weighted form carries with it an assumption that future samples will occur at similar rates to those given in **N**. This may not be so, and in any case one may be especially interested in identifying members of the rarely occurring groups, so then the last thing we want to do is to give rare groups low weights. Furthermore, allocation to a group using low-dimensional approximations is not ideal, as exact Mahalanobis distances in p dimensions are readily computed. Apart from the two cases just discussed, we may also choose $\mathbf{C} = \mathbf{I}$, in which case we retain the weighted centroid, which coincides with the centroid of all n samples, but do an unweighted PCA, with minor changes from the other unweighted case. The only effect of these variants is to give different projections, $\mathbf{VJV}'$ derived from (4.4), equivalently $\mathbf{MJM}^{-1}$ derived from (4.5), but the predicted fit $\hat{\mathbf{X}}$ given by (4.6) remains valid in all cases.

Often the researcher will make a scatterplot of the first two canonical variates. We show below that this plot can be enhanced by following the biplot format:

- representing both the means and information on the original variables;
- interpolating the original samples into the biplot;
- ensuring an aspect ratio of one to allow for visual appraisal of Mahalanobis distance;
- removing the scaffolding canonical axes scales but placing markers in the original units of measurement on the biplot axes.

We have seen that an observation in the canonical space is classified to its nearest (in the Mahalanobis sense) canonical mean. The CVA biplot is constructed in $r \leq m$ dimensions based on the transformation $\bar{\mathbf{Z}} = \bar{\mathbf{X}}\mathbf{M}_r$ and approximate r-dimensional convex nearest-neighbour classification regions are given by

$$C_{[t]j} = \left\{ \mathbf{z} \in \mathscr{R}^r : d_t(\mathbf{z} - \bar{\mathbf{z}}_j) < d_t(\mathbf{z} - \bar{\mathbf{z}}_h) \text{ for all } h \neq j \right\}, \tag{4.8}$$

where $d_t(.)$ denotes the Pythagorean distance of its argument calculated in dimension $t = r, r+1, \ldots, m$.

These neighbour regions are of two kinds, depending on whether $\bar{\mathbf{z}}_j$ and $\bar{\mathbf{z}}_h$ are in m or r dimensions. When $r = m$, (4.8) yields proper classification regions, but when $r < m$, (4.8) gives nearest-neighbour regions in the approximation space.

As well as classification regions, asymptotic confidence regions, which turn out to be circular, may be drawn around the canonical means. The canonical variables are uncorrelated with unit variance. If we assume that the initial, and hence also the canonical variables, originate from a normal distribution, it follows that the sum of squares of all points $(z_1, z_2, \ldots, z_r)$ in r-dimensional canonical space follows a χ_r^2 distribution. Thus, we may seek say a 95% confidence region for which $P(\chi_r^2 \leq z_1^2 + z_2^2 + \cdots + z_r^2) = 0.95$. This represents an r-dimensional sphere. For example, when $r = 2$, $\text{P}(\chi_2^2 \leq 5.9915) = 0.95$ so the radius of the 95% confidence circle is $\sqrt{5.9915} = 2.4478$. We may draw a circle of this radius around each canonical mean which then may be used to supplement the classification regions, when assigning a sample to its nearest group mean. If one is interested in an informal significance test for differences between the groups, the radius for the ith group should be divided by $\sqrt{n_i}$ and then one can inspect for the degree of overlap, if any, between the circles. When $r = 1$ the above confidence spheres reduce to confidence intervals of the form [canonical mean – $z_{\alpha/2}$, canonical mean + $z_{\alpha/2}$] or [canonical mean – $z_{\alpha/2}/\sqrt{n_i}$, canonical mean + $z_{\alpha/2}/\sqrt{n_i}$], where $z_{\alpha/2}$ is defined by $P(Z \leq z_{\alpha/2}) = 1 - \alpha/2$ with Z denoting the standard normal distribution. Equation (4.1) defines $\mathbf{W}$ as an SSP matrix, whereas the assumption of normality requires variances. Therefore, care has to be taken to scale down the radii by a factor $\sqrt{n - K}$, where $\mathbf{W}$ has $n - K$ degrees of freedom.

4.3 Geometric interpretation of the transformation to the canonical space

Let us consider three classes with ellipsoidal shaped distributions as shown in Figure 4.5. Each ellipsoid represents a set of sample points (not shown individually) and the whole is assumed, without loss of generality, to be mean-centred around the origin.

In step I we obtain the nonsingular matrix $\mathbf{L}$ such that the Mahalanobis distances between the class means in Figure 4.5 are Pythagorean distances in the canonical space. Let us consider geometrically how this is accomplished. Any matrix can be written in terms of its SVD, $\mathbf{L} = \mathbf{PDQ}'$. If we perform the transformation piecewise with three consecutive matrix multiplications, we have the transformations $\bar{\mathbf{X}} \rightarrow \bar{\mathbf{X}}\mathbf{P} \rightarrow \bar{\mathbf{X}}\mathbf{PD} \rightarrow \bar{\mathbf{X}}\mathbf{PDQ}' = \bar{\mathbf{X}}\mathbf{L}$. We notice that since both $\mathbf{P}$ and $\mathbf{Q}$ are orthogonal matrices, these transformations are rotations while $\mathbf{D}$ is a diagonal matrix resulting in a stretching or contracting of each dimension. This is displayed in Figure 4.6.

In the top left panel of Figure 4.6 we have the original data similar to Figure 4.5. The transformation implies $\bar{\mathbf{X}}\mathbf{W}^{-1}\bar{\mathbf{X}}' = \bar{\mathbf{X}}\mathbf{LL}'\bar{\mathbf{X}}' = \bar{\mathbf{X}}\mathbf{PDQ}'\mathbf{QDP}'\bar{\mathbf{X}}' = (\bar{\mathbf{X}}\mathbf{P})\mathbf{D}^2(\mathbf{P}'\bar{\mathbf{X}}')$, which

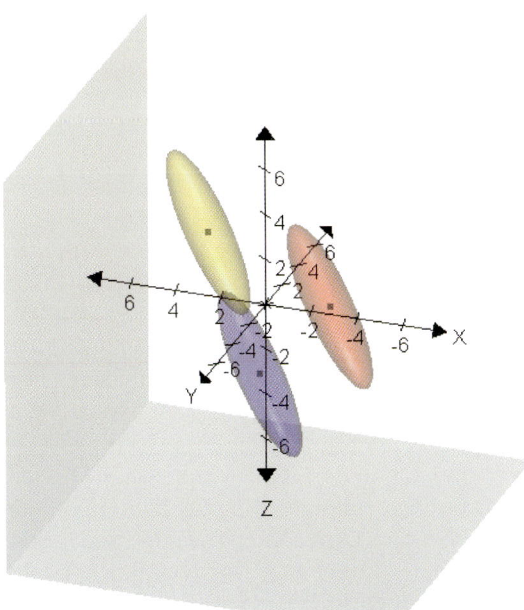

Figure 4.5 Three-class data set for illustrating the two-step transformation to the canonical space.

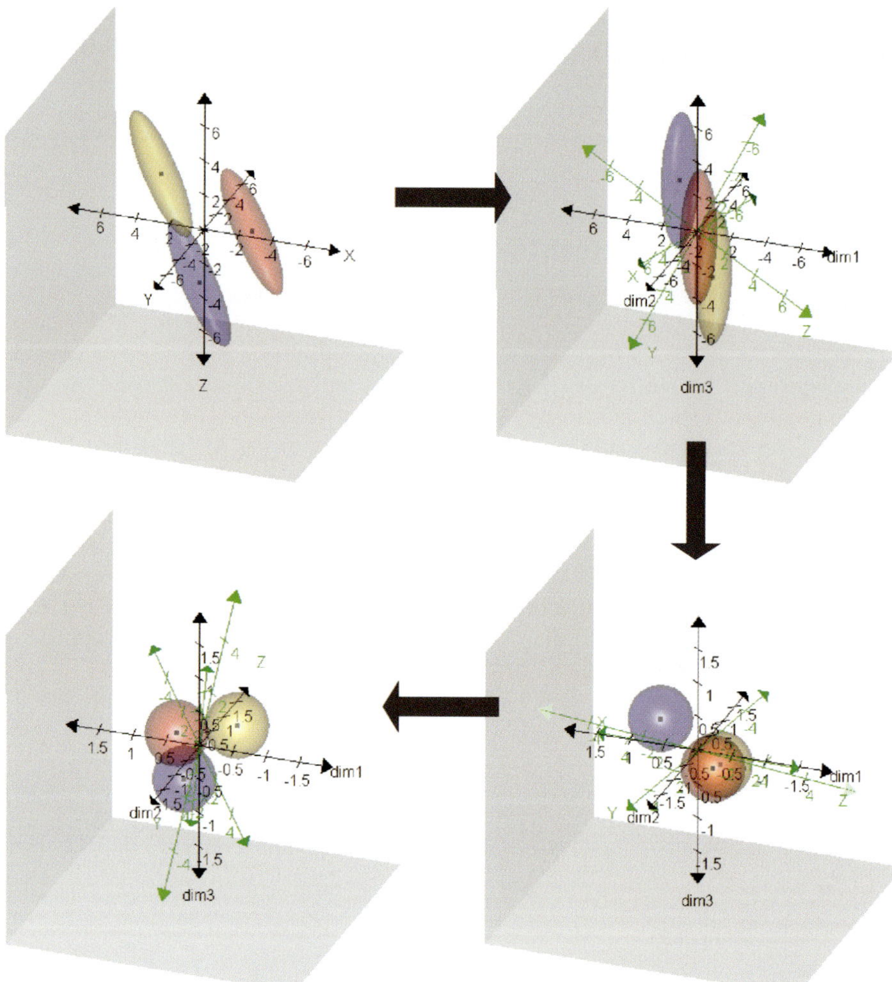

Figure 4.6 Step I illustrating the transformation to the canonical space.

shows that the distances in the top right panel have become additive weighted distances. The next action (see the bottom right panel) scales the dimensions such that we now have ordinary Pythagorean distances derived from the inner product $(\bar{\mathbf{X}}\mathbf{PD})(\mathbf{DP}'\bar{\mathbf{X}}')$. These distances are also obtained for any rotation by an orthogonal matrix multiplication $(\bar{\mathbf{X}}\mathbf{PD})\mathbf{R}'\mathbf{R}(\mathbf{DP}'\bar{\mathbf{X}}')$. We have chosen the scaling of $\mathbf{L}$ such that $\mathbf{L}'\mathbf{WL} = \mathbf{I}$, which now uniquely defines the final rotation to be given by the matrix $\mathbf{Q}'$. After this final stage in the transformation we obtain the configuration in the bottom left panel.

Step II now requires a PCA of $\bar{\mathbf{X}}\mathbf{L}$. In Figure 4.7(a) the three points (canonical means) are shown in three dimensions and the best-fitting two-dimensional PCA plane is added in Figure 4.7(b). Since we need to fit the plane to only three points, the fit is exact, rather than a two-dimensional approximation. In Chapter 3 we proceeded to construct

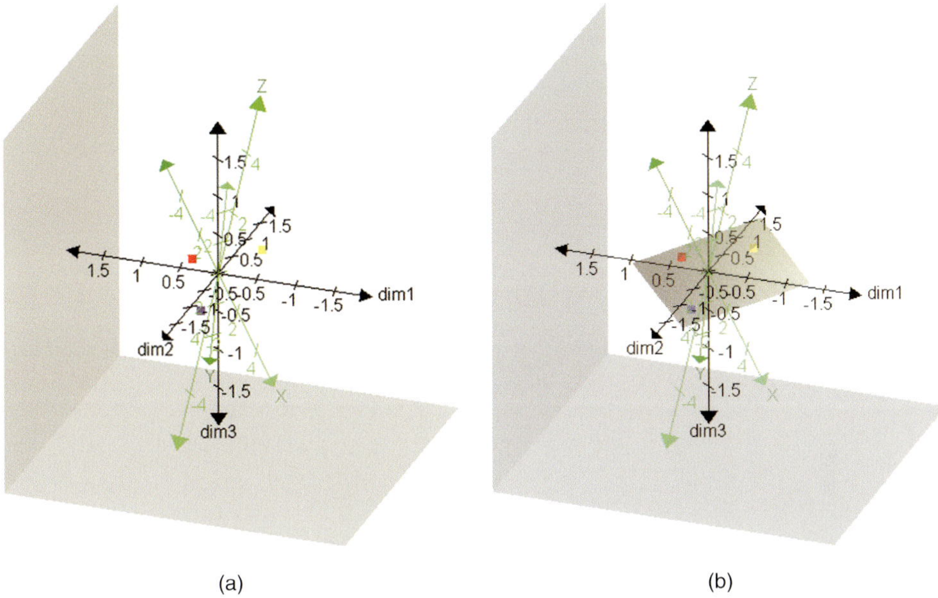

Figure 4.7 (a) Canonical means after the step I transformation by the matrix **L**. (b) PCA now implies fitting the least-squares plane of best fit to the canonical means in step II.

the biplot axes from Figure 3.9 which is equivalent to Figure 4.7(b). However, since rotations do not have any effect on the distances, we can rotate Figure 4.7(b) to have the two-dimensional PCA plane aligned with the first two dimensions. This is illustrated in Figure 4.8.

Figure 4.8 gives the final representation in the full canonical space. Any sample point $\mathbf{x}$ can be represented in the two-dimensional CVA biplot with the transformation $\mathbf{x}'\mathbf{LVJ}$ or $\mathbf{x}'\mathbf{MJ}$ which amounts to orthogonal projection onto the plane shown in Figure 4.8. The projection of the spheres amounts to their intersection with the above plane. Since there are only three classes in this example, the maximum number of dimensions needed to separate the three class means is two and then the two-dimensional CVA biplot is an exact representation of the three optimally separated class means. The three class means lie in the plane, even without orthogonal projection. In this case classification regions are just the nearest-neighbour regions for the means in the plane shown in Figure 4.8. With more than three means, a representation in two dimensions is not exact and the means are not contained in two dimensions. Then, the neighbour regions in two dimensions are given by those points that are nearest the true multidimensional means rather than to their projections (see (4.8)).

Notice that the axes representing the original variables X, Y and Z are embedded via a simple linear transformation into this space. In the next chapter we will see how to embed axes with more complex nonlinear transformations, but the underlying principle is the same.

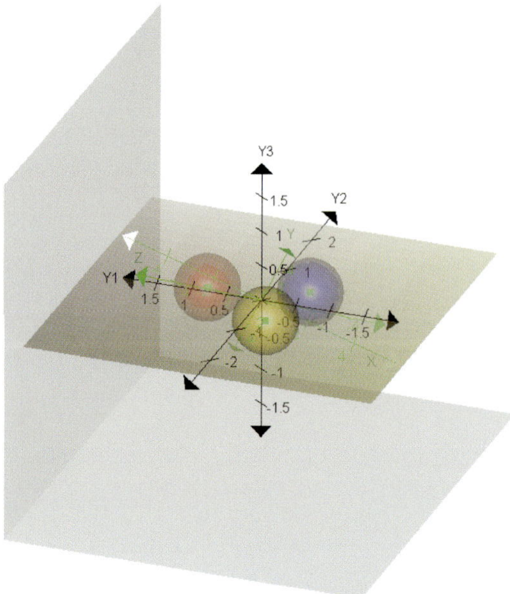

Figure 4.8 Biplot space aligned with the first two dimensions of the display.

4.4 CVA biplot axes

4.4.1 Biplot axes for interpolation

The transformation matrix $\mathbf{L}$ found in the first step of the two-step CVA procedure transforms the original space to the canonical space. The second step, as we have seen, is just an ordinary PCA in the canonical space, thus leading to the principal axes $\mathbf{Y} = \mathbf{LV}$ in the canonical space. Therefore, interpolative biplot axes can be constructed precisely as described in Section 3.2.2 for PCA biplots by simply doing first the same transformation on the original unit points $\mathbf{e}_k\,(k = 1, 2, \ldots, p)$ of the Cartesian axes of $\mathbf{X}$ to obtain $\mathbf{e}_k'\mathbf{L}$. This is valid because the unit points have the same form as the sample values and, indeed, are themselves putative data observations. Since the full transformation is defined by $\mathbf{y}' = \mathbf{x}'\mathbf{LV} = \mathbf{x}'\mathbf{M}$, and transformation to the r-dimensional biplot space by

$$\mathbf{z}' = \mathbf{x}'\mathbf{MJ}, \tag{4.9}$$

the direction and calibration of the marker μ on the kth interpolation biplot axis is $\mu\mathbf{e}_k'\mathbf{MJ}$, where $\mu \in (-\infty, \infty)$. Thus, unit markers for all the interpolation biplot axes are given by the rows of $\mathbf{MJ}$.

4.4.2 Biplot axes for prediction

We know that the direction of a predictive biplot axis in PCA is given by the orthogonal projection of the full axis onto the biplot space, with appropriate markers obtained by back-projection. The PCA prediction axes, so obtained, would allow us only to predict the values of the canonical variables, whereas we would like to predict values of the observed measurement scales of $\mathbf{X}$. The key is that a coordinate $\mathbf{y}$ plotted as a canonical sample point is obtained from an observed sample $\mathbf{x}$ by $\mathbf{y}' = \mathbf{x}'\mathbf{M}$ and therefore

$$\mathbf{x}' = \mathbf{y}'\mathbf{M}^{-1}.$$

This shows that the values of $\mathbf{x}$ may be predicted from the inner products of the sample point $\mathbf{y}$ in canonical space with the rows of $\mathbf{M}^{-1}$, which therefore represent the directions of axes for the original variables. These axes are not orthogonal but nevertheless the back-projection mechanism for PCA remains valid. In particular, the directions of the prediction axes in the r-dimensional approximation are obtained from $\mathbf{JM}^{-1}$. The kth of these axes has direction $\mathbf{JM}^{-1}\mathbf{e}_k$. It follows that in contrast to PCA where only the calibrations on predictive axes differ from those of interpolative axes, for CVA the directions $\mathbf{JM}^{-1}\mathbf{e}_k$ of prediction axes also differ from the directions $\mathbf{e}_k'\mathbf{MJ}$ of the corresponding interpolative axes.

It remains only to calibrate the prediction axes. We construct a plane $\mathscr{N}$ orthogonal to the kth embedded original axis as illustrated for $\boldsymbol{X}$ through the point $\mu = -4$. This is illustrated in Figure 4.9. Any point on the plane $\mathscr{N}$ is of the form (4.9) and predicts the value μ for the kth original axis, therefore $\mathbf{y}'\mathbf{M}^{-1}\mathbf{e}_k = \mu$. This defines the equation for the plane $\mathscr{N}$ and the intersection $\mathscr{L} \cap \mathscr{N}$ lies in $\mathscr{L}$ with equation $\mathbf{z}'\mathbf{JM}^{-1}\mathbf{e}_k = \mu$. We therefore have the biplot axis β_k in $\mathscr{L}$ with the marker μ a point on β_k. To facilitate orthogonal projection onto the biplot axes, we choose β_k to be orthogonal to $\mathbf{z}'\mathbf{JM}^{-1}\mathbf{e}_k = \mu$ and therefore of the form $\tau\mathbf{e}_k'(\mathbf{M}^{-1})'\mathbf{J}$ for $-\infty < \tau < \infty$. To predict the

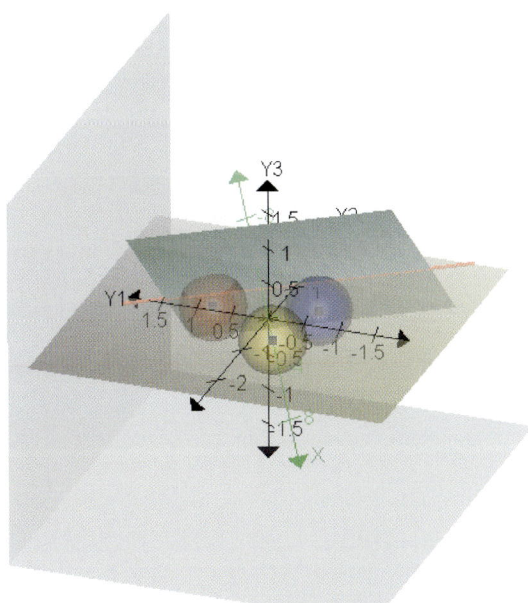

Figure 4.9 Construction of prediction biplot axes with the plane $\mathscr{N}$ orthogonal to original variable $\boldsymbol{X}$.

value μ we must have $(\tau \mathbf{e}_k'(\mathbf{M}^{-1})'\mathbf{J})(\mathbf{J}\mathbf{M}^{-1}\mathbf{e}_k) = \mu$ and therefore

$$\tau = \frac{\mu}{\mathbf{e}_k'(\mathbf{M}')^{-1}\mathbf{J}\mathbf{M}^{-1}\mathbf{e}_k}.$$

This shows that the position of the marker for predicting the value μ on the kth variable is given by

$$\frac{\mu}{\mathbf{e}_k'(\mathbf{M}')^{-1}\mathbf{J}\mathbf{M}^{-1}\mathbf{e}_k}\mathbf{e}_k'(\mathbf{M}^{-1})'\mathbf{J}.$$

4.5 Adding new points and variables to a CVA biplot

4.5.1 Adding new sample points

Similarly to PCA biplots, the vector-sum method may be used for adding new points to a CVA biplot. The process is illustrated in Figure 4.10. First we construct a CVA biplot with interpolation axes with the following function call:

```
CVAbipl(Ocotea.data[,3:5], X.new.samples = NULL,
    G = indmat(Ocotea.data[,2]), ax.type = "interpolative",
    alpha = 0.95, colours = c("red","blue","green"),
    n.int = c(10,10,10), pch.means = c(15,16,17),
    pch.samples = 0:2)
```

The function `vectorsum.interp` is then called and the values ***VesD*** = 134, ***VesL*** = 375 and ***FibL*** = 1170 selected on the respective interpolation axes to obtain the position of the specimen of unknown origin as indicated by the solid red circle in Figure 4.10. Note that we have increased the default `n.int` settings to obtain axis calibrations that allow accurate determination of the input values. As our preferred alternative we can use the incorporated algebraic calculation of (4.9) in the `CVAbipl` function with the following function call:

```
CVAbipl(Ocotea.data[,3:5], X.new.samples = matrix(c(134,375,1170),
    nrow = 1), G = indmat(Ocotea.data[,2]), ax.type = "predictive",
    alpha = 0.95, colours = c("red","blue","green"), pch.means =
    c(15,16,17), pch.samples = 0:2, pch.new.col = "black",
    pch.new = 8)
```

The latter call results in a CVA biplot exactly like Figure 4.4 but without the classification regions. We draw the reader's attention to the different directions of the interpolative axes of Figure 4.10 in comparison to the directions of the corresponding predictive axes of Figure 4.4.

4.5.2 Adding new variables

We saw in Section 3.5 how to use the regression method to add a new variable to a PCA biplot. The same method can also be used for adding a new variable to a CVA biplot. Let $\mathbf{x}^*$: $n \times 1$ denote a vector of sample values for a new variable. We assume

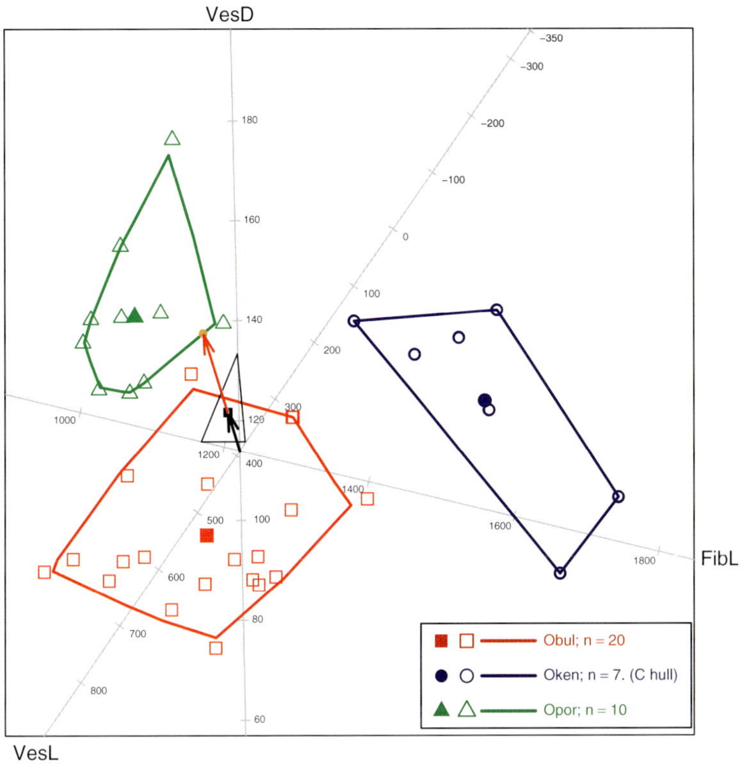

Figure 4.10 Vector-sum method for adding a new point to a CVA biplot. The biplot shown is similar to Figure 4.4, but with interpolation axes instead. The black triangle and red arrow illustrate the vector-sum method leading to the solid circle coinciding with the position of the star marking the position of the specimen of unknown origin in Figure 4.4.

that $\mathbf{x}^*$ has been centred in the same way as the other variables in $\mathbf{X}$ and that it has group means $\bar{\mathbf{x}}_k^*$ $(k = 1, \ldots, K)$. Then, we require the regression $\mathbf{b}$ of $\mathbf{C}^{1/2}\bar{\mathbf{x}}^*$ on the fitted values $\mathbf{C}^{1/2}\bar{\mathbf{X}}\mathbf{MJ}$. Thus,

$$\mathbf{b} = (\mathbf{JM}'\bar{\mathbf{X}}'\mathbf{C}\bar{\mathbf{X}}\mathbf{MJ})^{-1}\mathbf{JM}'\bar{\mathbf{X}}'\mathbf{C}\bar{\mathbf{x}}^* = (\mathbf{J}\mathbf{\Lambda}\mathbf{J})^{-1}\mathbf{JM}'\bar{\mathbf{X}}'\mathbf{C}\bar{\mathbf{x}}^*. \tag{4.10}$$

As a check, we examine what happens when $\bar{\mathbf{x}}^*$ is replaced by $\bar{\mathbf{x}}_k$, the means of the kth variable in $\mathbf{X}$. We now have $\mathbf{b} = (\mathbf{J}\mathbf{\Lambda}\mathbf{J})^{-1}\mathbf{JM}'\bar{\mathbf{X}}'\mathbf{C}\bar{\mathbf{x}}_k = (\mathbf{J}\mathbf{\Lambda}\mathbf{J})^{-1}\mathbf{JM}'\bar{\mathbf{X}}'\mathbf{C}\bar{\mathbf{X}}\mathbf{e}_k = (\mathbf{J}\mathbf{\Lambda}\mathbf{J})^{-1}\mathbf{J}\mathbf{\Lambda}\mathbf{M}^{-1}\mathbf{e}_k$ from the two-sided eigenvector equation (4.5) for CVA. Therefore $\mathbf{b} = \mathbf{JM}^{-1}\mathbf{e}_k$, agreeing with the expression for predictive CVA biplot axes derived in Section 4.4.2.

4.6 Measures of fit for CVA biplots

We saw in (4.1) and (4.2) that CVA is based on the decomposition $\mathbf{T} = \mathbf{B} + \mathbf{W}$. We shall write $\mathbf{H} = \mathbf{G}(\mathbf{G}'\mathbf{G})^{-1}\mathbf{G}'$ and note that $\mathbf{HX}$ gives a matrix of the group means (repeated as appropriate) and $(\mathbf{I} - \mathbf{H})\mathbf{X}$ the matrix of deviations from the means, that is, the within-group variation. First we note that the decomposition

$$\mathbf{X} = \mathbf{HX} + (\mathbf{I} - \mathbf{H})\mathbf{X} \tag{4.11}$$

satisfies $\mathbf{X}'\mathbf{X} = \mathbf{X}'\mathbf{HX} + \mathbf{X}'(\mathbf{I} - \mathbf{H})\mathbf{X}$, due to the idempotency of $\mathbf{H}$, showing that $\mathbf{T} = \mathbf{B} + \mathbf{W}$ has Type B orthogonality. However, Type A orthogonality does not hold for the decomposition (4.11) since $\mathbf{HXX}'(\mathbf{I} - \mathbf{H}) \neq 0$.

In the following, we are interested in the accuracy of the recovery of both the group means and the within-group values, and also to what extent Type A and Type B orthogonality are valid. To define between-group predictivity we proceed similarly to the discussion of measures of fit in Chapter 3 where we have a decomposition in terms of the PCA of the canonical means and the fitted values

$$\bar{\mathbf{X}}\mathbf{L} = \hat{\bar{\mathbf{X}}}\mathbf{L} + (\bar{\mathbf{X}} - \hat{\bar{\mathbf{X}}})\mathbf{L}, \tag{4.12}$$

where the fitted values $\hat{\bar{\mathbf{X}}}$ are given by (4.6). We would expect (4.12) to satisfy both Type A and Type B orthogonality. The centring matrix $\mathbf{C}$ can easily be incorporated by multiplying (4.12) from the left by $\mathbf{C}^{-1/2}$. Note that due to the diagonality of $\mathbf{N}$ and the idempotency of $\mathbf{I} - K^{-1}\mathbf{11}'$, we have that $\mathbf{C}^{1/2}$ is equal to $\mathbf{N}^{1/2}$ for the weighted case and, for the unweighted case, either to $\mathbf{I} - K^{-1}\mathbf{11}'$ or simply $\mathbf{I}$. Thus, for Type B orthogonality,

$$\mathbf{L}'\bar{\mathbf{X}}'\mathbf{C}\bar{\mathbf{X}}\mathbf{L} = \mathbf{L}'\hat{\bar{\mathbf{X}}}'\mathbf{C}\hat{\bar{\mathbf{X}}}\mathbf{L} + \mathbf{L}'(\bar{\mathbf{X}} - \hat{\bar{\mathbf{X}}})'\mathbf{C}(\bar{\mathbf{X}} - \hat{\bar{\mathbf{X}}})\mathbf{L}, \tag{4.13}$$

and for Type A orthogonality,

$$\mathbf{C}^{1/2}\bar{\mathbf{X}}\mathbf{LL}'\bar{\mathbf{X}}'\mathbf{C}^{1/2} = \mathbf{C}^{1/2}\hat{\bar{\mathbf{X}}}\mathbf{LL}'\hat{\bar{\mathbf{X}}}'\mathbf{C}^{1/2} + \mathbf{C}^{1/2}(\bar{\mathbf{X}} - \hat{\bar{\mathbf{X}}})\mathbf{LL}'(\bar{\mathbf{X}} - \hat{\bar{\mathbf{X}}})'\mathbf{C}^{1/2}. \tag{4.14}$$

Although (4.13) and (4.14) include the centring matrix $\mathbf{C}$, they essentially reproduce (3.12) and (3.14). To verify that they continue to be valid, we need to show that the corresponding cross product terms vanish. That this is so for Type B orthogonality follows from

$$\begin{aligned}\mathbf{L}'\hat{\bar{\mathbf{X}}}'\mathbf{C}(\bar{\mathbf{X}} - \hat{\bar{\mathbf{X}}})\mathbf{L} &= \mathbf{L}'\mathbf{M}'^{-1}\mathbf{JM}'\bar{\mathbf{X}}'\mathbf{C}\bar{\mathbf{X}}(\mathbf{I} - \mathbf{MJM}^{-1})\mathbf{L} \\ &= \mathbf{L}'\mathbf{M}'^{-1}\mathbf{J}[\mathbf{M}'\bar{\mathbf{X}}'\mathbf{C}\bar{\mathbf{X}}\mathbf{M}](\mathbf{I} - \mathbf{J})\mathbf{M}^{-1}\mathbf{L} \\ &= \mathbf{L}'\mathbf{M}'^{-1}\mathbf{J}\boldsymbol{\Lambda}(\mathbf{I} - \mathbf{J})\mathbf{M}^{-1}\mathbf{L} = \mathbf{0},\end{aligned}$$

since the term in square brackets simplify on recalling that $(\bar{\mathbf{X}}'\mathbf{C}\bar{\mathbf{X}})\mathbf{M} = \mathbf{WM}\boldsymbol{\Lambda}$ and $\mathbf{L}'\mathbf{WL} = \mathbf{M}'\mathbf{WM} = \mathbf{I}$. In the case of Type A orthogonality, we have

$$\begin{aligned}\mathbf{C}^{1/2}\hat{\bar{\mathbf{X}}}\mathbf{LL}'(\bar{\mathbf{X}} - \hat{\bar{\mathbf{X}}})'\mathbf{C}^{1/2} &= \mathbf{C}^{1/2}\bar{\mathbf{X}}\mathbf{MJM}^{-1}\mathbf{LL}'(\mathbf{I} - \mathbf{M}'^{-1}\mathbf{JM}')\bar{\mathbf{X}}'\mathbf{C}^{1/2} \\ &= \mathbf{C}^{1/2}\bar{\mathbf{X}}\mathbf{MJ}[\mathbf{M}^{-1}\mathbf{LL}'\mathbf{M}'^{-1}](\mathbf{I} - \mathbf{J})\mathbf{M}'\bar{\mathbf{X}}'\mathbf{C}^{1/2} \\ &= \mathbf{C}^{1/2}\bar{\mathbf{X}}\mathbf{MJ}(\mathbf{I} - \mathbf{J})\mathbf{M}'\bar{\mathbf{X}}'\mathbf{C}^{1/2} = \mathbf{0},\end{aligned}$$

since $\mathbf{W}^{-1} = \mathbf{LL}' = \mathbf{MM}'$.

Expressions (4.13) and (4.14) pertain to the canonical means, whereas we are concerned with predictivities for the original means. However, because **L** is nonsingular, it may be eliminated from the Type B result in (4.13), implying that

$$\overline{\mathbf{X}}'\mathbf{C}\overline{\mathbf{X}} = \hat{\overline{\mathbf{X}}}'\mathbf{C}\hat{\overline{\mathbf{X}}} + (\overline{\mathbf{X}} - \hat{\overline{\mathbf{X}}})'\mathbf{C}(\overline{\mathbf{X}} - \hat{\overline{\mathbf{X}}}). \tag{4.15}$$

Result (4.15) shows that Type B orthogonality remains available for the original means but since the matrix **L** cannot be eliminated from (4.14), Type A orthogonality is not available for the original variables. Because $\mathbf{LL}' = \mathbf{W}^{-1}$, the interpretation of Type B orthogonality in (4.14) is merely that Mahalanobis squared distances may be split into components in the approximation and residual spaces (Pythagoras).

The PCA measures of fit discussed in Section 3.3 and (4.15) enable us to define the following measures of fit for the class means in CVA. From (4.13) we may define

$$\textit{Overall quality} = \frac{\text{tr}(\mathbf{L}'\hat{\overline{\mathbf{X}}}'\mathbf{C}\hat{\overline{\mathbf{X}}}\mathbf{L})}{\text{tr}(\mathbf{L}'\overline{\mathbf{X}}'\mathbf{C}\overline{\mathbf{X}}\mathbf{L})}.$$

Now,

$$\begin{aligned}\text{tr}(\mathbf{L}'\hat{\overline{\mathbf{X}}}'\mathbf{C}\hat{\overline{\mathbf{X}}}\mathbf{L}) &= \text{tr}(\mathbf{V}'\mathbf{L}'\hat{\overline{\mathbf{X}}}'\mathbf{C}\hat{\overline{\mathbf{X}}}\mathbf{L}\mathbf{V}) = \text{tr}(\mathbf{M}'\hat{\overline{\mathbf{X}}}'\mathbf{C}\hat{\overline{\mathbf{X}}}\mathbf{M}) \\ &= \text{tr}(\mathbf{J}\mathbf{M}'\overline{\mathbf{X}}'\mathbf{C}\overline{\mathbf{X}}\mathbf{M}\mathbf{J}) = \text{tr}(\mathbf{J}\mathbf{\Lambda}\mathbf{J}) = \text{tr}(\mathbf{\Lambda}\mathbf{J}).\end{aligned}$$

The denominator is obtained in the same way but with **J** replaced by **I**. Thus

$$\textit{Overall quality} = \frac{\text{tr}(\mathbf{\Lambda}\mathbf{J})}{\text{tr}(\mathbf{\Lambda})} = \frac{\sum_{j=1}^{r}\lambda_j}{\sum_{j=1}^{p}\lambda_j}, \tag{4.16}$$

where, of course, at most the first $m = \min(K - 1, p)$ eigenvalues obtained from (4.4) may be nonzero.

It follows that when $r \geq m$, overall quality is always unity and $\hat{\overline{\mathbf{X}}} = \overline{\mathbf{X}}$. The measure (4.16) is the equivalent of the PCA measure (see (3.17)) and in CVA is concerned with the quality of fit in the canonical variables. Therefore, we denote the overall quality (4.16) as $\textit{Overall quality}_{\textit{Canvar}}$. We may be interested in a similar measure based directly on the original variables derived from the Type B orthogonality (4.15). This gives $\text{tr}(\hat{\overline{\mathbf{X}}}'\mathbf{C}\hat{\overline{\mathbf{X}}})/\text{tr}(\overline{\mathbf{X}}'\mathbf{C}\overline{\mathbf{X}})$. Now

$$\text{tr}(\hat{\overline{\mathbf{X}}}'\mathbf{C}\hat{\overline{\mathbf{X}}}) = \text{tr}\{\mathbf{M}'^{-1}\mathbf{J}[\mathbf{M}'\overline{\mathbf{X}}'\mathbf{C}\overline{\mathbf{X}}\mathbf{M}]\mathbf{J}\mathbf{M}^{-1}\} = \text{tr}\{\mathbf{M}'^{-1}\mathbf{J}[\mathbf{\Lambda}]\mathbf{J}\mathbf{M}^{-1}\} = \text{tr}\{(\mathbf{M}'\mathbf{M})^{-1}\mathbf{J}\mathbf{\Lambda}\mathbf{J}\}.$$

The denominator is obtained in the same way but with **J** replaced by **I**. Thus, we have

$$\textit{Overall quality}_{\textit{Origvar}} = \frac{\text{tr}(\hat{\overline{\mathbf{X}}}'\mathbf{C}\hat{\overline{\mathbf{X}}})}{\text{tr}(\overline{\mathbf{X}}'\mathbf{C}\overline{\mathbf{X}})} = \frac{\sum_{j=1}^{r}\lambda_j m^{jj}}{\sum_{j=1}^{p}\lambda_j m^{jj}}, \tag{4.17}$$

where m^{jj} represents the jth diagonal value of $(\mathbf{M}'\mathbf{M})^{-1}$. This would agree with (4.16) if we had $m^{jj} = 1$ for all j.

Using a similar argument leading to the definition of axis adequacies in the case of PCA (see (3.18), we define axis adequacy for CVA as the diagonal elements of the $p \times p$ matrix

$$\text{diag}(\mathbf{MJM}')[\text{diag}(\mathbf{MM}')]^{-1}. \tag{4.18}$$

Furthermore, for axis predictivity we proceed similar to before defining the diagonal elements of the matrix $\mathbf{\Pi} : p \times p$ where

$$\mathbf{\Pi} = \text{diag}(\hat{\overline{\mathbf{X}}}'\mathbf{C}\hat{\overline{\mathbf{X}}})[\text{diag}(\overline{\mathbf{X}}'\mathbf{C}\overline{\mathbf{X}})]^{-1}, \tag{4.19}$$

justified by (4.15). This applies to the predictivities of the original variables; a similar expression could be derived from (4.13) for the predictivities of the canonical variables. From (4.17) and (4.19) we have that

$$\textit{Overall quality} = \frac{\text{tr}[\mathbf{\Pi}(\overline{\mathbf{X}}'\mathbf{C}\overline{\mathbf{X}})]}{\text{tr}(\overline{\mathbf{X}}'\mathbf{C}\overline{\mathbf{X}})},$$

expressing overall quality as a weighted mean of the individual axis predictivities.

Class predictivity is defined by the diagonal elements of the matrix $\mathbf{\Psi} : K \times K$ where

$$\mathbf{\Psi} = \text{diag}(\mathbf{C}^{1/2}\hat{\overline{\mathbf{X}}}\mathbf{W}^{-1}\hat{\overline{\mathbf{X}}}'\mathbf{C}^{1/2})[\text{diag}(\mathbf{C}^{1/2}\overline{\mathbf{X}}\mathbf{W}^{-1}\overline{\mathbf{X}}'\mathbf{C}^{1/2})]^{-1} \tag{4.20}$$

pertaining to the canonical means and justified by (4.14). Result (4.20) simplifies to

$$\mathbf{\Psi} = \text{diag}(\hat{\overline{\mathbf{X}}}\mathbf{W}^{-1}\hat{\overline{\mathbf{X}}}')[\text{diag}(\overline{\mathbf{X}}\mathbf{W}^{-1}\overline{\mathbf{X}}')]^{-1}$$

for the weighted case after eliminating $\mathbf{C}$.

The above measures refer to class predictivity, rather than sample predictivity; information on the individual samples has not been used. For the samples, we look at predictivities within classes as discussed below, that is, the predictivities of the individual samples corrected for the class means, $(\mathbf{I} - \mathbf{H})\mathbf{X}$. These transform into canonical variables $(\mathbf{I} - \mathbf{H})\mathbf{XL}$ with the equivalent partition to (4.11),

$$(\mathbf{I} - \mathbf{H})\mathbf{XL} = (\mathbf{I} - \mathbf{H})\hat{\mathbf{X}}\mathbf{L} + (\mathbf{I} - \mathbf{H})(\mathbf{X} - \hat{\mathbf{X}})\mathbf{L},$$

where $\hat{\mathbf{X}} = \mathbf{XMJM}^{-1}$. These are standard orthogonal projections, so Type A and Type B orthogonality are satisfied: for Type B,

$$\mathbf{L}'\mathbf{X}'(\mathbf{I} - \mathbf{H})\mathbf{XL} = \mathbf{L}'\hat{\mathbf{X}}'(\mathbf{I} - \mathbf{H})\hat{\mathbf{X}}\mathbf{L} + \mathbf{L}'(\mathbf{X} - \hat{\mathbf{X}})'(\mathbf{I} - \mathbf{H})(\mathbf{X} - \hat{\mathbf{X}})\mathbf{L}, \tag{4.21a}$$

and for Type A,

$$(\mathbf{I} - \mathbf{H})\mathbf{XLL}'\mathbf{X}'(\mathbf{I} - \mathbf{H}) = (\mathbf{I} - \mathbf{H})\hat{\mathbf{X}}\mathbf{LL}'\hat{\mathbf{X}}'(\mathbf{I} - \mathbf{H}) + (\mathbf{I} - \mathbf{H})(\mathbf{X} - \hat{\mathbf{X}})(\mathbf{X} - \hat{\mathbf{X}})'(\mathbf{I} - \mathbf{H}). \tag{4.21b}$$

We may check these results by verifying that the relevant cross products vanish. That this is indeed so follows for Type B orthogonality from

$$\begin{aligned}\left\{(\mathbf{I}-\mathbf{H})\hat{\mathbf{X}}\mathbf{L}\right\}'(\mathbf{I}-\mathbf{H})(\mathbf{X}-\hat{\mathbf{X}})\mathbf{L} &= \mathbf{L}'(\mathbf{M}')^{-1}\mathbf{J}\mathbf{M}'\mathbf{X}'(\mathbf{I}-\mathbf{H})\mathbf{X}(\mathbf{I}-\mathbf{M}\mathbf{J}\mathbf{M}^{-1})\mathbf{L}\\ &= \mathbf{L}'(\mathbf{M}')^{-1}\mathbf{J}\mathbf{M}'\mathbf{W}\mathbf{M}(\mathbf{I}-\mathbf{J})\mathbf{M}^{-1}\mathbf{L}\\ &= \mathbf{L}'(\mathbf{M}')^{-1}\mathbf{J}(\mathbf{I}-\mathbf{J})\mathbf{M}^{-1}\mathbf{L}=\mathbf{0}.\end{aligned}$$

Type A orthogonality is the result of

$$\begin{aligned}(\mathbf{I}-\mathbf{H})\hat{\mathbf{X}}\mathbf{L}\mathbf{L}'(\mathbf{X}-\hat{\mathbf{X}})'(\mathbf{I}-\mathbf{H}) &= (\mathbf{I}-\mathbf{H})(\mathbf{X}\mathbf{M}\mathbf{J}\mathbf{M}^{-1})\mathbf{M}\mathbf{M}'(\mathbf{I}-(\mathbf{M}')^{-1}\mathbf{J}\mathbf{M}')\mathbf{X}'(\mathbf{I}-\mathbf{H})\\ &= (\mathbf{I}-\mathbf{H})\mathbf{X}\mathbf{M}\mathbf{J}(\mathbf{I}-\mathbf{J})\mathbf{M}'\mathbf{X}'(\mathbf{I}-\mathbf{H})=\mathbf{0}.\end{aligned}$$

Thus, as for the between-class measures, Type A and Type B orthogonality are satisfied within classes. Also $\mathbf{L}$ may be eliminated from the Type B result in (4.21a). The Type B result is trivial because $\mathbf{L}'\mathbf{X}'(\mathbf{I}-\mathbf{H})\mathbf{X}\mathbf{L}=\mathbf{L}'\mathbf{W}\mathbf{L}=\mathbf{I}$, so it simplifies to

$$\mathbf{I}=\mathbf{V}\mathbf{J}\mathbf{V}'+\mathbf{V}(\mathbf{I}-\mathbf{J})\mathbf{V}'$$

or, on eliminating $\mathbf{L}$, to

$$\mathbf{W}=(\mathbf{M}')^{-1}\mathbf{J}\mathbf{M}^{-1}+(\mathbf{M}')^{-1}(\mathbf{I}-\mathbf{J})\mathbf{M}^{-1}.$$

From (4.21a), after eliminating $\mathbf{L}$ we may define within-group axis predictivity as the diagonal elements of the matrix $\mathbf{\Pi}_{\mathrm{W}}: p\times p$ where

$$\mathbf{\Pi}_{\mathrm{W}}=\mathrm{diag}(\hat{\mathbf{X}}'(\mathbf{I}-\mathbf{H})\hat{\mathbf{X}})[\mathrm{diag}(\mathbf{X}'(\mathbf{I}-\mathbf{H})\mathbf{X})]^{-1}=\mathrm{diag}(\hat{\mathbf{X}}'(\mathbf{I}-\mathbf{H})\hat{\mathbf{X}})[\mathrm{diag}(\mathbf{W})]^{-1}. \tag{4.22}$$

Result (4.22) gives an overall measure of predictivity for each variable (column) of $(\mathbf{I}-\mathbf{H})\mathbf{X}$. A version could be constructed to give the axis predictivities within each group separately. From (4.21b), for Type A, we define within-group sample predictivity as the diagonal elements of the matrix $\mathbf{\Psi}_{\mathrm{W}}: n\times n$ where

$$\mathbf{\Psi}_{\mathrm{W}}=\mathrm{diag}((\mathbf{I}-\mathbf{H})\hat{\mathbf{X}}\mathbf{W}^{-1}\hat{\mathbf{X}}'(\mathbf{I}-\mathbf{H}))\{\mathrm{diag}((\mathbf{I}-\mathbf{H})\mathbf{X}\mathbf{W}^{-1}\mathbf{X}'(\mathbf{I}-\mathbf{H}))\}^{-1}. \tag{4.23}$$

Notice once again that the matrix $\hat{\mathbf{X}}$ in (4.22) as well as in (4.23) requires the appropriate matrix $\mathbf{M}$ obtained from the currently defined matrix $\mathbf{C}$.

Although the means of the K classes are exactly represented in m, or fewer, dimensions, the individual samples are generally not exactly represented in fewer than p dimensions. The within-group sample predictivities can be less than unity in dimensions $m+1,\ldots,p-1$. Section 4.2 shows that there is some degree of arbitrariness in the singular vectors corresponding to dimensions $m+1,\ldots,p$, making it invalid to examine each of the arbitrary dimensions individually. The best that can be done is to combine all the dimensions to give an overall within-group predictivity for the residual space orthogonal to the space of the group means. This can be done by subtracting the $(m-1)$-dimensional predictivities from unity (see (4.20)).

4.6.1 Predictivities of new samples and variables

We showed in Chapter 3 that new samples or variables added to a PCA biplot do not play any role in finding the scaffolding for constructing the biplot, but that it is still possible to calculate predictivities associated with the newly added samples or variables. Similar calculations are possible with CVA biplots. The matrix of within-group sample predictivities (4.23) can be written as

$$\boldsymbol{\Psi}_{\mathrm{W}} = \mathrm{diag}((\mathbf{I}-\mathbf{H})\mathbf{XMJM'X'}(\mathbf{I}-\mathbf{H}))\left\{\mathrm{diag}((\mathbf{I}-\mathbf{H})\mathbf{XMM'X'}(\mathbf{I}-\mathbf{H}))\right\}^{-1}, \qquad (4.24)$$

since $\hat{\mathbf{X}}\mathbf{W}^{-1}\hat{\mathbf{X}}' = (\mathbf{XMJM}^{-1})(\mathbf{MM}')(\mathbf{XMJM}^{-1})' = \mathbf{XMJM'X'}$. Given a new sample $\mathbf{x}$: $p \times 1$ with known values for all p original variables and belonging to group t, its within-group sample predictivity can be calculated as follows. Centre $\mathbf{x}$ by subtracting the same mean vector used in the column-centring of the original input data matrix $\mathbf{X}$. Let $\mathbf{x}^*$ denote the centred $\mathbf{x}$. The factor $(\mathbf{I}-\mathbf{H})\mathbf{X}$ in (4.24) is replaced by $\tilde{\mathbf{x}}' = \mathbf{x}^{*\prime}$ – row t of $(\mathbf{G'G})^{-1}\mathbf{G'X}$. Thus Within-group sample predictivity $= \tilde{\mathbf{x}}'\mathbf{MJM}'\tilde{\mathbf{x}}/\tilde{\mathbf{x}}'\mathbf{MM}'\tilde{\mathbf{x}}$.

If the group membership of the new sample is unknown, as was the case with the specimen of unknown origin in our *Ocotea* example, its group membership can be taken as suggested by the nearest group mean to its position in the biplot display. The above procedure generalizes in a straightforward way to an $s \times p$ matrix containing measurements of s new samples. Consider the p vector $\mathbf{C}^{1/2}\overline{\mathbf{x}}^*$ and its regression $\mathbf{b}$ introduced in equation (4.10). Similar to (3.24) and (3.25), we have the orthogonal decompositions

$$\mathbf{C}^{1/2}\overline{\mathbf{x}}^* = \mathbf{C}^{1/2}\overline{\mathbf{X}}\mathbf{MJb} + \mathbf{C}^{1/2}\overline{\mathbf{X}}\mathbf{M}(\mathbf{I}-\mathbf{J})\mathbf{b} + (\mathbf{C}^{1/2}\overline{\mathbf{x}}^* - \mathbf{C}^{1/2}\overline{\mathbf{X}}\mathbf{Mb}). \qquad (4.25)$$

In (4.25) the first two terms on the right-hand side represent the contribution of the regression in the canonical space, while the final term is the regression residual. Thus, in the full space we have $\mathbf{b} = \boldsymbol{\Lambda}^{-1}\mathbf{M}'\overline{\mathbf{X}}'\mathbf{C}\overline{\mathbf{x}}^*$, while $\mathbf{Jb} = \mathbf{J}\boldsymbol{\Lambda}^{-1}\mathbf{M}'\overline{\mathbf{X}}'\mathbf{C}\overline{\mathbf{x}}^*$ selects the r elements of $\mathbf{b}$ in the approximation space. From (4.25), axis predictivity for the newly added variable may be defined in two ways:

$$\Pi_{\mathrm{CVA.x}} = \frac{\mathbf{b}'\mathbf{JM}'\overline{\mathbf{X}}'\mathbf{C}\overline{\mathbf{X}}\mathbf{MJb}}{\overline{\mathbf{x}}^{*\prime}\mathbf{C}\overline{\mathbf{x}}^*} = \frac{\mathbf{b}'\mathbf{J}\boldsymbol{\Lambda}\mathbf{Jb}}{\overline{\mathbf{X}}^{*\prime}\mathbf{C}\overline{\mathbf{x}}^*}. \qquad (4.26)$$

This compares the regression fit in r dimensions of the canonical space with the sum of squares among the means of the new variable. Usually the residual term $(\mathbf{C}^{1/2}\overline{\mathbf{x}}^* - \mathbf{C}^{1/2}\overline{\mathbf{X}}\mathbf{Mb})$ in (4.25) vanishes because both the canonical means and $\overline{\mathbf{x}}^*$ will occupy $K-1$ dimensions. Then (4.26) simplifies to

$$\Pi_{\mathrm{CVA.x}} = \frac{\mathbf{b}'\mathbf{JM}'\overline{\mathbf{X}}'\mathbf{C}\overline{\mathbf{X}}\mathbf{MJb}}{\mathbf{b}'\mathbf{M}'\overline{\mathbf{X}}'\mathbf{C}\overline{\mathbf{X}}\mathbf{Mb}} = \frac{\mathbf{b}'\mathbf{J}\boldsymbol{\Lambda}\mathbf{Jb}}{\mathbf{b}'\boldsymbol{\Lambda}\mathbf{b}}. \qquad (4.27)$$

This compares the regression fit in r dimensions of the canonical space with the sum of squares in the remaining $p-r$ dimensions. However, when $\overline{\mathbf{x}}^*$ occupies fewer dimensions, as when $p < K-1$ or when there happen to be collinearities among the canonical means, then (4.27) excludes a nonzero regression residual in its denominator and so gives an overoptimistic indication of predictivity. Thus we prefer (4.26) to (4.27). Expressions (4.26) and (4.27) generalize in a straightforward way to t newly added variables.

4.7 Functions for constructing a CVA biplot

The main function for constructing CVA biplots is the function `CVAbipl`. Its usage and arguments are similar to those of the function `PCAbipl` as discussed in Section 3.7, with notable exceptions given in Section 4.7.1. There are also several functions closely related to `CVAbipl` for constructing specific types of CVA biplots or adding enhancements to an existing biplot. These functions are briefly introduced in Sections 4.7.2–4.7.7.

4.7.1 Function `CVAbipl`

This is the basic function for constructing CVA biplots. Provision is made for one-, two- and three-dimensional CVA biplots. Its call and arguments are almost identical to those of the function `PCAbipl` as discussed in Chapter 3, with the following exceptions:

Arguments

`G`	A required argument for `CVAbipl`. Indicator matrix defining K groups of samples in the data.
`scaled.mat`	Not available in `CVAbipl` since CVA is scale-invariant. If for some reason a CVA of scaled data is required it can be performed by assigning the scaled data to argument `X`.
`weightedCVA`	The default is `"weighted"`, specifying a weighted CVA to be performed. Other possible values are `"unweightedI"` or `"unweightedCent"` for specifying the different forms of unweighted CVA discussed in Section 4.2.
`conf.alpha`	Confidence coefficient, typically 0.95 or 0.99, for constructing confidence circles about each group mean. The factor $\sqrt{n_i}$, as described at the end of Section 4.2, is included in the calculations.

Value

This is a list with the following components:

`SSP.B`	The between sums of squares and products matrix.
`SSP.W`	The within sums of squares and products matrix.
`Lmat`	Matrix $\mathbf{L}$ that transforms the original space into the canonical space.
`Mmat`	Matrix $\mathbf{M} = \mathbf{LV}$.
`lambda`	The eigenvalues i.e. diagonal elements of $\mathbf{\Lambda}$.
`CVA.quality.canvar`	The overall quality of the display in terms of the canonical space.

`CVA.quality.origvar`	The overall quality of the display in terms of the original space.
`adequacy`	The axis adequacies.
`predictions`	Predictions of the means or of the samples as specified.
`Z.means.mat`	Matrix containing the canonical means.

`CVAbipl` is mainly called for its side effect, the plotting of the CVA biplot. The functions `CVA.predictions.mat` and `CVA.predictivities` provide extensive output of the various measures of fit and fitted values discussed in this chapter.

4.7.2 Function `CVAbipl.zoom`

A function for zooming into an interactively selected area of a one- or two-dimensional CVA biplot. In addition to the arguments of `CVAbipl`, this function has the argument `zoomval` for controlling the amount of zooming. Its usage is similar to that described for the function `PCAbipl.zoom` in Section 3.7.2.

4.7.3 Function `CVAbipl.density`

A function for overlaying two-dimensional CVA biplots on a density plot of a two-dimensional CVA approximation of the input matrix. Its usage and arguments additional to those of `CVAbipl` are identical to the description given for `PCAbipl.density` in Section 3.7.3.

4.7.4 Function `CVAbipl.density.zoom`

Analogous to `CVAbipl.zoom`, the function `CVAbipl.density.zoom` provides zooming functionality for `CVAbipl.density` through its argument `zoomval`.

4.7.5 Function `CVAbipl.pred.regions`

This function constructs nearest canonical means regions for a CVA biplot. In addition to the arguments of `CVAbipl`, it has the arguments described below. These latter arguments are passed to the function `pred.regions` which in turn calls the function `Colour.Pixel`. The functions `pred.regions` and `Colour.Pixel` are responsible for calculating the distance from a given grid point in the biplot space to each canonical mean in the full canonical space. The grid point is then coloured according to the canonical mean having the shortest distance to it. Additional arguments of `CVAbipl.pred.regions` for passing to functions `pred.regions` and `Colour.Pixel` are as follows:

`x.grid`	Defaults to 0.05. Determines the x-coordinates of a grid overlaying the biplot space.
`y.grid`	Defaults to 0.05. Determines the y-coordinates of a grid overlaying the biplot space.
`plot.symbol`	Defaults to 16. Plotting symbol for labelling a grid point.
`plot.symbol.size`	Defaults to 0.05. Size of plotting symbol for labelling a grid point.
`colours.pred.regions`	A k-vector specifying the colouring of each grid point according to the canonical mean with the shortest distance to it, where k denotes the number of canonical means.

4.7.6 Function `CVA.predictivities`

This is a function for calculating the various measures of fit discussed in Section 4.6. It has the five arguments discussed above for `CVAbipl`: `X`, `G`, `weightedCVA`, `X.new.samples` and `X.new.vars`. The various measures of fit are provided for dimensions $1, 2, \ldots, p$, and the output (value) of `CVA.predictivities` is a list with the following named components:

`OverAllMeans`	Means of the original data matrix.
`XBar.groups`	Group means of the uncentred data matrix.
`XcentBar.groups`	Group means of the column-centred data matrix.
`canon.group.means`	Canonical group means.
`CVA.quality.canvar`	Overall quality with respect to the canonical variables (4.16).
`CVA.quality.origvar`	Overall quality with respect to the original variables (4.17).
`Axis.Adequacies`	Adequacies associated with the variables (biplot axes) defined in (4.18).
`Axis.Predictivities`	Predictivities associated with the original variables (biplot axes) defined in (4.19).
`Class.Predictivities`	The predictivities for each group calculated according to (4.20).
`WithinGroup.axis.Predictivities`	The within-group axis predictivity for each variable (column) of $(\mathbf{I} - \mathbf{H})\mathbf{X}$, calculated according to (4.22).
`WithinGroup.Sample.Predictivities`	The within-group sample predictivities calculated according to (4.23).
`new.sample.pred`	The within-group sample predictivities for new samples calculated according to Section 4.6.1.
`Axis.predictivities.new.vars`	Axis predictivities for new variables calculated according to (4.26).

4.7.7 Function CVA.predictions.mat

The function `CVA.predictions.mat` takes the three arguments `X`, `G` and `weightedCVA` as described above for `CVAbipl`. Its return value is a list with the named components given below. Note that the various fitted values are given for dimensions 1, 2, ..., p.

`predictions.for.XcentBar.groups`	Fitted values for the group means of the column-centred data matrix.
`predictions.for.centredX`	Fitted values for the column centred data matrix.
`predictions.for.originalX`	Fitted values for the original data matrix.
`reconstr.error.Xbar`	Sum of squared differences between the elements of the group means of the column centred data matrix and its fitted values.
`reconstr.error.X`	Sum of squared differences between the elements of centred data matrix and its fitted values.

4.8 Continuing the *Ocotea* example

In Figures 4.2–4.4 we illustrated some of the functionality of the functions described in Section 4.7. Figure 4.4 results from a call to `CVAbipl.pred.regions`:

```
CVAbipl.pred.regions(X = Ocotea.data[,3:5],
   G = indmat(Ocotea.data[,2]), X.new.samples = matrix(c(134,375,
   1170), nrow = 1), alpha = 0.95, colours = c("red","blue",
   "green2"), pch.samples = 0:2, x.grid = 0.01, y.grid = 0.01,
   pch.new = 8, plot.symbol = 20, plot.symbol.size = 0.75,
   colours.pred.regions = c("coral","lightblue","lightgreen"),
   pch.new.cols = "black")
```

The classification regions in Figure 4.4 are calculated according to the nearest neigbour rule (4.8). The default procedure used in the above call to `CVAbipl.pred.regions` is to calculate the distances between a grid point in the biplot space and the canonical means in the full canonical space. In a subsequent example we will illustrate the effect of changes in this default procedure.

If we can assume that the *Ocotea* data come from a multivariate normal distribution, the CVA biplot can be equipped with the confidence circles introduced in Section 4.2. In the following call to `CVAbipl` we specify the argument `conf.alpha = 0.99` to request 99% confidence circles about each group mean. Furthermore, we specify `alpha = 0.99` for comparison purposes to obtain 0.99-bags for each of the groups. We remark that for this example the difference in the appearance of the biplots resulting from a weighted or unweighted CVA is negligible and therefore we give only the biplot for the weighted CVA in Figure 4.11.

```
CVAbipl(Ocotea.data[,3:5], X.new.samples = matrix(c(134,375,1170),
   nrow = 1), G = indmat(Ocotea.data[,2]), weightedCVA =
   "weighted", means.plot = TRUE, colours = c("red","blue",
   "green"), pch.samples = 0:2, pch.samples.size = 1.25,
   label = FALSE, pos = "Hor", line.type = rep(1,3),
   line.width = rep(2,3), offset = c(-0.2, 0.05, 0.1, 0.2),
   n.int = c(5,10,5), pch.means = c(15,16,17),
   pch.means.size = 1.5, side.label = rep("right",3),
   pos.m = c(4,4,4), offset.m = c(-0.1, -0.1, 0.1),
   parplotmar = c(3,3,3,3), alpha = 0.99, conf.alpha = 0.99,
   conf.type = "with.n.factor", specify.bags = 1:3,
   legend.type = c(TRUE,TRUE,TRUE), pch.new.col = "black",
   pch.new = 8)
```

Figure 4.11 clearly shows that the three confidence circles are well separated and that the specimen of unknown origin lies almost on the perimeter of the confidence circle around the mean of ***Opor***.

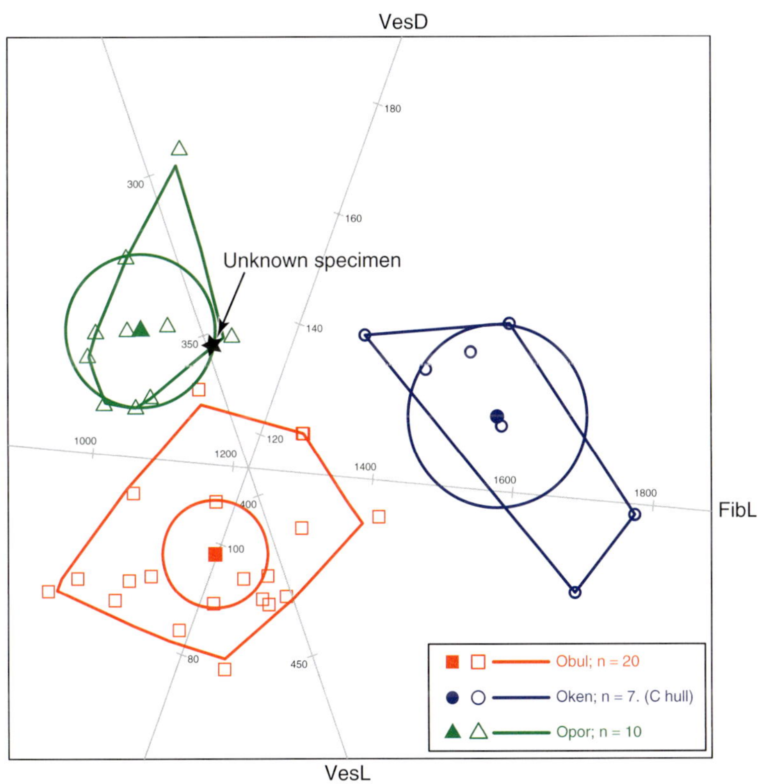

Figure 4.11 CVA biplot for the three-variable *Ocotea* data with 99% confidence circles drawn about each group mean. For comparison purposes 0.99-bags are also shown (convex hull for the ***Oken*** group).

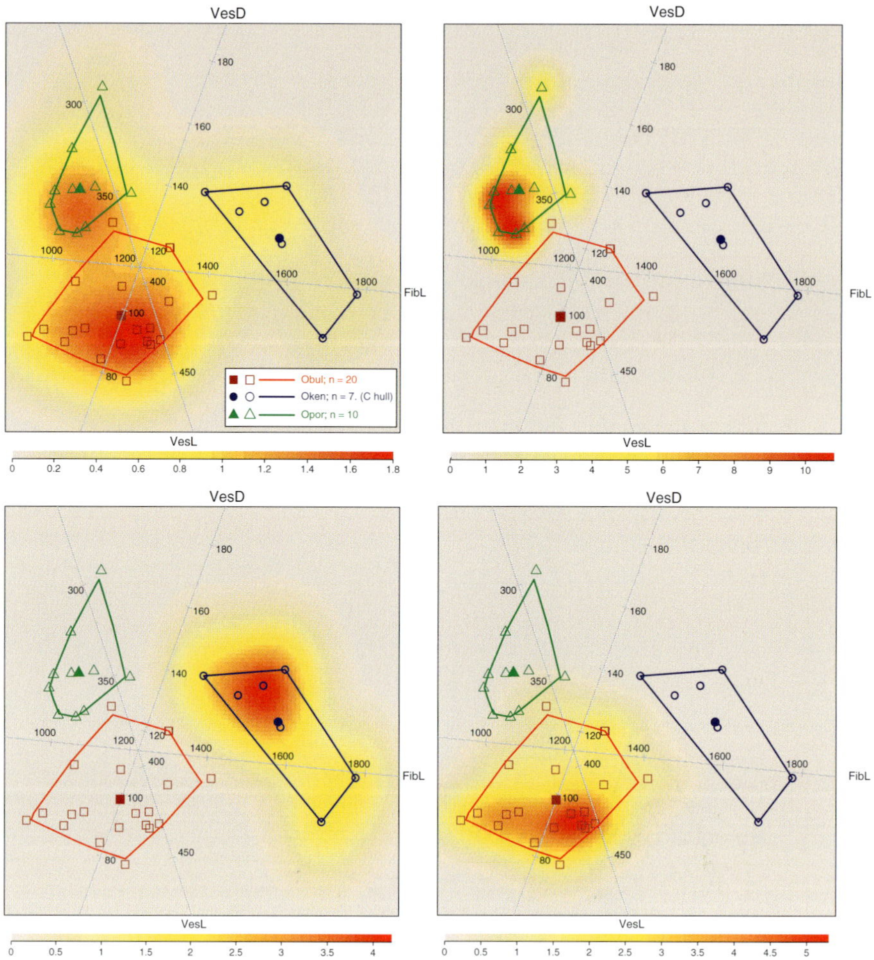

Figure 4.12 CVA biplots with density surfaces and 0.99-bags for each group: (top left) all samples used for the density surface; (top right) density surface for ***Opor***; (bottom left) density surface for ***Oken***; (bottom right) density surface for ***Obul***.

In Figure 4.12 we enhance the biplot with various density surfaces by consecutive calls to `CVAbipl.density`. In these calls we assign the argument `specify.density.class` the values `"allsamples"`, `"Opor"`, `"Obul"` and `"Oken"` to obtain the four biplots shown in Figure 4.12.

A biplot should always be considered together with the various measures of fit. When considering the measures of fit for the three variable *Ocotea* CVA biplots, we must keep in mind that, since we have three groups, the canonical means are exactly represented in a two-dimensional space. In Tables 4.4–4.11 we give the output of the following call to obtain detailed information to complement interpretation of the CVA biplots:

```
CVA.predictivities(X = Ocotea.data[,3:5],
   X.new.sample = c(134,375,1170), G = indmat(Ocotea.data[,2]),
   weightedCVA   = "weighted")
```

Table 4.4 Overall means, group means for three uncentred and centred variables of the *Ocotea* data, together with the canonical means calculated for the weighted and unweighted CVAs. The sample sizes are: ***Obul***, 20; ***Oken***, 7; ***Opor***, 10.

Overall means			
	VesD	VesL	FibL
	113.946	391.243	1221.811

Group means of original data matrix			
	VesD	VesL	FibL
Obul	98.100	412.000	1185.400
Oken	137.286	401.714	1568.857
Opor	129.300	342.400	1051.700

Group means of centred data matrix			
	VesD	VesL	FibL
Obul	−15.846	20.757	−36.411
Oken	23.340	10.471	347.046
Opor	15.354	−48.843	−170.111

Canonical group means: Weighted CVA			
Dimension	1	2	3
Obul	−0.074	−0.186	0.000
Oken	0.551	0.110	0.000
Opor	−0.237	0.294	0.000

Canonical group means: Unweighted **I** CVA			
Dimension	1	2	3
Obul	−0.078	−0.184	0.000
Oken	0.553	0.099	0.000
Opor	−0.232	0.299	0.000

Canonical group means: Unweighted cent. CVA			
Dimension	1	2	3
Obul	−0.063	−0.190	0.000
Oken	0.543	0.141	0.000
Opor	−0.254	0.280	0.000

Actually the call was also made by specifying `weightedCVA = "unweightedI"` and `weightedCVA = "unweightedCent"`. Recall that in Tables 4.4–4.11, reference to dimension 3, say, does not refer just to the third dimension but to the cumulative effects of all three dimensions.

Tables 4.4–4.11 are given mainly to show the kind of output available from our R functions. In this example they are not particularly interesting because, with only three groups, all group information is exact in two dimensions; hence the many 100% values shown for dimensions 2 and 3. Furthermore, only three variables have been used, again giving exact information on the variables in three dimensions. We have also given the

Table 4.5 CVA quality associated with CVA biplots of three variable *Ocotea* data.

	CVA quality with respect to canonical variables		
	Weighted	Unweighted I	Unweighted centred
Dim_1	63.013	73.277	74.837
Dim_2	100.000	100.000	100.000
Dim_3	100.000	100.000	100.000
	CVA quality with respect to original variables		
	Weighted	Unweighted I	Unweighted centred
Dim_1	96.459	97.614	98.232
Dim_2	100.000	100.000	100.000
Dim_3	100.000	100.000	100.000

Table 4.6 Adequacies associated with CVA biplots of three-variable *Ocotea* data.

	Axis adequacies								
	Weighted			Unweighted I			Unweighted centred		
	VesD	VesL	FibL	VesD	VesL	FibL	VesD	VesL	FibL
Dim_1	0.000	0.139	0.926	0.000	0.147	0.918	0.006	0.117	0.948
Dim_2	0.933	0.416	0.978	0.933	0.416	0.978	0.933	0.416	0.978
Dim_3	1.000	1.000	1.000	1.000	1.000	1.000	1.000	1.000	1.000

Table 4.7 Axis predictivities associated with CVA biplots of three-variable *Ocotea* data.

	Axis predictivities								
	Weighted			Unweighted I			Unweighted centred		
	VesD	VesL	FibL	VesD	VesL	FibL	VesD	VesL	FibL
Dim_1	0.190	0.170	0.995	0.297	0.225	0.995	0.219	0.335	1.000
Dim_2	1.000	1.000	1.000	1.000	1.000	1.000	1.000	1.000	1.000
Dim_3	1.000	1.000	1.000	1.000	1.000	1.000	1.000	1.000	1.000

Table 4.8 Class predictivities associated with CVA biplots of three-variable *Ocotea* data.

	Class predictivities								
	Weighted			Unweighted I			Unweighted centred		
	Obul	Oken	Opor	Obul	Oken	Opor	Obul	Oken	Opor
Dim_1	0.137	0.962	0.394	0.151	0.969	0.375	0.213	0.982	0.724
Dim_2	1.000	1.000	1.000	1.000	1.000	1.000	1.000	1.000	1.000
Dim_3	1.000	1.000	1.000	1.000	1.000	1.000	1.000	1.000	1.000

Table 4.9 Within group axis predictivities associated with CVA biplots of three-variable *Ocotea* data.

	Within-group axis predictivities								
	Weighted			Unweighted I			Unweighted centred		
	VesD	VesL	FibL	VesD	VesL	FibL	VesD	VesL	FibL
Dim_1	0.073	0.008	0.842	0.080	0.007	0.838	0.052	0.011	0.848
Dim_2	0.601	0.074	0.849	0.601	0.074	0.849	0.601	0.074	0.849
Dim_3	1.000	1.000	1.000	1.000	1.000	1.000	1.000	1.000	1.000

Table 4.10 The first and last five within-groups sample predictivities associated with CVA biplots of three-variable *Ocotea* data.

	Within-groups sample predictivities: Weighted CVA									
	S1	S2	S3	S4	S5	S33	S34	S35	S36	S37
Dim_1	0.941	0.839	0.723	0.366	0.831	0.005	0.077	0.015	0.548	0.918
Dim_2	0.985	0.864	0.788	0.407	0.999	0.980	0.379	0.380	0.558	0.923
Dim_3	1.000	1.000	1.000	1.000	1.000	1.000	1.000	1.000	1.000	1.000
	Within-groups sample predictivities: Unweighted I CVA									
	S1	S2	S3	S4	S5	S33	S34	S35	S36	S37
Dim_1	0.948	0.844	0.731	0.370	0.845	0.007	0.083	0.013	0.550	0.915
Dim_2	0.985	0.864	0.788	0.407	0.999	0.980	0.379	0.380	0.558	0.923
Dim_3	1.000	1.000	1.000	1.000	1.000	1.000	1.000	1.000	1.000	1.000
	Within-groups sample predictivities: Unweighted centred CVA									
	S1	S2	S3	S4	S5	S33	S34	S35	S36	S37
Dim_1	0.915	0.820	0.696	0.351	0.786	0.000	0.060	0.025	0.538	0.923
Dim_2	0.985	0.864	0.788	0.407	0.999	0.980	0.379	0.380	0.558	0.923
Dim_3	1.000	1.000	1.000	1.000	1.000	1.000	1.000	1.000	1.000	1.000

Table 4.11 New sample predictivities associated with CVA biplots of three-variable *Ocotea* data.

	New sample predictivities								
	Weighted			Unweighted I			Unweighted centred		
	Obul	Oken	Opor	Obul	Oken	Opor	Obul	Oken	Opor
Dim_1	0.000	0.941	0.898	0.000	0.932	0.890	0.003	0.964	0.916
Dim_2	0.991	0.996	0.938	0.991	0.996	0.938	0.991	0.996	0.938
Dim_3	1.000	1.000	1.000	1.000	1.000	1.000	1.000	1.000	1.000

measures for three kinds of centring: (i) weighted by the group sizes, (ii) ignoring group sizes and (iii) ignoring group sizes but passing through the weighted centroids of the group means. Perhaps because the sample sizes are not disparate, there is little to choose between the methods and, of course, with three groups they all give the same two-dimensional fit, merely distributing the group variation slightly differently between the first two dimensions – see Tables 4.5 and 4.7. The adequacies of the variables are given in Table 4.6. As we have explained, these are generally of less interest than predictivities. We see again that the adequacies do not depend greatly on the centring used and, slightly surprisingly, once we use two dimensions the type of centring has no effect whatsoever. This pattern persists in the within-group predictivity (Tables 4.9 and 4.10).

The measures given in the tables are meant to help researchers to detect which variables and samples are well or not well approximated, so that appropriate action can be taken. Normally, we would expect them to be displayed on the biplot maps. Though the R functions are available, this would be futile with the present example as the two-dimensional diagrams would be cluttered by useless 100% values, so it has not been done. Elsewhere in the book, this kind of information has been displayed (see, for example, Figure 4.1).

Finally, we show how to add new variables to a CVA biplot. We can almost use the exact procedure that allowed us to construct the PCA biplot of the full *Ocotea* data set in Figure 3.24 for adding the newly constructed variables ***VLDratio*** and ***RHWratio***. Instead of a call to `PCAbipl`, the call is made to `CVAbipl`. The only changes needed are omitting the argument `scaled.mat`, adding `G = indmat(Ocotea.data[,2])` and providing for plotting symbols and colours. The CVA biplot with the two newly added variables is shown in Figure 4.13. Note that we have also interactively shifted the origin. The reader can verify by calling `CVA.predictivities` that the axis predictivity for ***VLDratio*** is 0.0728 and that of ***RHWratio*** is 0.2326 in one dimension. In the two-dimensional biplot shown, both attain axis predictivity of unity.

4.9 CVA biplots for two classes

With only two classes, the points representing the two canonical means are necessarily collinear. The predictive biplot axes are then superimposed onto this single dimension and hence cannot be distinguished; a solution is to present them separately, as is shown in the example in Section 4.9.1. We may consider whether the dimensions orthogonal to the collinear means may be exploited. A potentially important use is for two-dimensional representations, where we have a spare unused dimension. A problem is that the scaling $\mathbf{L}'\mathbf{W}\mathbf{L} = \mathbf{I}$ implies the homogeneity of the dispersion of the canonical variables orthogonal to the space containing the means of the canonical variables. Then no preferential direction exists. We are investigating the possibility of finding a useful criterion that may be optimized in the orthogonal space to give points that may then be combined with the dimension holding the means.

4.9.1 An example of two-class CVA biplots

Gender remuneration inequalities at universities have been studied in various parts of the world (see, for example, Barbezat and Hughes, 2005; McNabb and Wass, 1997; Ward, 2001; Warman *et al.*, 2006). Although researchers were able to attribute part of the

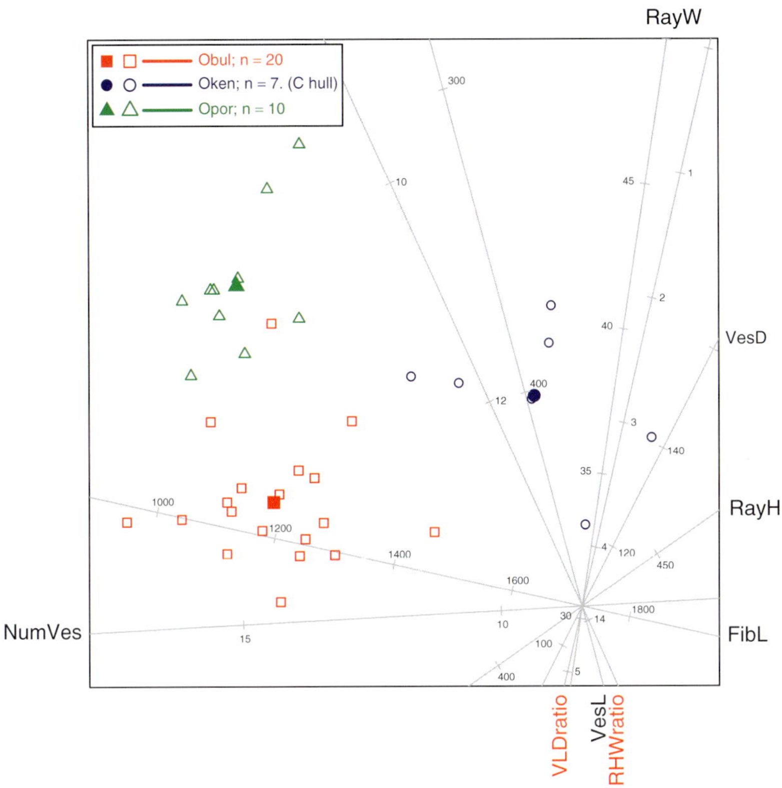

Figure 4.13 CVA biplot of the full *Ocotea* data set with the two newly constructed variables, ***VLDratio*** and ***RHWratio***, added as additional calibrated axes.

gender remuneration differentials to factors such as publication record (Ward, 2001), time allocation (Toutkoushian, 1999), rank and seniority (McNabb and Wass, 1997), institution type (Barbezat and Hughes, 2005), age and education (Bayer and Astin, 1975; Gordon *et al.*, 1974), an unexplained portion was generally found (see, for example, Toutkoushian, 1998; Barbezat and Hughes, 2005). This unexplained differential is commonly called the gender remuneration pay gap.

As far as the South African situation is concerned, South African higher education institutions were faced after 1994 with the reality of transforming to an organizational culture appropriate for the 'new' South Africa. In the case study presented here, we focus on the gender remuneration differentials for academic staff at Stellenbosch University for the years 2002–2005. Our data concern the permanent full-time academic staff at Stellenbosch University for 2002 and 2005. These data are available in the R dataframe `Remuneration.data` and are described in detail in Walters and Le Roux (2008). `Remuneration.data` consists of the following columns:

ID a coded identification number;

Remun the total cost of employment before deductions for December 2002 and 2005 (in units of R10 000) – inflation was not taken into account because it affected all staff equally during the study period;

Resrch	research output expressed on a continuous scale;
Rank	academic position or rank;
Age	in years;
Gender	male/female;
PQual	professional qualification (not used in the CVA discussed here);
AQual	academic qualification expressed on a continuous scale;
Faclty	integer value representing one of nine faculties included in the study;
Year	2002 or 2005.

The mean values of the samples in 2002 and two-dimensional CVA biplots, both grouped according to gender, are given in Table 4.12 and Figure 4.14, respectively.

The function `CVAbipl` was called four times with argument `e.vects` specified as `c(1,2)`, `c(1,3)`, `c(1,4)` and `c(1,5)` respectively to produce Figure 4.14. Arguments `predictions.mean = 1:2` and `ort.lty = 2` request the perpendicular dotted lines for predicting the mean values for the males and the females. The reader can verify that the predicted values obtained from any of the four biplots in Figure 4.14 correspond exactly to the values given in Table 4.12. This is so in spite of the large differences in the appearances of these biplots. Furthermore, a call to the function `CVA.predictivities` reveals that all biplot axes in the different biplots have axis predictivity of unity. Similarly, the class predictivities are all equal to unity (this is already true in the first dimension). Since we have only two groups, we know that the canonical means differ only in their first components and the call to `PCA.predictivities` returns the values for the canonical means given in Table 4.13.

The biplots in Figure 4.14 illustrate that although each of them provides exact predictions for all the mean values, incorrect conclusions can be drawn by judging from their appearance only. The reason for this is found in the form of the matrix $\mathbf{\Lambda}$ returned by the call to `CVAbipl`: all its diagonal elements, except the first one of 0.2906, are zero. Therefore, there is no uniquely defined second axis for constructing the scaffolding for the CVA biplot – only a one-dimensional CVA biplot is uniquely defined when dealing with two groups. In Figures 4.15 and 4.16 we give one-dimensional CVA biplots for the remuneration data.

Since in a one-dimensional biplot all the sample points and all the biplot axes are on the line extending through the means, we have overlaid the biplot in Figure 4.15 with separate density estimates for the two groups and also translated the axes arbitrarily

Table 4.12 Mean values of males and females for the variables in the remuneration data for 2002.

Variable	Female	Male
Remun	19.3128	26.1441
Resrch	0.2604	0.6319
Rank	2.4367	3.6046
Age	41.9755	47.9959
Aqual	6.7184	7.6304

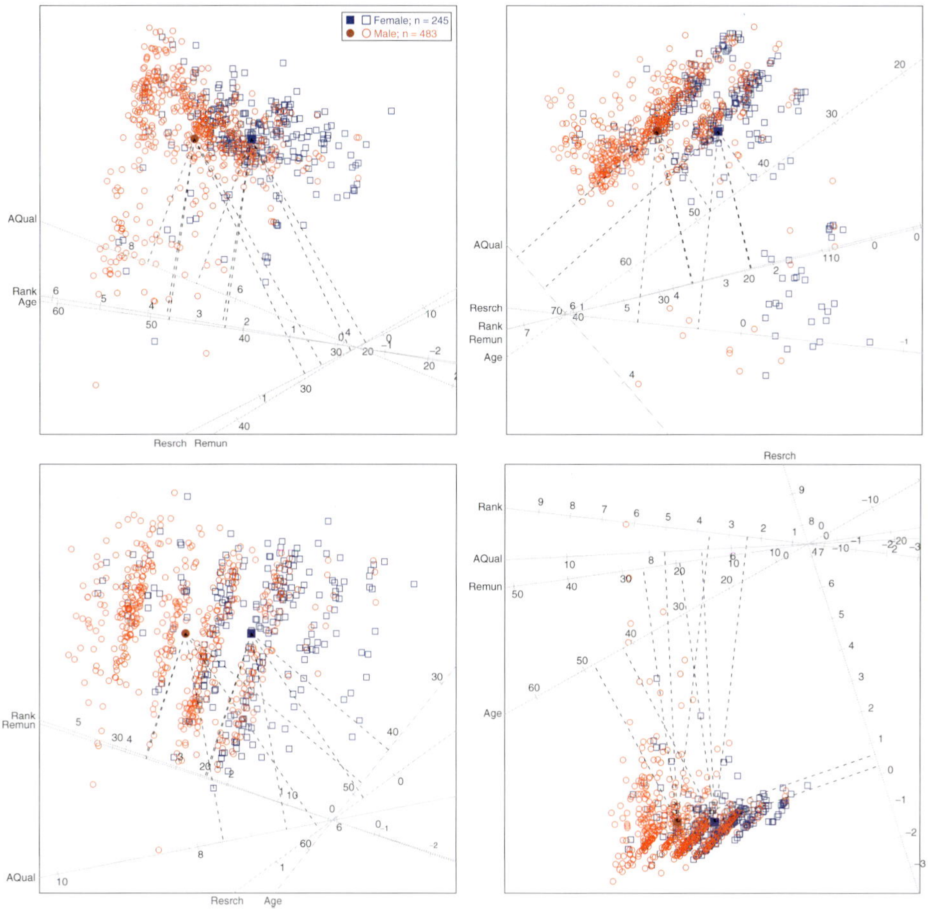

Figure 4.14 Two-dimensional CVA biplots for the 2002 remuneration data: (top left) eigenvectors 1 and 2; (top right) eigenvectors 1 and 3; (bottom left) eigenvectors 1 and 4; (bottom right) eigenvectors 1 and 5.

Table 4.13 Remuneration data: canonical means (weighted and unweighted CVA).

	1	2	3	4	5
Female	0.0281	0	0	0	0
Male	−0.0142	0	0	0	0

parallel to this line (argument `constant = -0.02` in the call below). The convention of labelling all biplot axes at the high end of the scale is maintained. In the bottom panel of Figure 4.15 a large-scale version of the biplot is given to ease the process of finding

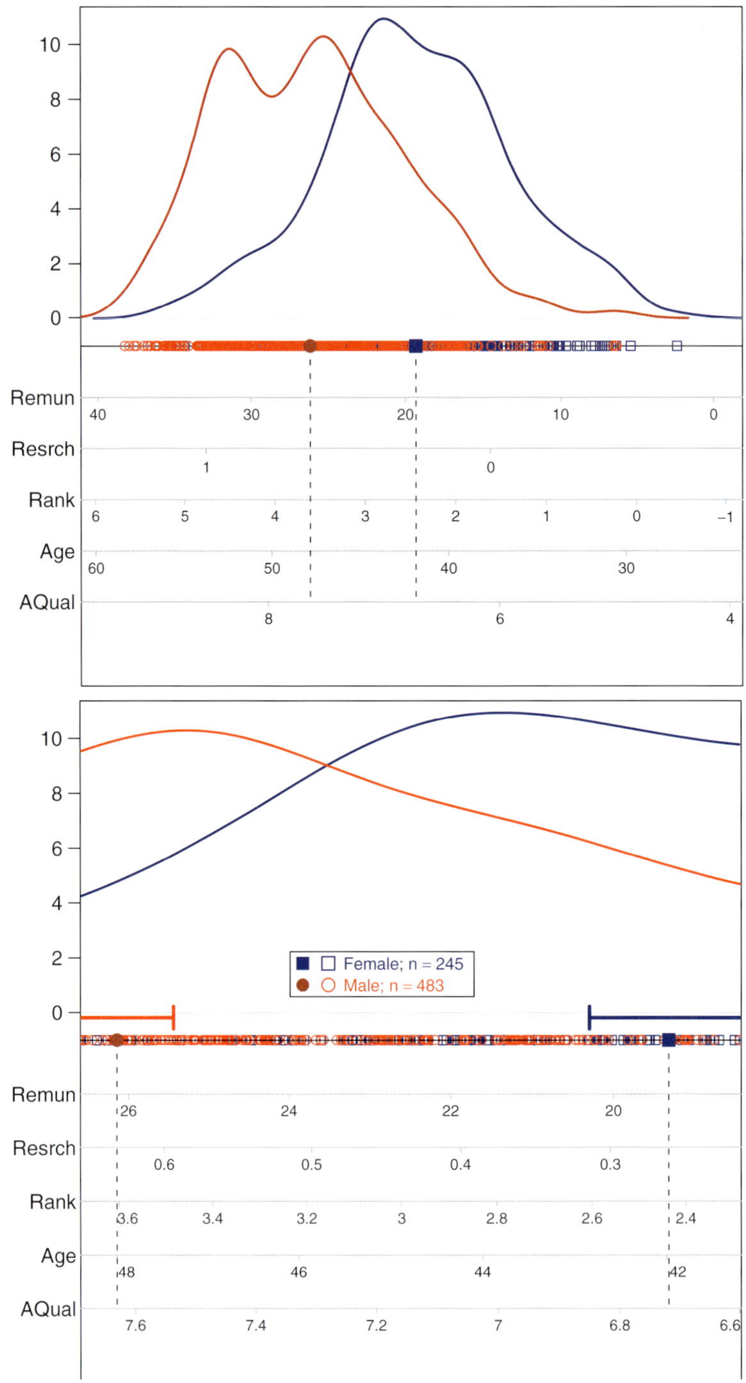

Figure 4.15 (Top) One-dimensional CVA biplot with density estimates for the 2002 remuneration data. (Bottom) Enlargement of the area occupied by the group means together with 99% confidence intervals about the group means.

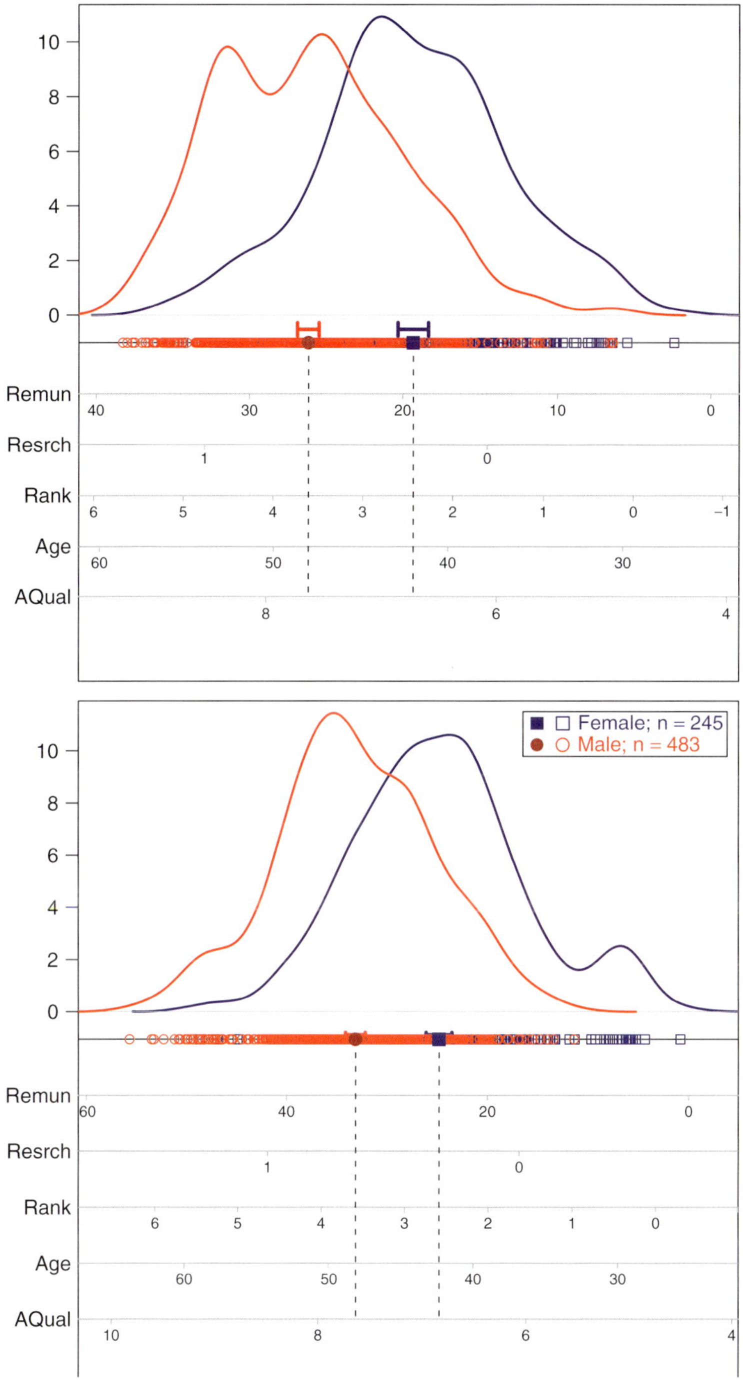

Figure 4.16 One-dimensional CVA biplot with density estimates and 99% confidence intervals for the remuneration data for: (top) 2002; (bottom) 2005.

predicted values for both means on all variables. These figures can be reproduced by the following function calls:

```
CVAbipl(Remuneration.data[Remuneration.data[,10] == 2002,
   c(2,3,4,5,8)], G = indmat (Remuneration.data
   [Remuneration.data[,10] == 2002,6]),
   dim.biplot = 1, legend.type = c(TRUE,TRUE,FALSE),
   n.int = c(5,10,5,5,5), predictions.mean = 1:2, pos = "Hor",
   ort.lty = 2, constant = -0.02, density.plot = "groups",
   line.width = 2)
```

Adding arguments `large.scale = T, conf.alpha = 0.99` to the above leads to the large-scale biplot in the bottom panel of Figure 4.15.

Although there is overlap between the females and the males, the extent of the remuneration gap between these groups is striking. It is also clear that the lower salaries of the females are related to a lower mean research output, a lower mean age, lower academic qualifications and the more junior positions.

The one-dimensional biplots resulting from an unweighted CVA look similar to those shown here for the weighted case. The within-class axis predictivities (see Table 4.14) are also similar. Note also that the within-class axis predictivities attain the maximum value of unity only in the full five-dimensional space.

In Table 4.15 (given as part of the output of a call to `CVA.predictivities` for selected samples) the within-class sample predictivities associated with the weighted CVA biplots given in Figure 4.15 are shown in comparison to those obtained in the unweighted analyses.

It follows from the various measures of fit discussed above that although one-dimensional CVA biplots provide exact representations of means in a two-groups application, the same is not true of the sample values. This may lead us to consider the question: are the samples equally well represented in the four biplots contained in Figure 4.14? In order to answer this question, we consider the sum of the squared differences between the observed and fitted values of the samples, a quantity we term the *squared reconstruction error*. The squared reconstruction error for the means is zero but for the individual samples it amounts to 52 306.4 for the top left, 49 662.5 for the top right, 28 648.1 for the bottom left and 54 383.8 for the bottom right biplot in Figure 4.14. In Figure 4.15 we have not used the second dimension available on a sheet of paper to give additional information, apart from plotting densities and exhibiting the one-dimensional predictive axes that otherwise would be superimposed. Other possibilities for using the second dimension are under consideration.

Table 4.14 Remuneration data: within-class axis predictivities (weighted and unweighted CVA).

	Remun	Resrch	Rank	Age	Aqual
Dim_1	0.6591	0.0923	0.8276	0.3034	0.4020
Dim_2	0.9344	0.1801	0.8390	0.3039	0.4985
Dim_3	0.9814	0.1892	0.9826	0.3085	0.8824
Dim_4	0.9994	0.3906	0.9965	0.9338	0.9068
Dim_5	1.0000	1.0000	1.0000	1.0000	1.0000

Table 4.15 Remuneration data: within-class sample predictivities for the first four and last four samples (weighted and unweighted CVA).

Weighted CVA								
	1	2	3	4	725	726	727	728
Dim_1	0.1530	0.5511	0.4751	0.4009	0.0494	0.1624	0.0028	0.3043
Dim_2	0.2323	0.5518	0.5650	0.8166	0.0670	0.4510	0.0042	0.3060
Dim_3	0.7094	0.7407	0.7884	0.8338	0.0980	0.6190	0.0055	0.3060
Dim_4	0.9646	0.9938	0.9997	0.9809	0.5536	0.7330	0.7394	0.4108
Dim_5	1.0000	1.0000	1.0000	1.0000	1.0000	1.0000	1.0000	1.0000
Unweighted I CVA								
	1	2	3	4	725	726	727	728
Dim_1	0.1530	0.5511	0.4751	0.4009	0.0494	0.1624	0.0028	0.3043
Dim_2	0.3126	0.6097	0.4794	0.5146	0.0757	0.2981	0.2204	0.3219
Dim_3	0.4500	0.8705	0.8620	0.9909	0.2665	0.5249	0.9377	0.3219
Dim_4	0.9709	0.9760	0.9970	0.9926	0.6204	0.8792	0.9904	0.5939
Dim_5	1.0000	1.0000	1.0000	1.0000	1.0000	1.0000	1.0000	1.0000
Unweighted I $- K^{-1}\mathbf{11}'$ CVA								
	1	2	3	4	725	726	727	728
Dim_1	0.1530	0.5511	0.4751	0.4009	0.0494	0.1624	0.0028	0.3043
Dim_2	0.1895	0.5534	0.5484	0.6952	0.2697	0.6369	0.0096	0.4169
Dim_3	0.2551	0.5580	0.5814	0.6953	0.2934	0.7123	0.2357	0.4315
Dim_4	0.4490	0.8590	0.9009	0.9941	0.3160	0.7649	0.9898	0.4943
Dim_5	1.0000	1.0000	1.0000	1.0000	1.0000	1.0000	1.0000	1.0000

Finally, we compare the 2002 data with the corresponding data in 2005 in the form of the CVA biplots given in Figure 4.16. A comparison of these biplots shows that the gender gap regarding all five variables persisted in 2005. The remuneration gap even shows an increase in absolute terms. It is also of interest that both groups show an increase in research output in 2005 as compared to 2002.

4.10 A five-class CVA biplot example

We introduced the copper froth data in Section 3.8.4, where we constructed PCA biplots. Here we exploit the group structure imposed on the data using the various forms of CVA introduced in Section 4.2. In a call similar to that given below, to `CVA.predictivities`, the output listed in Tables 4.16–4.22 is obtained. Figure 4.17 contains several snapshots of a three-dimensional CVA biplot of the Copper Froth data obtained with function call

```
CVAbipl(CopperFroth.data[,1:8], G =
   indmat(CopperFroth.data[,9]), weightedCVA = "unweightedCent",
   dim.biplot = 3, factor.x = 1.5, factor.y = 1.2,
   factor.3d.axes = 0, specify.classes = 1:5)
```

Table 4.16 Copper froth data: group means from uncentred **X** and centred **X** together with the canonical group means (weighted and unweighted CVA).

Group means calculated from uncentred **X**	X1	X2	X3	X4	X5	X6	X7	X8
Group 1	32.5178	6603.7	325.4417	−5.2152	10.5017	−9.4674	22084	1.0067
Group 2	32.1626	6993.7	322.5736	−5.2626	10.3304	−9.3068	19983	1.0581
Group 3	32.2835	6389.4	273.8918	−5.0399	5.8952	−5.6022	29354	1.1437
Group 4	32.2986	6341.1	276.1890	−5.0332	5.6658	−5.4059	29994	1.0522
Group 5	33.2088	6151.2	418.8002	−5.2480	12.1518	−10.7301	14645	0.9975

Group means calculated from centred **X**	X1	X2	X3	X4	X5	X6	X7	X8
Group 1	0.0447	140.7830	7.1893	−0.0751	2.1159	−1.8040	−2055.3	−0.0381
Group 2	−0.3105	530.7468	4.3212	−0.1225	1.9446	−1.6434	−4156.2	0.0133
Group 3	−0.1896	−73.5395	−44.3606	0.1002	−2.4906	2.0612	5215.1	0.0989
Group 4	−0.1744	−121.8450	−42.0634	0.1069	−2.7201	2.2575	5854.9	0.0074
Group 5	0.7358	−311.7450	100.5477	−0.1079	3.7659	−3.0667	−9493.5	−0.0473

Canonical means; weighted CVA	Dim_1	Dim_2	Dim_3	Dim_4	Dim_5
Group 1	0.1018	−0.0984	−0.0825	0.0007	0.0000
Group 2	0.1221	−0.1247	0.0403	0.0010	0.0000
Group 3	0.1082	0.0817	0.0019	0.0559	0.0000
Group 4	0.1043	0.0692	0.0021	−0.0111	0.0000
Group 5	−0.4478	−0.0035	0.0015	0.0002	0.0000

Canonical means; unweighted **I** CVA	Dim_1	Dim_2	Dim_3	Dim_4	Dim_5
Group 1	0.1065	−0.1027	−0.0701	0.0062	0.0000
Group 2	0.1258	−0.1142	0.0549	0.0141	0.0000
Group 3	0.1062	0.0907	−0.0084	0.0437	0.0000
Group 4	0.1015	0.0710	−0.0034	−0.0213	0.0000
Group 5	−0.4473	−0.0189	−0.0040	0.0110	0.0000

Canonical means; unweighted and centred CVA	Dim_1	Dim_2	Dim_3	Dim_4	Dim_5
Group 1	0.1066	−0.1013	−0.0711	0.0128	0.0000
Group 2	0.1259	−0.1146	0.0541	0.0135	0.0000
Group 3	0.1062	0.0919	−0.0044	0.0419	0.0000
Group 4	0.1014	0.0706	−0.0036	−0.0228	0.0000
Group 5	−0.4473	−0.0188	−0.0037	0.0118	0.0000

Table 4.17 Overall quality attained with CVA biplots of the copper froth data.

	Original variables			Canonical variables		
	Weighted	UnweightedI	UnweightedCent	Weighted	UnweightedI	UnweightedCent
Dim_1	55.1702	50.2295	51.3422	85.8998	83.8419	84.3715
Dim_2	99.4005	97.9858	98.0788	97.5865	96.3789	96.625
Dim_3	99.6814	98.8842	99.0975	99.4754	99.0842	99.2874
Dim_4	100.0000	100.0000	100.0000	100.0000	100.0000	100.0000

Table 4.18 Axis adequacies obtained with CVA biplots (weighted and unweighted) of the copper froth data.

	Weighted CVA							
	X1	X2	X3	X4	X5	X6	X7	X8
Dim_1	0.0120	0.0000	0.5236	0.3198	0.0194	0.0278	0.6527	0.0185
Dim_2	0.0124	0.0008	0.5441	0.4306	0.0800	0.0865	0.6706	0.1218
Dim_3	0.0541	0.3476	0.7010	0.5014	0.3734	0.3960	0.7103	0.1233
Dim_4	0.0653	0.4328	0.7352	0.5379	0.3766	0.3993	0.7162	0.9887
Dim_5	0.1236	0.7550	0.9795	0.9656	0.8678	0.9022	0.8612	0.9889
Dim_6	0.1250	0.7605	0.9799	0.9843	0.9896	0.9890	0.8612	0.9899
Dim_7	0.4895	0.8778	0.9807	1.0000	0.9956	0.9993	0.9605	0.9920
Dim_8	1.0000	1.0000	1.0000	1.0000	1.0000	1.0000	1.0000	1.0000
	Unweighted I CVA							
	X1	X2	X3	X4	X5	X6	X7	X8
Dim_1	0.0133	0.0001	0.5447	0.3421	0.0249	0.0343	0.6618	0.0168
Dim 2	0.0141	0.0001	0.5643	0.4430	0.1104	0.1179	0.6800	0.1923
Dim_3	0.0560	0.3639	0.7073	0.4964	0.3731	0.3959	0.7085	0.1945
Dim_4	0.0653	0.4328	0.7352	0.5379	0.3766	0.3993	0.7162	0.9887
Dim_5	0.0661	0.5086	0.7911	0.5822	0.8091	0.7703	0.7371	0.9890
Dim_6	0.1211	0.7771	0.9869	0.9329	0.9763	0.9666	0.9180	0.9896
Dim_7	0.3713	0.8911	0.9940	0.9509	0.9804	0.9689	0.9614	0.9996
Dim_8	1.0000	1.0000	1.0000	1.0000	1.0000	1.0000	1.0000	1.0000
	Unweighted Cent CVA							
	X1	X2	X3	X4	X5	X6	X7	X8
Dim_1	0.0134	0.0001	0.5450	0.3425	0.0250	0.0344	0.6620	0.0167
Dim_2	0.0139	0.0001	0.5618	0.4377	0.1049	0.1124	0.6798	0.2105
Dim_3	0.0533	0.3439	0.6982	0.4875	0.3673	0.3899	0.7106	0.2107
Dim_4	0.0653	0.4328	0.7352	0.5379	0.3766	0.3993	0.7162	0.9887
Dim_5	0.0697	0.4598	0.7766	0.5652	0.7615	0.7288	0.7248	0.9890
Dim_6	0.1225	0.7068	0.9187	0.9768	0.9779	0.9736	0.7707	0.9898
Dim_7	0.5635	0.7283	0.9876	0.9994	0.9902	0.9951	0.8789	0.9991
Dim_8	1.0000	1.0000	1.0000	1.0000	1.0000	1.0000	1.0000	1.0000

Table 4.19 Axis predictivities obtained with CVA biplots (weighted and unweighted) of the copper froth data.

	Weighted CVA							
	X1	X2	X3	X4	X5	X6	X7	X8
Dim_1	0.9478	0.3121	0.8361	0.2283	0.4331	0.4193	0.5522	0.3884
Dim_2	0.9478	0.9158	0.9992	0.9943	0.9929	0.9923	0.9942	0.4899
Dim_3	1.0000	0.9948	0.9999	0.9972	0.9971	0.9970	0.9968	0.6286
Dim_4	1.0000	1.0000	1.0000	1.0000	1.0000	1.0000	1.0000	1.0000
	Unweighted I CVA							
	X1	X2	X3	X4	X5	X6	X7	X8
Dim_1	0.9163	0.3877	0.8003	0.1607	0.3434	0.3284	0.5025	0.2645
Dim_2	0.9224	0.8455	0.9990	0.9773	0.9878	0.9871	0.9802	0.5856
Dim_3	0.9998	0.9837	0.9993	0.9889	0.9902	0.9899	0.9889	0.6538
Dim_4	1.0000	1.0000	1.0000	1.0000	1.0000	1.0000	1.0000	1.0000
	Unweighted Cent CVA							
	X1	X2	X3	X4	X5	X6	X7	X8
Dim_1	0.9185	0.3939	0.8067	0.1655	0.3553	0.3400	0.5137	0.2699
Dim_2	0.9232	0.8461	0.9990	0.9780	0.9890	0.9884	0.9811	0.6353
Dim_3	1.0000	0.9889	0.9993	0.9907	0.9910	0.9907	0.9910	0.7405
Dim_4	1.0000	1.0000	1.0000	1.0000	1.0000	1.0000	1.0000	1.0000

Note that the left-hand section of Table 4.17 combines estimates for the original variables and therefore is subject to commensurability problems, so it is not considered further. The right-hand section of Table 4.17 shows that a one-dimensional CVA fits quite well and two dimensions very well. Nevertheless, the two-dimensional CVA shown in Figure 4.18 shows that there is considerable overlap for two pairs of groups. The four different views of the three-dimensional CVA shown in Figure 4.17 reveal that although one group is well separated, the remaining four groups are not. The CVA biplot shown in Figure 4.18, its zoomed version in Figure 4.20 and the density plot of Figure 4.19, with 95% bags surrounding each group emphasize the problem, as do the numerical values of the canonical means given in Table 4.16. The different methods for centring make little difference so we shall focus our remarks on the weighted CVA analysis. Table 4.18 on axis adequacies need not detain us (see Sections 3.3 and 4.6). Table 4.19 on axis predictivities, the most important of these efficiency measures, shows that it requires two dimensions for good predictivity of the variables from the canonical means. Exceptionally, ***X8*** is poor even when we move to three dimensions; this draws attention to a feature that needs further examination even though it shows no abnormality on the biplot maps. As expected, the within-group predictivities of Table 4.20 are poor, except for ***X4***. Figure 4.21 shows variants of uncertainty regions. Figure 4.22 shows classification regions. In the top panel these represent the nearest canonical means in the full four-dimensional space; in the bottom panel they show regions nearest to the two-dimensional canonical means. Theoretically, the top panel would represent the optimal discriminatory rule for assignment to a class, but the bottom panel seems to reflect better the properties of the actual samples in the data. Now there is at least some separation of the samples between the overlapping groups.

Table 4.20 Within-class axis predictivities (weighted and unweighted CVA) obtained with CVA biplots of the copper froth data.

	Weighted CVA							
	X1	X2	X3	X4	X5	X6	X7	X8
Dim_1	0.1005	0.0240	0.1891	0.0317	0.0319	0.0332	0.0627	0.0055
Dim_2	0.1005	0.3654	0.4602	0.8144	0.3350	0.3669	0.4312	0.0161
Dim_3	0.3517	0.6419	0.4675	0.8328	0.3492	0.3838	0.4449	0.1054
Dim_4	0.3525	0.7071	0.4706	0.8965	0.3842	0.4228	0.5040	0.9667
Dim_5	0.3546	0.7462	0.4707	0.9025	0.3904	0.4325	0.5052	0.9967
Dim_6	0.5458	0.7966	0.4936	0.9242	0.9940	0.9973	0.5963	0.9973
Dim_7	0.9705	0.8640	0.8053	0.9650	0.9984	0.9995	0.8217	0.9987
Dim_8	1.0000	1.0000	1.0000	1.0000	1.0000	1.0000	1.0000	1.0000
	Unweighted **I** CVA							
	X1	X2	X3	X4	X5	X6	X7	X8
Dim_1	0.0957	0.0297	0.1709	0.0206	0.0235	0.0242	0.0508	0.0071
Dim_2	0.1000	0.2644	0.4546	0.7203	0.3177	0.3483	0.3737	0.0644
Dim_3	0.3508	0.5929	0.4560	0.7661	0.3228	0.3547	0.4008	0.1208
Dim_4	0.3525	0.7071	0.4706	0.8965	0.3842	0.4228	0.5040	0.9667
Dim_5	0.4937	0.7075	0.5159	0.9369	0.7796	0.7757	0.5395	0.9764
Dim_6	0.5599	0.7293	0.5911	0.9428	0.9722	0.9667	0.5442	0.9992
Dim_7	0.5778	0.9617	0.8293	0.9776	1.0000	0.9998	0.8713	1.0000
Dim_8	1.0000	1.0000	1.0000	1.0000	1.0000	1.0000	1.0000	1.0000
	Unweighted and centred CVA							
	X1	X2	X3	X4	X5	X6	X7	X8
Dim_1	0.0957	0.0298	0.1706	0.0204	0.0233	0.0240	0.0506	0.0071
Dim_2	0.0991	0.2654	0.4507	0.7114	0.3100	0.3398	0.3679	0.0731
Dim_3	0.3524	0.6080	0.4522	0.7612	0.3142	0.3450	0.3987	0.1605
Dim_4	0.3525	0.7071	0.4706	0.8965	0.3842	0.4228	0.5040	0.9667
Dim_5	0.5780	0.7090	0.5337	0.9408	0.8390	0.8312	0.5350	0.9737
Dim_6	0.5794	0.8646	0.5508	0.9472	0.9931	0.9955	0.6200	0.9883
Dim_7	0.5890	0.9819	0.7981	0.9809	0.9957	0.9970	0.8534	0.9953
Dim_8	1.0000	1.0000	1.0000	1.0000	1.0000	1.0000	1.0000	1.0000

4.11 Overlap in two-dimensional biplots

4.11.1 Describing the structure of overlap

In some instances it is not possible to completely separate the classes, as we saw in the five-group example above. All is not lost, however. With α-bags we can still deduce valuable conclusions even without separating the classes completely. In the data given as `archaeology.data` in the R library we have measurements on Stone Age artefacts for stone tools known as points. The complete data set of points and other tools known as blades is discussed in detail in Wurz *et al.* (2003). The data set contains three classes labelled ***MSAI***, ***MSAIIL*** and ***MSAIIU*** indicating different time periods in the Middle Stone Age. The oldest artefacts come from stage I,

Table 4.21 Class predictivities (weighted and unweighted CVA) obtained with CVA biplots of the copper froth data.

	Weighted CVA				
	Class 1	Class 2	Class 3	Class 4	Class 5
Dim_1	0.3859	0.4647	0.5444	0.6886	0.9999
Dim_2	0.7463	0.9493	0.8547	0.9919	1.0000
Dim_3	1.0000	1.0000	0.8549	0.9922	1.0000
Dim_4	1.0000	1.0000	1.0000	1.0000	1.0000
	Unweighted **I** CVA				
	Class 1	Class 2	Class 3	Class 4	Class 5
Dim_1	0.4228	0.4934	0.5249	0.6518	0.9975
Dim_2	0.8156	0.8997	0.9080	0.9707	0.9993
Dim_3	0.9986	0.9938	0.9113	0.9714	0.9994
Dim_4	1.0000	1.0000	1.0000	1.0000	1.0000
	Unweighted and centred CVA				
	Class 1	Class 2	Class 3	Class 4	Class 5
Dim_1	0.4970	0.5436	0.4863	0.5574	0.9999
Dim_2	0.8183	0.8800	0.9611	0.9381	1.0000
Dim_3	0.9999	0.9999	0.9612	0.9383	1.0000
Dim_4	1.0000	1.0000	1.0000	1.0000	1.0000

followed by stage II lower and the most recent from stage II upper. The variables are as follows:

platfang	platform angle;
platfwth	platform width;
platfthk	platform thickness;
pointlng	point length;
pointwdt	point width;
pointthk	point thickness;
Site	MSAI/MSAIIL/MSAIIU;
ratio	ratio of point length to platform thickness.

The biplots in Figure 4.23 show that there is a large amount of overlap even for 0.40-bags. This shows that structurally the data from the three groups cannot be separated. Even though we cannot separate the three time periods, these biplots provide the archaeologists with valuable information which can be related to practical considerations. We see that there is a large part of ***MSAI*** towards the right of each biplot which is completely separated from the ***MSAII*** classes. However, part of ***MSAI*** overlaps with the ***MSAII*** classes. The ***MSAII*** classes show more overlap, with ***MSAIIU*** tending towards the bottom. This can be interpreted as time progression. Tools evolve over time, with no clear cut-off between the periods. Originally the older ***MSAI*** tools had larger ratio values which reduced over time, that is, the ***pointlng*** reduced relative to ***platfthk***. The overlap indicates that the older ***MSAI*** methods were not yet extinct by the

Table 4.22 Within-class sample predictivities obtained for the first four and last four samples in CVA biplots (weighted and unweighted) of the copper froth data.

Weighted CVA	S1	S2	S3	S4	S494	S495	S496	S497
Dim_1	0.1043	0.0054	0.0822	0.1243	0.0028	0.0029	0.0086	0.2460
Dim_2	0.1928	0.2369	0.3715	0.2287	0.3854	0.0093	0.0198	0.2891
Dim_3	0.3814	0.5820	0.4569	0.3192	0.6559	0.0305	0.0571	0.3564
Dim_4	0.3848	0.6474	0.4962	0.3718	0.7384	0.0309	0.0710	0.3564
Dim_5	0.3851	0.7232	0.4962	0.3788	0.7543	0.3444	0.1034	0.3636
Dim_6	0.5295	0.7345	0.5900	0.3976	0.8901	0.8934	0.9570	0.4638
Dim_7	0.5934	0.9034	0.6785	0.7548	0.9956	0.9471	0.9729	0.9071
Dim_8	1.0000	1.0000	1.0000	1.0000	1.0000	1.0000	1.0000	1.0000

Unweighted **I** CVA	S1	S2	S3	S4	S494	S495	S496	S497
Dim_1	0.1057	0.0110	0.0883	0.1257	0.0080	0.0030	0.0104	0.2423
Dim_2	0.1634	0.2711	0.3483	0.2145	0.4116	0.0064	0.0224	0.2798
Dim_3	0.3846	0.5673	0.4793	0.3370	0.6274	0.0298	0.0568	0.3545
Dim_4	0.3848	0.6474	0.4962	0.3718	0.7384	0.0309	0.0710	0.3564
Dim_5	0.4340	0.6795	0.5225	0.3783	0.8935	0.9391	0.8913	0.4285
Dim_6	0.4415	0.7055	0.5313	0.3786	0.9123	0.9416	0.9945	0.4476
Dim_7	0.8523	0.7454	0.8317	0.5096	0.9144	0.9580	0.9968	0.4499
Dim_8	1.0000	1.0000	1.0000	1.0000	1.0000	1.0000	1.0000	1.0000

Unweighted Cent CVA	S1	S2	S3	S4	S494	S495	S496	S497
Dim_1	0.1058	0.0111	0.0884	0.1257	0.0081	0.0030	0.0104	0.2423
Dim_2	0.1671	0.2557	0.3571	0.2203	0.3924	0.0066	0.0212	0.2811
Dim_3	0.3835	0.5404	0.4769	0.3318	0.5974	0.0302	0.0534	0.3554
Dim_4	0.3848	0.6474	0.4962	0.3718	0.7384	0.0309	0.0710	0.3564
Dim_5	0.4802	0.6994	0.5576	0.4040	0.9242	0.9337	0.9449	0.3947
Dim_6	0.5253	0.7491	0.5879	0.4073	0.9250	0.9892	0.9864	0.4918
Dim_7	0.8728	0.8482	0.8469	0.5822	0.9295	0.9980	0.9985	0.5391
Dim_8	1.0000	1.0000	1.0000	1.0000	1.0000	1.0000	1.0000	1.0000

time some tools were made according to the ***MSAII*** methods. As the ***MSAIIU*** period followed after the ***MSAIIL*** period, the shape of ***MSAIIU*** towards the bottom shows the most modern trend. Alpha-bags are therefore not only useful for describing clouds of points, but can give important information on the structure of overlap in the data.

4.11.2 Quantifying overlap

Although the CVA biplot methodology is useful when classifying samples of unknown origin to a particular class, perhaps its main strength is its potency for describing the degree as well as the nature of the overlap among the various classes. We illustrate this

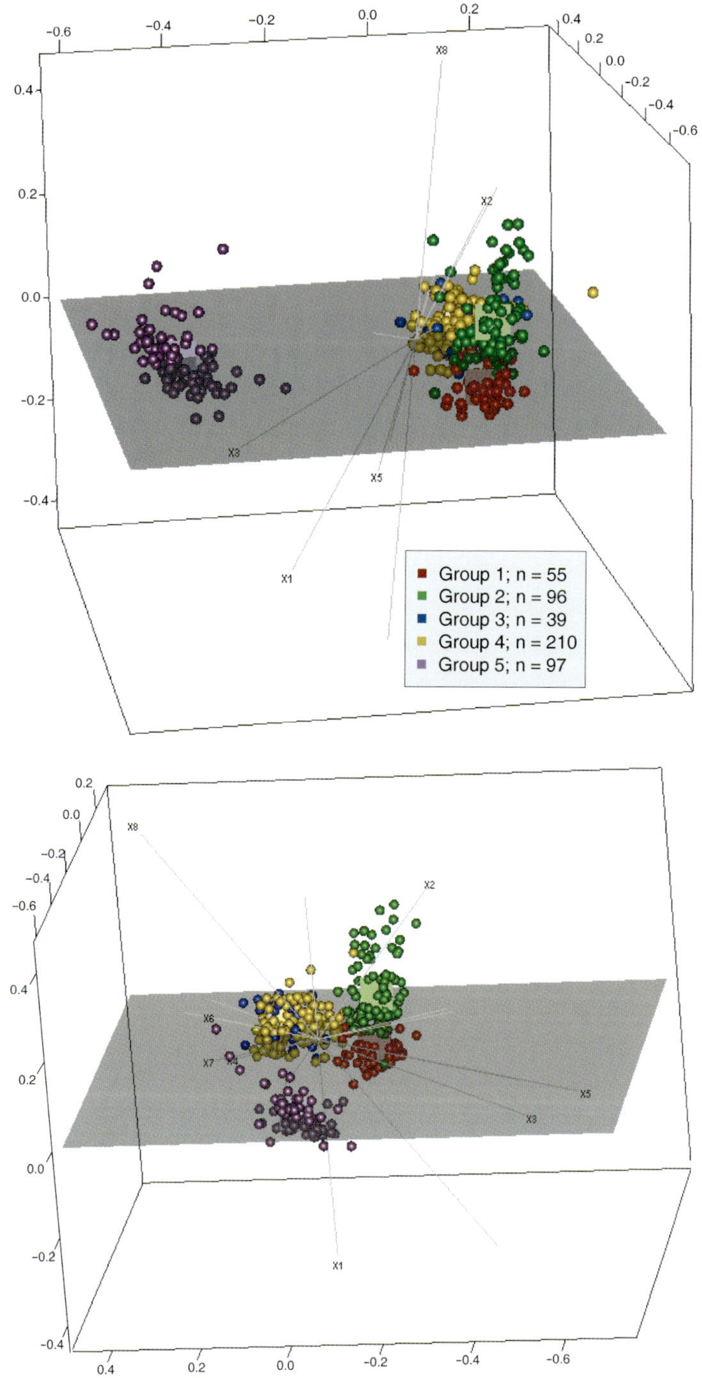

Figure 4.17 Four different views of a three-dimensional CVA biplot of the copper froth data. The best-fitting plane is also shown.

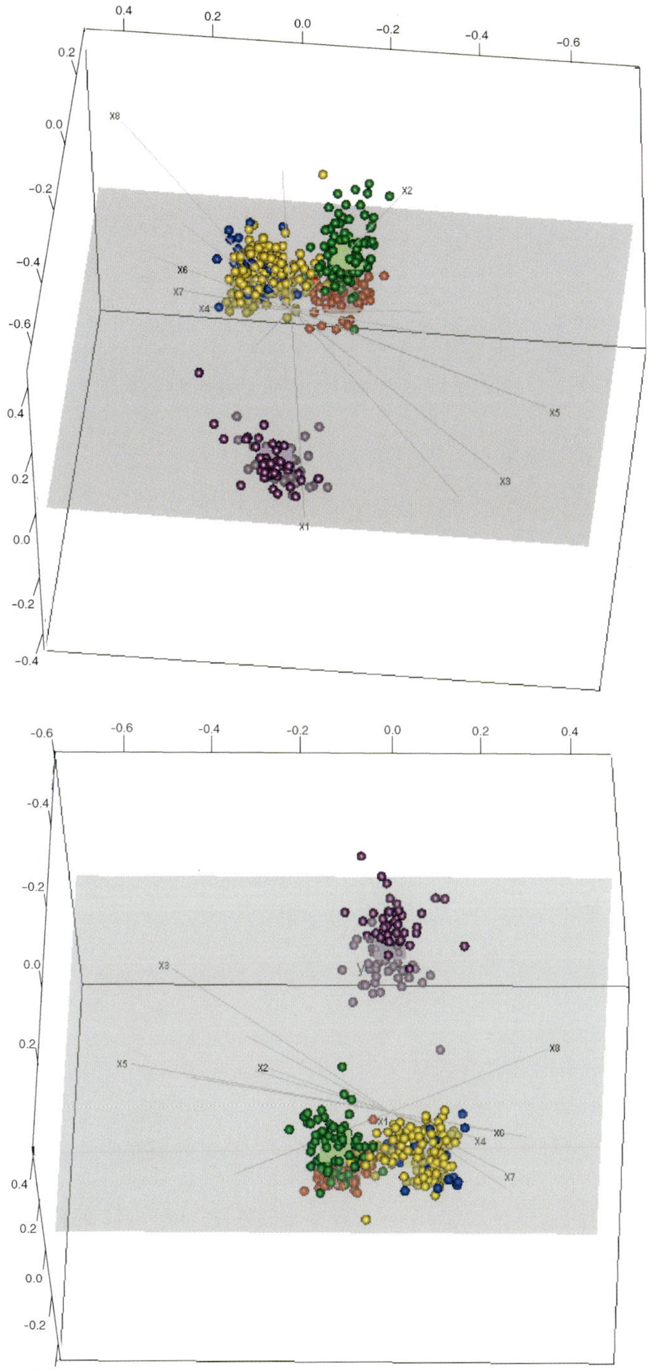

Figure 4.17 (*Continued*)

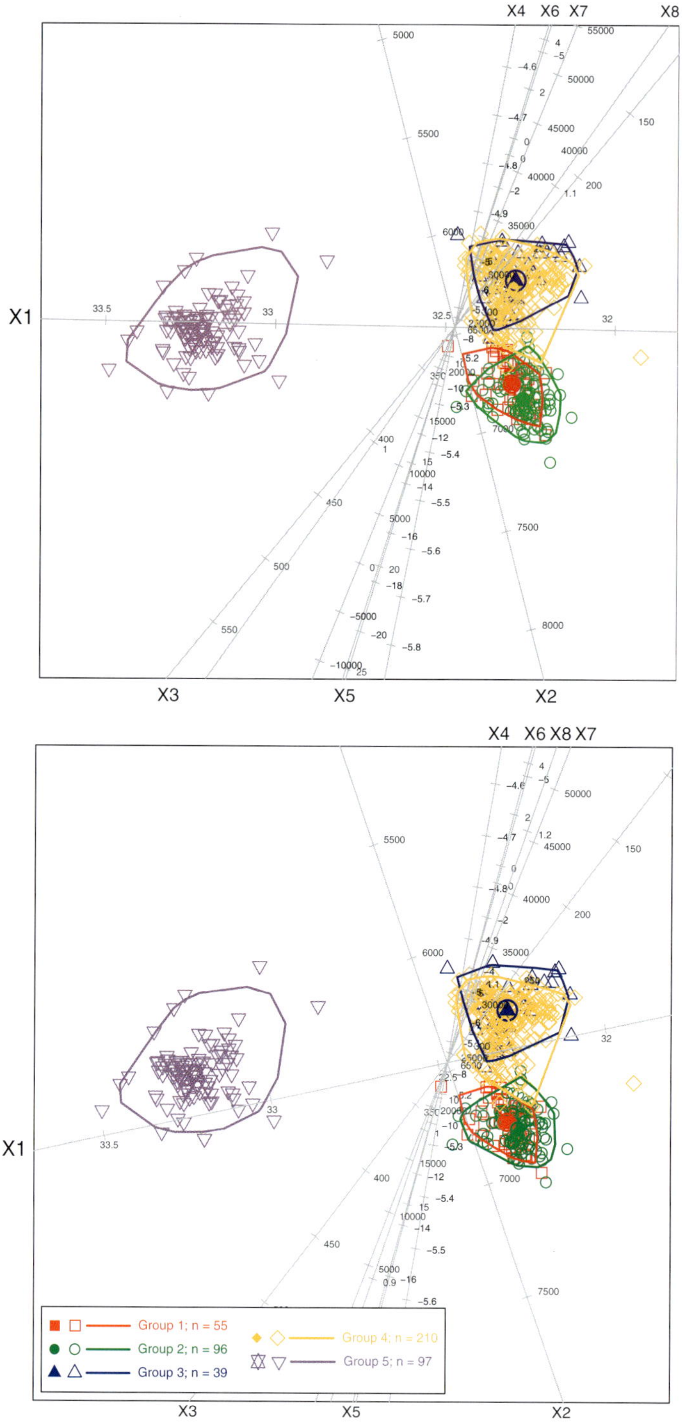

Figure 4.18 (Top) Weighted and (bottom) unweighted CVA biplot of copper froth data with 0.95-bags.

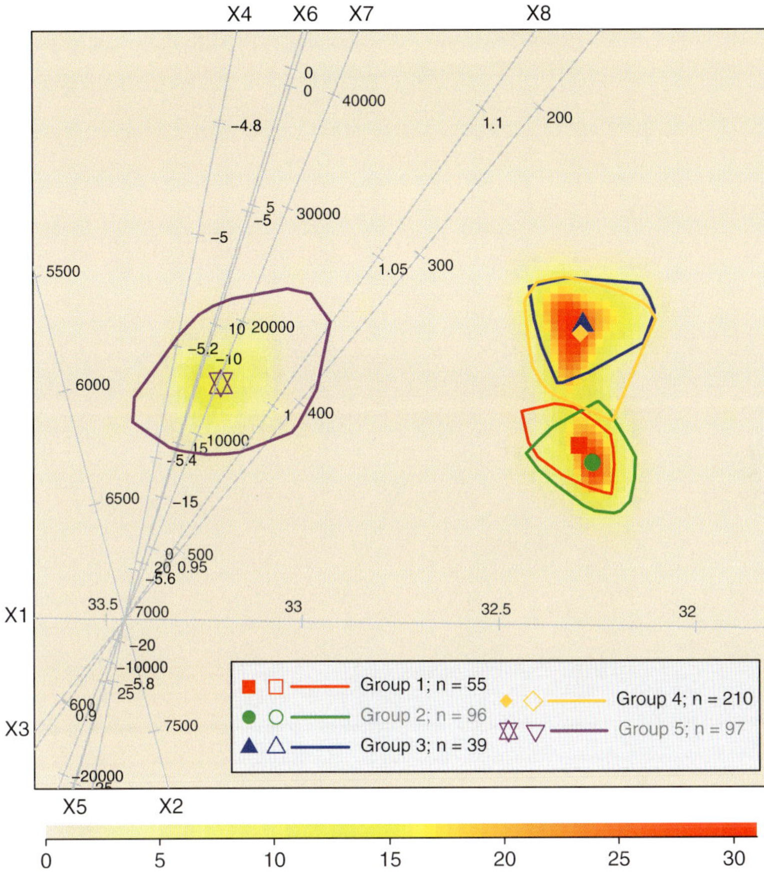

Figure 4.19 Two-dimensional CVA biplot of the copper froth data together with a density surface constructed from all the sample points regardless of their group membership. Class means and 0.95-bags are also shown. The origin has been interactively translated to a more convenient position.

ability with the education data set described by Van der Berg *et al.* (2002). This data set is available as the R dataframe `Education.data` and consists of measurements made in 1993 and again in 2006 for a sample of teenagers in the age group 13–18 years with respect to the following variables:

TScore total score on a Literacy Assessment Module where the level of the tests was set approximately to Grade 7 or age 12, providing a measure of educational attainment to complement the number of years of formal education completed (unfortunately this variable was not available in the 2006 survey);

EducY number of years of education completed by the respondent;

Age age of the respondent;

PcExpD decile of the monthly expenditure per household member;

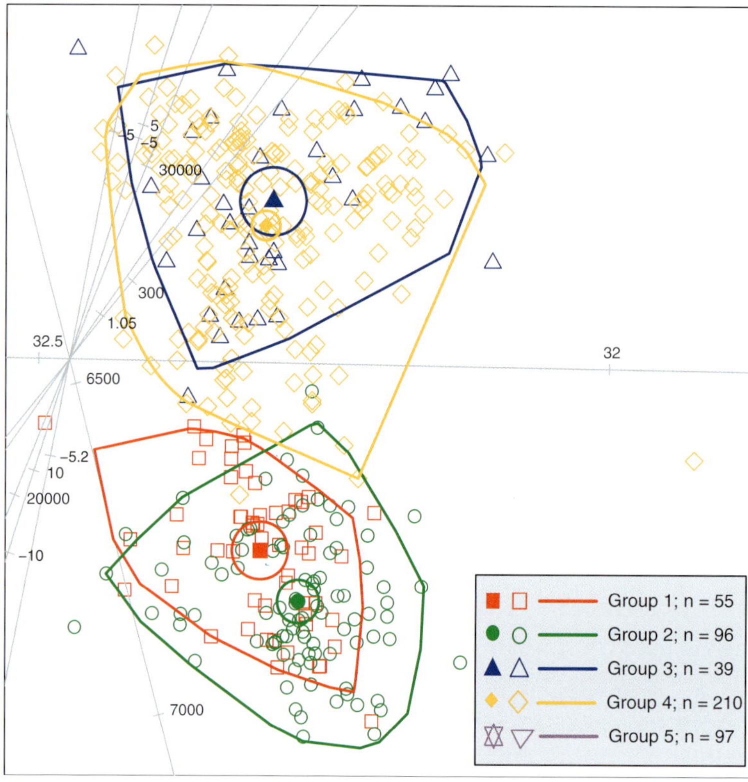

Figure 4.20 Zoomed version of the weighted CVA biplot in top panel of Figure 4.18 revealing clearly the 95% confidence circles about group centroids.

ParEdY number of years of education completed by the mother for the 1993 data or parent for the 2006 data;

Group race group – one of Black, Coloured, Indian or White;

Year 1993 or 2006 (initial data set consists only of the 1993 data).

The original data for 1993 formed part of a larger data set describing socio-economic differences among various race groups in South Africa prior to the democratic elections held in 1994. The original data set was supplemented by a similar set of data compiled in 2006. We first consider the following instruments for describing the nature and degree of overlap among the various groups in 1993 and then also for changes that took place between 1993 and 2006.

The CVA biplot of the 1993 education data in the top left panel of Figure 4.24 contains the means of the four groups with all the samples added. Since the differences between weighted and unweighted CVA biplots are negligible, we show only biplots for the weighted case. The weighted biplot has the following measures of overall quality: in terms of the canonical variables, 0.9970; and in terms of the original variables, 0.9992. Axis predictivities are shown in Tables 4.23 and 4.24, with the class predictivities given in Table 4.25.

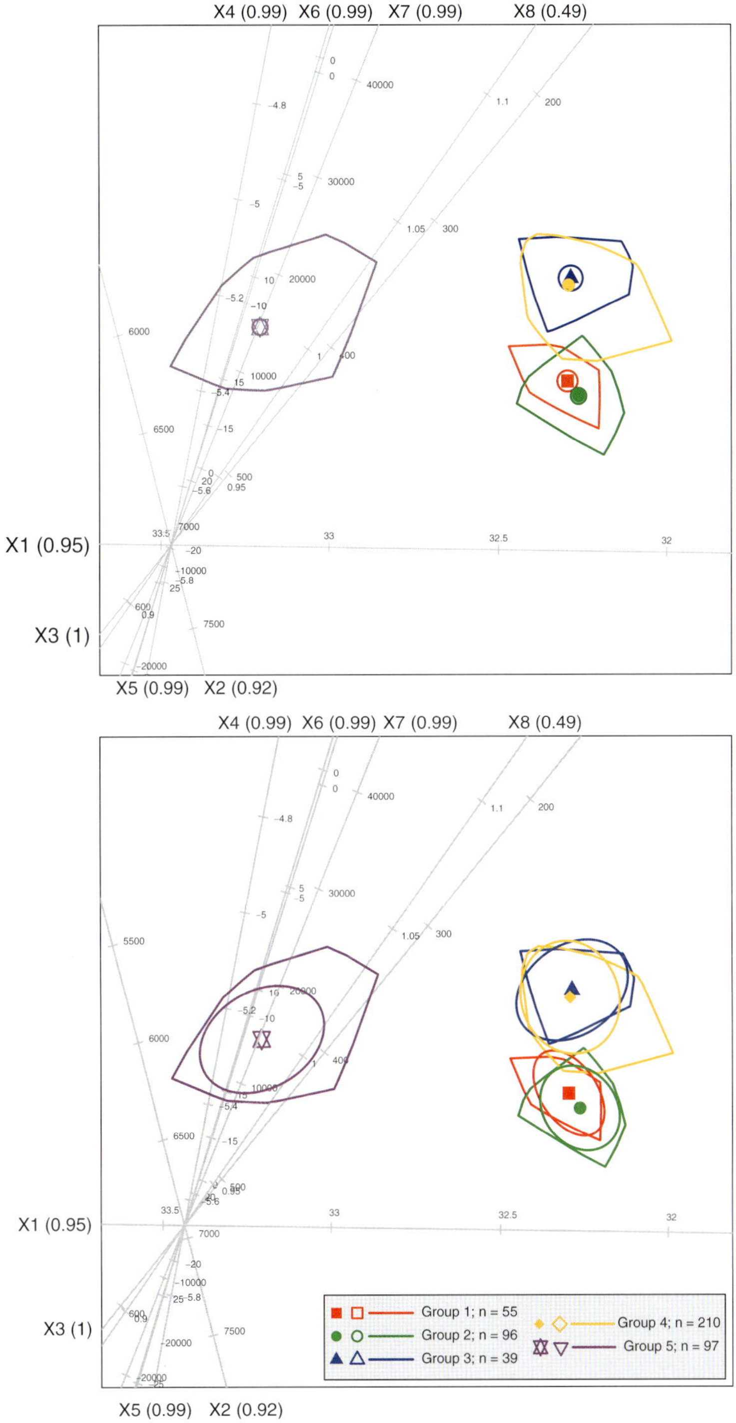

Figure 4.21 CVA biplot of copper froth data: 0.95-bags together with: (top) 95% confidence circles around the means; (bottom) concentration ellipses around each mean.

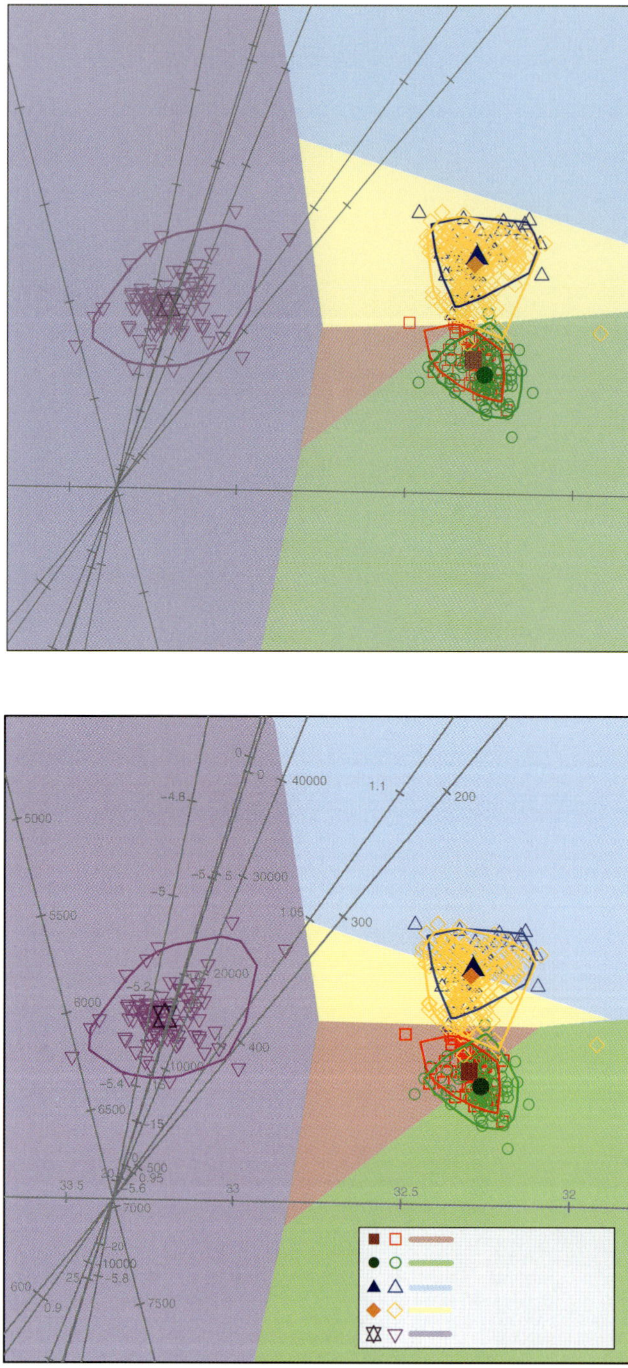

Figure 4.22 CVA biplot of copper froth data equipped with classification regions. (Top) Distance between point in biplot space and mean in full space. (Bottom) Distance between point in biplot space and mean also in biplot space.

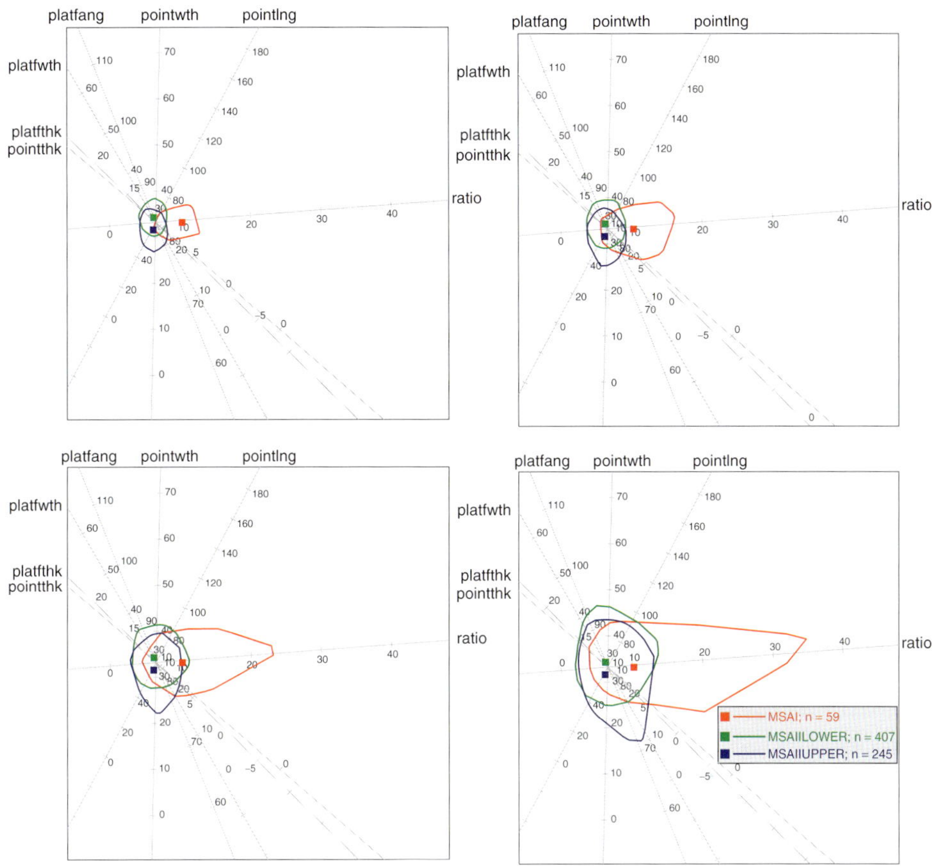

Figure 4.23 CVA biplot of the archaeological data set, fitted with: (top left) 0.40-bags; (top right) 0.60-bags; (bottom left) 0.80-bags; (bottom right) 0.95-bags.

It is clear from the biplots in the top panel of Figure 4.24 that there are large differences between the four groups with respect to all the variables with the exception of age. To gain insight into these differences, we equipped the biplot with α-bags for various values of alpha. The reader can verify that by choosing $\alpha = 0.50$, for example, bags are obtained such that there is no overlap between the black and the white groups, while the bag for the coloured group is situated between these two groups, overlapping with both. In order to quantify the separation among the different groups, we look for the smallest value of α such that all the groups overlap. The biplot in the top right panel of Figure 4.24 shows that this happens when $\alpha = 0.65$, thus enclosing the inner (with respect to the Tukey median) 65% of the multidimensional samples from each group. For this value of α the bags for the black and white groups barely touch each other. Instead of α-bags, κ-ellipses about each group mean are constructed in the biplot in the bottom left panel of Figure 4.24 where, for the purpose of comparison, we have chosen the value of κ such that the ellipse encloses 65% of the sample values. Because of the canonical transformation, we expect the kappa ellipses to be circular. This is true for

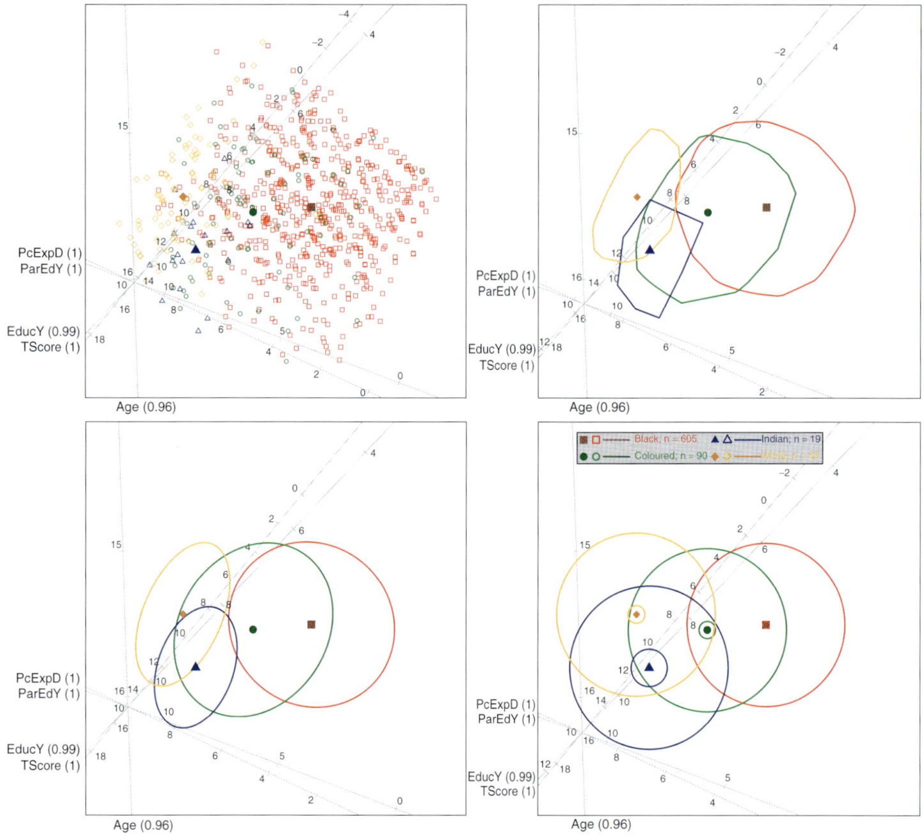

Figure 4.24 CVA biplots of the 1993 education data. (Top left) Means shown together with the individual samples obtained by calling `CVAbipl` with `specify.samples = 1:4, specify.bags = NULL, specify.ellipses = NULL, conf.alpha = NULL`. (Top right) 0.65-bags added to biplot obtained by calling `CVAbipl` with `specify.classes = NULL, alpha = 0.65, specify.bags = 1:4, specify.ellipses = NULL, conf.alpha = NULL`. (Bottom left) 0.65-ellipses added to biplot obtained by calling `CVAbipl` with `specify.classes = NULL, specify.bags = NULL, specify.ellipses = 1:4, ellipse.alpha = 0.65, conf.alpha = NULL`. (Bottom right) 65% confidence ellipses added to biplot obtained by calling `CVAbipl` with `specify.classes = NULL, specify.bags = NULL, specify.ellipses = NULL, conf.alpha = 0.65, conf.type="both"`.

the black group and to a lesser degree for the coloured group but not for the other two groups. This indicates a lack of homogeneity of within-groups covariance matrices. We notice also that the given α-bags and κ-ellipses are almost identical for the coloured and black groups but that they differ in the case of the two smaller groups. In the latter case the nonparametric nature of the α-bag results in a figure that follows the data more closely. Lastly, we add 65% confidence circles to the biplot as shown in the bottom right panel. The outer circles about each mean do not contain the division by the factor $\sqrt{n_i}$ (see Section 4.2) and it is clear that as n_i increases the diameter of the inner circle

Table 4.23 Axis predictivities of the CVA biplot of the education data.

	Tscore	EducY	Age	PcExpD	ParEdY
		Weighted CVA			
Dim_1	0.9633	0.9588	0.0384	0.9943	0.9960
Dim_2	1.0000	0.9871	0.9613	0.9999	0.9997
Dim_3	1.0000	1.0000	1.0000	1.0000	1.0000
		Weighted **I** CVA			
Dim_1	0.9527	0.9645	0.1752	0.9734	0.9839
Dim_2	1.0000	0.9902	0.9918	0.9999	0.9997
Dim_3	1.0000	1.0000	1.0000	1.0000	1.0000
		Weighted $\mathbf{I} - K^{-1}\mathbf{11}'$			
Dim_1	0.8988	0.9223	0.0727	0.9512	0.9716
Dim_2	1.0000	0.9805	0.9922	0.9998	0.9995
Dim_3	1.0000	1.0000	1.0000	1.0000	1.0000

Table 4.24 Within-groups axis predictivities of the CVA biplot of the education data.

	Tscore	EducY	Age	PcExpD	ParEdY
		Weighted CVA			
Dim_1	0.4186	0.0951	0.0001	0.7713	0.3014
Dim_2	0.9952	0.1968	0.0516	0.9279	0.3419
Dim_3	0.9989	0.6066	0.0707	0.9587	0.3749
Dim_4	0.9989	0.7072	0.0782	0.9925	1.0000
Dim_5	1.0000	1.0000	1.0000	1.0000	1.0000
		Weighted **I** CVA			
Dim_1	0.5605	0.1123	0.0008	0.6551	0.2697
Dim_2	0.9987	0.1593	0.0568	0.9360	0.3380
Dim_3	0.9989	0.6066	0.0707	0.9587	0.3749
Dim_4	0.9993	0.9381	0.3189	0.9681	0.7792
Dim_5	1.0000	1.0000	1.0000	1.0000	1.0000
		Weighted $\mathbf{I} - K^{-1}\mathbf{11}'$			
Dim_1	0.5365	0.1007	0.0006	0.6704	0.2797
Dim_2	0.9980	0.1493	0.0572	0.9322	0.3411
Dim_3	0.9989	0.6066	0.0707	0.9587	0.3749
Dim_4	0.9992	0.6070	0.3562	0.9999	0.8594
Dim_5	1.0000	1.0000	1.0000	1.0000	1.0000

Table 4.25 Class predictivities of the CVA biplot of the education data.

	Black	Coloured	Indian	White
		Weighted CVA		
Dim_1	0.9991	0.9534	0.8356	0.9902
Dim_2	0.9995	0.9683	0.9969	0.9997
Dim_3	1.0000	1.0000	1.0000	1.0000
		Weighted I CVA		
Dim_1	0.9836	0.9593	0.9266	0.9440
Dim_2	0.9988	0.9604	0.9999	0.9998
Dim_3	1.0000	1.0000	1.0000	1.0000
		Weighted $\mathbf{I} - K^{-1}\mathbf{11}'$		
Dim_1	0.9983	0.8844	0.7166	0.803
Dim_2	0.9983	0.9042	0.9996	0.999
Dim_3	1.0000	1.0000	1.0000	1.0000

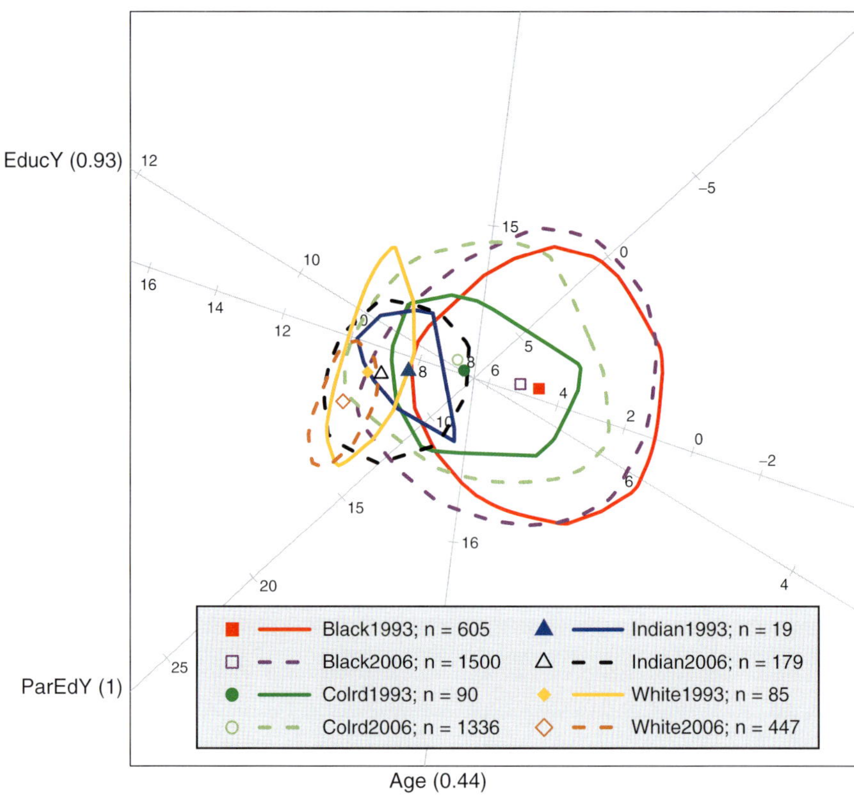

Figure 4.25 Quantifying overlap in the education data by using α-bags added to a CVA biplot.

approaches zero. It is also clear that confidence circles do not provide a good description of the overlap between the various groups.

The top right and bottom left CVA biplots in Figure 4.24 clearly show the degree of the disadvantage of the black group with respect to the Indian and white groups, with the coloured group occupying an in-between position. Thus these biplots provide an important baseline for studying progress in narrowing the gap between the four race groups with respect to socio-economical circumstances and the level of education attained. The education data also contain data collected in a survey in 2006. Unfortunately variable ***Tscore*** was not included in this survey. Figure 4.25 is a CVA biplot of the 1993 and 2006 data combined, while a zoomed version of this biplot is given in Figure 4.26. In producing these two figures a subsample of size 1500 was randomly selected from the 2006 black group. Notice that $\alpha = 0.65$ is the smallest bags where all four race groups overlap in 1993. There is evidence that the race groups are closer in 2006, since smaller α values will still have overlap between the 2006 bags. We invite the reader to experiment with the data by drawing κ-ellipses instead of the 0.65-bags shown here and also to experiment by taking different subsamples.

CVA is one of the most useful multivariate methods; its independence from the effects of measurement scales is a major advantage. The basic methodology has been known in applied mathematics and in statistics for many decades. Yet, new properties continue

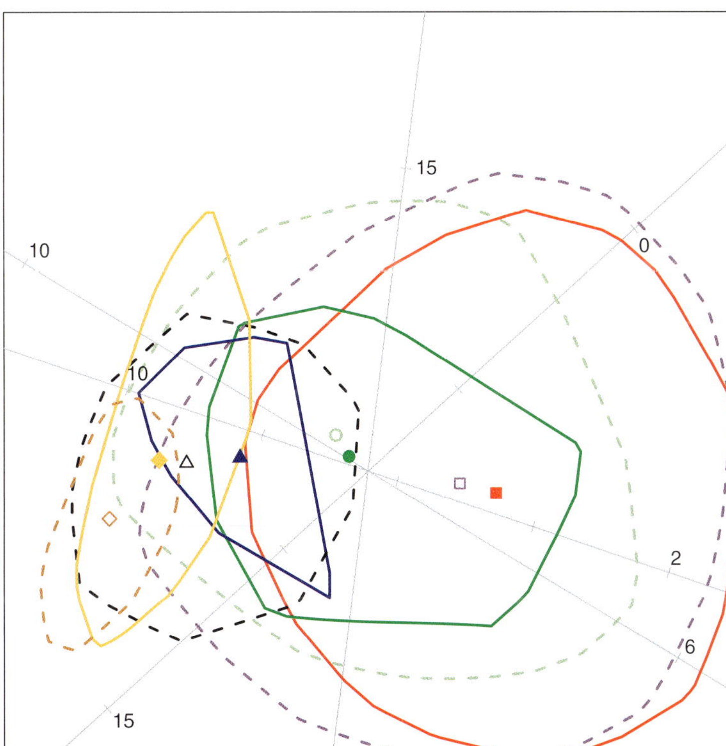

Figure 4.26 Zooming into the biplot of Figure 4.25 for a closer inspection of the overlap between the different groups.

to be discovered. Computing advances that allow density plotting enable large samples to be exhibited. Calibrated axes simplify use and classification regions, supplemented by uncertainty regions, readily allow the investigation of separation and discriminatory properties. We have emphasized, perhaps too much, the various measures of fit that are now available. These are best used in an interactive environment and one would not normally require these measures for all dimensionalities, especially when we know they must attain their optimal 100% values. Variables with low predictivity might be recognized and dropped interactively. However, we think that axis predictivities ought to be shown on most biplot maps to highlight any remaining ineffective variables. We have already shown (see Section 3.3) that usually adequacy is not a useful measure of the effectiveness of a variable. The scale invariance properties of canonical variate space extend to measures of fit in that space. However, one should be aware that fit measures concerning the original variables should be used with care and most safely by first normalizing as in PCA; this normalization has no effect on the CVA analysis itself. A happy exception to this warning is that axis predictivity for the original variables relies on ratios, so is invariant to scale. Finally, we have discussed three ways of centring the points representing the canonical means. These variants have had marginal effects on the examples discussed above, but could have major effects when group sizes are disparate. Only in the disparate case would one consider using more than one method.

5 Multidimensional scaling and nonlinear biplots

5.1 Introduction

So far we have been concerned with linear biplots and the use of Euclidean distances. With PCA the distance is Pythagorean (Chapter 3) and with CVA (Chapter 4) it is Mahalanobis distance, which is Euclidean embeddable. Later (Chapter 7) we shall encounter chi-squared distance, which again is Euclidean embeddable. The analyses accompanying these distances are all concerned with finding sets of points in a few dimensions whose inter-sample distances generate approximations to the data, generally using orthogonal projection in some form. There is a vast literature on a different approach, termed multidimensional scaling (MDS), which attempts to find points in a low-dimensional Euclidean space that generate distances that approximate distances given in an $(n \times n)$ symmetric matrix $\mathbf{D}^*$. Approximation may be measured in several ways, of which the two most important are:

$$stress = \sum_{i<j}^{n} (d_{ij} - \delta_{ij})^2, \tag{5.1}$$

$$sstress = \sum_{i<j}^{n} (d_{ij}^2 - \delta_{ij}^2)^2, \tag{5.2}$$

where d_{ij} are the distances given in $\mathbf{D}^*$ and δ_{ij} are the fitted distances. Here we use the notation $\mathbf{D}^*$ to distinguish this matrix form the matrix $\mathbf{D}$ containing *ddistances* used elsewhere in the text.

Minimizing these criteria is known as metric MDS and needs specialized algorithms and associated software. In nonmetric MDS, more sophisticated versions of these criteria exist in which the δ_{ij} are fitted to monotonic transformations of the observed d_{ij}

Understanding Biplots John Gower, Sugnet Lubbe and Niël le Roux

(see Chapter 10); Cox and Cox (2001) or Borg and Groenen (2005) may be consulted for overviews. In Chapter 3 we treated PCA as a matrix approximation method, but it may be brought into the purview of this chapter as a method for minimizing the criterion (5.1) where now δ_{ij} is constrained to be confined to distances between points arising from projecting $\mathbf{X}$ onto a low-dimensional space.

In MDS the distances in $\mathbf{D}^*$ may be observed directly or may be calculated from sample values given in a data matrix $\mathbf{X}$. In the former case information on which to base a biplot does not exist (see Chapter 10) but it does in the latter case.

In this chapter we shall see how biplots may be found in some cases of metric MDS. Gower *et al.* (1999) give an example of how to use the regression method to construct linear biplots with nonmetric scaling, the nonlinearity being absorbed by nonlinear calibration of the axes. Gower and Hand (1996) outlined, without implementation, a method for constructing so-called coherent nonlinear biplots, in the context of minimizing the metrics *stress* and *sstress*. Here, we shall be concerned with the regression method and nonlinear biplots.

It is important to understand that MDS methods are concerned with fitting δ_{ij} to d_{ij}; that is, δ_{ij} predicts d_{ij}, whereas biplots are concerned with the approximation of $\hat{\mathbf{X}}$ to $\mathbf{X}$ with associated graphical visualizations. This ambiguity of purpose raises some awkward issues: for example, the axis predictivities derived in Chapters 3 and 4 for the approximation of $\hat{\mathbf{X}}$ to $\mathbf{X}$ rest upon certain orthogonality relationships; these are unavailable in the MDS context of approximating distances.

5.2 The regression method

In PCA, we saw in (2.3) that $\mathbf{X} : n \times p = \mathbf{U\Sigma V}'$ is approximated by the rank r matrix

$$\hat{\mathbf{X}}_{[r]} = \mathbf{U\Sigma JV}' = \mathbf{UJ\Sigma V}' = \mathbf{UJ\Sigma JV}' = \mathbf{XVJV}',$$

so that the plotted points are given by $\mathbf{XV}_r = \mathbf{Z}$, say. If we calculate the regression $\mathbf{X} = \mathbf{Z\Gamma}$ of the data $\mathbf{X}$ on $\mathbf{Z}$ we obtain 'regression' coefficients

$$\mathbf{\Gamma} = \{(\mathbf{V}_r)'\mathbf{X}'\mathbf{XV}_r\}^{-1}(\mathbf{V}_r)'\mathbf{X}'\mathbf{X} = (\mathbf{\Lambda}_r)^{-1}\mathbf{\Lambda}_r(\mathbf{V}_r)' = (\mathbf{V}_r)', \tag{5.3}$$

where $\mathbf{X}'\mathbf{X} = \mathbf{V\Sigma}^2\mathbf{V}' = \mathbf{V\Lambda V}'$, with $\mathbf{V}_r$ defined as the matrix consisting of the first r columns of $\mathbf{V}$ (see Section 3.2) and $\mathbf{\Lambda}_r$ defined as the $r \times r$ diagonal matrix formed from the first r diagonal elements of $\mathbf{\Lambda}$.

In Section 3.5 we saw how to add new (linear) biplot axes to a PCA biplot using the regression method. The regression method also forms the basis for deriving the expression for calculating the axis predictivity of the newly added axis. Moreover, in Sections 4.5 and 4.6 we used the regression method for adding new biplot axes together with their axis predictivities in the case of a CVA biplot. The regression method has much wider applicability: the columns of $(\mathbf{V}_r)'$ in (5.3) are precisely the directions of the PCA biplot axes. This suggests that in other types of MDS with fitted coordinates $\mathbf{Z}$, we may use regression to provide linear biplot axes by setting $\mathbf{\Gamma} = (\mathbf{Z}'\mathbf{Z})^{-1}\mathbf{Z}'\mathbf{X}$. Then $\mathbf{x}_k = \mathbf{Z}\boldsymbol{\gamma}_k$, where $\boldsymbol{\gamma}_k$ is the kth column of $\mathbf{\Gamma}$, predicts the kth variable (i.e. the kth column of $\mathbf{X}$.) It follows that the unit marker on the kth axis is given by

$$\boldsymbol{\gamma}_k/\boldsymbol{\gamma}_k'\boldsymbol{\gamma}_k. \tag{5.4}$$

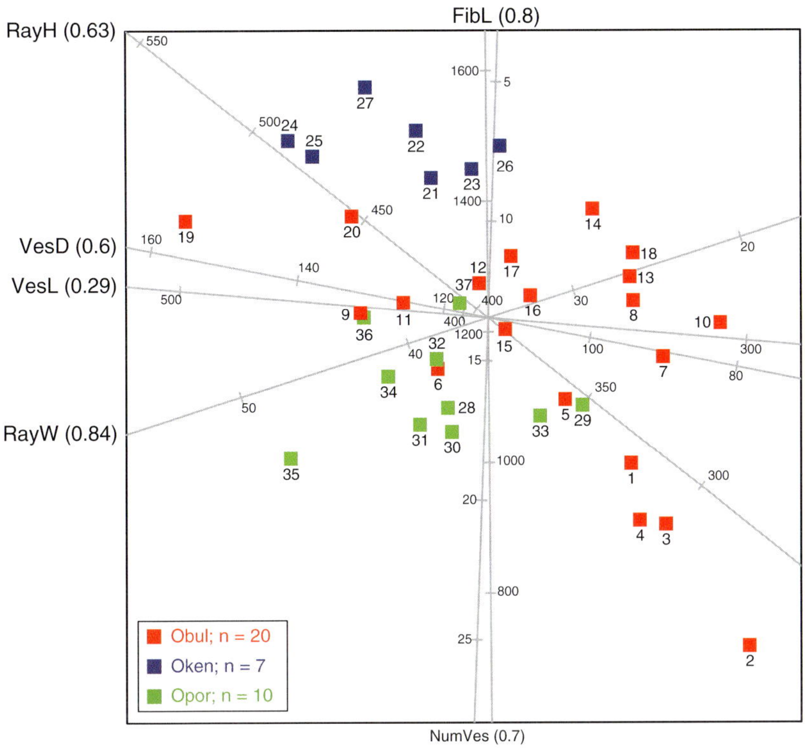

Figure 5.1 PCA biplot of the normalized *Ocotea* data. Axis predictivity is shown in brackets after the axis label.

Although the above has been concerned with predicting the values of **X**, the regression method may be used to add axes for completely new variables to an existing biplot, as we already have illustrated for PCA and CVA biplots in Figures 3.24 and 4.13, respectively.

In nonlinear situations it is an open question how well the regression will work. In the next section we shall see why it could be unsatisfactory.

As an example of an MDS biplot we again consider the *Ocotea* data. Its PCA biplot, given in Figure 4.1, is reproduced in rotated form in Figure 5.1 to aid visual comparison with the MDS biplot given in Figure 5.2.

The MDS biplot given in Figure 5.2 is obtained by calling our function `MDSbipl`. This function accepts a data matrix $\mathbf{X}$: $n \times p$ as input. Various distance metrics are available for calculating the distances between the rows of $\mathbf{X}$ to form an $n \times n$ distance matrix $\mathbf{D}^*$. For Figure 5.2 we used the default Pythagorean distance. The coordinates of the sample points in two dimensions are needed for producing the δ_{ij} such that the stress criterion (5.1) is minimized. The default method of `MDSbipl` is to use an implementation of the SMACOF algorithm as explained in Borg and Groenen (2005). For further details on SMACOF, see De Leeuw (1977) and Heiser and De Leeuw (1977).

In the construction of Figure 5.2 we initialize the SMACOF algorithm with an $n \times 2$ matrix of randomly selected normal (0, 1) values. The function `MDSbipl` allows several random starts of the SMACOF algorithm. The coordinates resulting in the smallest value

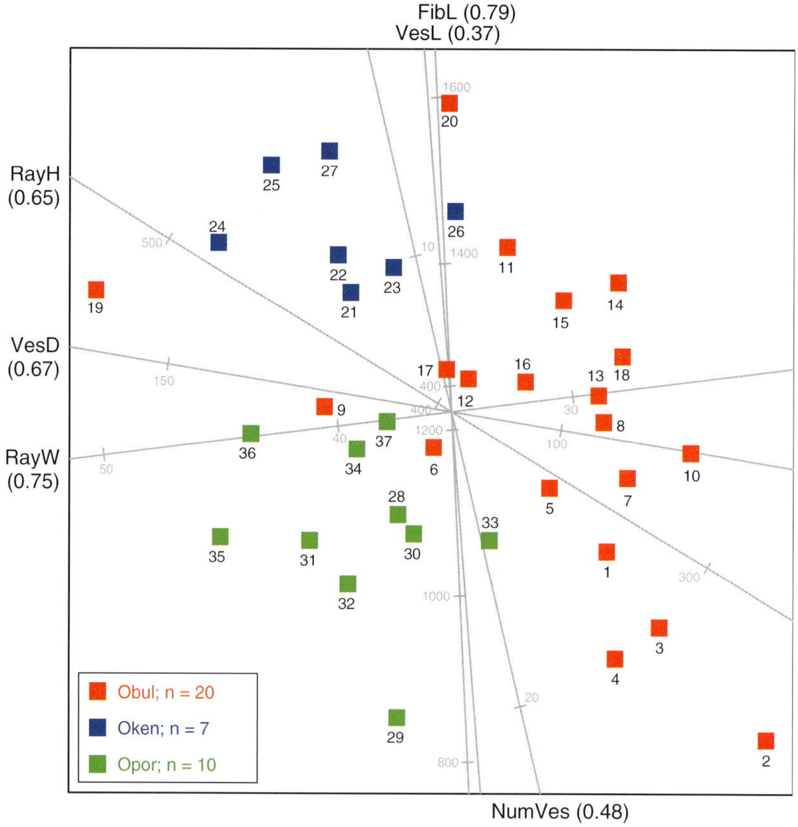

Figure 5.2 MDS biplot of the normalized *Ocotea* data. 'Regression predictivities' have been added to the names of the biplot axes.

of (5.1) are then used for the MDS map of the points. Finally, biplot axes are constructed using (5.3) and (5.4) of the regression method described above. The axes are calibrated according to the procedure described in Chapter 2. The minimum value for the stress (5.1) found in 250 random starts of the SMACOF algorithm was 1586.576. This was achieved after 712 iterations. If we normalize the raw stress value by dividing by $\mathbf{1}'\mathbf{D}\mathbf{1}/2$ we obtain a value of 0.7327. We compute 'regression predictivities' using the formula $\text{diag}(\mathbf{\Gamma}'\mathbf{Z}'\mathbf{Z}\mathbf{\Gamma})/\text{diag}(\mathbf{X}'\mathbf{X})$ and add these values to the names of the axes in Figure 5.2. Note that, apart from the variable with the low axis predictivity (***VesL***), the MDS biplot of Figure 5.2 looks remarkably similar to the PCA biplot in Figure 5.1.

5.3 Nonlinear biplots

There is one case where the effects of nonlinearity can be analysed and represented by nonlinear biplot axes. This is when the distances in $\mathbf{D}$ can be generated as Pythagorean distances by a set of n points in $n-1$ dimensions. When this is so, the distances are said to be *Euclidean embeddable*.

Gower and Legendre (1986) discuss various Euclidean embeddable distance measures. The square root of the Manhattan distance,

$$d_{ij} = \sqrt{\sum_{k=1}^{p} |x_{ik} - x_{jk}|}, \tag{5.5}$$

is one example; another is Clark's distance (Gower and Ngouenet, 2005), defined for nonnegative values x_{ik}, x_{jk} by

$$d_{ij} = \sqrt{\sum_{k=1}^{p} \left(\frac{x_{ik} - x_{jk}}{x_{ik} + x_{jk}} \right)^2}. \tag{5.6}$$

The nonlinear biplot (Gower and Harding, 1988) is a generalization of the PCA biplot, providing for distance measures other than Pythagorean distance. Let $\mathbf{X}$: $n \times p$ be a matrix of n samples with observations on p variables and let d_{ij} indicate the distance between samples $\mathbf{x}_i$ and $\mathbf{x}_j$. The matrix $\mathbf{D} = \{-\frac{1}{2}d_{ij}^2\}$ is used to define the double-centred matrix

$$\mathbf{B} = \left(\mathbf{I} - \tfrac{1}{n}\mathbf{11}'\right) \mathbf{D} \left(\mathbf{I} - \tfrac{1}{n}\mathbf{11}'\right). \tag{5.7}$$

If $\mathbf{B}$ is positive semi-definite it can be expressed as $\mathbf{B} = \mathbf{Y}^*\mathbf{Y}^{*\prime}$. The rows of $\mathbf{Y}^*$ provide coordinates that generate the distances d_{ij} and therefore a sufficient condition for Euclidean embeddability is that $\mathbf{B}$ be positive semi-definite. It is also a necessary condition. A slight generalization is that a necessary and sufficient condition for Euclidean embeddability is that $\mathbf{B} = (\mathbf{I} - \mathbf{1s}')\mathbf{D}(\mathbf{I} - \mathbf{s1}')$ be positive semi-definite for any $\mathbf{s}$ such that $\mathbf{s}'\mathbf{1} = 1$ and $\mathbf{s}'\mathbf{D} \neq \mathbf{0}'$. For a proof, see Gower (1982). The choice of $\mathbf{s}$ centres $\mathbf{Y}^*$ so that $\mathbf{s}'\mathbf{Y}^* = \mathbf{0}'$; in particular, when $\mathbf{s} = n^{-1}\mathbf{1}$, as in (5.7), we have $\mathbf{1}'\mathbf{Y}^* = \mathbf{0}'$ so the origin of $\mathbf{Y}^*$ is at its centroid.

The matrix $\mathbf{B}$ defined by (5.7) is invariant with respect to orthogonal transformations applied to $\mathbf{Y}^*$ expressing the well-known property of invariance of distances to orthogonal rotations. In principal coordinate analysis, $\mathbf{Y}^*$ is given by the eigenvectors satisfying $\mathbf{BY} = \mathbf{Y\Lambda}$ scaled so that $\mathbf{Y}'\mathbf{Y} = \mathbf{\Lambda}$, that is, the SVD $\mathbf{B} = \mathbf{V\Lambda V}'$ provides $\mathbf{Y} = \mathbf{V\Lambda}^{1/2}$. Then, because $\mathbf{Y}'\mathbf{Y} = \mathbf{\Lambda}$ is diagonal, $\mathbf{Y}$ is referred to principal axes through the centroid, as in PCA. As a simple illustration of Euclidean embeddability, let us consider the data matrix

$$\mathbf{X} = \begin{bmatrix} 6 & 4 \\ 4 & 8 \\ 4 & 2 \\ 2 & 2 \end{bmatrix} \begin{matrix} \text{a} \\ \text{b} \\ \text{c} \\ \text{d} \end{matrix} \tag{5.8}$$

graphically represented in Figure 5.3. If, instead of calculating ordinary Pythagorean distances between the four samples, we calculate Clark's distance, we obtain the following distances $\{d_{ij}\}$:

$$\mathbf{D}^* = \begin{bmatrix} 0 & 0.39 & 0.39 & 0.60 \\ 0.39 & 0 & 0.60 & 0.69 \\ 0.39 & 0.60 & 0 & 0.33 \\ 0.60 & 0.69 & 0.33 & 0 \end{bmatrix}.$$

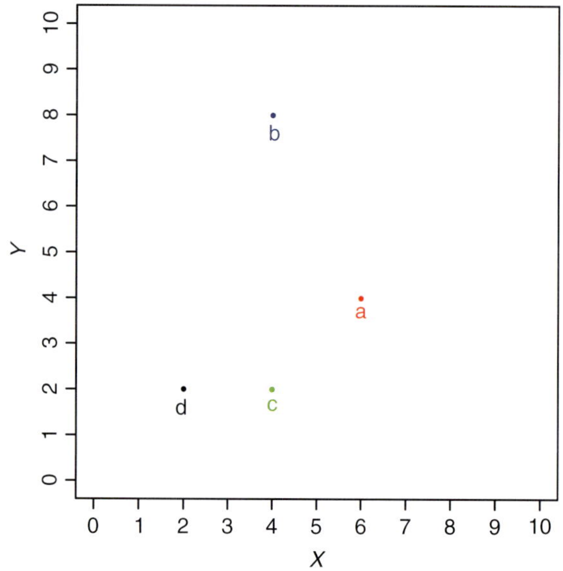

Figure 5.3 Two-dimensional graphical representation of the data matrix $\mathbf{X}$ in (5.8).

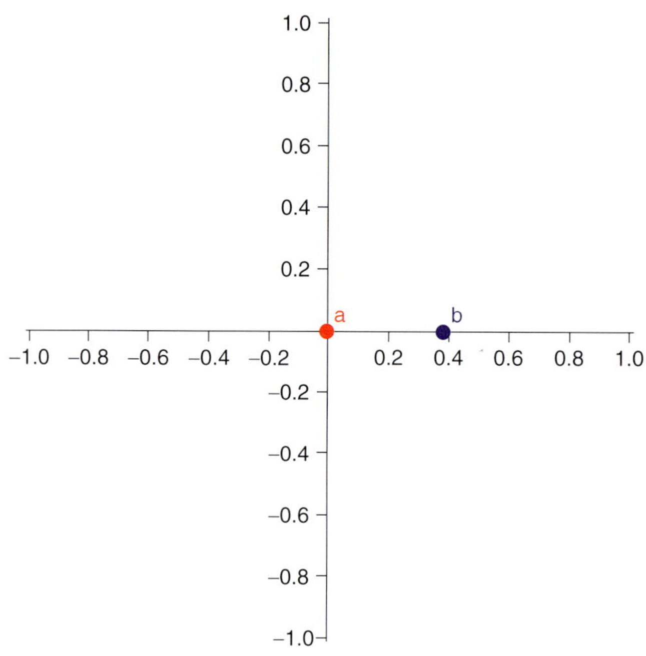

Figure 5.4 Euclidean space for samples $\boldsymbol{a}$ and $\boldsymbol{b}$.

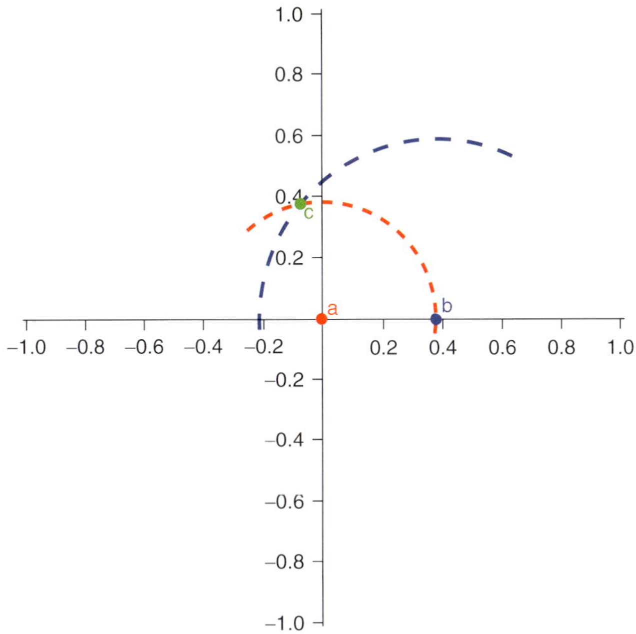

Figure 5.5 Euclidean space for samples $\boldsymbol{a}$, $\boldsymbol{b}$ and $\boldsymbol{c}$.

To embed our data matrix, $\mathbf{X}$, into a Euclidean space, we need to find a coordinate representation that will reproduce the matrix $\mathbf{D}^*$ when calculating the inter-sample Pythagorean distances. A simple way to accomplish this is as follows: we construct a two-dimensional Euclidean space and place $\boldsymbol{a}$ at the origin. Sample $\boldsymbol{b}$ now has to be a distance of 0.39 away from $\boldsymbol{a}$. This is accomplished by placing $\boldsymbol{b}$ at coordinates (0.39, 0) as illustrated in Figure 5.4.

Next, $\boldsymbol{c}$ has to be embedded in the Euclidean space such that it is at a distance 0.39 from $\boldsymbol{a}$ and 0.60 from $\boldsymbol{b}$. Finding the intersection (strictly one of the two possible intersections) of the circles with radii 0.39 and 0.60 respectively, $\boldsymbol{c}$ is placed in the two-dimensional Euclidean space as shown in Figure 5.5.

To embed $\boldsymbol{d}$ into a Euclidean space such that it has distances 0.60, 0.69 and 0.33 from $\boldsymbol{a}$, $\boldsymbol{b}$ and $\boldsymbol{c}$ respectively, it is necessary to turn to a three-dimensional Euclidean space. The embedding of $\boldsymbol{d}$ is illustrated in Figure 5.6. In general, n samples are embedded in an $(n-1)$-dimensional Euclidean space. The process described here can be continued further, but cannot be visually represented in dimensions higher than 3. A more straightforward algebraic way of finding a representation $\mathbf{Y}$ embedded in the Euclidean space is to solve the eigenvector problem $\mathbf{B} = \mathbf{YY}'$ imposing the scaling $\mathbf{Y}'\mathbf{Y} = \mathbf{\Lambda}$. This is known as *principal coordinate analysis* (PCO) or *classical scaling*. Of course, if $\mathbf{B}$ is not positive semi-definite, it will not be possible to find such a configuration $\mathbf{Y}$. As an example, if the distances between $\boldsymbol{a}$, $\boldsymbol{b}$ and $\boldsymbol{c}$ were such that the circles centred at $\boldsymbol{a}$ and $\boldsymbol{b}$ did not intersect, there would be no Euclidean representation where $\boldsymbol{c}$ can be embedded.

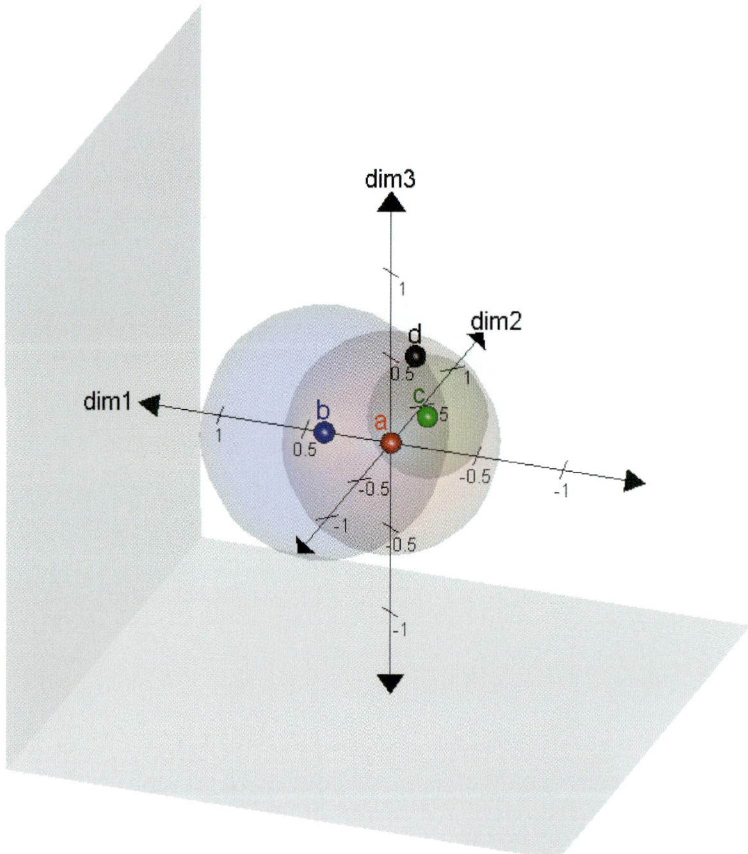

Figure 5.6 Euclidean space for samples $\boldsymbol{a}$, $\boldsymbol{b}$, $\boldsymbol{c}$ with $\boldsymbol{d}$ at the intersection of three spheres.

The above shows how to do embedding in a simple case but fails when there are more than four samples. Then we have to resort to approximations given by PCO, which amounts to the same thing as doing PCA on the points embedded by the simple method. Performing a PCO on a Euclidean embeddable distance matrix $\mathbf{D}$, the representation of $\mathbf{Y}$ in $\mathscr{R}$ is given in Figure 5.7. This representation is exact and approximation is not relevant. The best-fitting r-dimensional subspace $\mathscr{L}$ is obtained from the first r principal components given by the first r columns of $\mathbf{Y}$ denoted by $\mathbf{Z} = \mathbf{Y}_r$, as shown in the form of a biplot in Figure 5.8.

5.4 Providing nonlinear biplot axes for variables

A major difference between MDS methods and the methods of previous chapters is that, as a consequence of working with distances, there is no provision for representing the original variables in $\mathscr{R}$, only the sample points. This is a serious drawback because biplots are concerned with approximating the variables. The way forward is to note that

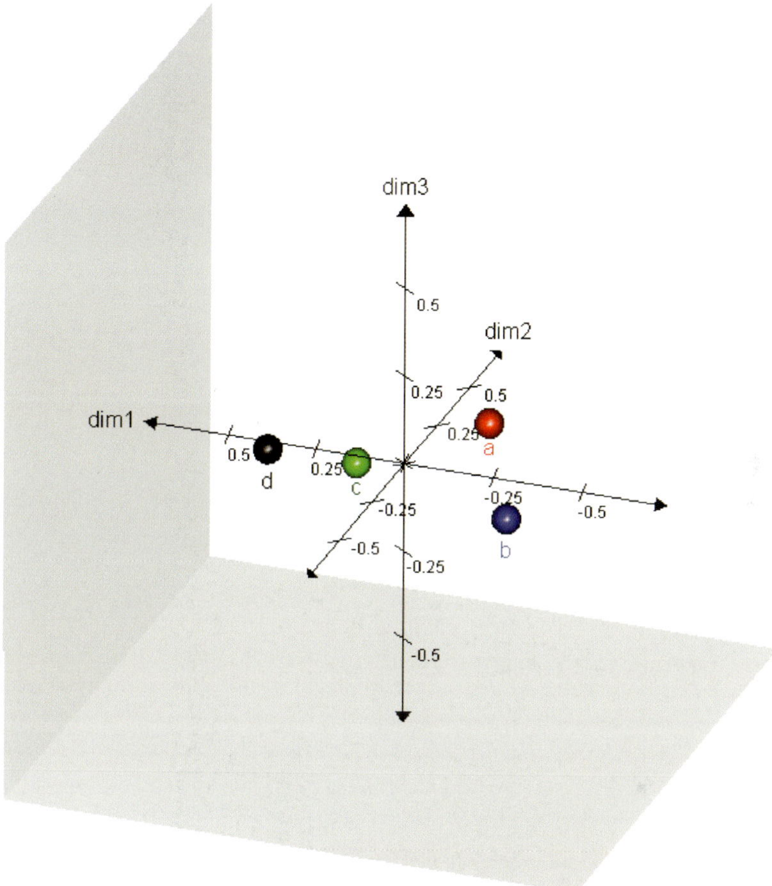

Figure 5.7 Euclidean embedded representation of the matrix **X** in (5.8).

a kth Cartesian axis may be regarded as the locus of the *pseudo-sample* $\mu\mathbf{e}_k$ as μ varies. It follows that if we can place $\mu\mathbf{e}_k$ in the MDS space then we have defined a kth 'axis'. To emphasize that there is no guarantee that this 'axis' is linear, we shall refer to it as a *trajectory*. All p trajectories must have a common point of concurrency at $\mu = 0$. We will denote this point of concurrency by O.

To embed a new point $\mathbf{x}^*$ into $\mathscr{R}$, the vector $\mathbf{d}_{n+1} = \{-\frac{1}{2}d_{n+1,\mathrm{i}}^2\}$ of *ddistances* between the samples $\mathbf{x}_i$ and the new point $\mathbf{x}^*$ must be calculated according to the chosen Euclidean embeddable distance measure. Gower (1968) shows that for embedding this sample into the m-dimensional Euclidean space $\mathscr{R}$, a further $(m+1)$th dimension is needed where m is generally defined as $n-1$. The (nonzero) coordinates are given by

$$\mathbf{y}^{*\prime} = (\mathbf{y}' : 1 \times m,\ y_{m+1}),$$

with

$$\mathbf{y} = \mathbf{\Lambda}^{-1}\mathbf{Y}'\left(\mathbf{d}_{n+1} - \tfrac{1}{n}\mathbf{D}\mathbf{1}\right) \tag{5.9}$$

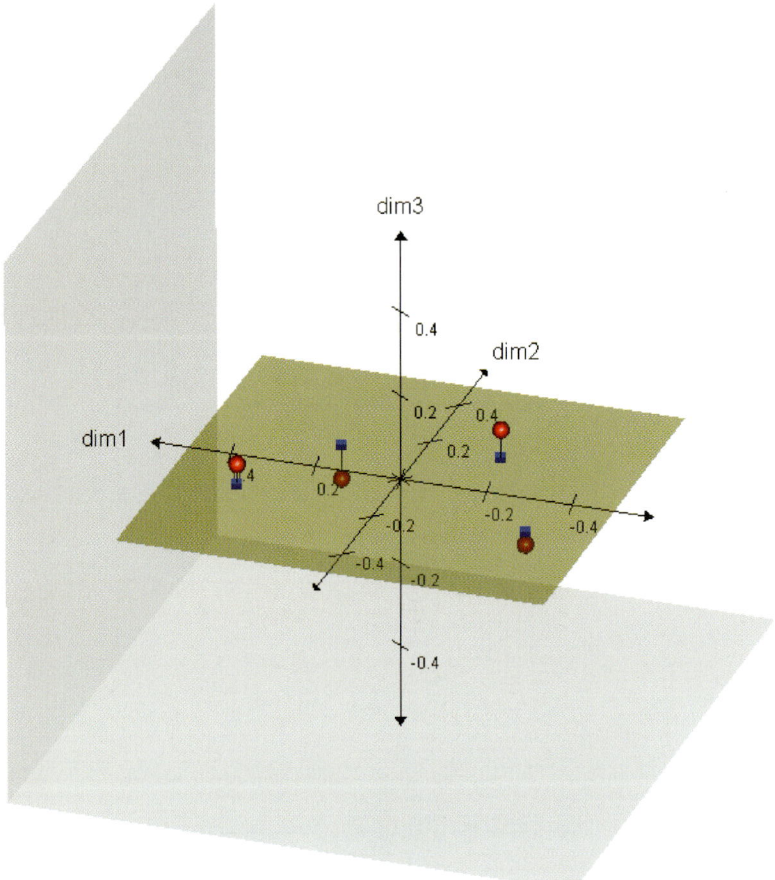

Figure 5.8 Two-dimensional space with projected samples contained in the (5.8) data matrix **X** being shown.

and

$$y_{m+1}^2 = \tfrac{1}{n^2}\mathbf{1}'\mathbf{D}\mathbf{1} - \tfrac{2}{n}\mathbf{1}'\mathbf{d}_{n+1} - \mathbf{y}'\mathbf{y}. \tag{5.10}$$

Note that in (5.9) and (5.10) the matrix $\mathbf{\Lambda}$: $m \times m$ excludes the last singular value of zero as well as the corresponding column of $\mathbf{Y}$, that is, the $\mathbf{Y}'$ in (5.9) is of size $m \times n$.

Although every embedded sample creates a new extra dimension, Gower and Hand (1996) give arguments why we may proceed as if there were only one extra dimension; the version of $\mathscr{R}$ augmented with this extra dimension will be denoted by $\mathscr{R}^+$. Distances calculated between any point in $\mathscr{R}$ and a point in $\mathscr{R}^+$ will be correct, but distances between points that are both in $\mathscr{R}^+$ might not be. Since the extra dimension is orthogonal to the first m dimensions, the orthogonal projection of the new sample onto the m-dimensional subspace $\mathscr{R} \subset \mathscr{R}^+$ is given by (5.9).

The original variables, axes $\boldsymbol{X}$ and $\boldsymbol{Y}$, in our example in Figure 5.3 are embedded into $\mathscr{R}$ by calculating the distance vector $\mathbf{d}_{n+1}$ from the pseudo-sample $\mathbf{x}^* = \mu\mathbf{e}_k$ for

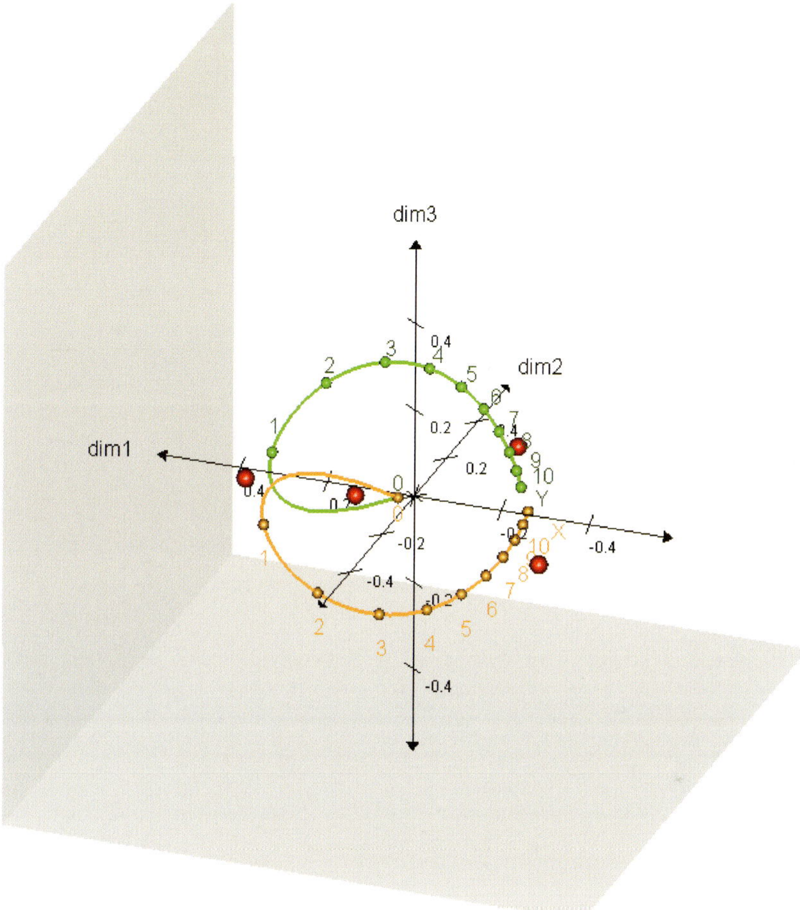

Figure 5.9 Embedded nonlinear axes of the data matrix **X** given in (5.8) when Clark's distance (5.6) is used for expressing inter-sample distances.

a series of values of μ and for $k = 1, 2$. The embedded axes for Clark's distance are shown in Figure 5.9 and are seen to be highly nonlinear.

5.4.1 Interpolation biplot axes

Since the linear Cartesian axes X and Y above become nonlinear trajectories when embedded with a non-Pythagorean distance measure, the projected interpolation biplot axes are also nonlinear. Their method of use remains the vector-sum method illustrated in Figure 3.11 for PCA and Figure 4.10 for CVA biplots. The projection of the embedded original axes X and Y onto the biplot space $\mathscr{L}$ gives the representation of the interpolation biplot axes as shown in Figure 5.10. In Figure 5.11 the biplot space representation of the sample points, together with interpolation biplot axes of X and Y, is shown.

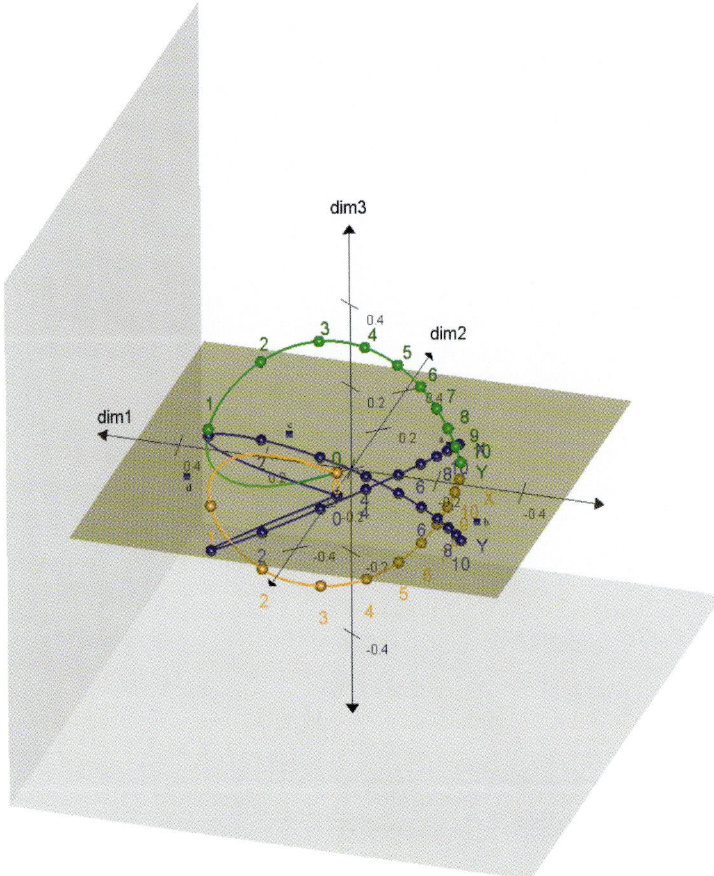

Figure 5.10 Interpolation of embedded nonlinear axes as nonlinear interpolation axes in the biplot space.

In order to derive an explicit expression for interpolating new samples we make the assumption that the variables contribute independently to the squared distance between a new sample and one of the original samples – that is, that the distance measure is additive. It is clear that Pythagorean distance as well as the distances (5.5) and (5.6) are additive, but Mahalanobis distance is not. Consider the pseudo-sample $(0, \ldots, 0, x_k, 0, \ldots, 0) = x_k \mathbf{e}'_k$ with associated vector $\mathbf{d}_{n+1}$ denoted by $\mathbf{d}_{n+1}(x_k)$. The squared distance from this pseudo-sample to each of the original samples is of the form

$$d^2_{n+1,i}(x_k) = \sum_{h=1}^{p} d^2(x_{ih}, 0) + d^2(x_{ik}, x_k) - d^2(x_{ik}, 0).$$

Therefore,

$$\sum_{k=1}^{p} d^2(x_{ik}, x_k) = \sum_{k=1}^{p} d^2_{n+1,i}(x_k) + \sum_{k=1}^{p} d^2(x_{ik}, 0) - p \sum_{h=1}^{p} d^2(x_{ih}, 0). \tag{5.11}$$

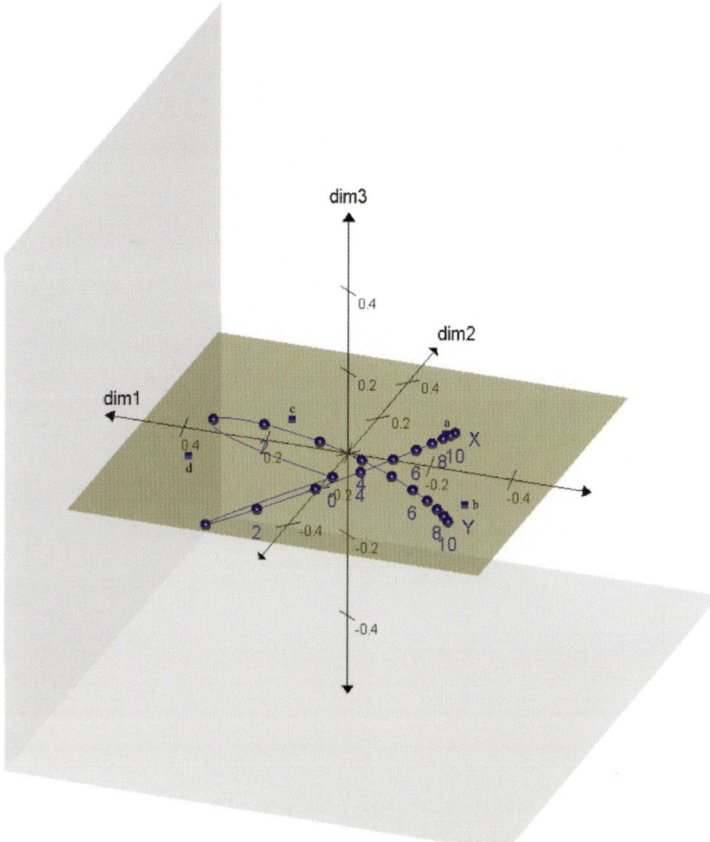

Figure 5.11 Representation of sample points in biplot space together with the nonlinear interpolation biplot axes projected onto the two-dimensional biplot space.

It follows from (5.11) that for a new sample $\mathbf{x} = \sum_{k=1}^{p} x_k \mathbf{e}_k$ we have

$$d_{n+1,i}^2(\mathbf{x}) = \sum_{k=1}^{p} d^2(x_{ik}, x_k) = \sum_{k=1}^{p} d_{n+1,i}^2(x_k) - (p-1)\sum_{h=1}^{p} d^2(x_{ih}, 0)$$

so that

$$\mathbf{d}_{n+1}(\mathbf{x}) = \sum_{k=1}^{p} \mathbf{d}_{n+1}(x_k) - (p-1)\sum_{h=1}^{p} \{-\tfrac{1}{2}d^2(x_{ih}, 0)\}. \qquad (5.12)$$

Substituting (5.12) into (5.9) leads to the expression for vector-sum interpolation:

$$\begin{aligned}\mathbf{y} &= \mathbf{\Lambda}^{-1}\mathbf{Y}'\left[\sum_{k=1}^{p} \mathbf{d}_{n+1}(x_k) - (p-1)\sum_{h=1}^{p} \{-\tfrac{1}{2}d^2(x_{ih}, 0)\} - \tfrac{1}{n}\mathbf{D1}\right] \\ &= \mathbf{\Lambda}^{-1}\mathbf{Y}'\left[\sum_{k=1}^{p} \left(\mathbf{d}_{n+1}(x_k) - \tfrac{1}{n}\mathbf{D1}\right) - (p-1)\left(\sum_{h=1}^{p} \{-\tfrac{1}{2}d^2(x_{ih}, 0)\} - \tfrac{1}{n}\mathbf{D1}\right)\right]. \qquad (5.13)\end{aligned}$$

The first summation in (5.13) gives the vector sum for the markers $x_1, x_2, \ldots, x_p$ of the pseudo-samples, while the second summation gives the interpolant of the mean $(0, 0, \ldots, 0)$, that is, the point of concurrency of the trajectories.

5.4.2 Prediction biplot axes

To construct prediction biplot axes in $\mathscr{L}$ for our example, the process follows similar principles to those described for the linear PCA and CVA biplots in Sections 3.2.3 and 4.4.2, respectively. To predict the marker μ on the kth biplot axis, a plane $\mathscr{N}$ is constructed at the point corresponding to $\mu\mathbf{e}_k$. The plane $\mathscr{N}$ is normal to the embedded Cartesian axes in $\mathscr{R}^+$. All points in this plane predict the value μ for the kth variable. This plane, $\mathscr{N}$, intersects the (in general) r-dimensional approximation space, $\mathscr{L}$, in an $(r-1)$-dimensional linear subspace $\mathscr{L} \cap \mathscr{N}$. Since embedding the marker $\mu\mathbf{e}_k$ in

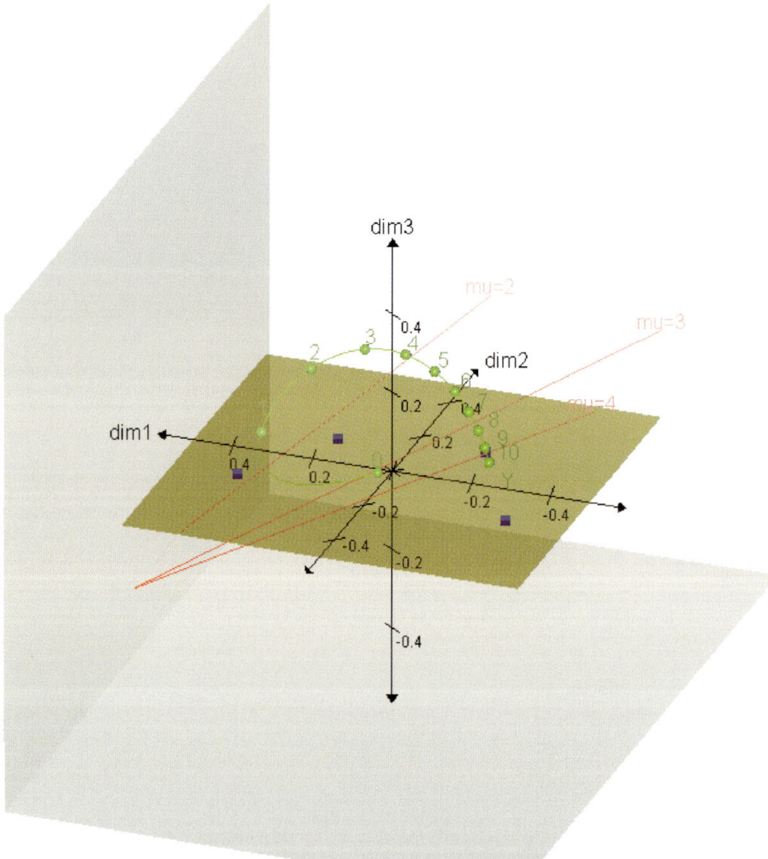

Figure 5.12 Intersection spaces $\mathscr{L} \cap \mathscr{N}$ for the original variable $\boldsymbol{Y}$ at the points $\mu = 2, 3, 4$.

our three-dimensional example in Figure 5.9 induces an extra fourth dimension, it is not possible to visually represent the marker and the plane $\mathscr{N}$ normal to the marker at this point. However, the intersection spaces for the markers $\mu = 2$, 3 and 4 of the embedded axis of original variable $\boldsymbol{Y}$ can be shown, as is illustrated in Figure 5.12. Since the embedded original axis $\boldsymbol{Y}$ is nonlinear, it follows that the intersection spaces $\mathscr{L} \cap \mathscr{N}$ as shown in Figure 5.12 are not parallel as was the case for PCA (Section 3.2.3). This, in turn, leads to nonlinear prediction biplot trajectories.

Gower and Ngouenet (2005) give details of three different methods to obtain the prediction biplot trajectories. The different trajectories are constructed by identifying different 'optimal' points on the same intersection spaces $\mathscr{L} \cap \mathscr{N}$. However, the prediction methods identify the same intersection and hence all will yield the same predictions.

In order to proceed we need the equation of the intersection space for a given marker μ which we denote by $\mathscr{L} \cap \mathscr{N}(\mu)$. The space $\mathscr{N}(\mu)$ is normal to the embedded kth biplot trajectory $\boldsymbol{\xi}_k^+$ at the point μ. Therefore, from (5.8) and (5.9), it passes through the point

$$\boldsymbol{\xi}_k^+(\mu) = \begin{bmatrix} \mathbf{\Lambda}^{-1}\mathbf{Y}'\left(\mathbf{d}_{n+1}(\mu) - \frac{1}{n}\mathbf{D1}\right) \\ \xi_{k,m+1}(\mu) \end{bmatrix},$$

where

$$\xi_{k,m+1}^2(\mu) = \frac{1}{n^2}\mathbf{1}'\mathbf{D1} - \frac{2}{n}\mathbf{1}'\,\mathbf{d}_{n+1}(\mu) - \boldsymbol{\xi}_k'(\mu)\boldsymbol{\xi}_k(\mu),$$

and is orthogonal to $\frac{d}{d\mu}\boldsymbol{\xi}_k^+(\mu)$, which implies that

$$\left\{\frac{d}{d\mu}\boldsymbol{\xi}_k^+(\mu)\right\}'\begin{bmatrix} \mathbf{y} - \mathbf{\Lambda}^{-1}\mathbf{Y}'\left(\mathbf{d}_{n+1}(\mu) - \frac{1}{n}\mathbf{D1}\right) \\ y_{m+1} - \xi_{k,m+1}(\mu) \end{bmatrix} = 0,$$

that is,

$$\left\{\frac{d}{d\mu}\boldsymbol{\xi}_k(\mu)\right\}'\{\mathbf{y} - \mathbf{\Lambda}^{-1}\mathbf{Y}'(\mathbf{d}_{n+1}(\mu) - \tfrac{1}{n}\mathbf{D1})\} + \frac{d}{d\mu}\xi_{k,m+1}(\mu)\{y_{m+1} - \xi_{k,m+1}(\mu)\} = 0. \tag{5.14}$$

The equation for $\mathscr{N}(\mu)$ is found by solving (5.14): now,

$$\frac{d}{d\mu}\boldsymbol{\xi}_k(\mu) = \mathbf{\Lambda}^{-1}\mathbf{Y}'\left\{\frac{d}{d\mu}\mathbf{d}_{n+1}(\mu)\right\}$$

and

$$2\xi_{k,m+1}(\mu)\left\{\frac{d}{d\mu}\xi_{k,m+1}(\mu)\right\} = -\frac{2}{n}\mathbf{1}'\left\{\frac{d}{d\mu}\mathbf{d}_{n+1}(\mu)\right\} - 2\boldsymbol{\xi}_k'(\mu)\left\{\frac{d}{d\mu}\boldsymbol{\xi}_k(\mu)\right\}.$$

Substituting into (5.14) and simplifying leads to

$$\left(\left\{\frac{d}{d\mu}\mathbf{d}_{n+1}(\mu)\right\}'\mathbf{Y\Lambda}^{-1},\quad \frac{d}{d\mu}\xi_{k,m+1}(\mu)\right)\begin{bmatrix}\mathbf{y}\\ y_{m+1}\end{bmatrix} = -\frac{1}{n}\mathbf{1}'\left\{\frac{d}{d\mu}\mathbf{d}_{n+1}(\mu)\right\}. \tag{5.15}$$

In the case of a two-dimensional biplot the intersection $\mathscr{L} \cap \mathscr{N}(\mu)$ follows by setting all elements of $(\mathbf{y}', y_{m+1})$ except the first two equal to zero. Denote these two nonzero coordinates as $\mathbf{z}' = (z_1, z_2)$. Then

$$\left\{ \frac{d}{d\mu} \mathbf{d}_{n+1}(\mu) \right\}' \left(\frac{1}{\lambda_1} \mathbf{y}_1 \quad \frac{1}{\lambda_2} \mathbf{y}_2 \right) \mathbf{z} = -\frac{1}{n} \mathbf{1}' \left\{ \frac{d}{d\mu} \mathbf{d}_{n+1}(\mu) \right\}. \tag{5.16}$$

Thus the intersection space becomes a line with equation of the form $\mathbf{a}'\mathbf{z} = t^*$. Multiplying the coefficient vector $\mathbf{a}$ and t^* with any constant will result in an equally valid equation for the intersection space $\mathscr{L} \cap \mathscr{N}(\mu)$. Therefore we reparameterize the equation as $\boldsymbol{l}'(\mu)\mathbf{z} = t$ such that $l_1^2(\mu) + l_2^2(\mu) = 1$.

5.4.2.1 Normal projection

With normal projection, we find a trajectory that is normal to all intersection spaces $\mathscr{L} \cap \mathscr{N}$. If we denote the kth two-dimensional biplot trajectory at the marker μ by

$$\boldsymbol{\beta}_k(\mu) = \begin{bmatrix} b_1(\mu) \\ b_2(\mu) \end{bmatrix},$$

then the direction of the tangent at $\boldsymbol{\beta}_k(\mu)$ is defined by

$$\begin{bmatrix} \frac{d}{d\mu} b_1(\mu) \\ \frac{d}{d\mu} b_2(\mu) \end{bmatrix}$$

and the gradient of the tangent is given by

$$\frac{\frac{d}{d\mu} b_2(\mu)}{\frac{d}{d\mu} b_1(\mu)}.$$

The tangent must be orthogonal to the direction of the intersection space $\mathscr{L} \cap \mathscr{N}(\mu)$ with gradient $-l_1(\mu)/l_2(\mu)$, therefore

$$-\frac{l_1(\mu)}{l_2(\mu)} = -\frac{\frac{d}{d\mu} b_1(\mu)}{\frac{d}{d\mu} b_2(\mu)}.$$

Furthermore, since $\boldsymbol{\beta}_k(\mu)$ lies on the intersection space, it can be expressed as

$$\boldsymbol{l}(\mu)' \boldsymbol{\beta}_k(\mu) = t(\mu),$$

which after differentiation becomes

$$\left\{ \frac{d}{d\mu} l_1(\mu) \right\} b_1(\mu) + \left\{ \frac{d}{d\mu} b_1(\mu) \right\} l_1(\mu) + \left\{ \frac{d}{d\mu} l_2(\mu) \right\} b_2(\mu) + \left\{ \frac{d}{d\mu} b_2(\mu) \right\} l_2(\mu) = \frac{d}{d\mu} t(\mu).$$

After some algebraic manipulation it follows that

$$\frac{d}{d\mu} \left(\frac{b_1(\mu)}{l_2(\mu)} \right) = l_1(\mu) \left\{ \frac{d}{d\mu} \left(\frac{t(\mu)}{l_2(\mu)} \right) \right\}. \tag{5.17}$$

For a fixed μ_0 we write

$$\boldsymbol{\beta}'_k(\mu_0) = \begin{pmatrix}\alpha_1 & \alpha_2\end{pmatrix}$$

and, integrating (5.17), we have

$$\frac{b_1(\mu)}{l_2(\mu)} = \int_{\mu_0}^{\mu} l_1(\mu)\left\{\frac{d}{d\mu}\left(\frac{t(\mu)}{l_2(\mu)}\right)\right\} d\mu + k. \tag{5.18}$$

To solve for the constant k, we have

$$\alpha_1 = b_1(\mu_0) = l_2(\mu_0)\left[\int_{\mu_0}^{\mu_0} l_1(\mu)\left\{\tfrac{d}{d\mu}\left(\frac{t(\mu)}{l_2(\mu)}\right)\right\} d\mu + k\right],$$

which simplifies to $\alpha_1 = l_2(\mu_0)(0 + k)$, implying

$$k = \frac{\alpha_1}{l_2(\mu_0)}.$$

Similarly,

$$b_2(\mu) = l_1(\mu)\left[\int_{\mu_0}^{\mu} l_2(\mu)\left\{\tfrac{d}{d\mu}\left(\frac{t(\mu)}{l_1(\mu)}\right)\right\} d\mu + \frac{\alpha_2}{l_1(\mu_0)}\right].$$

For convenience we choose $(\alpha_1 \quad \alpha_2)$ as the origin of $\mathscr{L}$, which implies that $t(\mu_0) = 0$. Then the coordinates in $\mathscr{L}$ for the marker μ on the kth biplot trajectory are given by

$$\boldsymbol{\beta}_k = \begin{bmatrix} b_1(\mu) \\ b_2(\mu) \end{bmatrix} = \begin{bmatrix} l_2(\mu)\displaystyle\int_{\mu_0}^{\mu} l_1(\mu)\left\{\frac{d}{d\mu}\left(\frac{t(\mu)}{l_2(\mu)}\right)\right\} d\mu \\ l_1(\mu)\displaystyle\int_{\mu_0}^{\mu} l_2(\mu)\left\{\frac{d}{d\mu}\left(\frac{t(\mu)}{l_1(\mu)}\right)\right\} d\mu \end{bmatrix}. \tag{5.19}$$

To construct the biplot with our R function `Nonlinbipl`, a series of μ values are selected and the trajectory is computed by substituting the symbolic differentiation into the integrand which is then solved by numerical integration (see Section 5.6 for details of the method employed). The symbolic differentiation of $t(\mu)/l_i(\mu)$ contains the first- and second-order derivatives of $\mathbf{d}_{n+1}(\mu)$. This follows since

$$l_i(\mu) = \frac{a_i(\mu)}{\sqrt{a_1^2(\mu) + a_2^2(\mu)}} \quad \text{and} \quad t(\mu) = \frac{t^*(\mu)}{\sqrt{a_1^2(\mu) + a_2^2(\mu)}},$$

where

$$\mathbf{a}(\mu) = \{\tfrac{d}{d\mu}\mathbf{d}_{n+1}(\mu)\}'(\tfrac{1}{\lambda_1}\mathbf{y}_1 \quad \tfrac{1}{\lambda_2}\mathbf{y}_2) \quad \text{and} \quad t^*(\mu) = -\tfrac{1}{n}\mathbf{1}'\{\tfrac{d}{d\mu}\mathbf{d}_{n+1}(\mu)\}.$$

This means that for normal projection the distance function must not allow second-order derivatives to vanish.

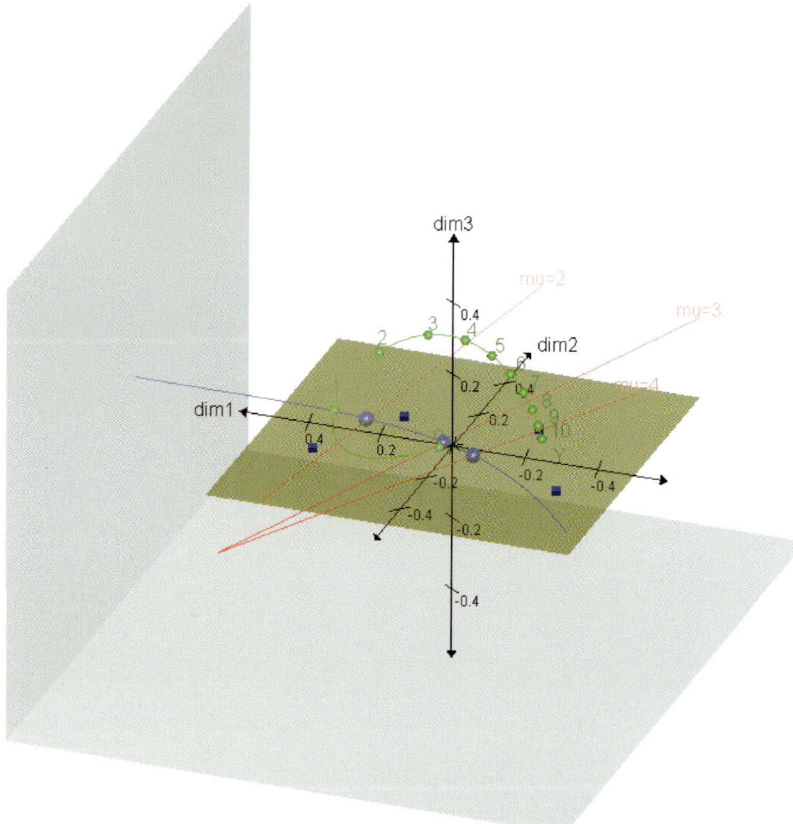

Figure 5.13 Normal projection biplot axis with markers for $\mu = 2, 3$ and 4 for original variable $\boldsymbol{Y}$.

In Figure 5.13 the markers $\mu = 2, 3$ and 4 for original variable $\boldsymbol{Y}$ are shown on the $\mathscr{L} \cap \mathscr{N}$ intersection spaces $\boldsymbol{l}(2)'\mathbf{z} = m(2)$, $\boldsymbol{l}(3)'\mathbf{z} = m(3)$ and $\boldsymbol{l}(4)'\mathbf{z} = m(4)$. The normal prediction biplot axis calculated according to (5.19) for a series of μ values is shown in Figure 5.14.

The two nonlinear prediction biplot trajectories for normal projection are shown in the biplot in Figure 5.15. Orthogonal projection from point P onto the axis $\boldsymbol{X}$ results in a value of `5.6` and for $\boldsymbol{Y}$ we read off the value `2`.

In Section 5.6 we discuss the implementation of normal projection in our $\mathscr{R}$ function `Nonlinbipl`.

5.4.2.2 Circular projection

An alternative to orthogonal projection is to define the marker μ on the intersection space $\mathscr{L} \cap \mathscr{N}$ as the orthogonal projection of the origin O (intersection of the scaffolding axes) onto $\mathscr{L} \cap \mathscr{N}$. This is illustrated in Figure 5.16. The equation of the intersection space is $\boldsymbol{l}(\mu)'\mathbf{z} = t(\mu)$ with intercept $t(\mu)/l_2(\mu)$. To obtain the point P, the orthogonal

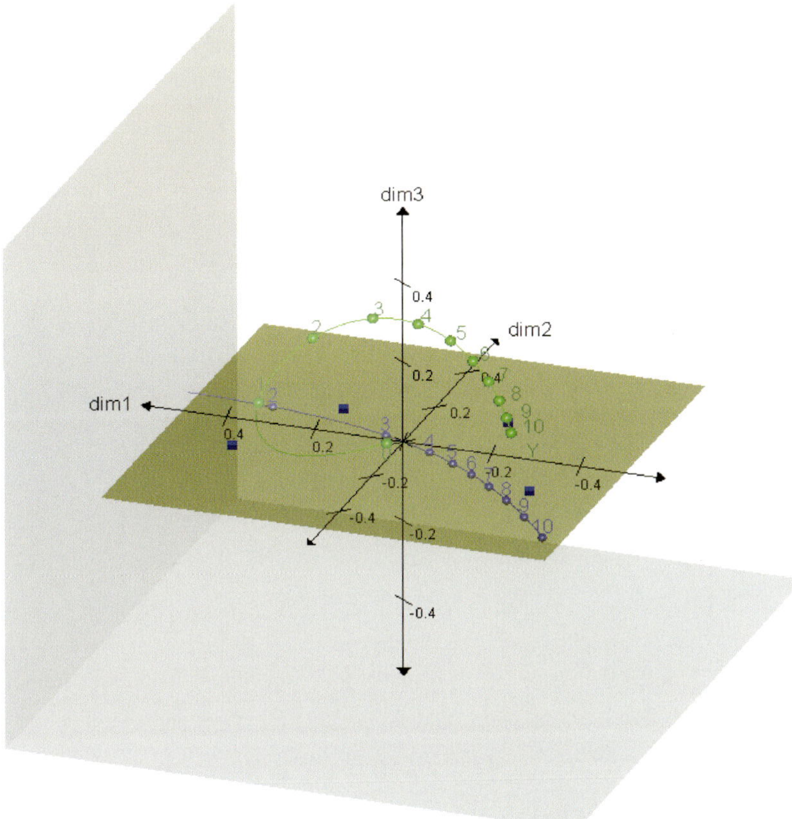

Figure 5.14 Calibrated normal projection prediction biplot axis for original variable $\boldsymbol{Y}$.

projection of O onto the intersection space, we first calculate the orthogonal projection, S, of the point R onto the line $\boldsymbol{l}^*(\mu)$, where $\boldsymbol{l}^*(\mu)$ has been translated vertically by $-t(\mu)/l_2(\mu)$ units to pass through the origin. The coordinates of the point R are given by $(0, -t(\mu)/l_2(\mu))$ and the line $\boldsymbol{l}^*(\mu)$ is generated by the vector $(l_2(\mu), -l_1(\mu))$, so that the projection S is given by $(l_1(\mu)t(\mu), -l_1^2(\mu)t(\mu)/l_2(\mu))$. The point P representing the marker μ on the kth prediction biplot trajectory is given by

$$\left(l_1(\mu)t(\mu),\quad -\tfrac{l_1^2(\mu)t(\mu)}{l_2(\mu)}\right) + \left(0,\quad \tfrac{t(\mu)}{l_2(\mu)}\right) = \left(l_1(\mu)t(\mu),\quad l_2(\mu)t(\mu)\right).$$

This is illustrated for the intersection spaces where $\mu = 2, 3$ and 4 for the original variable $\boldsymbol{Y}$ in Figure 5.17 in $\mathscr{R}$ with three dimensions and in Figure 5.18 in $\mathscr{L}$ with two dimensions. If we let μ vary continuously from 0 to 10, a series of intersection spaces is formed and, connecting the projections of O onto each of these, produces the circular prediction biplot trajectory as shown in Figure 5.19.

To predict the value of original variable $\boldsymbol{Y}$ for the point P in the approximation space, a circle is constructed with OP as diameter. The diameter OP subtends a right angle at

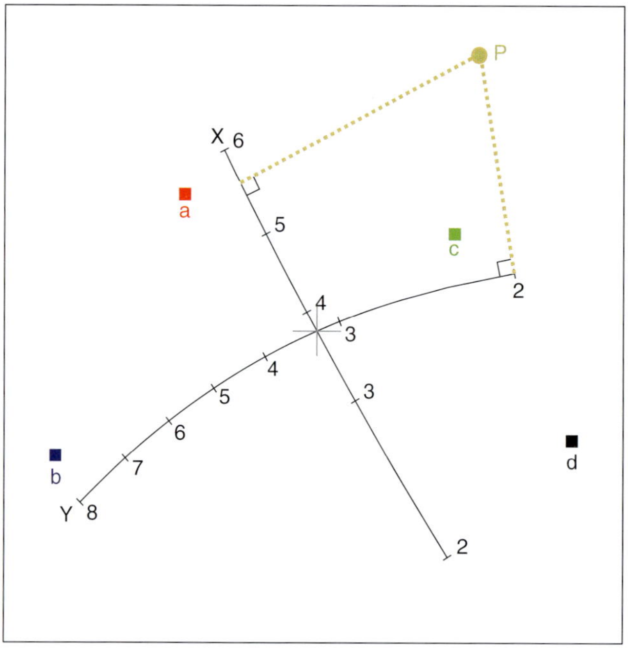

Figure 5.15 Nonlinear biplot with normal projection prediction biplot trajectories.

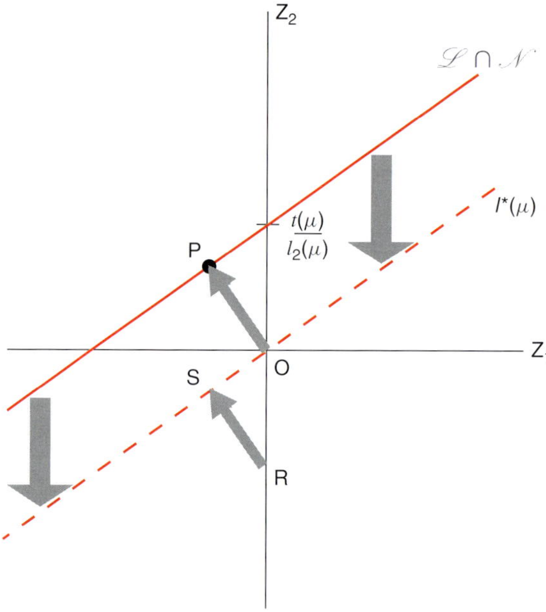

Figure 5.16 Orthogonal projection of the origin (O) onto the intersection space $\mathscr{L} \cap \mathscr{N}$.

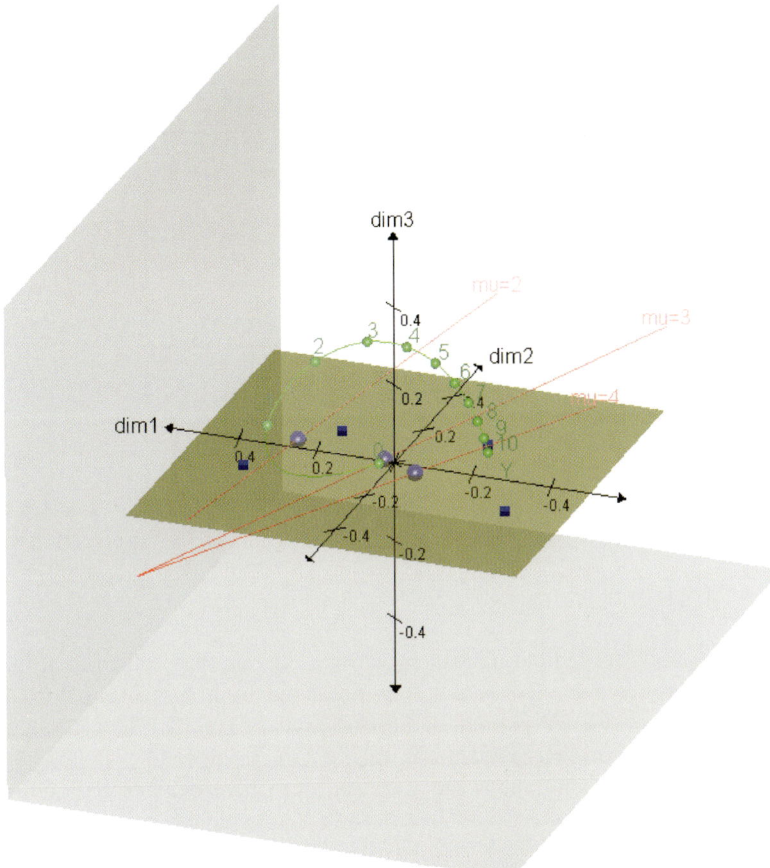

Figure 5.17 Circle projection prediction markers are found by orthogonal projection of the origin in $\mathscr{R}$ onto the intersection spaces $\mathscr{L} \cap \mathscr{N}$.

the point $\boldsymbol{\beta}_k(\mu)$ where the circle intersects the biplot axis ensuring that P is orthogonally projected onto the line $\mathrm{O}\boldsymbol{\beta}_k(\mu)$. The predicted value of the original variable $\boldsymbol{Y}$ for point P is therefore the marker corresponding to the point $\boldsymbol{\beta}_k(\mu)$ on the biplot trajectory. This process is illustrated in Figure 5.20. Once p prediction biplot trajectories are fitted to the nonlinear biplot, the predictions for all p original variables are obtained by the one circle with diameter OP where the circle intersects each of the biplot trajectories. If a biplot trajectory intersects the circle more than once, we follow the convention of selecting the intersection point closest to O. At first sight this may seem an elaborate process, but it may be helpful to note that when the biplot trajectories happen to be linear, circle projection and normal projection are the same thing, as we saw in Section 3.8.8.

Notice that once the prediction biplot trajectory is constructed in $\mathscr{L}$, there is no need to consider $\mathscr{R}$ or $\mathscr{R}^+$ nor for the embedded representation of the original axis in these spaces. All that is available to the user is the biplot space $\mathscr{L}$ which contains

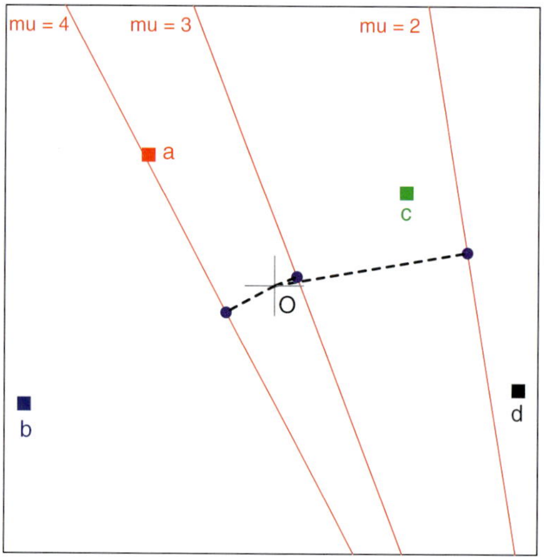

Figure 5.18 Orthogonal projection of point O onto each of the intersection spaces for $\mu = 2, 3, 4$.

everything necessary for prediction or interpolation. The circular prediction of Figure 5.20 is illustrated in Figure 5.21. The point P is predicted by constructing the circle with diameter OP and the value '2' is predicted for original variable $\boldsymbol{Y}$. This circle also gives the predicted value for original variable $\boldsymbol{X}$ as '5.6'.

In Section 5.4 we discuss how the user can perform circular prediction with the function `Nonlinbipl`.

5.4.2.3 Back-projection

A third possibility for obtaining prediction biplot trajectories is based on finding the point on $\mathscr{L} \cap \mathscr{N}$, nearest to the marker μ on the embedded original variable axis, that is, utilizing the back-projection mechanism we discussed when deriving the linear prediction biplot axes for PCA and CVA biplots. The nearest point is found by projecting the embedded axis $(\boldsymbol{\xi}_k'(\mu), \xi_{k,m+1}(\mu))$ onto $\mathscr{L} \cap \mathscr{N}(\mu)$. Similar to Figure 5.16, we translate the intersection space to the origin and project

$$\left(\boldsymbol{\xi}_k'(\mu), \quad \xi_{k,m+1}(\mu)\right) + \left(0 \quad -\frac{t(\mu)}{l_2(\mu)} \quad 0 \quad \ldots \quad 0\right)$$

onto $(l_2(\mu) \quad -l_1(\mu) \quad \mathbf{0}')$ to obtain

$$\begin{bmatrix} l_2^2(\mu)\xi_{k,1}(\mu) - l_1(\mu)l_2(\mu)\xi_{k,2}(\mu) + l_1(\mu)t(\mu) \\ -l_1(\mu)l_2(\mu)\xi_{k,1}(\mu) + l_1^2(\mu)\xi_{k,2}(\mu) - \dfrac{l_1^2(\mu)t(\mu)}{l_2(\mu)} \end{bmatrix}$$

and, after translating back with $t(\mu)/l_2(\mu)$ units in the second direction, the marker μ is given by $(\mathbf{I}_r - \boldsymbol{l}(\mu)\boldsymbol{l}'(\mu))\boldsymbol{\xi}_{k,r}(\mu) + t(\mu)\boldsymbol{l}(\mu)$.

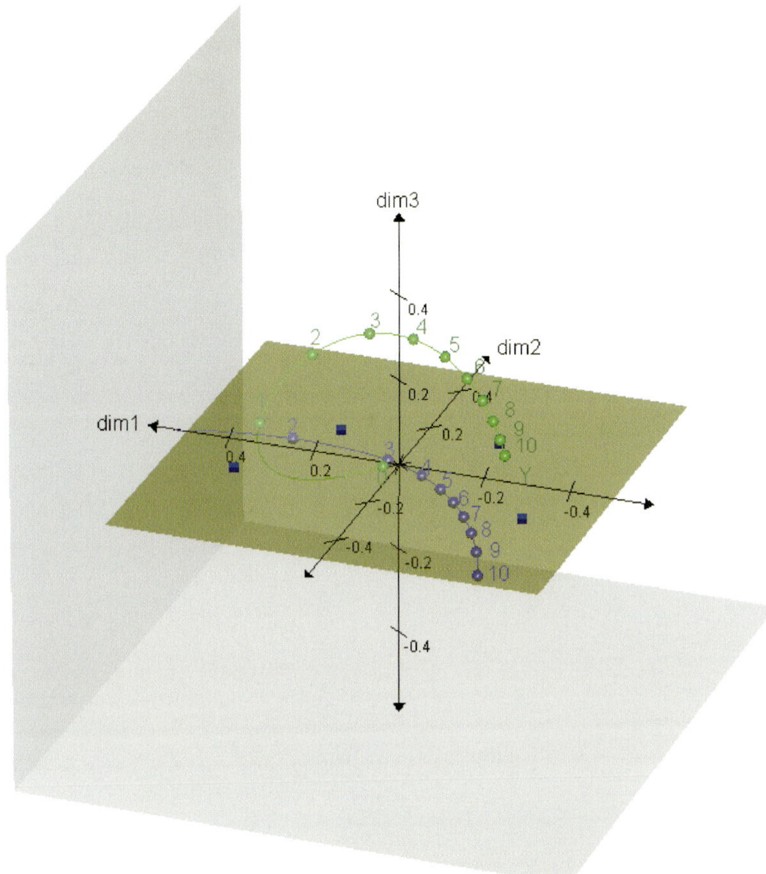

Figure 5.19 Illustration of circle projection prediction axis for the original variable $\boldsymbol{Y}$.

In Figure 5.22 the markers $\mu = 2, 3$ and 4 on the intersection spaces $\mathscr{L} \cap \mathscr{N}$ are shown for the original variable $\boldsymbol{Y}$. As μ varies along the embedded original variable axis, a nonlinear biplot trajectory is traced in the approximation space, as shown in Figure 5.23. The shortest distance property is attractive but unfortunately there is currently no known way of predicting from a back-projection trajectory. In the linear case, however, prediction axes derived from circle projection, normal projection and back-projection are identical.

Although we do not have a method for predicting the values of a point P from the back-projection prediction biplot trajectories, the biplot is shown in Figure 5.24 and the procedure is also implemented in the function `Nonlinbipl`.

5.5 A PCA biplot as a nonlinear biplot

When the matrix **D** is based on Pythagorean distances, then the nonlinear biplot reduces to a PCA biplot and the nonlinear biplot trajectories become linear. All three methods of

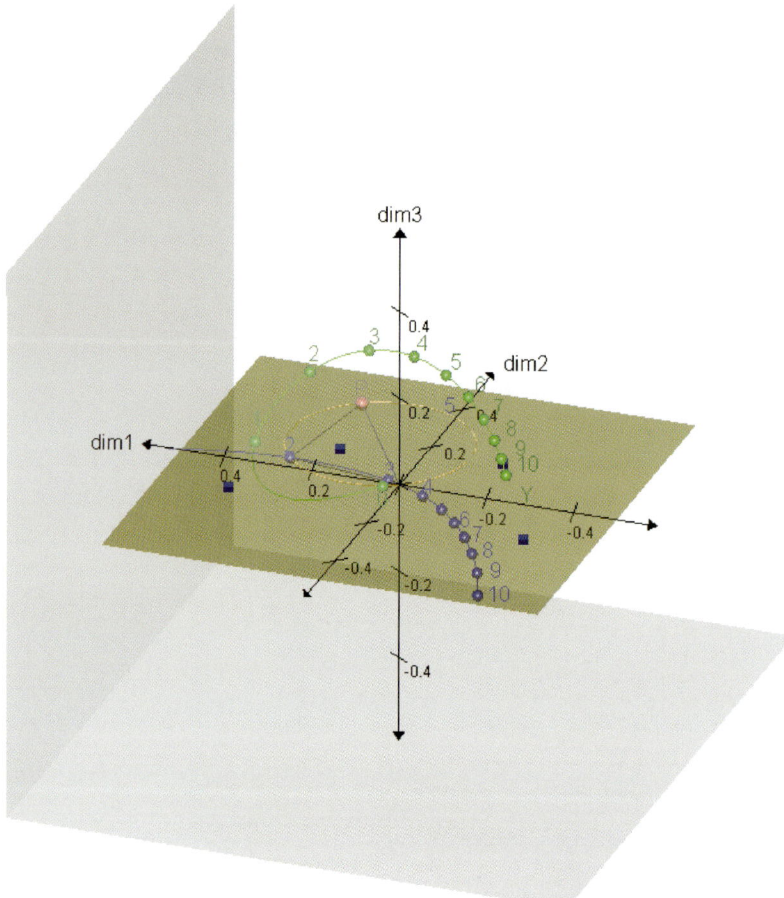

Figure 5.20 Circular prediction of the point P with the prediction biplot axis for original variable $\boldsymbol{Y}$.

constructing prediction biplot axes then coincide with the prediction biplot axes described in Chapter 3.

For Pythagoras distance, it follows that

$$\mathbf{D} = -\tfrac{1}{2}[\mathbf{11}'\{\text{diag}(\mathbf{XX}')\} + \{\text{diag}(\mathbf{XX}')\}\mathbf{11}' - 2\mathbf{XX}'], \tag{5.20}$$

and the matrix $\mathbf{B}$ defined in (5.7) can be written as

$$\begin{aligned}\mathbf{B} &= (\mathbf{I} - \tfrac{1}{n}\mathbf{11}')(\mathbf{XX}' - \tfrac{1}{2}\mathbf{11}'\{\text{diag}(\mathbf{XX}')\} - \tfrac{1}{2}\{\text{diag}(\mathbf{XX}')\}\mathbf{11}')\left(\mathbf{I} - \tfrac{1}{n}\mathbf{11}'\right)\\ &= \mathbf{XX}' - (\mathbf{1}\bar{\mathbf{x}}')\mathbf{X}' - \mathbf{X}(\bar{\mathbf{x}}\mathbf{1}') + \mathbf{1}\bar{\mathbf{x}}'\bar{\mathbf{x}}\mathbf{1}'.\end{aligned}$$

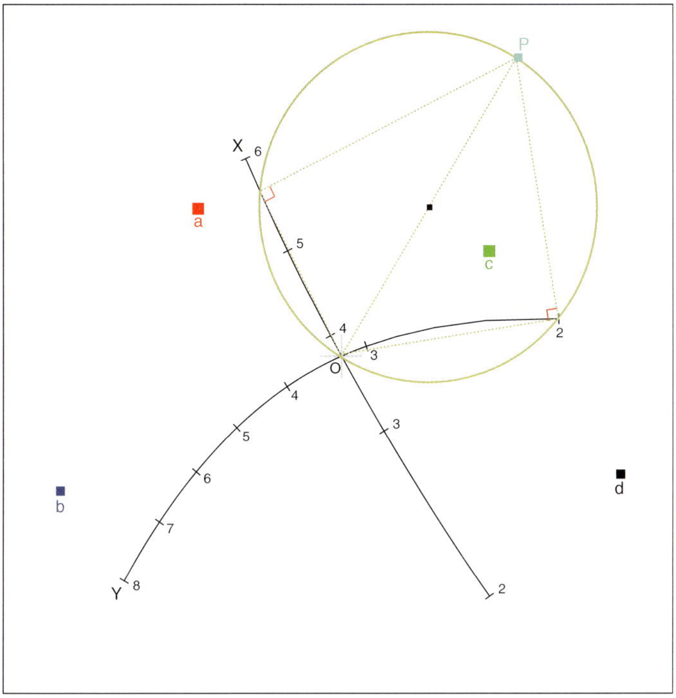

Figure 5.21 Circular prediction of the point P for the example data set: a circle with diameter the dotted line extending from O to P is drawn and the intersection of each of the biplot axes with the circle gives the predictions for point P. Note the difference in the positions of the rectangle indicating the Y-prediction of point P for (a) circular prediction and (b) normal prediction (Figure 5.15). Since the X-axis is almost linear the rectangles indicating the circular and normal X-predictions of P are nearly identical.

With the standard assumption that our data matrix $\mathbf{X}$ for PCA is centred, $\bar{\mathbf{x}} = \mathbf{0}$, it follows that $\mathbf{B} = \mathbf{XX}'$. This is of the same form as $\mathbf{B} = \mathbf{YY}'$, showing that the representation in Euclidean space is already obtained from the matrix $\mathbf{X}$ and no embedding needs to be applied. Alternatively, the embedding step can be viewed as a linear identity transformation. Once the Euclidean representation $\mathbf{X}$ (or $\mathbf{Y}$) is obtained, the principal axes are computed for optimal representation in r dimensions (see Section 5.6.1).

5.6 Constructing nonlinear biplots

Our main function for constructing nonlinear biplots is the function `Nonlinbipl`. In this section we provide details of its usage and capabilities. `Nonlinbipl` has many features in common with those of `PCAbipl` and `CVAbipl` as discussed in Chapters 3 and 4. Thus, in this section we concentrate on what is specific to `Nonlinbipl`.

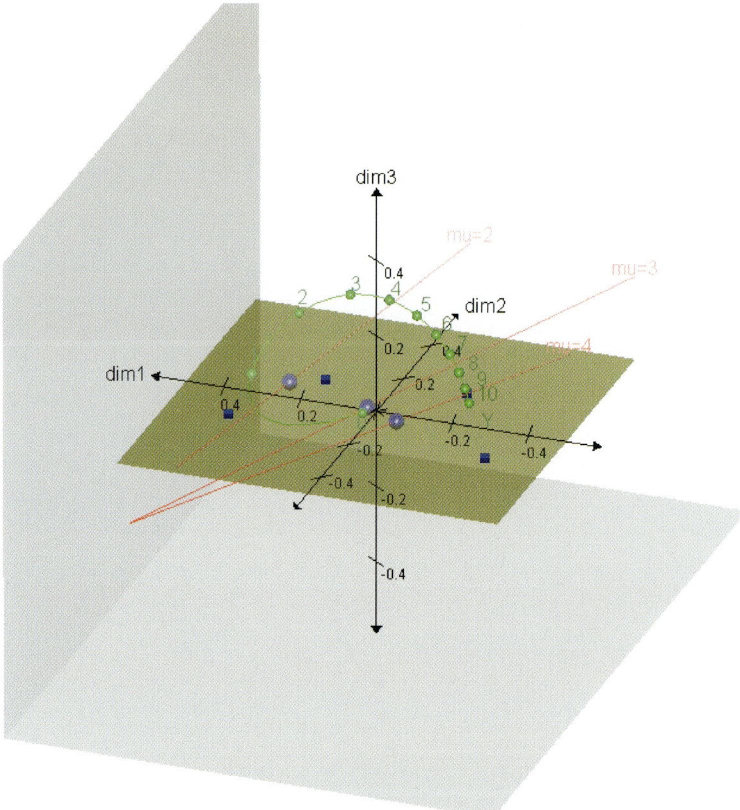

Figure 5.22 Back-projection points on the intersection spaces of the planes orthogonal to the original variable $\boldsymbol{Y}$ at $\mu = 2, 3$ and 4 with the biplot space.

5.6.1 Function `Nonlinbipl`

This function is used for constructing two-dimensional nonlinear biplots as described in Sections 5.3–5.5. Its call and usage, as far as the following arguments are concerned, are similar to `PCAbipl` as described in Section 3.7.1:

```
X               between.columns   line.size          pch.samples.size
X.new.samples   char.legend.size  line.width         predictions.sample
scaled.mat      c.hull.n          marker.size        specify.bags
e.vects         colour.scheme     max.num            specify.classes
alpha           colours           n.int              text.width.mult
ax              columns           parlegendmar       Title
ax.name.col     exp.factor        parplotmar         Tukey.median
ax.name.size    label             plot.class.means   zoomval
ax.type         label.size        pch.means
ax.col          legend.type       pch.means.size
between         line.length       pch.samples
```

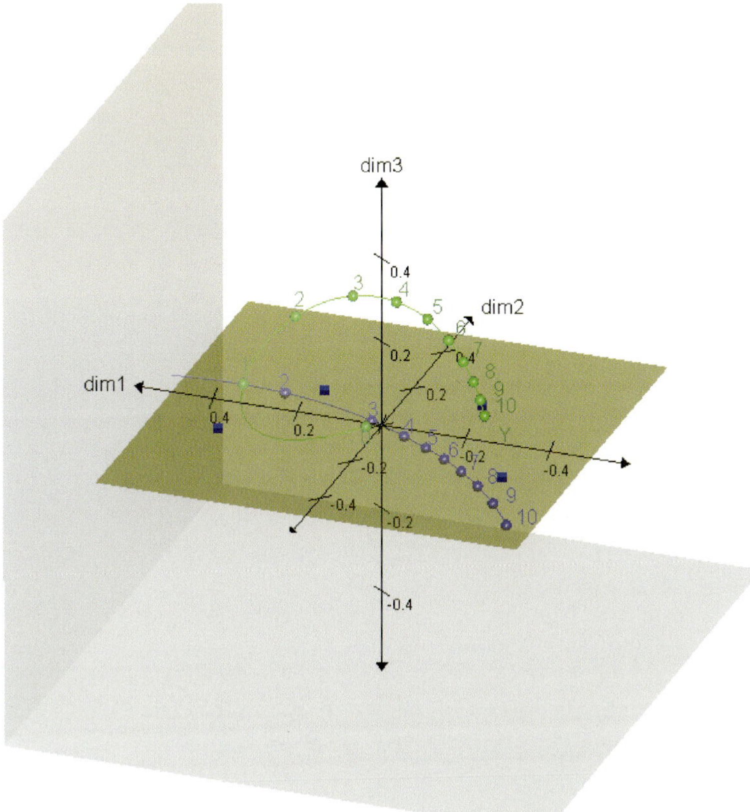

Figure 5.23 Back-projection prediction biplot axis for the original variable $\boldsymbol{Y}$.

A number of arguments are specific to `Nonlinbipl` and need special consideration.

Arguments

`dist`	The distance function for calculating the **D** matrix. In the current implementation, one of `"Pythagoras"` , `"Clark"` or `"SqrtL1"`. Can easily be extended to any qualifying distance metric, as described below. Default is `"Pythagoras"`.
`num.points`	Default is 25. A larger value (100 or 200) results in better-looking nonlinear biplot axes, but slows down the construction when `prediction.type = "normal"` is chosen.
`prediction.type`	One of `"circle"`, `"normal"` or `"back"`. Default is `"circle"`.
`scale.axes`	Scaling factor incorporated in graphical interpolation. Default is FALSE. See discussion below.
`straight`	If set to TRUE when `dist = "Pythagoras"`, sample predictions are obtained as with `PCAbipl`. Should be set to FALSE for all other distance metrics.

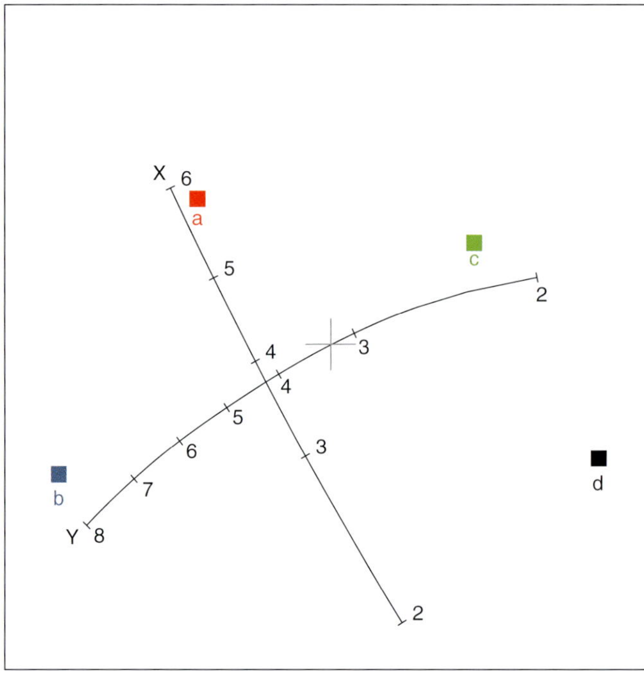

Figure 5.24 Nonlinear biplot with back-projection prediction biplot trajectories.

Details

In principle, `Nonlinbipl` will accept any allowable Euclidean embeddable distance measure that is additive. Nearly all that is needed is to add to `Nonlinbipl` a one-line specification of a typical term of the distance function. The following specifications serve as examples of how this should be done for Pythagoras distance, square root of Manhattan (5.5) and Clark distance (5.6):

```
{ if (dist == "Pythagoras") dist.expr <- function (xik, xjk)
   { (xik-xjk)^2 }
if (dist == "SqrtL1") dist.expr <- function (xik, xjk)
   { abs(xik-xjk) }
if (dist == "Clark") dist.expr <- function (xik, xjk)
   { ((xik-xjk)/(xik+xjk))^2 }
}
```

`Nonlinbipl` then uses the above specification to calculate the matrix **D** to allow the computations required by (5.7), (5.9), (5.13) and (5.16). In order to perform the integration in (5.19) for the currently included distance functions, the derivative in (5.16) is first symbolically expressed in terms of an R expression with the following function defined within `Nonlinbipl`:

```
ddmu.expr <- function (x)
 {  if (dist == "Pythagoras" || dist == "Clark")
     { dist.text <- deparse(dist.expr)[3]
       dist.text <- paste ("-0.5", dist.text, sep = "*")
       dist.text <- gsub ("xik", x, dist.text)
       dist.text <- gsub ("xjk", "mu", dist.text)
       dist.expression <- parse (text = dist.text)
       AA <- D(dist.expression,"mu")
     }
    if(dist == "SqrtL1") AA <- expression(-(sign(x-mu)))
    return(AA)
 }
```

The `derivative` of any newly defined distance function must also be specified in a function `ddmu.expr` according to the example code given above. The built-in R function `D` is then called recursively to obtain the expression for the second-order derivative as discussed in Section 5.4.2.1 before the integral (5.16) is numerically solved using `my.integrate`, a slightly modified version of the built-in R function `integrate`.

Value

A list with the following components:

`D`	The $n \times n$ matrix $\{-\frac{1}{2}d_{ij}^2\}$.
`Z`	A matrix with n rows and five columns containing the details of each sample point to be plotted: two-dimensional coordinates, plotting character, colour, line type.
`Z.axes`	A list with each component a matrix containing the details of a biplot axis to be plotted: two-dimensional coordinates, marker value, indicator if value is shown as a calibration on the biplot axis.
`e.vals`	All the eigenvalues (squared singular values).
`quality`	Overall quality of two-dimensional biplot display.
`predictions`	Sample predictions for samples specified in `predictions.samples`. Only available if `straight = TRUE` when `dist = "Pythagoras"`. Use function `CircularNonLinear.predictions` described below to obtain sample predictions for nonlinear biplots in general.
`O.co`	Coordinates of the point of concurrency of biplot trajectories.

5.6.2 Function `CircularNonLinear.predictions`

This function calculates the predictions on all variables for any point in the display space of a two-dimensional nonlinear biplot.

Arguments

`ToProject`	A two-component vector consisting of the coordinates of the point to be projected. The coordinates must be in terms of the scaffolding axes, e.g. the first two elements in a row of the element `Z` returned by `Nonlinbipl`.
`g`	The point of concurrency of the axes in the form of a two-component vector consisting of the coordinates in terms of the scaffolding axes.
`Biplot.axis`	A list with elements in the form of matrices: one matrix for each trajectory. Each of these matrices has two columns representing the two-dimensional coordinates for constructing the trajectory.
`Biplot.markers`	A list with elements in the form of vectors: one vector for each trajectory. Each of these vectors contains the calibrations associated with a trajectory. Calibrations correspond to the rows of the respective matrices in `Biplot.axis`.

Value

A list with the following two elements:

`Markers`	A p-vector, where p represents the number of biplot trajectories. Each element in the vector is the prediction for the (sample) for that particular variable.
`Coordinates`	The coordinates on the circle predictive biplot trajectory of the corresponding prediction in `Markers`.

5.7 Examples

5.7.1 A PCA biplot as a nonlinear biplot

In Section 5.5 we showed under what circumstances a PCA biplot can be regarded as a nonlinear biplot. Here, we use the *Ocotea* data of Table 3.9 to illustrate how to obtain a PCA biplot with `Nonlinbipl` in comparison with the output of `PCAbipl`. These biplots are given in Figure 5.25 and are constructed with the function calls

```
PCAbipl(X = Ocotea.data[,3:8], scaled.mat = TRUE,
   colours = "blue", pch.samples = 15, pos = "Hor", offset =
   c(-0.3, 0.1, 0.1, 0.1), offset.m = rep(-0.2, 6),
   reflect = "x")
Nonlinbipl(X = Ocotea.data[,3:8], dist = "Pythagoras",
   scaled.mat = TRUE, colours = "blue", pch.samples = 15)
```

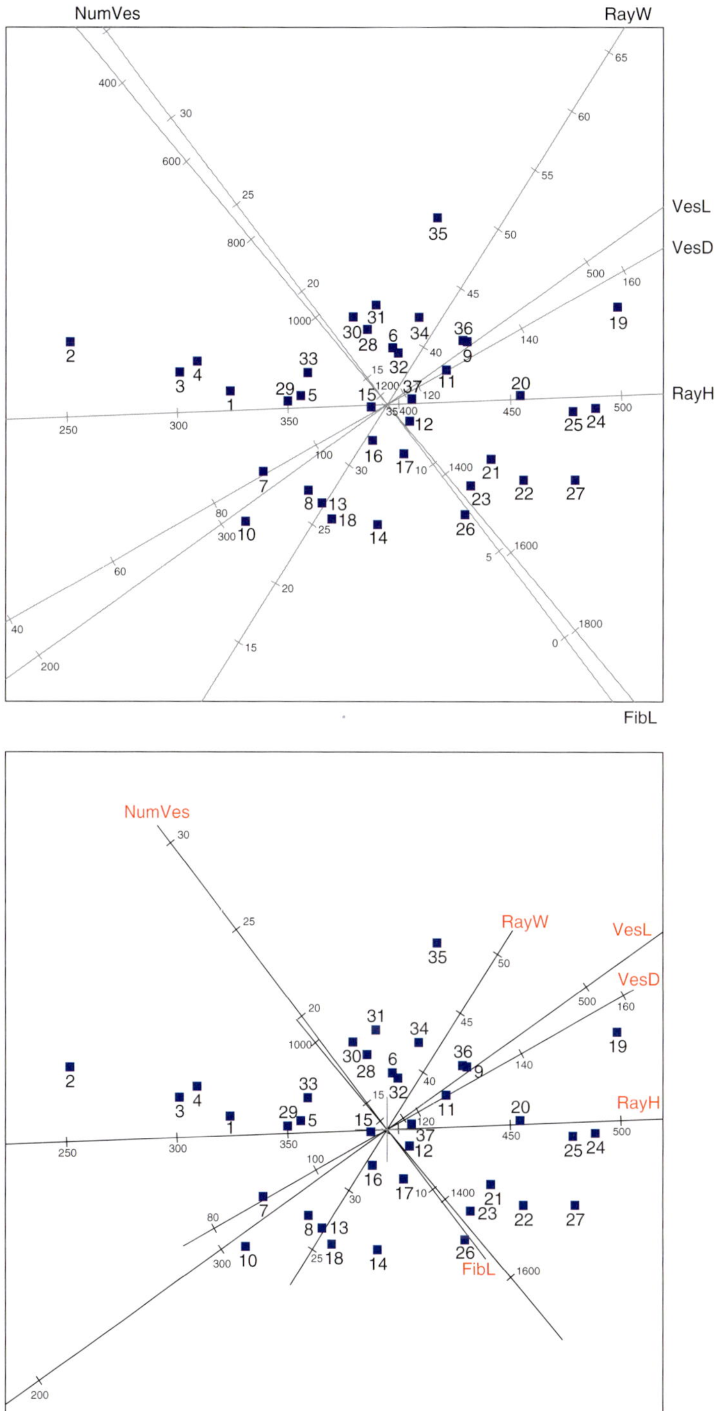

Figure 5.25 Predictive PCA biplot of the *Ocotea* data: (top) output of `PCAbipl`; (bottom) output of `Nonlinbipl`.

All that is needed to obtain a PCA biplot with `Nonlinbipl` is to specify the distance as `"Pythagoras"`. The reader can verify that changing `prediction.type` from `"circle"` to `"normal"` or to `"back"` does not change the appearance of the biplot in the bottom panel of Figure 5.25. Although both the above functions can be used for PCA biplots, we recommend the use of `PCAbipl` because of its superior facilities for fine-tuning a PCA biplot.

5.7.2 Nonlinear interpolative biplot

Calling `Nonlinbipl` with `X = aircraft.data[,2:5]` and `ax.type = "interpolative"` while `dist` is assigned `"Pythagoras"` or `"SqrtL1"` or `"Clark"`, the nonlinear interpolative biplots given in Figure 5.26 are obtained.

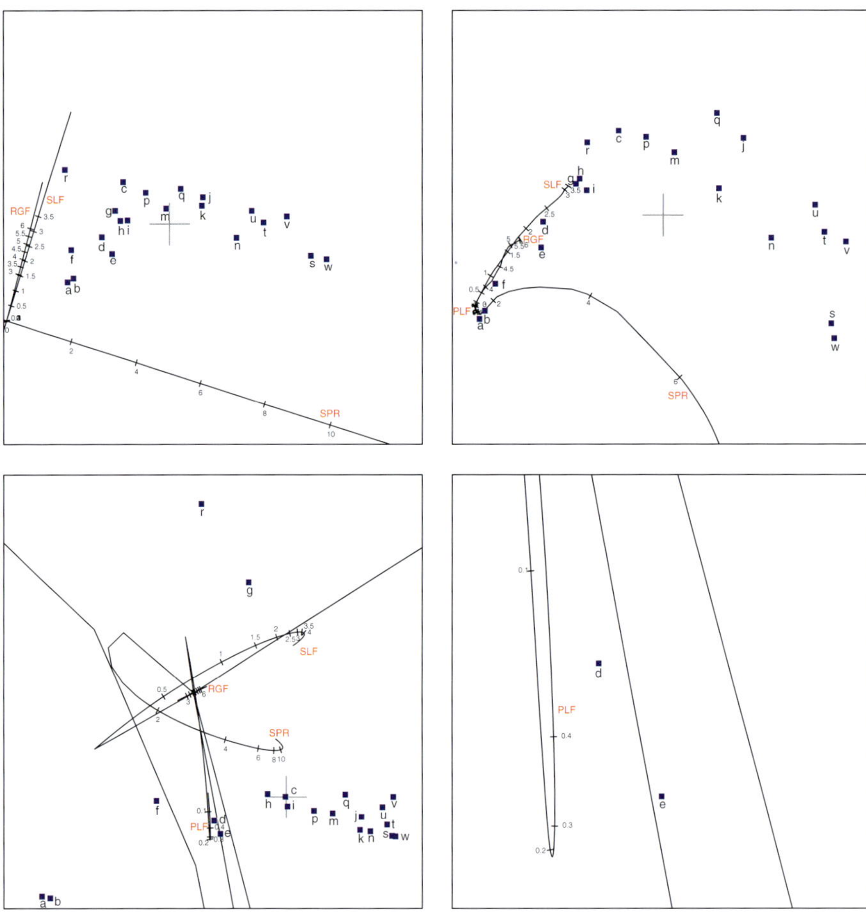

Figure 5.26 Interpolative nonlinear biplots of the aircraft data: (top left) Pythagorean distance; (top right) square root of Manhattan distance; (bottom left) Clark's distance (after adding 0.0001 to the data matrix); (bottom right) zooming into the positions of aircraft `d` and `e` in the bottom left panel biplot.

Although the interpolation axes provide us with a method of graphical interpolation, the highly nonlinear nature of the axes for Clark's distance makes the use of these axes quite difficult in practice. As mentioned before, it is intuitive to use axes to read off values and therefore we recommend that these biplots are fitted with prediction biplot axes. In the following section we do provide an example of the vector-sum interpolation method with a different data set.

5.7.3 Interpolating a new point into a nonlinear biplot

To illustrate the application of the vector-sum method for interpolating a new sample into a nonlinear biplot we consider the soils data described by Gower and Harding (1988). This data set consists of 12 trace elements (in the logarithm of parts per million) measured at 15 sites in Glamorganshire. Interpolative biplots of this data set are shown in Figure 5.27. The function calls for constructing these biplots are:

```
Nonlinbipl(soils.data, ax.type = "interpolative",
    dist = "SqrtL1")
Nonlinbipl(soils.data, ax.type = "interpolative",
    dist = "SqrtL1", zoomval = 0.1)
Nonlinbipl(soils.data, ax.type = "interpolative",
    dist = "SqrtL1", ax = c(5,7,9,12), scale.axes = TRUE)
Nonlinbipl(soils.data, ax.type = "interpolative", dist = "SqrtL1",
    ax = c(5,7,9,12), scale.axes = TRUE, zoomval = 0.15)
```

We illustrate the vector-sum method for interpolating a new point into a nonlinear biplot in Figure 5.28. Notice that the blue arrow originates in the point of concurrency, O, and not in the position of the grey cross. This process involves the R code given below. Note that the coordinates of the point of concurrency are returned by `Nonlinbipl`. These coordinates are assigned to the argument `from` in the call to `vectorsum.interp`. Note also that the argument `p` in the latter call is assigned the value 1 and not 4, since the factor of 4 has already been included in the construction of the biplot.

```
> out <- Nonlinbipl(soils.data, ax.type = "interpolative",
  dist = "SqrtL1", scale.axes = TRUE, ax = c(5,7,9,12))
> draw.arrow(col = "green", angle = 30, length = 0.3, lwd = 2.5)
#(four times)
> vectorsum.interp(vertex.points = 4, p = 1, from = out$O.co,
  pch.centroid = "", col.centroid = "blue",
  pch.interp = "", border = "red", length = 0.25, angle = 30)
> draw.text(string = "P", cex = 1.5, col = "blue")
```

5.7.4 Nonlinear predictive biplot with Clark's distance

From the distance function

$$d_{ij}^2 = \sum_{k=1}^{p} \left(\frac{x_{ik} - x_{jk}}{x_{ik} + x_{jk}} \right)^2$$

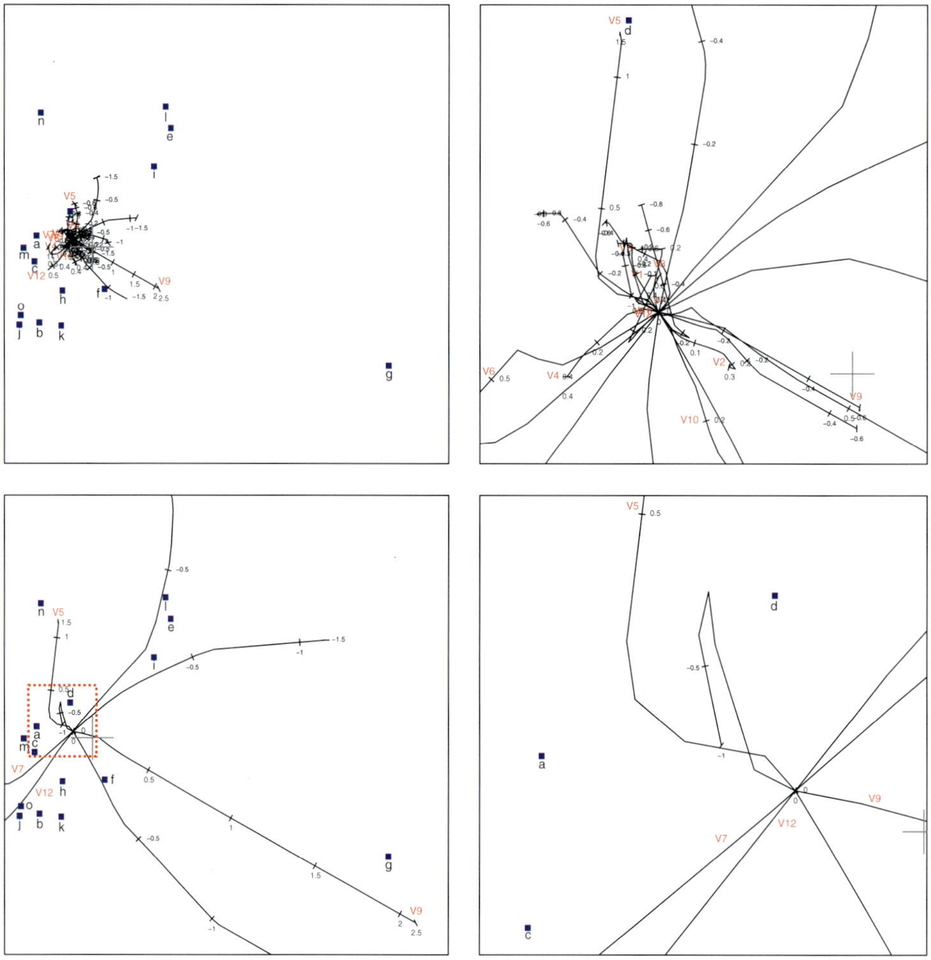

Figure 5.27 Interpolative nonlinear biplots of the soils data (see Gower and Harding, 1988). Distance metric: square root Manhattan. (Top left) Nonlinear biplot showing all 12 biplot axes. (Top right) Using the zooming facility of `Nonlinbipl` to zoom into the marked rectangle in the top left panel. (Bottom left) Interpolative biplot based upon all 12 variables but showing only the four axes that are not judged to be ineffective by Gower and Harding (1988). Trajectories are scaled from the point of concurrency, O, by a factor of 4, the number of variables shown. The grey cross marks the interpolant of the mean of **Y**, i.e. the point $(0, 0, \ldots, 0)$. (Bottom right) Enlargement of red rectangle shown in the bottom left panel.

it is clear that Clark's distance performs a scaling in the inter-sample distances in that smaller sample values are given heavier weights when dividing by small values $(x_{ik} + x_{jk})$. However, this distance is only defined for strictly positive values. The aircraft data given in Table 1.1 show a zero value for sample $\boldsymbol{g}$ on $\boldsymbol{PLF}$. The variable $\boldsymbol{PLF}$ in general has much smaller values than the other variables. Unscaled, this variable contributes far less

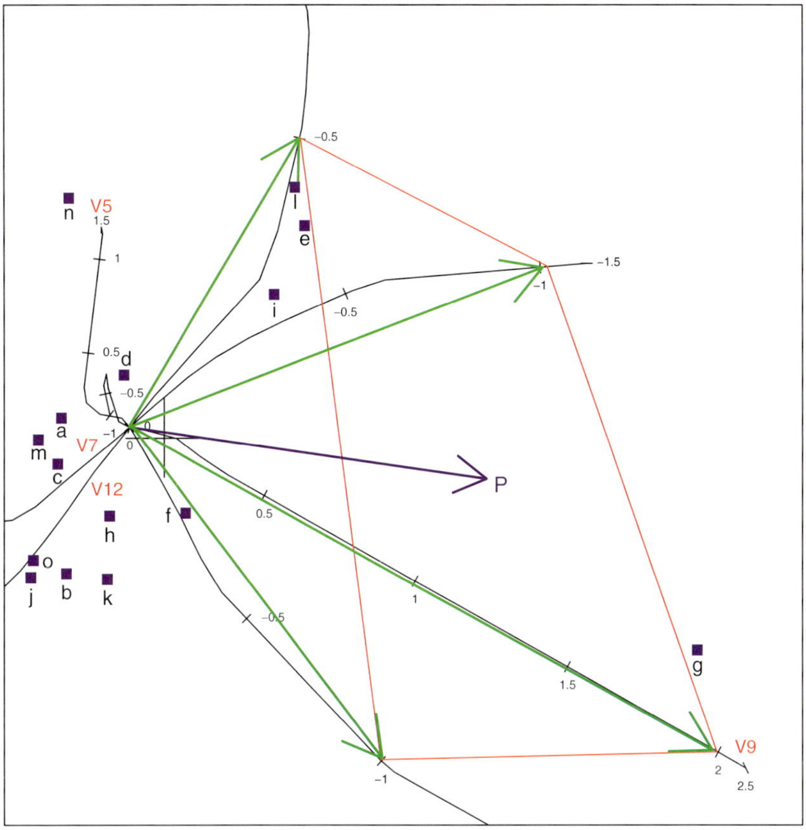

Figure 5.28 The interpolative nonlinear biplot of the soils data as constructed in the bottom left panel of Figure 5.27, illustrating the graphical interpolation of a point P with values ***V5*** $= -1.0$, ***V7*** $= -1.1$, ***V9*** $= 2$ and ***V12*** $= -0.5$.

to the inter-sample distances than the other three variables. By way of illustration we translate the data set by adding 0.2 to each value for each variable, thus resolving any potential problems with a zero value, and then calculate the squared inter-sample Clark's distances as given in Table 5.1.

The biplots for normal projection and circle projection are shown in Figures 5.29 and 5.30, respectively. The function call for constructing the biplot in Figure 5.29 is

```
Nonlinbipl(X = aircraft.data[,2:5]+0.2, ax.type = "predictive",
   dist = "Clark", prediction.type = "normal", colours = "blue",
   pch.samples = 15, num.points = 50)
```

Changing the argument `prediction.type = "circle"` leads to the biplot in Figure 5.30. Although the biplot trajectories appear different, the different projection methods result in equivalent predictions. In Figure 5.30 we show the circle predictions for aircraft ***f*** and ***g*** by calling `circle.projection.interactive(col = "green", cent.line = TRUE)` and selecting the required position. The function

Table 5.1 Squared inter-sample Clark's distances for the aircraft data when 0.2 is added to the data matrix.

	a	b	c	d	e	f	g	h	i	j	k	m	n	p	q	r	s	t	u	v	w
a	0.00	0.00	0.74	0.46	0.45	0.33	0.75	0.63	0.69	0.97	0.97	0.84	0.99	0.80	0.92	0.96	1.13	1.13	1.09	1.16	1.16
b	0.00	0.00	0.72	0.44	0.43	0.34	0.73	0.60	0.67	0.93	0.94	0.80	0.94	0.77	0.88	0.96	1.10	1.09	1.03	1.11	1.12
c	0.74	0.72	0.00	0.18	0.22	0.29	0.11	0.06	0.03	0.12	0.14	0.06	0.20	0.02	0.09	0.41	0.32	0.27	0.22	0.28	0.34
d	0.46	0.44	0.18	0.00	0.01	0.08	0.28	0.09	0.14	0.36	0.37	0.25	0.35	0.22	0.32	0.59	0.47	0.42	0.44	0.54	0.52
e	0.45	0.43	0.22	0.01	0.00	0.11	0.32	0.12	0.15	0.35	0.35	0.25	0.32	0.23	0.33	0.71	0.44	0.40	0.44	0.53	0.49
f	0.33	0.34	0.29	0.08	0.11	0.00	0.29	0.17	0.24	0.55	0.58	0.41	0.55	0.37	0.49	0.43	0.71	0.71	0.65	0.74	0.74
g	0.75	0.73	0.11	0.28	0.32	0.29	0.00	0.08	0.09	0.21	0.29	0.13	0.28	0.13	0.14	0.34	0.46	0.48	0.28	0.33	0.46
h	0.63	0.60	0.06	0.09	0.12	0.17	0.08	0.00	0.02	0.15	0.21	0.08	0.18	0.07	0.11	0.44	0.33	0.32	0.22	0.30	0.36
i	0.69	0.67	0.03	0.14	0.15	0.24	0.09	0.02	0.00	0.10	0.12	0.03	0.14	0.02	0.08	0.47	0.27	0.25	0.20	0.26	0.29
j	0.97	0.93	0.12	0.36	0.35	0.55	0.21	0.15	0.10	0.00	0.04	0.02	0.03	0.05	0.01	0.64	0.10	0.09	0.03	0.05	0.11
k	0.97	0.94	0.14	0.37	0.35	0.58	0.29	0.21	0.12	0.04	0.00	0.04	0.05	0.06	0.08	0.74	0.09	0.07	0.10	0.12	0.10
m	0.84	0.80	0.06	0.25	0.25	0.41	0.13	0.08	0.03	0.02	0.04	0.00	0.06	0.01	0.02	0.57	0.16	0.14	0.08	0.12	0.17
n	0.99	0.94	0.20	0.35	0.32	0.55	0.28	0.18	0.14	0.03	0.05	0.06	0.00	0.11	0.06	0.73	0.03	0.05	0.03	0.04	0.04
p	0.80	0.77	0.02	0.22	0.23	0.37	0.13	0.07	0.02	0.05	0.06	0.01	0.11	0.00	0.04	0.51	0.22	0.18	0.14	0.19	0.23
q	0.92	0.88	0.09	0.32	0.33	0.49	0.14	0.11	0.08	0.01	0.08	0.02	0.06	0.04	0.00	0.56	0.17	0.15	0.05	0.08	0.17
r	0.96	0.96	0.41	0.59	0.71	0.43	0.34	0.44	0.47	0.64	0.74	0.57	0.73	0.51	0.56	0.00	0.87	0.89	0.70	0.73	0.86
s	1.13	1.10	0.32	0.47	0.44	0.71	0.46	0.33	0.27	0.10	0.09	0.16	0.03	0.22	0.17	0.87	0.00	0.03	0.07	0.07	0.00
t	1.13	1.09	0.27	0.42	0.40	0.71	0.48	0.32	0.25	0.09	0.07	0.14	0.05	0.18	0.15	0.89	0.03	0.00	0.09	0.11	0.05
u	1.09	1.03	0.22	0.44	0.44	0.65	0.28	0.22	0.20	0.03	0.10	0.08	0.03	0.14	0.05	0.70	0.07	0.09	0.00	0.01	0.07
v	1.16	1.11	0.28	0.54	0.53	0.74	0.33	0.30	0.26	0.05	0.12	0.12	0.04	0.19	0.08	0.73	0.07	0.11	0.01	0.00	0.06
w	1.16	1.12	0.34	0.52	0.49	0.74	0.46	0.36	0.29	0.11	0.10	0.17	0.04	0.23	0.17	0.86	0.00	0.05	0.07	0.06	0.00

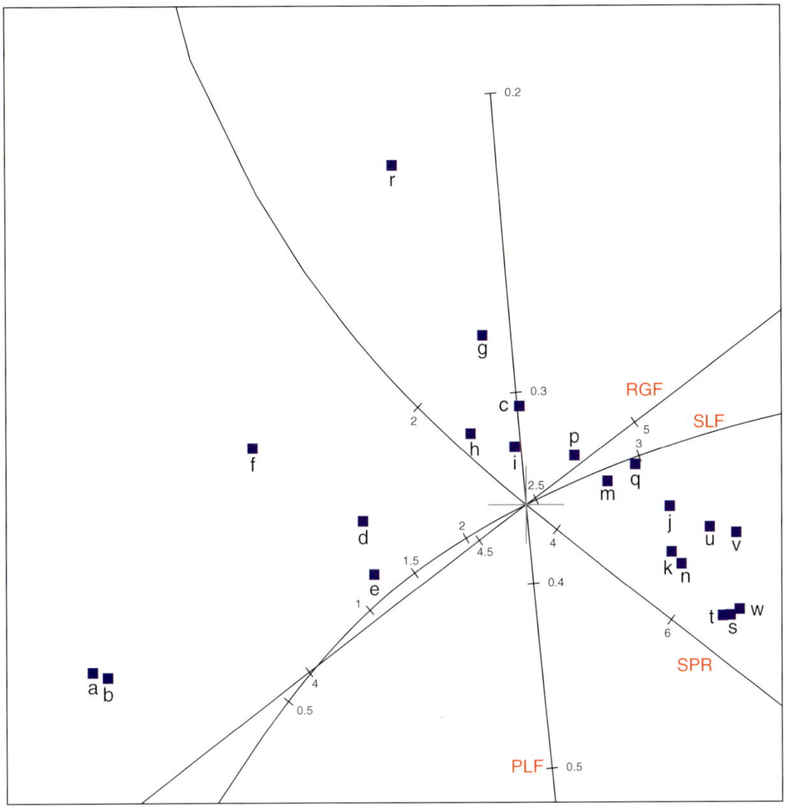

Figure 5.29 Nonlinear biplot based on Clark's distance for aircraft data with normal projection prediction biplot trajectories.

`CircularNonLinear.predictions` can be called as shown below for aircraft $\boldsymbol{f}$ and $\boldsymbol{g}$ for returning the calculated predictions of any desired point:

```
> out <- Nonlinbipl(X = aircraft.data[,2:5]+0.2, ax.type =
  "predictive", dist = "Clark", prediction.type = "circle",
  colours = "blue", pch.samples = 15, num.points = 50)
> CircularNonLinear.predictions(out$Z[7,1:2], Biplot.axis =
  lapply(out$Z.axes, FUN = function(x)x[,1:2]),
  Biplot.markers = lapply(out$Z.axes,
  FUN = function(x)x[,3]))$Markers
```

The return value (i.e. the predictions for aircraft $\boldsymbol{f}$ and $\boldsymbol{g}$) of the above call is:

	SPR	RGF	PLF	SLF
f	1.6002	4.1932	0.3157	1.1097
g	1.9815	4.8125	0.2744	2.5983

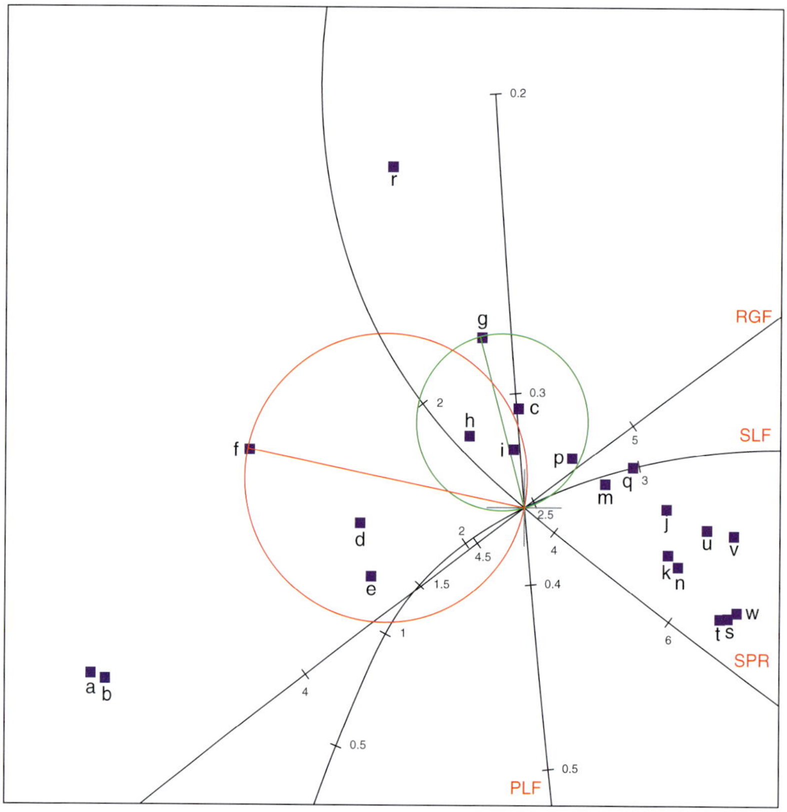

Figure 5.30 Nonlinear biplot based on Clark's distance for aircraft data with circle projection prediction biplot trajectories.

Although we currently do not know how to obtain predictions from back-projection nonlinear prediction trajectories, we can construct such a biplot by assigning `prediction.type = "back"` in the above call. The resulting biplot is shown in Figure 5.31.

5.7.5 Nonlinear predictive biplot with square root of Manhattan distance

Finally, we give an example of predictive biplots when square root of Manhattan distances are used. This distance function differs from our previous examples in that (a) it has a discontinuity at zero and (b) its second-order derivative vanishes. The last property means that normal projection prediction trajectories cannot be constructed with `Nonlinbipl`. We again use the aircraft data to obtain circle and back-projection prediction trajectories. The function calls are similar to those encountered in Section 5.7.4 except for the changes `dist = "SqrtL1"`, `prediction.type = "circle"` or `"back"`. The biplots are shown in Figure 5.32. The effect of the absolute value can clearly be seen.

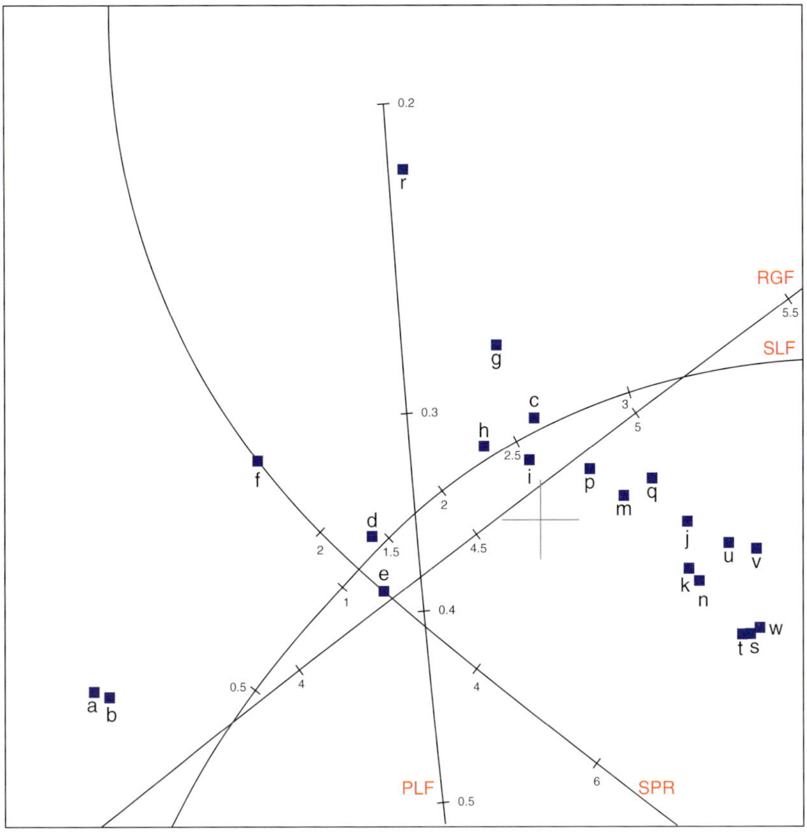

Figure 5.31 Nonlinear biplot based on Clark's distance for aircraft data with back-projection prediction biplot trajectories.

At every sample point there is an abrupt change – that is, every time we calculate (5.16) for $\mu = x_{ik}$. As μ varies, we see the sharp bends in the prediction biplot axes at each of the points where $\mu = x_{ik}$.

5.8 Analysis of distance

We saw in Chapter 4 how CVA gives a useful method for describing and assessing the differences between the means of K groups. In this chapter, so far we have seen how the ideas of PCA can be generalized to cope with nonlinear distances. In this section we show how a similar generalization can be made when handling grouped data. In CVA a key concept has been the use of Mahalanobis distance. In Section 4.1 the PCA biplot of the *Ocotea* data was compared to the CVA biplot which takes the grouping into account (see Figures 4.1 and 4.2). A fairer comparison would have been to perform a PCA on the group means and interpolate the individual samples into this space. This is illustrated in Figure 5.33.

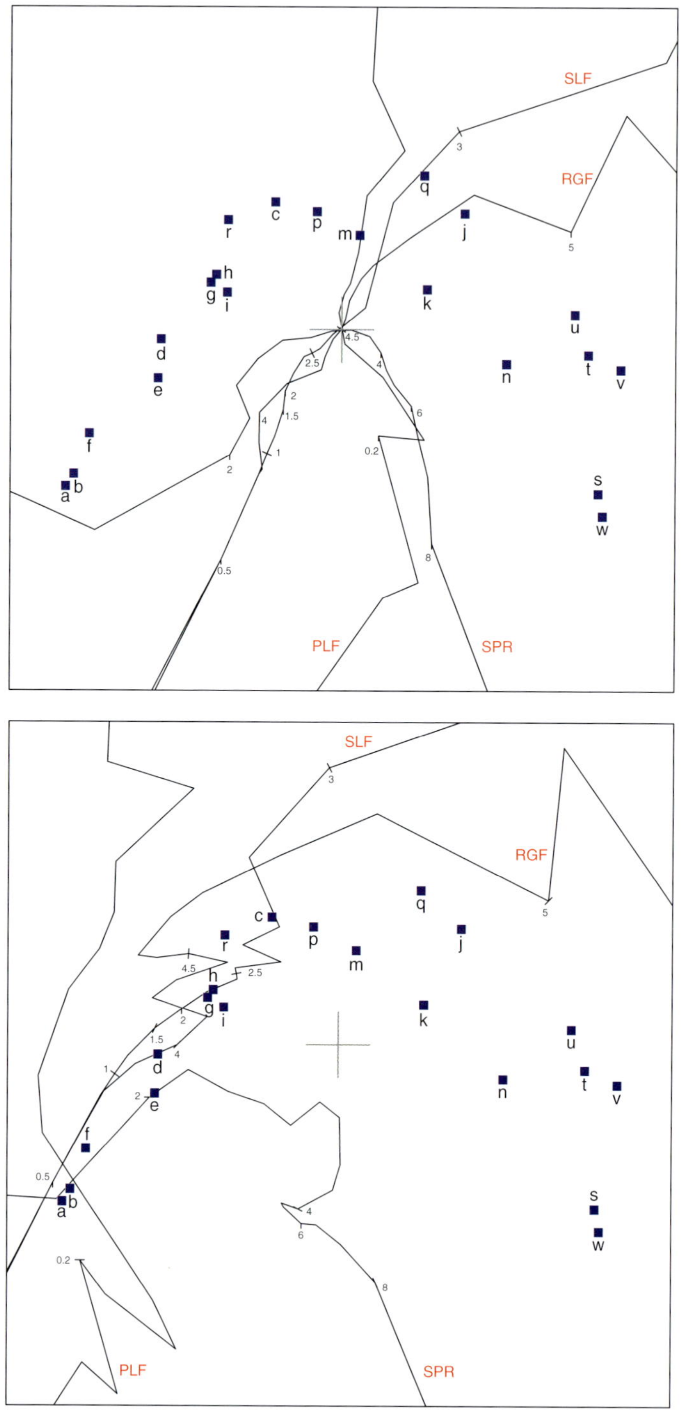

Figure 5.32 Nonlinear prediction biplots based on the square root of the Manhattan distance of the aircraft data: (top) circle prediction trajectories; (bottom) back-projection prediction trajectories.

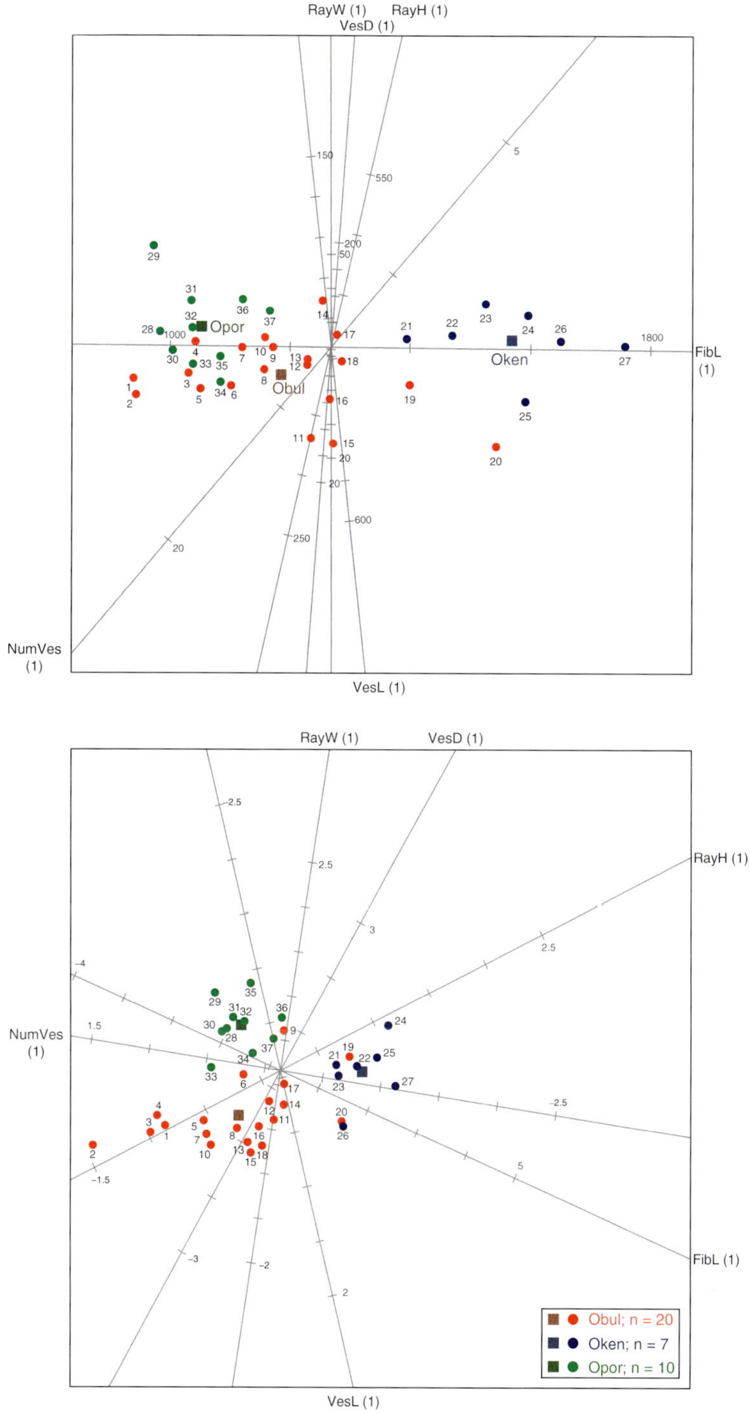

Figure 5.33 PCA biplot of *Ocotea* group means with all samples interpolated into the biplot: (top) unnormalized data; (bottom) data normalized before calculating the matrix of group means which is used as input to `PCAbipl` as described in the text.

The biplot in the bottom panel of Figure 5.33 results from the following function call:

```
PCAbipl(X = means.mat.scaled, G = indmat(1:3),
   pch.samples = rep(15,3), pch.samples.size = 1.25,
   pch.new.labels.size = 0.6, pch.new.size = 1,
   colours = UBcolours[(1:3)+12], X.new.samples = data,
   pch.new = 16, pch.new.labels = 1:37,
   pch.new.cols = rep(UBcolours[1:3],c(20,7,15)), exp.factor = 2,
   markers = FALSE, pos = "Hor", offset = c(0.02, 0.3, 0.1, 0.1),
   offset.m = rep(-0.1, 6), n.int = rep(2,6))
```

First we compare this representation with that given in Figure 4.1. The top panel is concerned with unnormalized data which, as we have seen, can have a profound effect on PCA, as is verified by comparing with the bottom panel showing an unstructured but normalized PCA. The normalization may be regarded as a first step towards removing the incommensurabilities that are fully handled by CVA itself (Figure 4.2). In Figure 5.33 (bottom panel) the three means are represented exactly in two dimensions, as in CVA, roughly at the vertices of an equilateral triangle and, as might be expected, with a different orientation than in Figure 4.1 (bottom panel). The between-group dispersion is much clearer in Figure 5.33 and has much less overlap than in Figure 4.1 (bottom panel). The grouping given by CVA in Figure 4.2 is even clearer but looks remarkably similar to Figure 5.33. Apart from ***Numves***, the corresponding biplot axes in Figure 4.2 and the bottom panel of Figure 5.33 are almost identical. The PCA of the group means with added within-group dispersion has worked very well and might be considered as a model for further development.

We could proceed as in the above PCA example by doing a PCO of $\mathbf{D}$. Then we would evaluate the group means to produce a map of the K group means by using PCA. Finally, we would rotate all n samples so that the group means occupy the first $K - 1$ dimensions and show the within-group samples in this space. A problem with this approach is that n may be very large, entailing a massive eigendecomposition. This problem can be avoided by using the methodology described below, which requires the whole of $\mathbf{D}$ but the eigenstructure of only a $K \times K$ matrix.

The method followed in constructing the biplots in Figure 5.33 may be generalized to an analysis of distance (AoD) where the *ddistances* between all pairs of samples are available in the form of an $n \times n$ matrix $\mathbf{D} = \{-\frac{1}{2}d_{ij}^2\}$. Distances may be defined very generally, though it is desirable that they be Euclidean embeddable as we assume here. In addition, as above, we have grouping information available in $\mathbf{G}$ (see Section 4.2) with associated partitioning of $\mathbf{D}$ conveniently written as

$$\begin{bmatrix} \mathbf{D}_{11} & \mathbf{D}_{12} & \cdots & \mathbf{D}_{1K} \\ \mathbf{D}_{21} & \mathbf{D}_{22} & \cdots & \mathbf{D}_{2K} \\ \vdots & \vdots & \cdots & \vdots \\ \mathbf{D}_{K1} & \mathbf{D}_{K2} & \cdots & \mathbf{D}_{KK} \end{bmatrix},$$

though there is no requirement in the following that the n samples be presented in the implied group-by-group order.

We may evaluate an average $K \times K$ matrix,

$$\overline{\mathbf{D}} = \mathbf{N}^{-1}\mathbf{G}'\mathbf{D}\mathbf{G}\mathbf{N}^{-1}, \tag{5.21}$$

from which we may derive

$$\Delta_{hk} = -\tfrac{1}{2}(\overline{\mathbf{D}}_{kk} + \overline{\mathbf{D}}_{hh} - 2\overline{\mathbf{D}}_{hk}). \tag{5.22}$$

Gower and Hand (1996, p. 249) show that the quantity in parentheses is the squared distance between the centroids of groups h and k. Defining $\mathbf{g}_k$ to be the kth column of $\mathbf{G}$, then

$$\overline{\mathbf{D}}_{hk} = \frac{1}{n_h n_k}\mathbf{g}_h'\mathbf{D}\mathbf{g}_k$$

gives the average of the *ddistances* between the members of the hth and kth groups; when $h = k$ the zero diagonals and repeated symmetric values are all included.

The Δ_{hk} in (5.22) may be assembled into a *ddistance* matrix $\boldsymbol{\Delta} = \{\Delta_{hk}\} : K \times K$ which may be approximated by any method of multidimensional scaling (see above) to give a map of the group means analogous to the map of canonical means given by CVA. In the following we shall use PCO. This completes the between-group map. We show in Section 5.5 that when we restrict the distances to be Pythagorean, a PCO of $\boldsymbol{\Delta}$ results in a PCA of the group means, as illustrated for the *Ocotea* data in Figure 5.33. If in $\mathbf{D}$ we defined the Mahalanobis distance between all pairs of samples, rather than just the sample means, PCO would recover the CVA of the canonical means. With other choices of embeddable distance and MDS, different analyses and representations will ensue.

So far, we have been concerned with between-group structure but when PCO is used to represent $\boldsymbol{\Delta}$, it is relatively easy to add points representing the individual samples. To do this we repeatedly use the technique described in Section 5.4.1 for adding a point P to a PCO. This requires the distances from the new point to all the n original points. We assume that these are given in a column vector $\mathbf{d}$: $n \times 1$ of elements

$$\mathbf{d}' = (\mathbf{d}_1', \mathbf{d}_2', \ldots, \mathbf{d}_K'), \tag{5.23}$$

where $\mathbf{d}_k$ is a vector of size n_k giving the *ddistances* of the new point from the samples in the kth group. Note that $\mathbf{d}$ may represent a completely new sample, but often it will be one of the columns of $\mathbf{D}$. Denoting the centroids of the groups by $G_1, G_2, \ldots, G_K$, $\boldsymbol{\Delta}$ gives the *ddistances* between all pairs of centroids. We need the squared distances of P from every centroid and the *ddistances* of every centroid from the overall centroid G, say. The latter is given by

$$(\mathbf{1}'\boldsymbol{\Delta}\mathbf{1})\mathbf{1}/K^2 - 2\boldsymbol{\Delta}\mathbf{1}/K. \tag{5.24}$$

The squared distance of P from G_k is given by

$$\delta_k^2 = \overline{D}_{kk} - \frac{2}{n_k}\mathbf{1}'\mathbf{d}_k, \tag{5.25}$$

which may be assembled into a *ddistance* vector $\boldsymbol{\delta}' = (-\frac{1}{2}\delta_1^2, -\frac{1}{2}\delta_2^2, \ldots, -\frac{1}{2}\delta_K^2)$.

Suppose $\mathbf{Y}$ are the coordinates given by a PCO of $\boldsymbol{\Delta}$ – that is,

$$\mathbf{YY}' = (\mathbf{I} - \mathbf{11}'/K)\boldsymbol{\Delta}(\mathbf{I} - \mathbf{11}'/K), \tag{5.26}$$

where $\mathbf{Y}'\mathbf{Y} = \boldsymbol{\Lambda}$. Then the coordinates $\mathbf{y}$ of P are (Gower and Hand, 1996, p. 250)

$$\mathbf{y} = \boldsymbol{\Lambda}^{-1}\mathbf{Y}'(\boldsymbol{\delta} - \boldsymbol{\Delta}\mathbf{1}/K), \tag{5.27}$$

where the final term derives from (5.24) after taking into account that $\mathbf{Y}'\mathbf{1} = \mathbf{0}$, because of the centring of $\mathbf{Y}$. Equation (5.27) gives the coordinates of the added point in $K - 1$ or fewer dimensions; in r-dimensional approximations, only the first r columns of $\mathbf{Y}$ and the first r eigenvalues will be needed. A further coordinate in a Kth dimension is necessary for exact representations, but this is rarely needed – see Section 5.4 for further details. It may be verified that if $\boldsymbol{\delta}$ is successively taken to be the n_k columns pertaining to the kth group, then the centroids of the inserted points are at the same position as the kth group mean in $\mathbf{Y}$, as is shown in Section 5.8.1.

We may write $\boldsymbol{\Delta}$ in full matrix form as

$$\boldsymbol{\Delta} = \overline{\mathbf{D}} - \tfrac{1}{2}[\text{diag}(\overline{\mathbf{D}})\mathbf{11}' + \mathbf{11}'\text{diag}(\overline{\mathbf{D}})],$$

whence

$$\mathbf{n}'\boldsymbol{\Delta}\mathbf{n} = \mathbf{n}'\overline{\mathbf{D}}\mathbf{n} - n[\mathbf{n}'\text{diag}(\overline{\mathbf{D}})\mathbf{1}]. \tag{5.28}$$

From (5.28) we have

$$\mathbf{n}'\boldsymbol{\Delta}\mathbf{n} = \mathbf{1}'\mathbf{D}\mathbf{1} - n\left[\sum_{k=1}^{K} n_k \overline{D}_{kk}\right]$$

which rearranges to

$$\frac{\mathbf{1}'\mathbf{D}\mathbf{1}}{n} = \frac{\mathbf{n}'\boldsymbol{\Delta}\mathbf{n}}{n} + \sum_{k=1}^{K} \frac{\mathbf{g}_k'\mathbf{D}\mathbf{g}_k}{n_k}. \tag{5.29}$$

Recalling that $-\mathbf{1}'\mathbf{D}\mathbf{1}/n$ is the total sum of squares and $-\mathbf{g}_k'\mathbf{D}\mathbf{g}_k/n_k$ is the sum of squares within the kth group, we see that, apart from sign, the analysis of distance (5.29) is an expression of the CVA orthogonal analysis of variance:

Total sum of squares = Between-group sum of squares + Within-group sum of squares.

Thus, from (5.29) we may form an AoD table in which the contributions between and within groups are exhibited. Furthermore, we may break this down into the contribution arising from different dimensions and sets of dimensions, especially the r fitted dimensions and the remaining residual dimensions. Note that with K groups, the means fit into $K - 1$ or fewer dimensions so the remaining 'residual' dimensions for the group means are null.

We have represented within-group variation by choosing $\mathbf{d}$ in (5.23) as the successive columns of $\mathbf{D}$. However, $\mathbf{d}$ may refer to a genuine new sample, in which case (5.27) interpolates that sample into the map. Also, $\mathbf{d}$ may also be chosen as a pseudo-sample (see Section 5.4) and used to plot predictive trajectories for the variables. In this way the AoD of the individual samples may be enhanced to include information on the variables to give a true biplot. The regression method (Chapter 4) may be used to give approximate linear biplot axes. Krzanowski (2004) suggests a half-way house where a limited number of pseudo-samples (say 10) are fitted with linear axes. In principle, the formulae for adding a point may also be used to construct the convex prediction regions of generalized biplots, as described in Chapter 9.

The cloud of points surrounding each centroid may be enclosed in any region that expresses spread, analogously to the confidence circles of CVA. Thus we may use minimal covering circles or ellipses enclosing, say, all or 95% of the points, or we may use α-bags, or we may use convex hulls (Section 2.9.3). Furthermore, a nonparametric testing procedure can be used for testing as illustrated in the example below.

Finally, we note that the above uses unweighted centroids. Again, analogously to CVA, we may use centroids weighted by sample sizes. The starting point is the weighted PCO of $\mathbf{\Delta}$, where now (5.26) is replaced by

$$\mathbf{YY}' = \left(\mathbf{I} - \frac{\mathbf{1n}'}{n}\right)\mathbf{\Delta}\left(\mathbf{I} - \frac{\mathbf{n1}'}{n}\right). \tag{5.30}$$

Because $\mathbf{n}'\mathbf{Y} = \mathbf{0}'$, G, the centroid of the samples, is now at the weighted centroid of the group centroids. As with CVA, the use of a weighted centroid does not affect the distances between the individual centroids; but in approximations, groups with smaller sample size will be less well represented than those with the larger sample sizes.

5.8.1 Proof of centroid property for interpolated points in AoD

We now show that the centroids of the interpolated samples from the kth group coincide with the kth column of $\mathbf{Y}$, the coordinates of the group means. Of course, this is a trivial result in the linear case, but it is not obvious that it carries over to the nonlinear case and therefore requires proof.

We begin with the basic interpolation formula (5.27):

$$\mathbf{y} = \mathbf{\Lambda}^{-1}\mathbf{Y}'(\boldsymbol{\delta} - \mathbf{\Delta 1}/K),$$

where $\boldsymbol{\delta} = \{-\frac{1}{2}\delta_i^2\}$ with $\delta_i^2 = \overline{D}_{ii} - \frac{2}{n_i}\mathbf{1}'\mathbf{d}_i$ for $i = 1, 2, \ldots, K$. To interpolate all the samples from the kth group requires the n_k columns of $\mathbf{D}$ pertaining to the kth group. This gives n_k columns of $(\boldsymbol{\delta} - \mathbf{\Delta 1}/K)$ as follows:

$$-\frac{1}{2}\left[\begin{pmatrix} \overline{D}_{11}\mathbf{1}'_{n_k} \\ \overline{D}_{22}\mathbf{1}'_{n_k} \\ \vdots \\ \overline{D}_{KK}\mathbf{1}'_{n_k} \end{pmatrix} - 2\begin{pmatrix} \frac{1}{n_1}\mathbf{1}'\mathbf{D}_{1k} \\ \frac{1}{n_2}\mathbf{1}'\mathbf{D}_{2k} \\ \vdots \\ \frac{1}{n_K}\mathbf{1}'\mathbf{D}_{Kk} \end{pmatrix}\right] - \frac{\mathbf{\Delta 1}}{K}\mathbf{1}'_{n_k}.$$

Multiplying by $\mathbf{1}_{n_k}$ and dividing by n_k to obtain the coordinates of the centroid of the kth interpolated group gives its coordinates as

$$\mathbf{y}_k = \mathbf{\Lambda}^{-1}\mathbf{Y}'\left(-\frac{1}{2}\left[\begin{pmatrix}\overline{D}_{11}\\ \overline{D}_{22}\\ \vdots\\ \overline{D}_{KK}\end{pmatrix} - 2\begin{pmatrix}\overline{D}_{1k}\\ \overline{D}_{2k}\\ \vdots\\ \overline{D}_{Kk}\end{pmatrix}\right] - \frac{\mathbf{\Delta 1}}{K}\right).$$

Now, by definition $\Delta_{hk} = -\frac{1}{2}(\overline{D}_{hh} + \overline{D}_{kk} - 2\overline{D}_{hk})$, so the above may be written as

$$\mathbf{y}_k = \mathbf{\Lambda}^{-1}\mathbf{Y}'\left(\begin{pmatrix}\Delta_{1k} + \frac{1}{2}\overline{D}_{kk}\\ \Delta_{2k} + \frac{1}{2}\overline{D}_{kk}\\ \vdots\\ \Delta_{Kk} + \frac{1}{2}\overline{D}_{kk}\end{pmatrix} - \frac{\mathbf{\Delta 1}}{K}\right) = \mathbf{\Lambda}^{-1}\mathbf{Y}'\left(\begin{pmatrix}\Delta_{1k}\\ \Delta_{2k}\\ \vdots\\ \Delta_{Kk}\end{pmatrix} - \frac{\mathbf{\Delta 1}}{K}\right)$$

where the constant column is eliminated because of the centring $\mathbf{Y}'\mathbf{1} = \mathbf{0}$.

The dimensions of the centroids $\mathbf{y}_k$ are $r \times 1$, and if we combine all the K interpolated centroids into $\mathbf{Y}_K$ we obtain

$$\mathbf{Y}_K = \mathbf{\Lambda}^{-1}\mathbf{Y}'\left(\mathbf{\Delta} - \frac{\mathbf{\Delta 11}'}{K}\right),$$

which may be written, again because of the centring, as

$$\mathbf{Y}_K = \mathbf{\Lambda}^{-1}\mathbf{Y}'\left(\mathbf{I} - \frac{\mathbf{11}'}{K}\right)\mathbf{\Delta}\left(\mathbf{I} - \frac{\mathbf{11}'}{K}\right).$$

Finally,

$$\left(\mathbf{I} - \frac{\mathbf{11}'}{K}\right)\mathbf{\Delta}\left(\mathbf{I} - \frac{\mathbf{11}'}{K}\right) = \mathbf{YY}',$$

therefore we have

$$\mathbf{Y}_K = \mathbf{\Lambda}^{-1}\mathbf{Y}'(\mathbf{YY}') = \mathbf{\Lambda}^{-1}(\mathbf{Y}'\mathbf{Y})\mathbf{Y}' = \mathbf{Y}'.$$

So the interpolated centroids coincide with the group average positions given by $\mathbf{Y}$.

5.8.2 A simple example of analysis of distance

Note that in this example, where we use Pythagorean distance, the AoD reduces to a PCA of the group means. In this example, similar to our introductory AoD example, we now have five groups and so no exact two-dimensional group mean representation. The AoD methodology is actually redundant, although we can still use it as a motivation for performing the permutation testing procedure discussed below.

Clarke *et al.* (2003) reported on an investigation conducted at a South African wood mill of the underlying relationships between genetic (species) and physiological factors of wood as well as pulp quality. The data were placed in the public domain with the intention of inviting researchers to perform their own statistical analyses. The data are also described by Gardner *et al.* (2005) and are represented here in the form of an R dataframe `Pine.data`. This dataframe consists of the following measurements:

species	Five pine species: *Pinus elliottii, P. kesiya, P. maximinoi, P. patula* and *P. taeda*
TotYield	Total pulp yield expressed as a percentage of the original mass
Alkali	Percentage alkali consumption
Density	Wood density in kg m^{-3}
TEA	Tensile energy absorption in mJ g^{-1}
Tensile	Tensile index
Tear	Tearing index in mN m^2 g^{-1}
Burst	Burst index in kPa m^2 g^{-1}
Growth	Height in metres at 11 years

A preliminary statistical analysis of `Pine.data` shows that not only are there large differences between the within-species covariance matrices but also, due to small sample sizes, some of these covariance matrices are singular. The AoD described above needs only the assumption that a distance function exists such that inter-sample distances can be calculated between each pair of the 37 rows of `Pine.data`. In this example we restrict ourselves to Pythagorean distances between the samples (after normalizing the `Pine.data` to unit column variances) so that, as we have seen above, the AoD biplot is simply a PCA biplot of the class means with the individual samples interpolated onto the display. This biplot is shown in Figure 5.34.

The AoD biplot shows that the five group means lie at quite a distance from one another, with ***P.kes*** and ***P.tae*** closest. The latter two species are almost similar with respect to ***Tensile*** and ***Growth***; they differ mostly on ***Tear***, ***Burst*** and ***TEA***. ***P.max*** and ***P.pat*** are also almost similar on ***Tensile*** and ***Growth***, showing small values in contrast to the large values of ***P.kes*** and ***P.tae***. ***P.ell*** has moderate ***Tensile*** and ***Growth*** values but has the smallest ***Alkali*** values. If pulp is needed for a product requiring high ***Tensile*** values, then ***P.kes*** and ***P.tae*** are candidates to consider; if the product requires high density values, then ***P.max*** is better avoided, although it gives maximum ***TotYield***.

Are the differences between the five group means statistically significant? In order to answer this question we can turn to the breakdown of the total sum of squared distances that we have seen in Section 5.8 to be equivalent to the usual analysis of variance breakdown of a total sum of squares. For the normalized pine data this becomes T = 280; B = 84.6722 and W = 195.3278. A permutation testing procedure making no distributional or homogeneous covariance matrix assumptions can be employed to test the null hypothesis of no significant differences between the group means. Under the null hypothesis the B term above should remain approximately constant over random permutations of the 37 samples to form the five groups fixing the sample sizes. However, under the

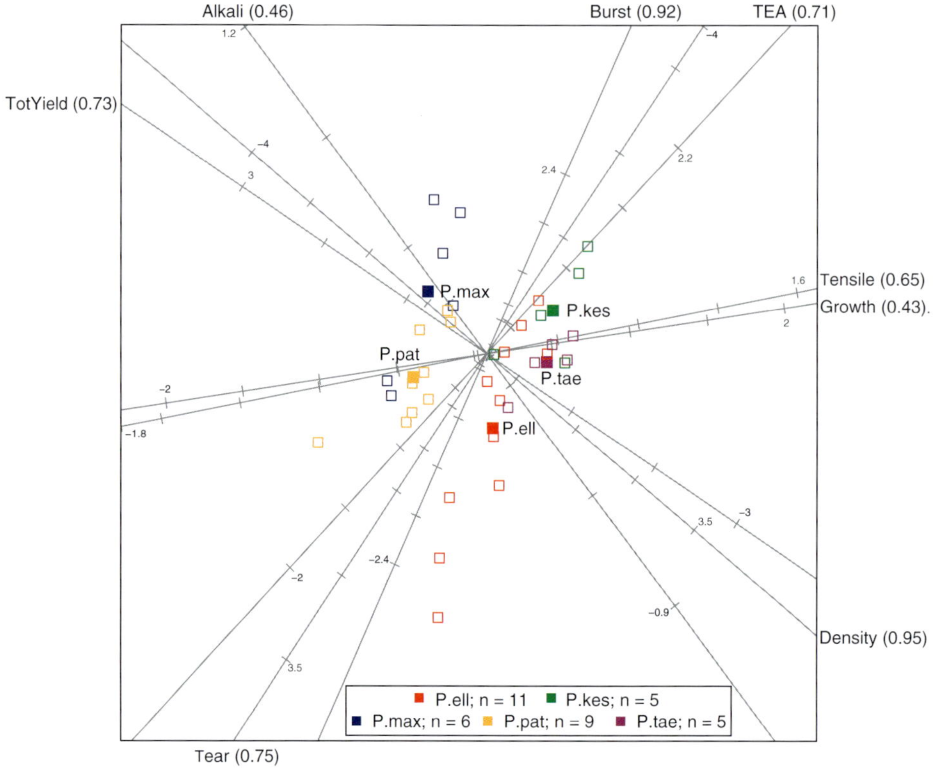

Figure 5.34 AoD biplot of the pine data. After normalizing the data, `PCAbipl` was called to construct a PCA biplot of the group means. Subsequently the colour-coded samples were interpolated into the biplot. Axis predictivities are given in parentheses.

alternative hypothesis permutation of the sampling units should tend to increase the B-value. The algorithm provided by Good (2000) is implemented in our function `PermutationAnova` to randomly permute n samples into K groups of fixed sample sizes summing to n. Applying `PermutationAnova` to the normalized pine data resulted in an achieved significance level of 0.0001.

As a final example we present in Figure 5.35 an AoD plot of the pine data based on Clark's distance. Comparing Figure 5.35 with Figure 5.34, we see that the use of Clark's distance resulted in a shift of ***P.tae*** away from ***P.kes***, while ***P.ell*** shifted to a more central position to end up closer to ***P.tae***. Figure 5.35 is the result of the following function call to `AODplot`:

```
AODplot(X = Pine.data[,-1], dist = "Clark", X.new.samples =
   Pine.data[,-1], group.vec = Pine.data[,1], exp.factor = 2.5,
   label.size = 0.6, pch.samples = rep(0,length(unique
   (Pine.data[,1]))), pch.samples.col = c("red",
   "green","blue","orange","purple"), pch.samples.size = 0.9,
   pch.means = rep(15, length(unique(Pine.data[,1]))),
   pch.means.col = c("red","green","blue","orange","purple"),
   pch.means.size = 1.5, pos.labs = 1, weight = "unweighted")
```

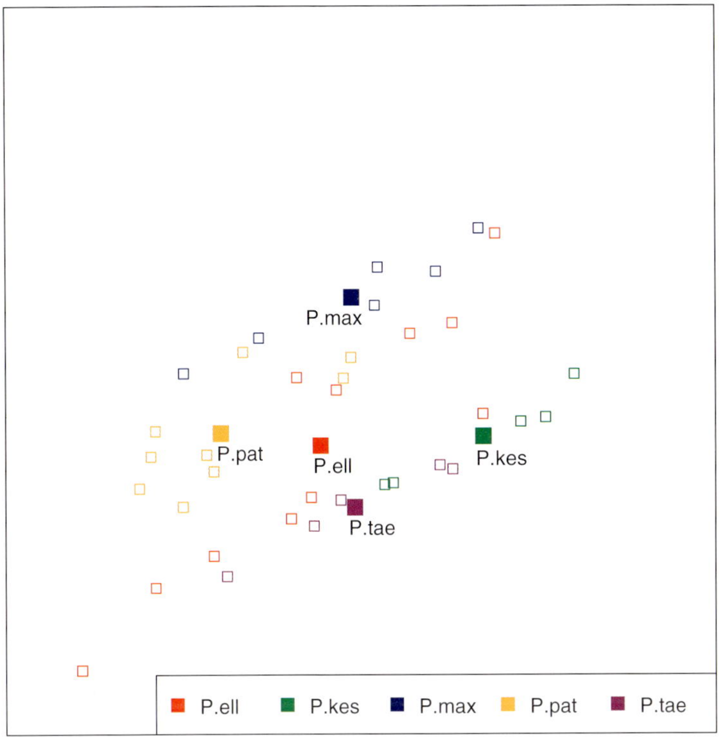

Figure 5.35 AoD plot of the pine data using the function `AODplot` with Clark's distance. Sum of squares breakdown: T = 0.5963, B = 0.1908 and W = 0.4054.

5.9 Functions `AODplot` and `PermutationAnova`

5.9.1 Function `AODplot`

This function is used for constructing the two-dimensional AoD plots described in Section 5.8. Its call and usage are similar to those of `Nonlinbipl` with which it shares the arguments:

`dist`	`label`	`pch.means.size`	`Title`
`e.vects`	`label.size`	`pch.samples`	
`exp.factor`	`pch.means`	`pch.samples.size`	

A number of arguments are specific to `AODplot` or need special consideration:

Arguments

`X` The $n \times p$ data matrix *after* performing any prescaling, e.g. centring or normalization.

`X.new.samples` A $t \times p$ matrix containing the data on t new samples *after* being subjected to the same prescaling performed on `X`. If NULL, only group means are plotted.

`group.vec`	n-vector specifying the groups.
`dim.biplot`	Implemented only for `dim.biplot = 2`.
`pch.means.col`	K-vector specifying the colour of each group's plotting character.
`pch.samples.col`	K-vector specifying the colour of plotting character of the samples in each of the K groups.
`pos.labs`	Position of the group labels.
`weight`	Either `"unweighted"` (the default) or `"weighted"`, requesting an unweighted or weighted AoD.

Value

An AoD plot as discussed in Section 5.8 is constructed and a list with the following components is returned:

`T.ss`	The total sum of squares, T.
`B.ss`	The between sum of squares, B.
`W.ss`	The within sum of squares, W.
`nvec`	The K-vector containing the group sizes.
`Dmat`	The $n \times n$ *ddistance* matrix **D**.
`Dbar`	The matrix $\overline{\mathbf{D}}$ defined in (5.21).
`Deltabar`	The $K \times K$ matrix with elements defined by (5.22).

5.9.2 Function `PermutationAnova`

This function performs the permutation testing procedure described in Section 5.8. It shares the following three arguments with `AODplot`: `X`, `group.vec` and `dist`. It has a fourth argument `n.perm` for specifying the number of permutations required.

Value

A list with the following three components

`TandB.ss.null`	A vector with first element the total sum of squares T and second element the between sum of squares B obtained in the initial AoD.
`out.perm.TandB`	Matrix with `n.perm` rows and two columns containing the total sum of squares T and the between sum of squares B obtained for each random permutation cycle of the AoD.
`asl`	The achieved significance level, i.e. the proportion of times a random permutation cycle resulted in a value of B larger than the B obtained in the initial AoD.

6 *Two-way tables: biadditive biplots*

6.1 Introduction

Previous chapters have been concerned with biplots for a variety of forms of data matrix where, typically, the rows refer to n samples and the columns to p variables. As we have seen, samples and variables are very different concepts, entailing different kinds of statistical treatment. In this chapter we shall be concerned with two-way tables with p rows and q columns which refer to similar entities – hence our change of notation. The body of the table may contain

(i) the numerical values of a single variable or

(ii) counts, so defining a contingency table, or

(iii) the values of a single categorical variable.

This chapter concerns problem (i); contingency tables are covered by Chapter 7, while issue (iii) is covered in Chapter 8. Crucially, in these chapters the body of the table is regarded as a dependent variable, with the rows and column classifiers treated as independent variables. In case (i) we shall see that additive or biadditive models may be fitted and, of particular interest for biplots, additional multiplicative terms included. Case (iii) may be handled by replacing the categories with optimal scores and then treated as in (i), but can also be handled by multiple correspondence analysis with three variables – that labelling the rows, that labelling the columns and, thirdly, the categorical variable in the body of the table – but then the dependency relationship is ignored (see Chapter 8).

Understanding Biplots John Gower, Sugnet Lubbe and Niël le Roux

6.2 A biadditive model

Consider the data given in Table 6.1. The measurements are the yields in grams per square metre (g/m^2) of 12 varieties of winter wheat grown at seven sites at each of two levels (low and high) of nitrogen. Varieties *Cappelle* (***Cap***), *Ranger* (***Ran***), *Huntsman* (***Hun***), *Templar* (***Tem***) and *Kinsman* (***Kin***) are classified as conventional varieties, while *Fundin* (***Fun***), *Durin* (***Dur***), *Hobbit* (***Hob***), *Sportsman* (***Spo***), *TJB 259/95* (***T95***), *TJB 325/464* (***T64***) and *TJB 368/268* (***T68***) are classified as semi-dwarf varieties. These data are described in detail by Blackman *et al.* (1978) and are available as the R matrix object `wheat.data`. An early example of a biadditive biplot using these data was given by Kempton (1984).

Apart from the column of row means and the row consisting of the column means, the data contained in Table 6.1 appear to have the same form as the *samples* $\times$ *variables* data matrix $\mathbf{X}$: $n \times p$ encountered in previous chapters. However, the body of the table is quantitative, all elements referring to the same variable, unlike a data matrix whose columns refer to different variables. Therefore it would be wrong to normalize the columns as is often justifiable for a data matrix. Indeed, the two-way table is concerned with only three variables: the row and column factors and the values in the body of the table. Although there is possible interest in main effects, there is at least as much, if not more, interest in the interaction of varieties with sites. The two-way table will be written as $\mathbf{X}$: $p \times q$, where p denotes the number of levels of the row factor (the sites in our example) and q the number of levels of the column factor (the varieties in our example). The measurement x_{ij} can be thought of as a sample observation generated by the model

$$X_{ij} = \mu + \alpha_i + \beta_j + \sum_{k=1}^{r} \gamma_{ik}\delta_{jk} + \varepsilon_{ij}, \quad \text{for } i = 1,\dots,p \text{ and } j = 1,\dots,q. \tag{6.1}$$

The $\{\varepsilon_{ij}\}$ in (6.1) are assumed to be independently distributed with equal variances. It is often convenient to write (6.1) in matrix form:

$$\mathbf{X} = \mu\mathbf{11}' + \boldsymbol{\alpha}\mathbf{1}' + \mathbf{1}\boldsymbol{\beta}' + \sum_{k=1}^{r} \boldsymbol{\gamma}_k\boldsymbol{\delta}'_k + \boldsymbol{\varepsilon}. \tag{6.2}$$

6.3 Statistical analysis of the biadditive model

Two-way tables such as Table 6.1 may be analysed by fitting the biadditive model (6.1) to the data contained therein. So, unlike PCA, we now have a dependent variable X with $p + q$ independent dummy variables labelling the rows and columns in addition to $p + q$ dummy variables for each of r multiplicative interaction terms. The parameters $\boldsymbol{\alpha}, \boldsymbol{\beta}, \boldsymbol{\gamma}_k$, and $\boldsymbol{\delta}_k$ as well as the constant term μ are to be estimated. In its simplest form the multiplicative terms may be omitted and the additive terms, known as *main effects*, estimated by

$$\hat{\mu} = \bar{x}, \ \hat{\alpha}_i = \bar{x}_{i.} - \bar{x}_{..}, \ \hat{\beta}_j = \bar{x}_{.j} - \bar{x}_{..}, \quad \text{for } i = 1,\dots,p \text{ and } j = 1,\dots,q, \tag{6.3}$$

Table 6.1 Yields (g/m^2) of $q = 12$ varieties of winter wheat grown at seven sites at each of two levels (low and high) of nitrogen, giving $p = 14$ (From Blackman *et al.*, 1978.).

	Cap	Ran	Hun	Tem	Kin	Fun	Dur	Hob	Spo	T95	T64	T68	Mean
Cra L	321.00	285.00	287.00	346.00	356.00	278.00	314.00	332.00	369.00	325.00	293.00	322.00	319.00
Cra H	411.00	436.00	399.00	445.00	441.00	348.00	431.00	493.00	473.00	482.00	414.00	447.00	435.00
Beg L	317.00	328.00	316.00	360.00	312.00	266.00	318.00	374.00	381.00	336.00	289.00	339.00	328.00
Beg H	429.00	450.00	442.00	552.00	517.00	446.00	442.00	544.00	576.00	561.00	482.00	569.00	500.83
Fow L	364.00	418.00	341.00	455.00	442.00	329.00	457.00	490.00	522.00	447.00	395.00	373.00	419.42
Fow H	464.00	532.00	384.00	453.00	506.00	464.00	513.00	583.00	533.00	528.00	450.00	520.00	494.17
Tru L	408.00	409.00	382.00	413.00	407.00	401.00	466.00	487.00	426.00	441.00	457.00	453.00	429.17
Tru H	434.00	462.00	484.00	403.00	505.00	493.00	552.00	568.00	538.00	522.00	479.00	546.00	498.83
Box L	419.00	404.00	438.00	394.00	415.00	423.00	483.00	525.00	509.00	429.00	437.00	480.00	446.33
Box H	492.00	392.00	434.00	370.00	460.00	491.00	499.00	533.00	514.00	518.00	479.00	513.00	474.58
Ear L	496.00	523.00	532.00	519.00	600.00	579.00	601.00	646.00	579.00	579.00	523.00	522.00	558.25
Ear H	448.00	481.00	496.00	536.00	514.00	526.00	512.00	563.00	569.00	512.00	491.00	525.00	514.42
Edn L	566.00	747.00	633.00	707.00	838.00	749.00	767.00	850.00	764.00	760.00	750.00	751.00	740.17
Edn H	625.00	740.00	615.00	644.00	764.00	667.00	708.00	859.00	694.00	689.00	745.00	704.00	704.50
Mean	442.43	471.93	441.64	471.21	505.50	461.43	504.50	560.50	531.93	509.21	477.43	504.57	490.19

where

$$\bar{x} = \frac{1}{pq}\sum_{i=1}^{p}\sum_{j=1}^{q} x_{ij},\ \bar{x}_{i.} = \frac{1}{q}\sum_{j=1}^{q} x_{ij} \text{ and } \bar{x}_{.j} = \frac{1}{p}\sum_{i=1}^{p} x_{ij}.$$

Note that any constant may be added to each $\hat{\alpha}_i$ and $\hat{\beta}_j$ but, provided $\hat{\mu}$ is accordingly adjusted, fitted values remain unchanged. The identification constraints adopted give the simple unique parameterization (6.3) in which the main effects sum to zero. When the multiplicative terms are included, the main effects continue to be estimated by (6.3) and the multiplicative terms, representing the interaction between the row and column factors, are estimated by least squares from the singular value decomposition of the residual matrix (also called the interaction matrix) $\mathbf{Z}$: $p \times q$ with elements

$$z_{ij} = x_{ij} - \hat{\mu} - \hat{\alpha}_i - \hat{\beta}_j = x_{ij} - \bar{x}_{i.} - \bar{x}_{.j} + \bar{x}, \quad \text{for } i = 1, \ldots, p \text{ and } j = 1, \ldots, q. \tag{6.4}$$

Defining centring matrices $\mathbf{P} = \mathbf{11}'/p$ and $\mathbf{Q} = \mathbf{11}'/q$, we may write (6.1) in matrix terms and in terms of the above estimates as

$$\mathbf{X} = \mathbf{PXQ} + (\mathbf{I} - \mathbf{P})\mathbf{XQ} + \mathbf{PX}(\mathbf{I} - \mathbf{Q}) + (\mathbf{I} - \mathbf{P})\mathbf{X}(\mathbf{I} - \mathbf{Q}) \tag{6.5}$$

from which the following orthogonal analysis of variance is immediate:

$$||\mathbf{X} - \mathbf{PXQ}||^2 = ||(\mathbf{I} - \mathbf{P})\mathbf{XQ}||^2 + ||\mathbf{PX}(\mathbf{I} - \mathbf{Q})||^2 + ||(\mathbf{I} - \mathbf{P})\mathbf{X}(\mathbf{I} - \mathbf{Q})||^2 \tag{6.6}$$

In particular,

$$\mathbf{Z} = (\mathbf{I} - \mathbf{P})\mathbf{X}(\mathbf{I} - \mathbf{Q}) \tag{6.7}$$

which may be compared with the similar formula (7.2) arising in correspondence analysis.

The matrix $\mathbf{Z}$ may be analysed in the usual way by appealing to the Eckart-Young theorem to give estimates of the multiplicative terms, presented in order of decreasing importance according to the size of their corresponding singular values. Because this computational process is shared, fitting biadditive models is often regarded as a version of PCA. Above, we have stressed the fundamental statistical differences between fitting a biadditive model and PCA, not the least of which is that one includes a dependent variable and the other does not. Yet both depend on the same algebraic underpinning, and this carries over into common algorithmic procedures.

Table 6.2 gives the values of the residuals or interactions (6.4), together with the estimated main effects.

The whole analysis may be summarized in the orthogonal analysis of variance (6.6) with entries for the main effects and each multiplicative term, together with degrees of freedom. This is shown in Table 6.3 for the data of Table 6.1.

Our main purpose here is to discuss biplots associated with two-way tables like Table 6.1, but first we make a few remarks commenting on Table 6.3. From this table it is clear that both main effects are substantial relative to the Total Sites × Varieties interaction. However, the first two dimensions of the interaction account for about 62% of the total interaction and merit closer examination. The mean sum of squares for the remaining nine terms with 99 degrees of freedom is 730.45 and can be considered as

Table 6.2 The residuals **Z** and main effects derived from Table 6.1.

	Cap	Ran	Hun	Tem	Kin	Fun	Dur	Hob	Spo	T95	T64	T68	Sites Main effects
Cra L	49.8	−15.7	16.5	46.0	21.7	−12.2	−19.3	−57.3	8.3	−13.0	−13.2	−11.4	−171.19
Cra H	23.8	19.3	12.5	29.0	−9.3	−58.2	−18.3	−12.3	−3.7	28.0	−8.2	−2.4	−55.19
Beg L	36.8	18.3	36.5	51.0	−31.3	−33.2	−24.3	−24.3	11.3	−11.0	−26.2	−3.4	−162.19
Beg H	−24.1	−32.6	−10.3	70.1	0.9	−26.1	−73.1	−27.1	33.4	41.1	−6.1	53.8	10.64
Fow L	−7.7	16.8	−29.9	54.6	7.3	−61.7	23.3	0.3	60.8	8.6	−11.7	−60.8	−70.77
Fow H	17.6	56.1	−61.6	−22.2	−3.5	−1.4	4.5	18.5	−2.9	14.8	−31.4	11.5	3.98
Tru L	26.6	−1.9	1.4	2.8	−37.5	0.6	22.5	−12.5	−44.9	−7.2	40.6	9.5	−61.02
Tru H	−17.1	−18.6	33.7	−76.9	−9.1	22.9	38.9	−1.1	−2.6	4.1	−7.1	32.8	8.64
Box L	20.4	−24.1	40.2	−33.4	−46.6	5.4	22.4	8.4	20.9	−36.4	3.4	19.3	−43.86
Box H	65.2	−64.3	8.0	−85.6	−29.9	45.2	10.1	−11.9	−2.3	24.4	17.2	24.0	−15.61
Ear L	−14.5	−17.0	22.3	−20.3	26.4	49.5	28.4	17.4	−21.0	1.7	−22.5	−50.6	68.06
Ear H	−18.7	−15.2	30.1	40.6	−15.7	40.3	−16.7	−21.7	12.8	−21.4	−10.7	−3.8	24.23
Edn L	−126.4	25.1	−58.6	−14.2	82.5	37.6	12.5	39.5	−17.9	0.8	22.6	−3.5	249.98
Edn H	−31.7	53.8	−41.0	−41.5	44.2	−8.7	−10.8	84.2	−52.2	−34.5	53.3	−14.9	214.31
Varieties Main effects	−47.76	−18.26	−48.55	−18.98	15.31	−28.76	14.31	70.31	41.74	19.02	−12.76	14.38	

Table 6.3 Analysis of variance for a biadditive model fitted to the data of Table 6.1.

Source	Sum of squares	DF	Mean SS
Sites Main Effect	2200767.00	13	169289.80
Varieties Main Effect	196211.90	11	17837.45
Total S × V Interaction	188245.10	143	1316.40
Multiplicative term 1	65718.31	23	2857.32
Multiplicative term 2	50211.31	21	2391.02
Multiplicative term 3	21394.60	19	1126.03
Multiplicative term 4	15905.03	17	935.59
Multiplicative term 5	11115.25	15	741.02
Multiplicative term 6	7904.39	13	608.03
Multiplicative term 7	5753.79	11	523.07
Multiplicative term 8	3891.06	9	432.34
Multiplicative term 9	3669.92	7	524.27
Multiplicative term 10	2427.90	5	485.58
Multiplicative term 11	253.53	3	84.51
Mean	40368170.00	1	
Mean corrected Total	2585224.00	167	
Total	42953390.00	168	

residual variation. The sums of squares for the interactions are given by the squares of the singular values of $\mathbf{Z}$ (see Section 6.4). Note that the degrees of freedom for successive interaction terms decrease by two and sum to the total of 143. This is a commonly used approximation that deals sensibly with the degrees of freedom, but which ultimately derives from canonical analysis (see Rao, 1952) and appears to have no firm theoretical basis for use in other contexts.

6.4 Biplots associated with biadditive models

In what follows we assume that rank$(\mathbf{X}) = q \le p$. There is no loss of generality because if $q > p$ we can take $\mathbf{X}'$ since rows and columns of two-way tables have the same status.

It often happens that main effects swamp the interactions. Nevertheless the interactions may be commercially important, showing for example that certain varieties perform better than others at certain sites. The structure of the interactions may be presented as a biplot of the interaction matrix $\mathbf{Z}$ based on the decomposition

$$\mathbf{Z} = \mathbf{U\Sigma V}'. \tag{6.8}$$

The biplot may be presented in several equivalent forms:

(i) Plot $\mathbf{U\Sigma}^{1/2}\mathbf{J}$ for the rows and $\mathbf{V\Sigma}^{1/2}\mathbf{J}$ for the columns.

(ii) Plot $\mathbf{U\Sigma J}$ for the rows and $\mathbf{VJ}$ for the columns.

(iii) Plot $\mathbf{UJ}$ for the rows and $\mathbf{V\Sigma J}$ for the columns.

Of the above, (i) treats rows and columns equally, so respecting the symmetric nature of $\mathbf{X}$ and $\mathbf{Z}$. Unfortunately, it gives two sets of points and therefore suffers from the usual difficulties associated with the visual inspection of inner products from pairs of points (see Section 2.3). Methods (ii) and (iii) merely interchange the roles of rows and columns, so it suffices to discuss (ii). In (ii), rows are represented by p points and the columns by q axes. As with PCA, axes may be calibrated and the devices of axis lambda-scaling (Section 2.3) and axis shifts (Section 2.4) are available for improving presentation. In addition, the axis calibrations may be adjusted to include and mark the relevant main effects. To assess how well the columns of the r-dimensional approximation account for the full representation in $\mathbf{Z}$ we may calculate

$$\text{Column predictivities} = \{\text{diag}(\hat{\mathbf{Z}}'\hat{\mathbf{Z}})\}\{\text{diag}(\mathbf{Z}'\mathbf{Z})\}^{-1} \tag{6.9}$$

and associate these values with the axes. Similarly, for (iii) we may calculate

$$\text{Row predictivities} = \{\text{diag}(\hat{\mathbf{Z}}\hat{\mathbf{Z}}')\}\{\text{diag}(\mathbf{Z}\mathbf{Z})'\}^{-1}. \tag{6.10}$$

Note that (6.9) and (6.10) remain valid for all of (i), (ii) and (iii), but (ii) provides a convenient means for displaying (6.9) and (iii) a convenient means for displaying (6.10).

6.5 Interpolating new rows or columns

The requirement to interpolate a new row or column into $\mathbf{X}$ is rather unlikely; usually it would be simpler and better to redo the whole analysis on the enlarged $\mathbf{X}$. However, it is not an impossible requirement, so here we give the details of how to do it. The process proceeds similarly to PCA, but modified to handle the main effects.

Given a new row $\mathbf{z}'$ of $\mathbf{Z}$, we merely have to calculate $\mathbf{z}'\mathbf{V}\mathbf{J}, \mathbf{z}'\mathbf{V}\boldsymbol{\Sigma}^{-1/2}\mathbf{J}$ or $\mathbf{z}'\mathbf{V}\boldsymbol{\Sigma}^{-1}\mathbf{J}$ according to how we have scaled the row coordinates in the display of $\mathbf{Z}$. When λ-scaling has been used, this too must be taken into account. Thus, we have only to decide how to calculate $\mathbf{z}'$, given a new row $\mathbf{x}'$ of $\mathbf{X}$. Clearly we have to eliminate the main effect associated with the new row and also to adjust for the main effects associated with the q columns. From (6.7) this gives

$$\mathbf{z}' = \mathbf{x}'(\mathbf{I} - \mathbf{Q}) - \tfrac{1}{p}\mathbf{1}'\mathbf{X}(\mathbf{I} - \mathbf{Q}). \tag{6.11}$$

Because $\mathbf{V}$ is a singular vector of $\mathbf{Z}$, which has zero row and column totals, it follows that $\mathbf{1}'\mathbf{V}\mathbf{J} = \mathbf{0}'$ and thence $\mathbf{Q}\mathbf{V}\mathbf{J} = \mathbf{0}$. Thus, when calculating $\mathbf{z}'\mathbf{V}\mathbf{J}$, (6.11) implies that

$$\mathbf{z}'\mathbf{V}\mathbf{J} = \left(\mathbf{x}' - \tfrac{1}{p}\mathbf{1}'\mathbf{X}\right)\mathbf{V}\mathbf{J}. \tag{6.12}$$

It follows that the predictivity of the new row $\mathbf{x}'$: $1 \times q$ is given by

$$\frac{\mathbf{z}'\mathbf{V}\mathbf{J}\mathbf{V}'\mathbf{z}}{\mathbf{z}'\mathbf{z}},$$

which, on using (6.11) and (6.12), may be written as

$$\frac{(\mathbf{x}' - \frac{1}{p}\mathbf{1}'\mathbf{X})\mathbf{V}\mathbf{J}\mathbf{V}'(\mathbf{x}' - \frac{1}{p}\mathbf{1}'\mathbf{X})'}{(\mathbf{x}' - \frac{1}{p}\mathbf{1}'\mathbf{X})(\mathbf{I} - \mathbf{Q})(\mathbf{x}' - \frac{1}{p}\mathbf{1}'\mathbf{X})'}. \tag{6.13}$$

Note that predictivity is a function of $\mathbf{z}$ and does not depend on the particular factorization used to give the biplot display.

Similarly, adding a new column $\mathbf{x}$: $p \times 1$ requires

$$\mathbf{Uz} = \mathbf{U}\left(\mathbf{x} - \tfrac{1}{q}\mathbf{X1}\right) \tag{6.14}$$

with predictivity

$$\frac{(\mathbf{x} - \frac{1}{q}\mathbf{X1})'\mathbf{UJU}'(\mathbf{x} - \frac{1}{q}\mathbf{X1})}{(\mathbf{x} - \frac{1}{q}\mathbf{X1})'(\mathbf{I} - \mathbf{P})(\mathbf{x} - \frac{1}{q}\mathbf{X1})}. \tag{6.15}$$

6.6 Functions for constructing biadditive biplots

Our main function for constructing biadditive biplots is the function `biadbipl`. In Section 6.6.1 we give a detailed description of `biadbipl`. The reader is encouraged to study and experiment with the arguments of `biadbipl`. As with our other main functions like `PCAbipl` and `CVAbipl` introduced in previous chapters, several functions are called by `biadbipl` for drawing the biplot, adding features to it and changing its appearance, that are generally not directly called by the user. In addition to these functions, the functions `biad.ss` and `biad.predictivities` may be called in order to obtain additional information about the biplots constructed with `biadbipl`. These two functions are briefly introduced in Sections 6.6.2 and 6.6.3.

6.6.1 Function `biadbipl`

This is a function for constructing biadditive biplots associated with a two-way table. Provision is made for approximating the two-way table itself, the overall mean-corrected two-way table as well as the two-way table of interactions. The function `biadbipl` allows for one-, two- or three-dimensional biadditive biplots. Several enhancements to the basic biadditive biplots are available.

Usage

`biadbipl` uses the same calling conventions as `PCAbipl` and shares the following arguments with it:

```
alpha.3d        dim.biplot     markers.size          predictivity.print
aspect.3d       exp.factor     n.int                 reflect
ax              e.vects        offset                rotate.degrees
ax.col.3d       factor.x       offset.m              select.origin
ax.name.col     factor.y       ort.lty               side.label
ax.name.size    font.3d        parplotmar            size.ax.3d
cex.3d          ID.labs        pos                   size.points.3d
col.plane.3d    ID.3d          pos.m                 Title
col.text.3d     line.length    predictions.3d        Titles.3d
constant        markers        predictions.sample
```

Arguments specific to `biadbipl`

Argument	Description
`X`	Required argument. Two-way table in the form of a matrix of size $p \times q$ representing observations on a response variable depending on two factors.
`X.new.rows`	Optional argument. Matrix of size $k \times q$ containing the measurements of k new rows to be interpolated into the biplot.
`X.new.columns`	Optional argument. Matrix of size $p \times m$ containing the measurements of m new columns to be interpolated into the biplot.
`add.maineffect`	Logical value requesting or suppressing adding of main effects to calibrations on axes of an interaction biplot. Defaults to FALSE.
`show.maineffect`	Logical value indicating whether main effects are shown on the biplot axes of a biadditive biplot of the interaction matrix. Defaults to FALSE.
`biad.variant`	One of `"InteractionMat"`, `"Xmat"`, `"XminMeanMat"` specifying whether to use the interaction matrix (the default), the two-way table itself or the two-way table with the overall mean subtracted. As discussed in Sections 6.1–6.5, biadditive biplots are mainly constructed from the interaction matrix. The other two alternatives above are rarely used but are available if needed.
`SigmaHalf`	Logical value indicating whether $\mathbf{V}$ or $\mathbf{V\Sigma}^{1/2}$ is to be used in plotting the rows.
`circle.proj`	NULL or an integer-valued vector specifying circles to be drawn with diameter from a specified row point to the origin. Intersections of the circle with the biplot axes provide the predicted values for the specified row point.
`expand.markers`	NULL or scalar value for use together with `n.int` to ensure at least two marker points appear on a biplot axis.
`axis.col`	A single integer or character value specifying the colour of all biplot axes. A vector of integers or character values can be given to allow different colours for the different biplot axes.
`column.points.col`	Vector specifying colours of plotting symbols for the columns. Defaults to `"red"`, i.e. `rep("red", ncol(X))`.
`column.points.size`	Value specifying size of plotting symbol for the columns. Defaults to 1.
`column.points.label.size`	Value specifying size of labels for the column points. Defaults to 0.8.
`column.points.text`	Logical value specifying whether labels of column points are shown. Defaults to TRUE.

`lambda`	Logical value requesting lambda-scaling. Defaults to FALSE.
`legend.show`	Accepts logical TRUE or FALSE for requesting a legend on a separate graph window. The top panel is the legend for the rows and the bottom panel for the columns.
`legend.columns`	A two-component integer valued vector: the first element specifies the number of columns for the rows legend; the second element the number of columns for the columns legend. Defaults to `c(1,1)`.
`line.length`	A numeric vector consisting of two components: the first component determines the length of the tick marks on all biplot axes; the second component determines the distance of the scale marker to the tick mark. The default is `c(1,1)`.
`marker.col`	Specifies the colour used for markers on biplot axes.
`pch.row.points`	Vector of length the number of rows, specifying the plotting character for each row point.
`pch.col.points`	Vector of length the number of columns specifying the plotting character for each column point.
`plot.col.points`	Logical TRUE or FALSE requesting or suppressing plotting of the columns as points.
`predictions.allsamples.onaxis`	NULL or an integer value specifying the biplot axis onto which all row points are projected.
`propshift`	Numerical value used together with constant for translating and shifting biplot axes in a one-dimensional biplot. Defaults to 0.
`row.points.col`	Vector specifying colours of plotting symbols for the rows. Defaults to `"green"`, i.e. `rep("green", nrow(X))`.
`row.points.size`	Value specifying size of plotting symbol for the rows. Defaults to 1.
`row.points.label.size`	Value specifying size of labels for the row points. Defaults to 0.8
`row.points.text`	Logical value specifying whether labels of row points are shown. Defaults to TRUE.
`samples.plot`	Logical TRUE or FALSE requesting or suppressing plotting of the rows as points.
`text.pos`	Two-component vector. The first (second) element specifies position of the labels for a row (column) point: values of 1, 2, 3 and 4, respectively indicate positions below, to the left of, above and to the right of the specified coordinates. Defaults to `c(1,1)`.
`tick.marker.col`	Specifies the colour for tick marks on a biplot axis.
`output`	Specifies output to be printed to the screen. See VALUE below.
`X.new.columns.pch`	Plotting character for plotting of new columns as points.

`X.new.columns.col`	Colour of plotting character for plotting of new columns as points.
`X.new.columns.pch.cex`	Size of plotting character for plotting of new columns as points.
`X.new.columns.labels`	Labels of new columns as points.
`X.new.columns.labels.cex`	Size of labels of new columns as points.
`X.new.rows.pch`	Plotting character for plotting new row points.
`X.new.rows.col`	Colour of plotting character for plotting new row points.
`X.new.rows.pch.cex`	Size of plotting character for plotting new row points.
`X.new.rows.labels`	Labels of new row points.
`X.new.rows.labels.cex`	Size of labels of new row points.
`zoomval`	Specify `zoomval = NULL` for no zooming; a value less than unity for zooming in and a value larger than unity for zooming out. See Section 3.7.2 for usage.

Value

The function `biadbipl` constructs the specified biadditive biplot and returns a list with the following named components:

`out`	Predictions for specified row points or new row points on all the columns (variables).
`X.mat`	Input $\mathbf{X}$ matrix.
`lambda`	The computed lambda value if lambda-scaling is requested.
`Quality`	Quality of display.
`predictions`	Three-dimensional predictions; NULL in the case of one- or two-dimensional biplots.
`Axis.predictivities`	The axis predictivities for the displayed biplot.
`main.effects.cols`	The column main effects.
`main.effects.rows`	The row main effects.
`SVDmat`	Matrix that is subjected to an SVD.
`Umat`	The $\mathbf{U}$ of the SVD $\mathbf{U\Sigma V}'$.
`Vmat`	The $\mathbf{V}$ of the SVD $\mathbf{U\Sigma V}'$.
`Sigma`	The $\mathbf{\Sigma}$ of the SVD $\mathbf{U\Sigma V}'$.

6.6.2 Function `biad.predictivities`

Description

This function is for for calculating row and column adequacies, row and column predictivities, overall quality, main effects and predictions associated with biadditive biplots. This function is intended for use together with `biadbipl`.

Usage

```
biad.predictivities (X = wheat.data, e.vects = 1:ncol(X),
  X.new.rows = NULL, X.new.columns = NULL, add.maineffect =
  FALSE, biad.variant = c("Xmat", "XminMeanMat",
  "InteractionMat"), predictions.dim = c("All", 1:min(nrow(X),
  ncol(X))))
```

Arguments

`X`	Required argument. It must be in the form of a $p \times q$ matrix representing a two-way table of size $p \times q$ being observations (measurements) on a response variable depending on two factors. It is assumed that $p \geq q$. If this is not the case **X** must be transposed.
`e.vects`	Vector specifying the eigenvectors to be used for the quality calculations. The default is to use all eigenvectors.
`X.new.rows`	If not NULL it must be a matrix of size $k \times q$ representing k new rows interpolated into the biadditive biplot.
`X.new.columns`	If not NULL it must be a matrix of size $p \times m$ representing m new columns (or axes) interpolated into the biadditive biplot.
`add.maineffect`	Logical value requesting or suppressing adding of main effects for calculated predicted values when biplotting the interaction matrix. Defaults to FALSE.
`biad.variant`	One of `"Xmat"`, `"XminMeanMat"`, `"InteractionMat"` specifying whether to use the two-way table itself, the two-way table with the overall mean subtracted or the interaction matrix.
`predictions.dim`	`"All"` or integer value requesting predictions in dimension `"integer"`. Default of `"All"` requests predictions in all dimensions.

Value

`biad.predictivities` returns a list with the following named components:

`Quality`	Overall quality in $1, 2, \ldots, q$ dimensions.
`Column.adequacies`	Column adequacies in $1, 2, \ldots, q$ dimensions.
`Column.predictivities`	Column predictivities in $1, 2, \ldots, q$ dimensions.
`Row.adequacies`	Row adequacies in $1, 2, \ldots, q$ dimensions.
`Row.predictivities`	Row predictivities in $1, 2, \ldots, q$ dimensions.
`main.effects.columns`	The column main effects.
`main.effects.rows`	The row main effects.

`predictions.rows`	By default a list is returned. Each element of the list is a matrix of the same size as the input two-way table containing the specified predicted values. The first matrix contains the predictions in the first dimension, the second one in the first two dimensions, and so on.
`New.Row.Predictivities`	Predictivities in $1, 2, \ldots, q$ dimensions for each of the k rows given in the matrix `X.new.rows`.
`New.Column.Predictivities`	Predictivities in $1, 2, \ldots, q$ dimensions for each of the m columns given in the matrix `X.new.columns`.

6.6.3 Function `biad.ss`

Description

This function is for calculating the partitioning of the total sum of squares associated with a two-way table. This function is intended for use together with `biadbipl`. It is called with one argument, the input $\mathbf{X}$ matrix to `biadbipl`.

Value

The function `biad.ss` returns a list with the following named components:

`X`	Input $\mathbf{X}$.
`X.hat`	The estimated (approximated) $\mathbf{X}$ in the two-dimensional biplot space. Main effects are estimated according to (6.3), while the interactions defined by (6.4) and assembled into the matrix $\mathbf{Z}$ are approximated by the singular vector pairs associated with the two largest singular values of $\mathbf{Z}$.
`Interaction.mat`	Interaction matrix.
`C.mat`	Matrix $\mathbf{U\Sigma}^{1/2}$ where the interaction matrix is $\mathbf{U\Sigma V}'$.
`D.mat`	Matrix $\mathbf{V\Sigma}^{1/2}$ where the interaction matrix is $\mathbf{U\Sigma V}'$.
`SS.Interaction.Total`	Total sum of squares for interaction.
`SS.Interaction.components`	The break-down of `SS.Interaction.Total` into two degrees of freedom components.
`cumsum`	Cumulative sum of the `SS.Interaction.components`.
`ANOVA.Table`	Complete ANOVA table with components as in Table 6.3.

6.7 Examples of biadditive biplots: the wheat data

We begin by presenting a biplot of Table 6.1. This biplot (see Figure 6.2), approximating the overall mean, both main effects and interaction part of the model, is not very exciting

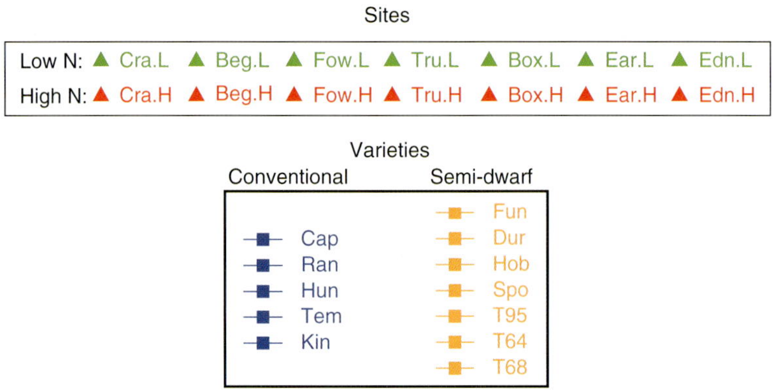

Figure 6.1 Legend to accompany biplots related to the wheat data set.

but serves as an introduction to the usage of `biadbipl`. The legend given in Figure 6.1 is obtained by adding the specifications

```
legend.show = TRUE, legend.columns = c(7,1)
```

in our call to `biadbipl`, given below. This legend applies not only to Figure 6.2 but to all the figures in this section and will therefore not be repeated for later figures.

In our first example of a biadditive biplot we examine a biplot of Table 6.1 after eliminating the overall mean, obtained by specifying `biad.variant = "Xminus-MeanMat"`. Thus the biplot will approximate both the main effects and interaction part of the model. This biplot is obtained in the usual way, from the SVD $\mathbf{X}_{\text{adj}} = \mathbf{U\Sigma V}'$, and is shown in Figure 6.3, where the first two dimensions of $\mathbf{U\Sigma} = \mathrm{X}_{\text{adj}}\mathbf{V}$ are plotted for the sites and of $\mathbf{V}$ for the varieties in the top panel, as well as in the bottom left panel. In the bottom right panel the roles of the sites and varieties have been switched by specifying `X = t(wheat.data)`. The full function call to produce the biplot in the top left panel of Figure 6.2 is

```
biadbipl(wheat.data, lambda = F, ax = 1:12, biad.variant =
   "XminMeanMat", show.origin = F, SigmaHalf = FALSE,
   predictivity.print = TRUE, pch.row.points = 17,
   row.points.col = rep(c("green","red"),2),
   row.points.size = 1.25, ax.name.col =
   c(rep("blue",5),rep("orange",7)), pch.col.points = 15,
   column.points.col = c(rep("blue",5),rep("orange",7)),
   column.points.size = 1.25, pos = "Orthog",
   axis.col = c(rep("blue",5),rep("orange",7)),
   column.points.text = TRUE, offset = c(2.5, 2.8, 0.1, 0.1),
   offset.m = rep(-0.1, 14))))
```

The quality of the display has the high value of 95.3%, and Table 6.4 shows that the two-dimensional row and column predictivities are equally impressive.

Nevertheless, this does not guarantee a satisfactory biplot: the biplot in the top left panel of Figure 6.2 is evidently unsatisfactory in many ways. In particular, the points for the varieties very nearly coincide. We can adjust for this by using lambda-scaling as in the top right panel. The result is far more satisfactory. A striking feature of the figure

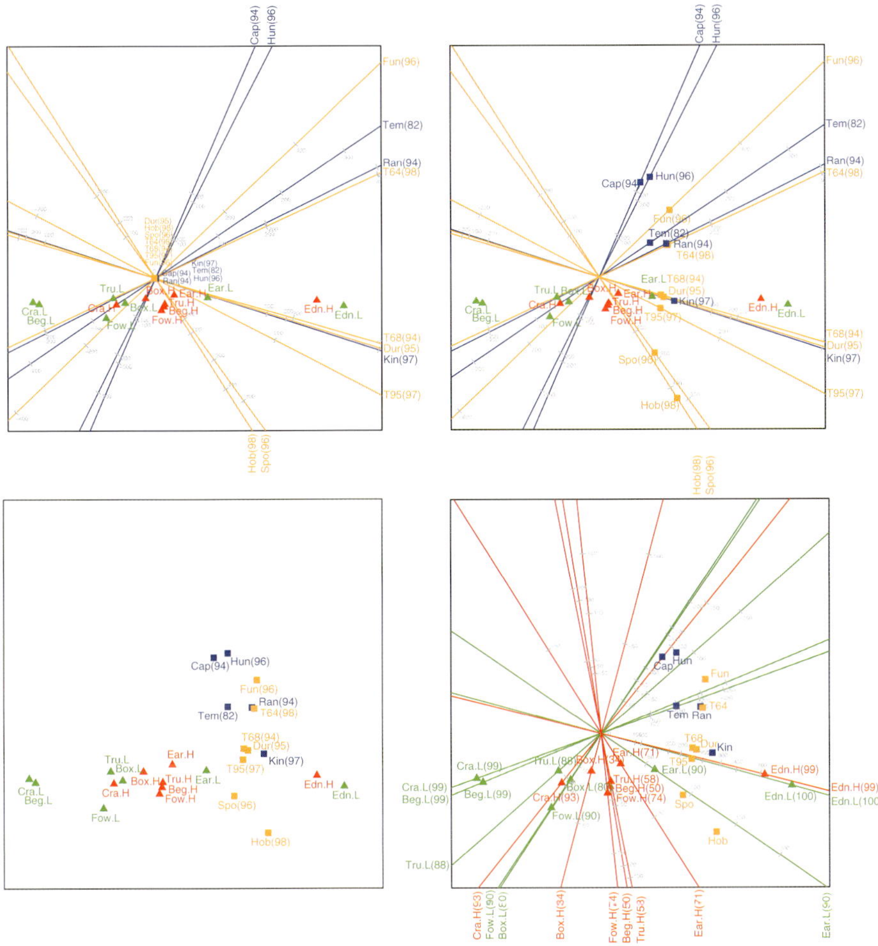

Figure 6.2 Biadditive biplots of wheat data (Table 6.1) after eliminating the overall mean: (top left) $\mathbf{X}_{\text{adj}}\mathbf{V}$ used for plotting the sites and $\mathbf{V}$ for the varieties; varieties almost on top of each other; (top right) biplot with lambda-scaling; (bottom left) $\mathbf{U\Sigma}^{1/2}$ used for plotting the sites and $\mathbf{V\Sigma}^{1/2}$ for the varieties; (bottom right) biplot showing sites as biplot axes with $\mathbf{U\Sigma}^{1/2}$ used for plotting the varieties and $\mathbf{V\Sigma}^{1/2}$ for the sites. Calibrations on axes are in terms of deviations from the overall mean.

is that the two sets of points are approximately orthogonal. This is a diagnostic feature that suggests that a main effects model suffices – we comment briefly on diagnostic biplots towards the end of this chapter. We could have achieved a similar rescaling by basing the plot on $\mathbf{U\Sigma}^{1/2}$ and $\mathbf{V\Sigma}^{1/2}$. The biplot in the top left panel of Figure 6.2 is distorted because $\mathbf{\Sigma}$ incorporates all information on size. This suggests that plots for biadditive models should always be based on the $\mathbf{\Sigma}^{1/2}$-scaling, possibly followed by further lambda-scaling to improve presentation. because with our data p and q are of similar size, additional lambda-scaling will have little effect. We emphasize that none of these cosmetic changes affect the approximating inner product.

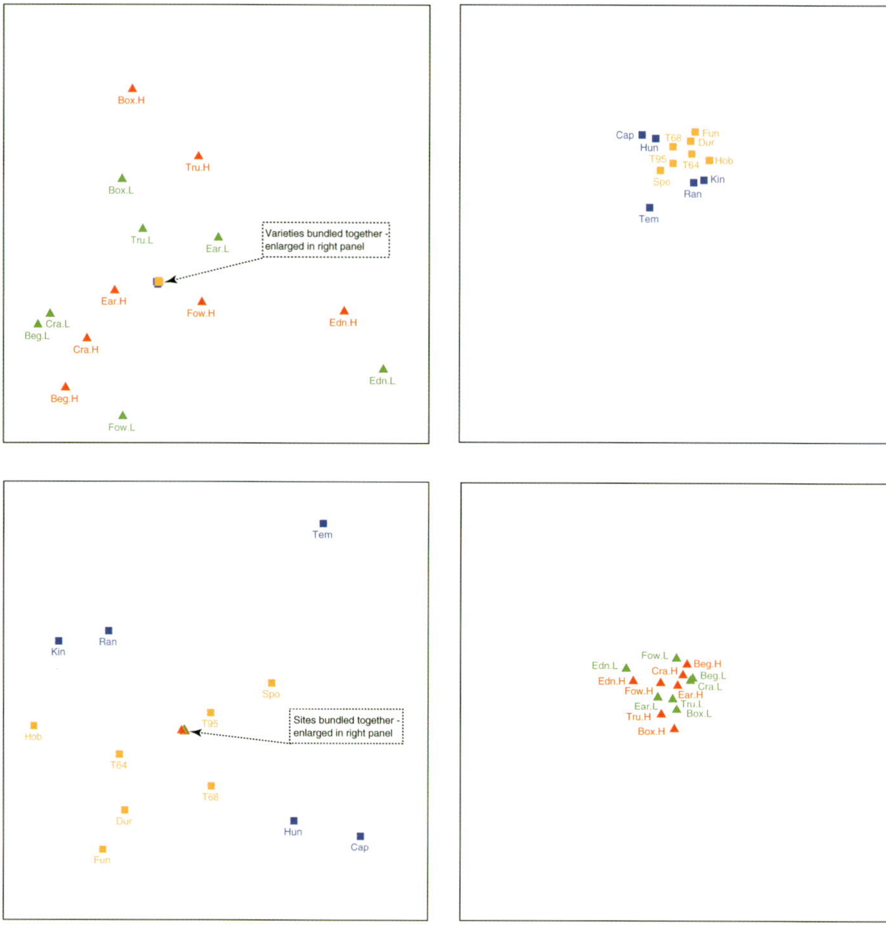

Figure 6.3 Biadditive biplots of the interaction matrix $\mathbf{Z} = \mathbf{U\Sigma V}'$ of the wheat data (Table 6.2): (top left) $\mathbf{ZV}$ used for plotting the sites and $\mathbf{V}$ for the varieties; (top right) zooming with `zoomval = 0.025` into positions of the bundled varieties on the left; (bottom left) carry out the SVD $\mathbf{Z}' = \mathbf{U\Sigma V}'$ to use $\mathbf{Z}'\mathbf{V}$ for plotting the varieties and $\mathbf{V}$ for the sites; (bottom right) zooming with `zoomval = 0.025` into positions of the bundled varieties on the left. The origin is marked with a black cross.

The high quality of Figure 6.2 may suggest a main effects model. However, a glance at the analysis of variance shows that interaction, although relatively small, cannot be ignored. For a farmer planning to grow wheat it is important to know which varieties give the best yields in his location. From here on we shall be concerned with biplots of the interaction matrix $\mathbf{Z}$ as shown in Table 6.2. The least-squares estimates of the multiplicative parameters are found from the SVD $\mathbf{Z} = \mathbf{U\Sigma V}'$ with an associated biplot of $\mathbf{U\Sigma}$ for the rows and $\mathbf{V}$ for the columns, as with PCA. Indeed, the methodology of biadditive biplots is so close to that of PCA biplots that the two are often confused. We may proceed as with PCA by identifying either the rows or the columns of $\mathbf{Z}$ as 'variables'. Treating the rows (the 14 sites) of Table 6.2 as variables, Kempton (1984),

Table 6.4 Row and column predictivities for the biplot of the wheat data when specifying `biad.variant = "XminusMeanMat"`.

Predictivities for the sites

	Cra.L	Cra.H	Beg.L	Beg.H	Fow.L	Fow.H	Tru.L	Tru.H	Box.L	Box.H	Ear.L	Ear.H	Edn.L	Edn.H
Dim_1	0.953	0.643	0.947	0.043	0.554	0.026	0.726	0.069	0.484	0.068	0.796	0.411	0.974	0.969
Dim_2	0.987	0.925	0.992	0.503	0.899	0.742	0.885	0.580	0.797	0.342	0.902	0.708	0.995	0.987
Dim_3	0.989	0.948	0.995	0.643	0.946	0.743	0.909	0.903	0.924	0.869	0.917	0.721	0.997	0.987
Dim_4	0.989	0.952	0.996	0.923	0.959	0.810	0.924	0.918	0.925	0.902	0.917	0.859	0.997	0.994
Dim_5	0.990	0.970	0.996	0.998	0.976	0.848	0.933	0.925	0.926	0.924	0.976	0.912	0.998	0.997
Dim_6	0.992	0.970	0.997	0.998	0.976	0.916	0.937	0.927	0.988	0.957	0.979	0.946	0.998	0.998
Dim_7	0.994	0.970	0.998	0.998	0.978	0.984	0.946	0.949	0.990	0.983	0.980	0.953	0.998	0.999
Dim_8	0.997	0.975	0.998	0.999	0.989	0.984	0.970	0.958	0.992	0.996	0.983	0.954	1.000	0.999
Dim_9	0.999	0.977	0.999	0.999	0.997	0.986	0.982	0.968	0.992	0.997	0.988	0.964	1.000	1.000
Dim_10	0.999	0.998	1.000	1.000	0.998	1.000	0.987	0.996	0.995	0.999	0.990	0.996	1.000	1.000
Dim_11	1.000	1.000	1.000	1.000	1.000	1.000	1.000	1.000	0.999	0.999	1.000	1.000	1.000	1.000
Dim_12	1.000	1.000	1.000	1.000	1.000	1.000	1.000	1.000	1.000	1.000	1.000	1.000	1.000	1.000

Predictivities for the varieties

	Cap	Ran	Hun	Tem	Kin	Fun	Dur	Hob	Spo	T95	T64	T68
Dim_1	0.647	0.918	0.722	0.786	0.965	0.893	0.942	0.813	0.828	0.945	0.959	0.929
Dim_2	0.939	0.937	0.961	0.817	0.974	0.962	0.950	0.985	0.965	0.967	0.977	0.936
Dim_3	0.952	0.956	0.966	0.988	0.982	0.979	0.971	0.988	0.967	0.967	0.977	0.943
Dim_4	0.961	0.989	0.972	0.994	0.982	0.984	0.979	0.993	0.981	0.974	0.979	0.964
Dim_5	0.978	0.990	0.978	0.994	0.986	0.991	0.993	0.993	0.984	0.978	0.983	0.991
Dim_6	0.982	0.991	0.991	0.996	0.991	0.994	0.993	0.996	0.984	0.990	0.986	0.992
Dim_7	0.984	1.000	0.991	0.996	0.993	0.994	0.993	0.996	0.985	0.990	0.996	0.995
Dim_8	0.997	1.000	0.993	0.996	0.995	0.995	0.993	0.997	0.994	0.992	0.997	1.000
Dim_9	0.999	1.000	0.993	0.996	0.996	0.996	1.000	0.999	0.994	0.995	0.999	1.000
Dim_10	1.000	1.000	1.000	0.998	0.999	1.000	1.000	0.999	0.995	0.997	1.000	1.000
Dim_11	1.000	1.000	1.000	1.000	1.000	1.000	1.000	1.000	1.000	1.000	1.000	1.000
Dim_12	1.000	1.000	1.000	1.000	1.000	1.000	1.000	1.000	1.000	1.000	1.000	1.000

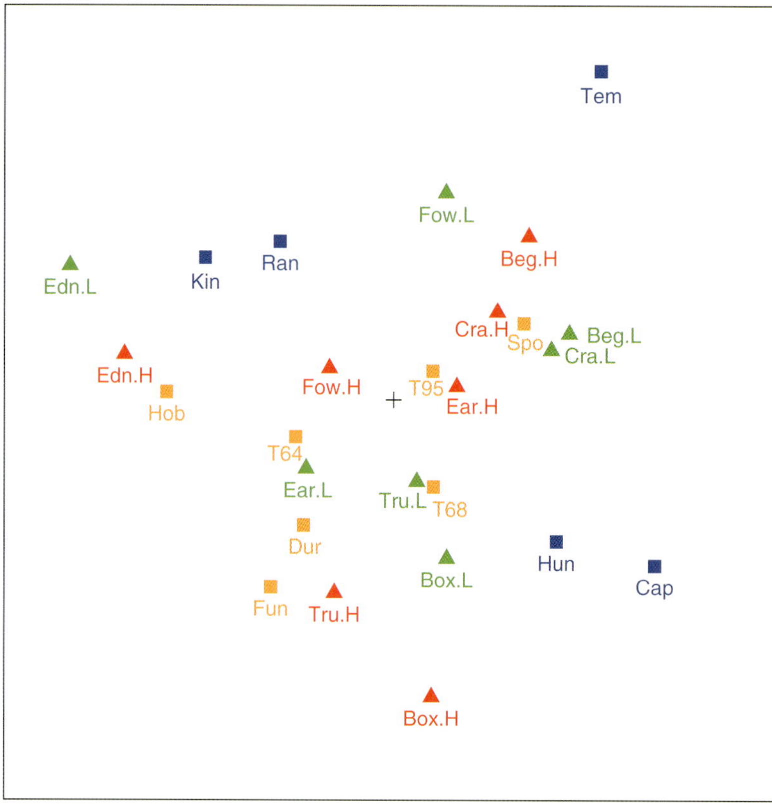

Figure 6.4 Biadditive biplot of the interaction matrix $\mathbf{Z} = \mathbf{U\Sigma V}'$ of the wheat data (Table 6.2). Sites are plotted from first two dimensions of $\mathbf{U\Sigma}^{1/2}$ and varieties from first two dimensions of $\mathbf{V\Sigma}^{1/2}$ using our R function `biadbipl` with arguments `X = wheat.data, biad.variant = "InteractionMat", ax = NULL, SigmaHalf = TRUE`. Note that exactly the same graph is obtained with the setting `X = t(wheat.data)`. The origin is marked with a black cross.

in his Figure 4, gives an inner product type biplot where the variables are represented as vectors. Here, Figure 6.6 gives the equivalent biplot with calibrated axes.

In Figures 6.3 and 6.4 we show a biadditive biplot of the interaction matrix $\mathbf{Z}$, where we have used points to represent both varieties and sites, leaving users to attempt to evaluate inner products. Previously we have emphasized the value of representing at least one of the factors by a set of calibrated biplot axes. This we show in the top panel of Figure 6.5, where the sites are shown as axes. To obtain approximate predictions for the variety yields expected at each site, we merely project each variety onto the selected site axis and read off the calibration. We might wish to do things the other way round and predict which site is best for each variety. This requires that the varieties be represented as axes as in the bottom panel of Figure 6.5. In Figure 6.5 the two-dimensional row and column predictivities are shown. These are not as good as for the biplots of the original data table nor for those of the mean-adjusted data. Full details of the predictivities are

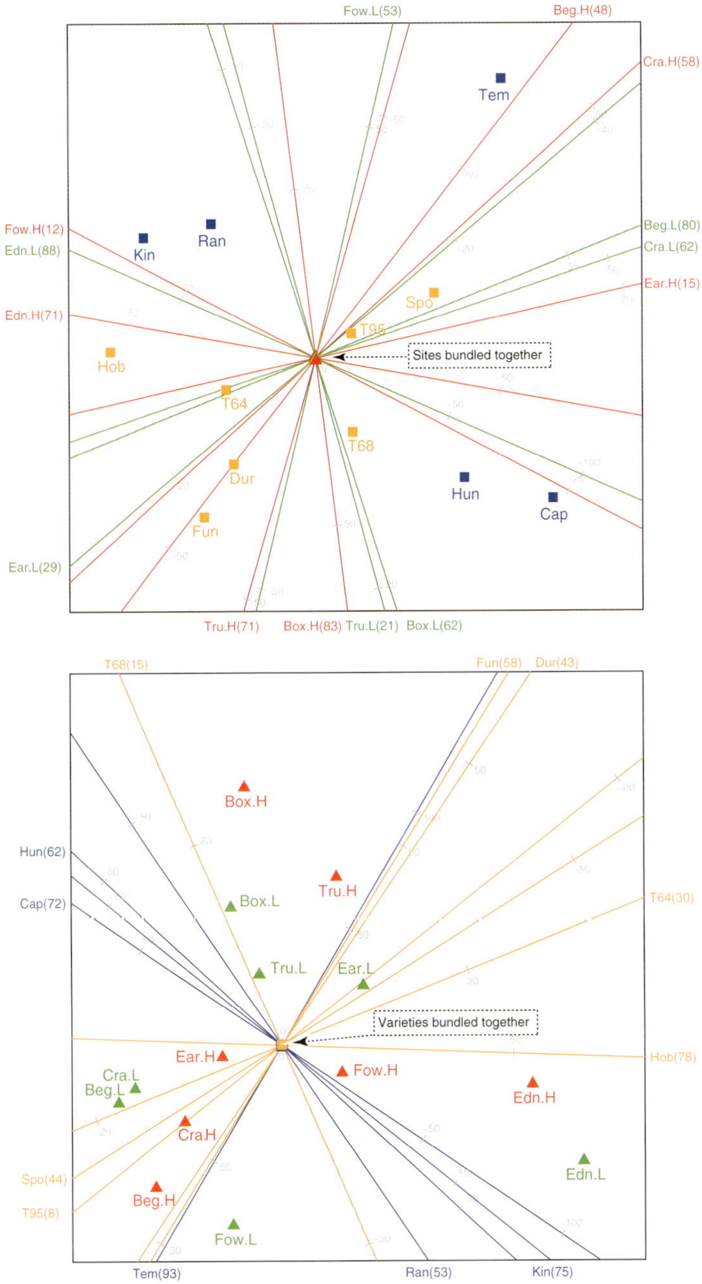

Figure 6.5 Biadditive biplots of the interaction matrix $\mathbf{Z} = \mathbf{U\Sigma V}'$ of the wheat data (Table 6.2) with sites as biplot axes in the top panel and varieties as biplot axes in the bottom panel. The quality of the display is 61.58%. The scales on the axes are in terms of interactions (see Table 6.2). Note that all axes pass through zero, coinciding almost with the position of the sites in the top panel and the varieties in the bottom panel.

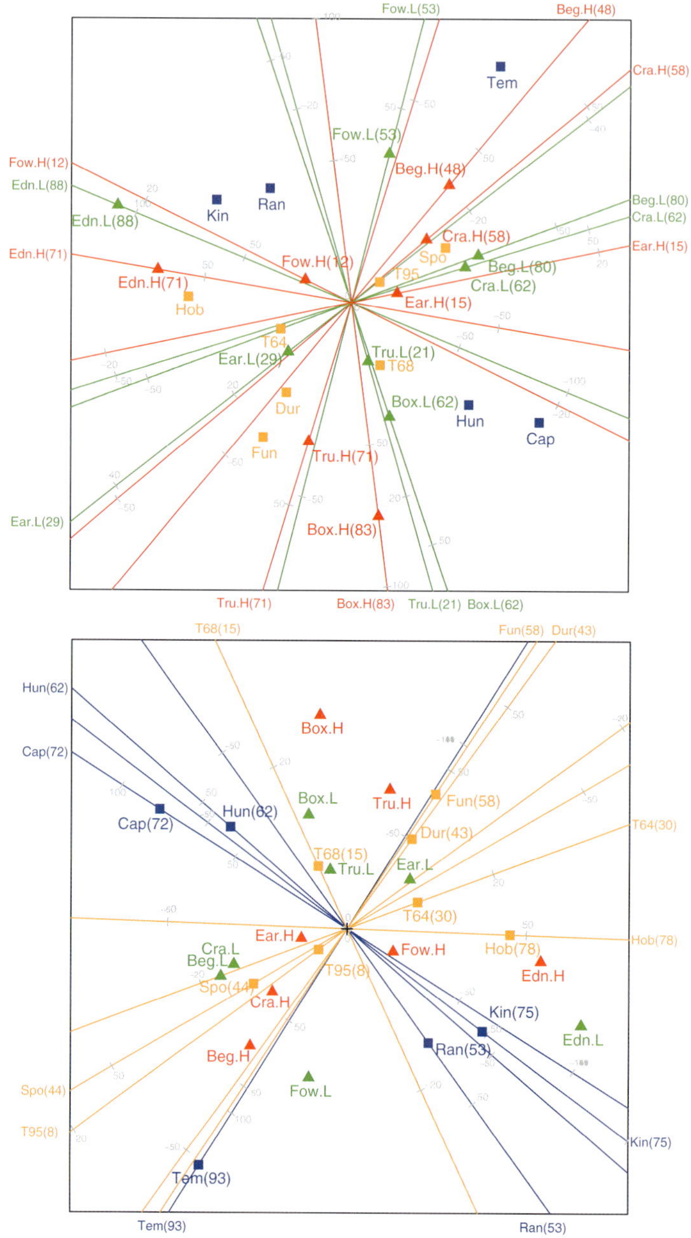

Figure 6.6 Extension of Figure 6.5. SVD of $\mathbf{Z} = \mathbf{U\Sigma V}'$. In the top panel varieties are plotted from first two dimensions of $\mathbf{U\Sigma}^{1/2}$ and sites from first two dimensions of $\mathbf{V\Sigma}^{1/2}$ using our R function `biadbipl` with arguments `X = wheat.data, biad.variant = "InteractionMat", SigmaHalf = TRUE`. Sites are also shown as calibrated axes. Calibrations are in terms of interactions. Notice that each axis passes through zero and one of the sites. Using a similar function call with `X = t(wheat.data)` results in the biplot in the bottom panel. Note that in these biplots neither column points nor row points are lumped together.

Table 6.5 Row and column predictivities for the biplot of the wheat data when specifying `biad.variant = "InteractionMat"`.

Predictivities for the sites

	Cra.L	Cra.H	Beg.L	Beg.H	Fow.L	Fow.H	Tru.L	Tru.H	Box.L	Box.H	Ear.L	Ear.H	Edn.L	Edn.H
Dim_1	0.569	0.362	0.714	0.216	0.038	0.099	0.019	0.072	0.076	0.017	0.189	0.140	0.771	0.698
Dim_2	0.616	0.576	0.797	0.477	0.535	0.119	0.209	0.712	0.621	0.827	0.285	0.146	0.880	0.715
Dim_3	0.618	0.743	0.853	0.738	0.607	0.378	0.343	0.759	0.626	0.830	0.328	0.538	0.989	0.853
Dim_4	0.652	0.793	0.863	0.982	0.748	0.485	0.377	0.759	0.648	0.872	0.801	0.677	0.989	0.880
Dim_5	0.676	0.793	0.916	0.983	0.897	0.706	0.603	0.802	0.656	0.914	0.802	0.830	0.990	0.961
Dim_6	0.873	0.793	0.928	0.985	0.914	0.723	0.608	0.894	0.944	0.957	0.900	0.836	0.990	0.961
Dim_7	0.874	0.794	0.990	0.988	0.991	0.959	0.654	0.902	0.945	0.985	0.908	0.895	0.990	0.967
Dim_8	0.878	0.897	0.998	0.990	0.998	0.996	0.839	0.934	0.947	0.986	0.936	0.954	0.992	0.979
Dim_9	0.880	0.986	0.998	0.993	1.000	0.999	0.995	0.981	0.990	0.997	0.937	0.980	0.996	1.000
Dim_10	1.000	0.998	0.999	0.998	1.000	1.000	0.999	1.000	0.995	0.999	0.995	0.991	1.000	1.000
Dim_11	1.000	1.000	1.000	1.000	1.000	1.000	1.000	1.000	1.000	1.000	1.000	1.000	1.000	1.000
Dim_12	1.000	1.000	1.000	1.000	1.000	1.000	1.000	1.000	1.000	1.000	1.000	1.000	1.000	1.000

Predictivities for the varieties

	Cap	Ran	Hun	Tem	Kin	Fun	Dur	Hob	Spo	T95	T64	T68
Dim_1	0.542	0.202	0.381	0.296	0.508	0.197	0.164	0.781	0.346	0.052	0.270	0.031
Dim_2	0.722	0.526	0.617	0.926	0.748	0.578	0.427	0.782	0.445	0.076	0.302	0.151
Dim_3	0.930	0.862	0.654	0.940	0.798	0.823	0.491	0.867	0.500	0.133	0.317	0.266
Dim_4	0.934	0.864	0.769	0.950	0.808	0.853	0.773	0.870	0.531	0.323	0.424	0.885
Dim_5	0.934	0.866	0.873	0.983	0.810	0.859	0.849	0.871	0.798	0.708	0.641	0.886
Dim_6	0.991	0.872	0.914	0.984	0.910	0.909	0.872	0.906	0.899	0.777	0.642	0.945
Dim_7	0.991	0.988	0.916	0.984	0.914	0.950	0.882	0.906	0.910	0.785	0.968	0.972
Dim_8	0.991	0.989	0.991	0.996	0.964	0.984	0.917	0.918	0.911	0.804	0.994	0.973
Dim_9	0.999	0.994	0.999	0.997	0.965	0.995	0.990	0.956	0.992	0.907	0.994	0.976
Dim_10	1.000	0.996	1.000	0.999	1.000	1.000	0.998	0.997	0.998	0.999	0.995	0.998
Dim_11	1.000	1.000	1.000	1.000	1.000	1.000	1.000	1.000	1.000	1.000	1.000	1.000
Dim_12	1.000	1.000	1.000	1.000	1.000	1.000	1.000	1.000	1.000	1.000	1.000	1.000

obtained with a call to `biad.predictivities` and are given in Table 6.5. The overall quality of the biplots in Figure 6.4 is 61.58%, showing that the interactions are less well represented in two dimensions than are the main effects. The two biplots in Figure 6.5 are shown separately; we could have incorporated both sets of axes in a single biplot, but it does not help with intelligibility and therefore is not recommended.

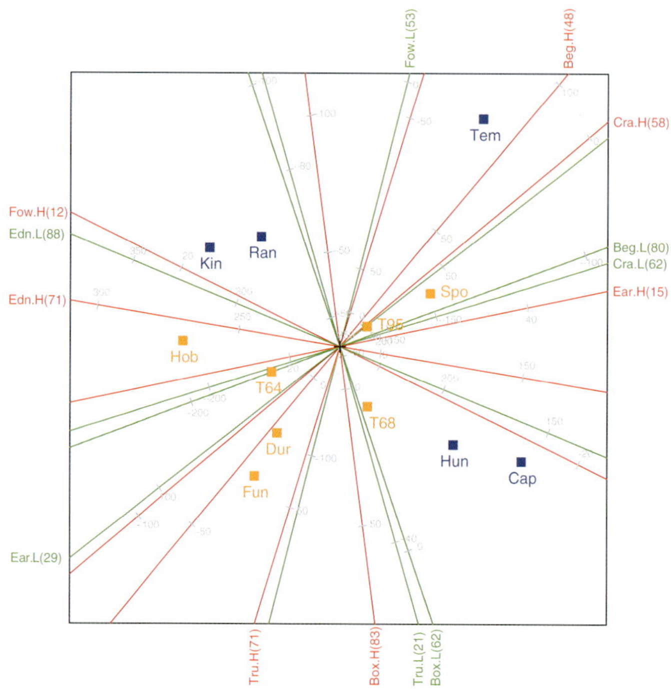

(a)

Figure 6.7 Similar to the biplot in the top panel of Figure 6.6 but with sites main effects added by calling the function `biadbipl` with argument `add.maineffects = TRUE`. Note that the scales on the biplot axes have changed to reflect the role of the main effects. The main effects can also be shown on the biplot axes. This can be implemented by setting argument `show.maineffects = TRUE` when calling `biadbipl`. All main effects are shown as solid black circles on the respective biplot axes. In (a) these black circles lie on top of one another, coinciding with the origin, and are therefore not explicitly visible. In the biplot in (b) the point of concurrency of the axes has been translated in order to improve the display of the biplot. Specifying in `biadbipl` the argument `select.origin = TRUE` allows the user to indicate by means of the mouse the preferred point of concurrency of the axes. The position of the origin can be marked with a cross by specifying `show.origin = TRUE`. Notice how the black circles indicating the respective main effects have moved together with the axes and are now readily appreciated. Unlike the biplot in the top panel of Figure 6.6, only the row points and columns as axes are shown; plotting of the column points is suppressed by specifying `plot.col.points = FALSE` and `col.points.text = FALSE` in the call to `biadbipl`.

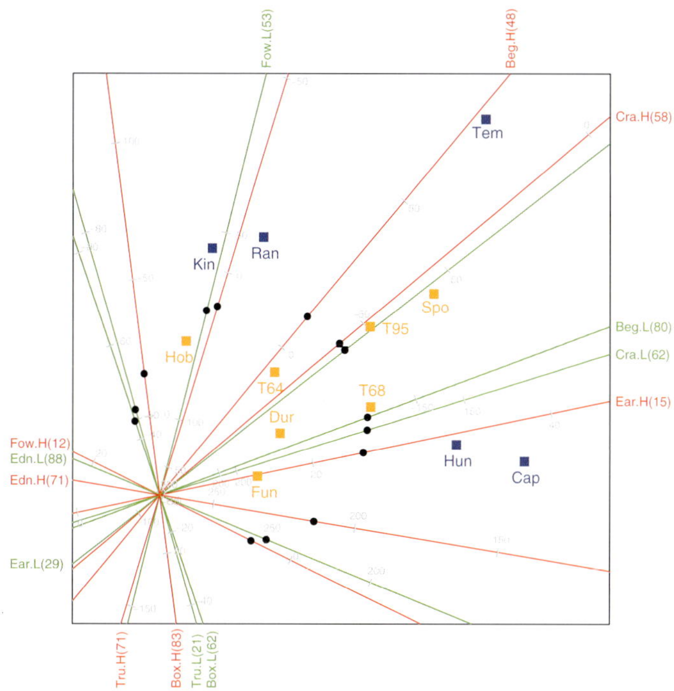

(b)

Figure 6.7 (*Continued*)

Knowledge of the size of the interaction is useful, but plant breeders will want to know the absolute value of yield at each site. This is easily obtained by adding the site main effect to the interaction, which can be readily achieved by adjusting the markers on the site biplot axes. When this is done the actual main effect itself may be marked by a special symbol. This is done in Figure 6.7, where we have also taken the opportunity to shift the axes to a position where the points representing the varieties are not obscured by the axes and labels. Of course, the same could be done for the biplot showing variety axes.

We have demonstrated that, just as we may evaluate axes predictivities for PCA, so may we evaluate row and column predictivities for two-way tables – all we have to do is to find the ratios of the sum of squares for the rows (or columns) of $\hat{\mathbf{X}}$ to the same sum of squares for the same rows (or columns) of $\mathbf{X}$. In practice we would first remove the overall mean from $\mathbf{X}$. We have seen that it is also useful to remove row and column main effects to provide for the residual matrix $\mathbf{Z}$ with its own row and column predictivities. By projecting the variety points onto an environmental axis, we obtain estimates of the interaction term. To get the total environmental effect, one may wish to include the contribution of the environmental main effect in the biplot. To do so, one merely has to shift the scale-markers by an amount equal to the magnitude of the main effect. Thus, the total environmental contribution (main effects + interaction) may be represented in a calibrated biplot. This has been done in Figure 6.7(b) where, at the same time we have shifted axes (see Chapter 2) to a more convenient position that interferes less with the plotted information.

The main effects for the varieties can be added to the biplot in Figure 6.7 by calling the function `biadbipl` with the argument `add.maineffects = TRUE`. This has been done in Figure 6.8.

The function `biadbipl` allows simultaneously all the interaction predictions for any given site to be obtained. This is achieved by using circle projection as described in Chapter 2 to draw a circle with diameter the line connecting the site with the origin. Predictions are given where the circle intersects the biplot axes. This is shown in Figure 6.9 for site ***Fow.L*** by calling `biadbipl` with arguments `X = wheat.data,`

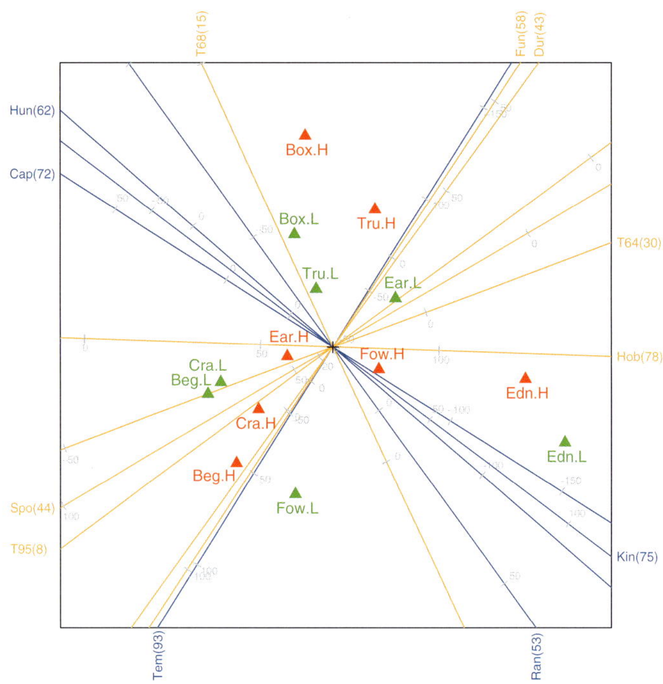

(a)

Figure 6.8 Similar to the biplot in the bottom panel of Figure 6.6 but with varieties main effects added by calling function `biadbipl` with argument `add.maineffects = TRUE`. Note that the scales on the biplot axes have changed to reflect the role of the main effects. The main effects can also be shown on the biplot axes. This can be implemented by setting argument `show.maineffects = TRUE` when calling `biad-bipl`. All main effects are shown as solid black circles on the respective biplot axes. In (a) of Figure 6.8 these black circles lie on top of one another coinciding with the origin and is therefore not explicitly visible. However, in the biplot in (b) where the origin has been translated the black circles indicating the respective main effects are readily appreciated. Note that plotting the varieties as points has been suppressed by specifying `plot.col.points = FALSE` and `col.points.text = FALSE` in the call to `biadbipl`.

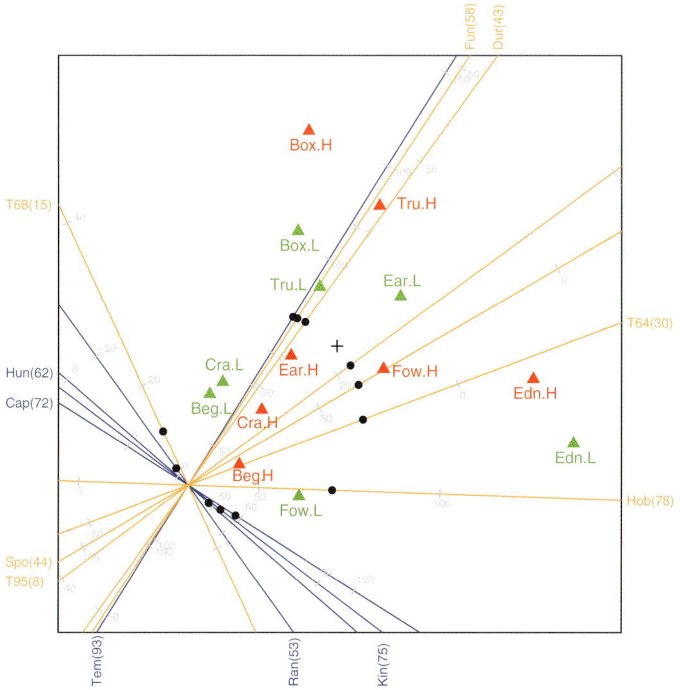

(b)

Figure 6.8 *(Continued)*

`circle.proj = 5`. Similarly, finding for any given variety the interaction predictions at all sites is achieved by construct a circle with diameter the line connecting the variety with the origin. This is shown for variety ***Ran*** in Figure 6.10 by calling `biadbipl` with arguments `X = t(wheat.data), circle.proj = 2`. The reader can verify that the predictions obtained from Figures 6.9 and 6.10 are as given in Table 6.6. Note that the predictions in Table 6.6 are readily obtainable by utilizing the argument `predictions.sample` when calling `biadbipl`. The reader can verify that this argument allows connecting lines to be drawn to the predictions in the circles shown in Figures 6.9 and 6.10.

If preferred, the origin of the biplot axes in Figures 6.9 and 6.10 can be interactively shifted to a more convenient position by setting `select.origin = TRUE` in the call to `biadbipl`. Examples are shown in Figure 6.11.

A biadditive biplot showing the predictions at all sites for the same selected variety or the predictions for all varieties at any selected site can easily be constructed by calling `biadbipl` with arguments `X` or `t(X)` together with an appropriate value given for the argument `predictions.allsamples.onaxis`. This is illustrated in Figures 6.12–6.14. The reader may check the predictions from Figures 6.12 and 6.13 by calling `biad.predictivities` appropriately. In Figure 6.14 we show how to obtain predictions (in terms of interactions with main effects added) in a one-dimensional biadditive biplot.

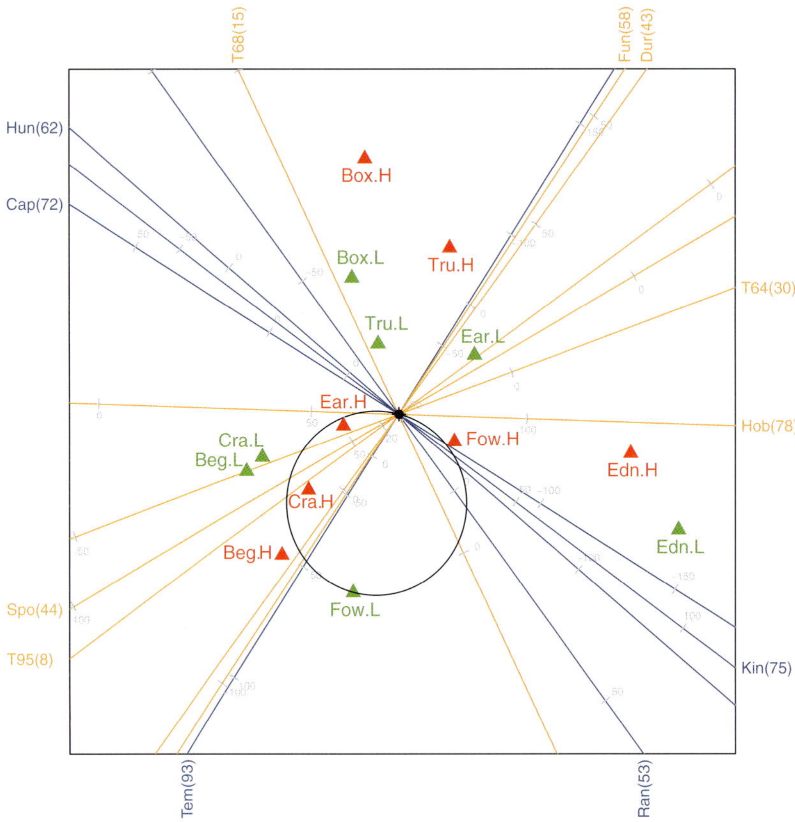

Figure 6.9 Using circular projection to show all interaction predictions pertaining to site ***Fow.L***. Note that the main effects, indicated by black solid circles, are all superimposed at the origin.

It is clear from Table 6.5 that by adding a further dimension to the biplots in Figures 6.4–6.13 several predictivities will be substantially increased. Furthermore, from the eigenvalues returned by `biadbipl` it follows that the overall quality increases from 61.6% in two dimensions to 72.9% in three dimensions. Therefore it might be advantageous to consider a three-dimensional biplot of the interaction matrix in Table 6.2. We provide in Figure 6.15 two screenshots of three-dimensional biadditive biplots. The screenshot in the top panel was made after the following call to `biadbipl`:

```
biadbipl(t(wheat.data), lam = F, biad.variant = "InteractionMat",
   show.origin = FALSE, SigmaHalf = TRUE, dim.biplot = 3,
   pch.row.points = 15, row.points.col = c(rep("blue",5),
   rep("orange",7)), row.points.size = 1.25,  pch.col.points = 17,
   plot.col.points = FALSE, column.points.col =
   rep(c("green","red"),7), column.points.size = 1.25,
   col.text.3d = rep(c("green","red"),7),
   axis.col = rep(c("green","red"),7), column.points.text = TRUE)
```

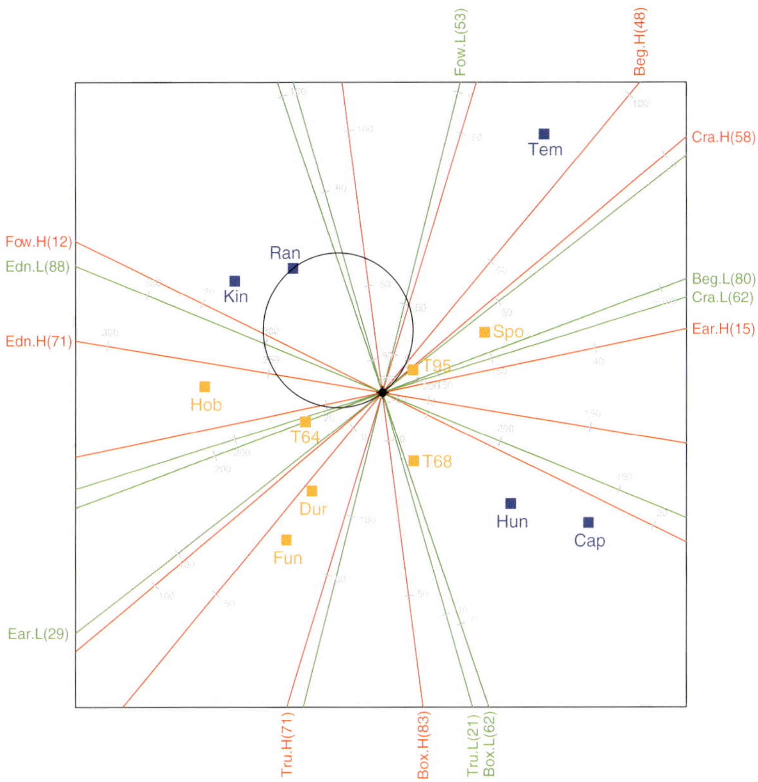

Figure 6.10 Using circular projection to show all interaction predictions pertaining to variety ***Ran***. Note that the main effects, indicated by black solid circles, are all superimposed at the origin.

Table 6.6 Interaction predictions for site ***Fow.L*** and variety ***Ran***.

Interaction predictions for variety ***Ran***		Interaction predictions for site ***Fow.L***	
Cra.L(62)	−180.51	Cap(72)	−64.42
Cra.H(58)	−54.01	Ran(53)	3.96
Beg.L(80)	−171.14	Hun(62)	−65.54
Beg.H(48)	18.71	Tem(93)	47.61
Fow.L(53)	−48.55	Kin(75)	31.28
Fow.H(12)	14.65	Fun(58)	−66.92
Tru.L(21)	−74.23	Dur(43)	−11.56
Tru.H(71)	−11.07	Hob(78)	61.20
Box.L(62)	−70.30	Spo(44)	61.07
Box.H(83)	−58.91	T95(8)	25.86
Ear.L(29)	67.77	T64(30)	−23.69
Ear.H(15)	19.60	T68(15)	1.17
Edn.L(88)	300.55		
Edn.H(71)	247.45		

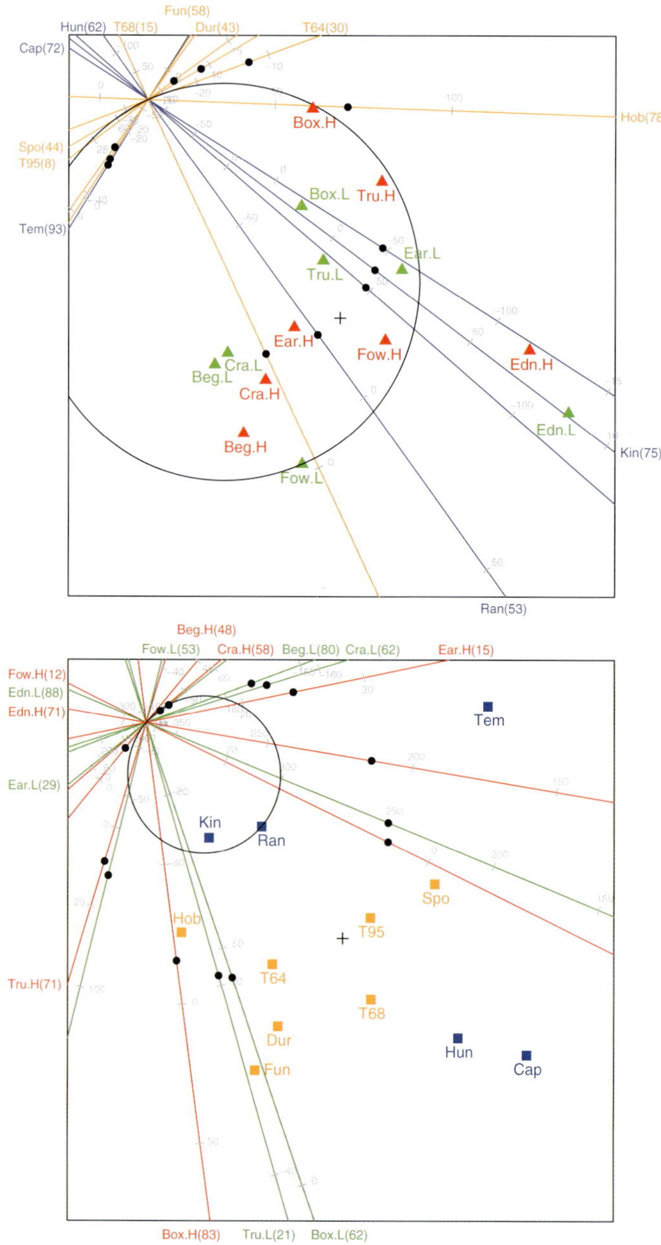

Figure 6.11 Similar to Figures 6.9 and 6.10, but with axes shifted as a result of setting `select.origin = TRUE` when calling `biadbipl`. Note that all predictions are exactly as in Figures 6.9 and 6.10. This figure illustrates why it can be advantageous to have the point of concurrency at or near the origin: all intersections of constructed circles with axes would tend to appear on the biplot. Note also that the black solid circles representing the main effects have moved with the axes and are no longer superimposed at the origin that is marked with a black cross.

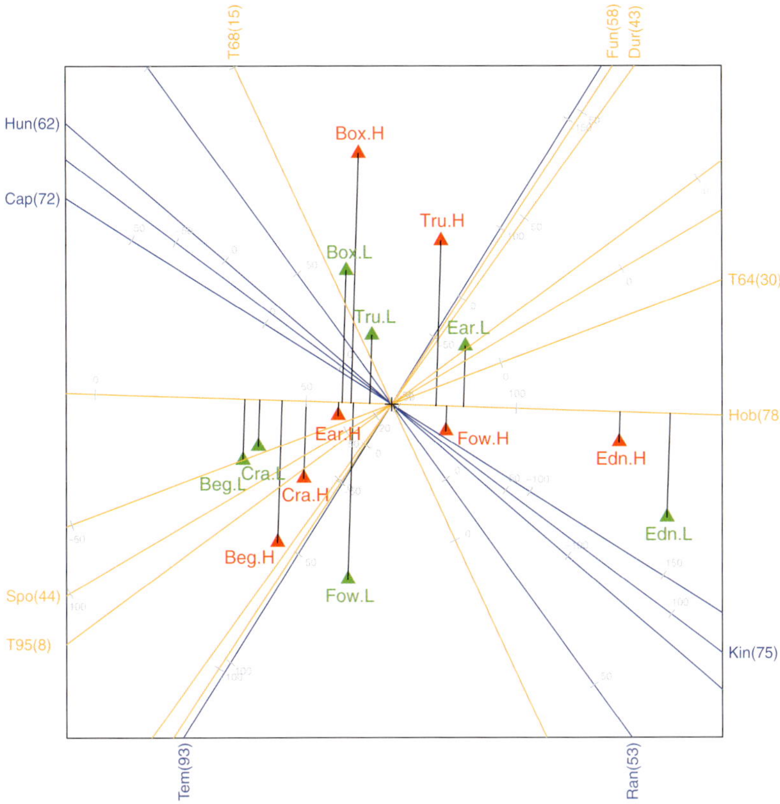

Figure 6.12 Predicting the values (in terms of interactions with main effects added) for all sites for the same variety (***Hob***). Accomplished by calling `biadbipl` with arguments `X = wheat.data, predictions.allsamples.onaxis = 8`.

The reader is encouraged to reproduce the three-dimensional biadditive biplots and to use the mouse buttons to interactively explore the three-dimensional biplot display.

As a final exercise, the reader is asked to reproduce the two-dimensional biplots of the interaction matrix of the wheat data with extra points added representing the average yield of the high nitrogen and low nitrogen at each site together with the predictivities of these extra points. First, read Section 6.5 carefully. Recall arguments `X.new.rows` and `X.new.columns` of `biadbipl`. Recall the output of function `biad.predictivities`.

If an estimate of the true error variance is available, biadditive biplots may also be embellished with confidence ellipses. In this regard, the reader is referred to Gower and Hand (1996) and the references contained therein.

6.8 Diagnostic biplots

The foregoing gives the most important use of biplots for a quantitative two-way table. Another, less used, application is to diagnose what model to fit, chosen from within the

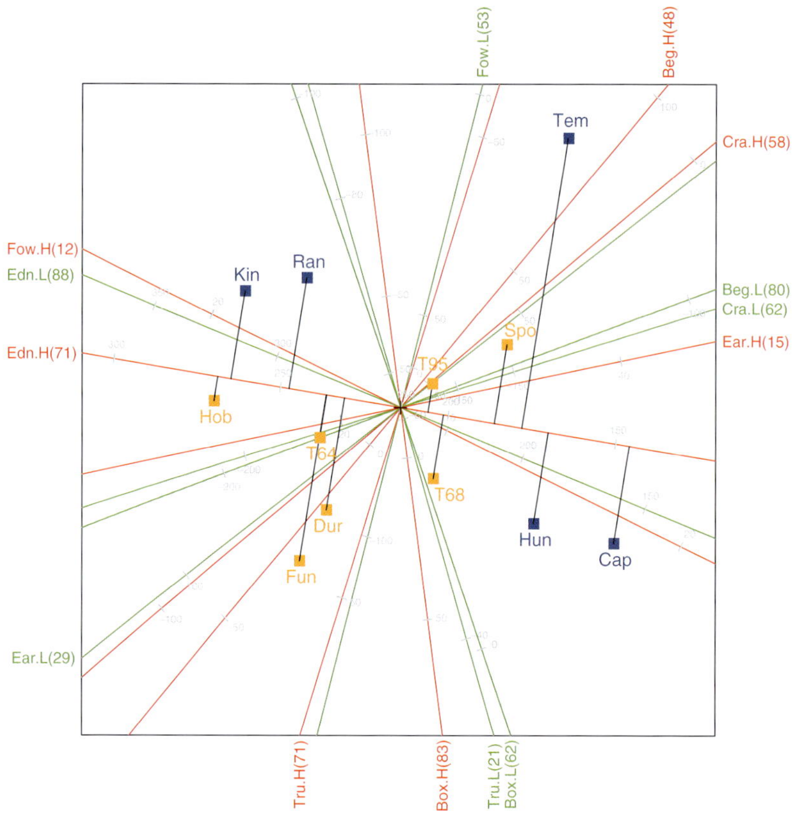

Figure 6.13 Similar to Figure 6.12 but predicting the values (in terms of interactions with main effects added) for all varieties at the same site (***Edn.H***). Accomplished by calling `biadbipl` with arguments `X = t(wheat.data), predictions.allsamples.onaxis = 12`.

biadditive family. Rather than operating on the interaction matrix $\mathbf{Z}$, we now operate on the original data $\mathbf{X}$ and use its SVD, $\mathbf{X} = \mathbf{U\Sigma V}'$. When the simple additive model holds, the matrix form of (6.1) is $\mathbf{X} = \mu\mathbf{11}' + \boldsymbol{\alpha}\mathbf{1}' + \mathbf{1}\boldsymbol{\beta}'$ or, on rewriting, $\mathbf{X} = (\mu\mathbf{1} + \boldsymbol{\alpha})\mathbf{1}' + \mathbf{1}\boldsymbol{\beta}'$, which, being the sum of two products, shows that $\mathbf{X}$ is of rank 2. Thus, if we make a biplot with coordinates $(\mu\mathbf{1} + \boldsymbol{\alpha}, \mathbf{1})$ and $(\mathbf{1}, \boldsymbol{\beta})$ we get two lines at right angles. The values of $\mu, \boldsymbol{\alpha}, \boldsymbol{\beta}$ are unknown but the SVD of $\mathbf{X}$ will exhibit the same orthogonal structure. The converse is also true, that if the SVD exhibits this structure, then the underlying model must be of simple additive form. This result may be used as a diagnostic for the additive model (Bradu and Gabriel, 1978). In applications there will be variations in the data, modelled by the usual error term, and the geometry will not be exact, but nevertheless it remains useful.

In the general case where there is one biadditive term, $\mathbf{X} = \mu\mathbf{11}' + \boldsymbol{\alpha}\mathbf{1}' + \mathbf{1}\boldsymbol{\beta}' + \boldsymbol{\gamma}\boldsymbol{\delta}'$, $\mathbf{X}$ is of rank 3 and the two-dimensional biplot derived from the SVD exhibits no structure. Then we would infer that there was at least one biadditive term. There are two

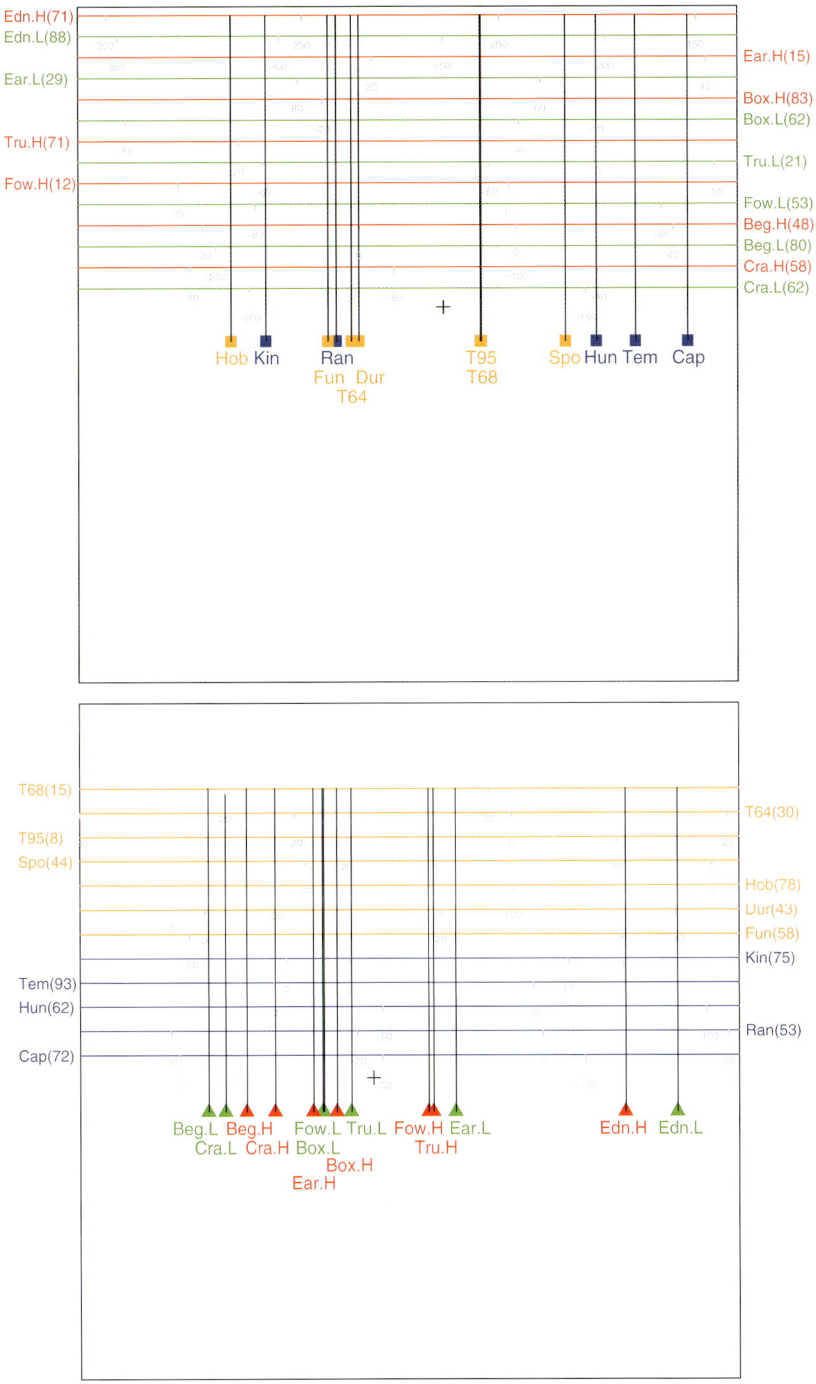

Figure 6.14 Similar to Figures 6.12 and 6.13 but using a one-dimensional biadditive biplot for predicting the values (in terms of interactions with main effects added) for: (top) all varieties at the same site; (bottom) at all sites for the same variety.

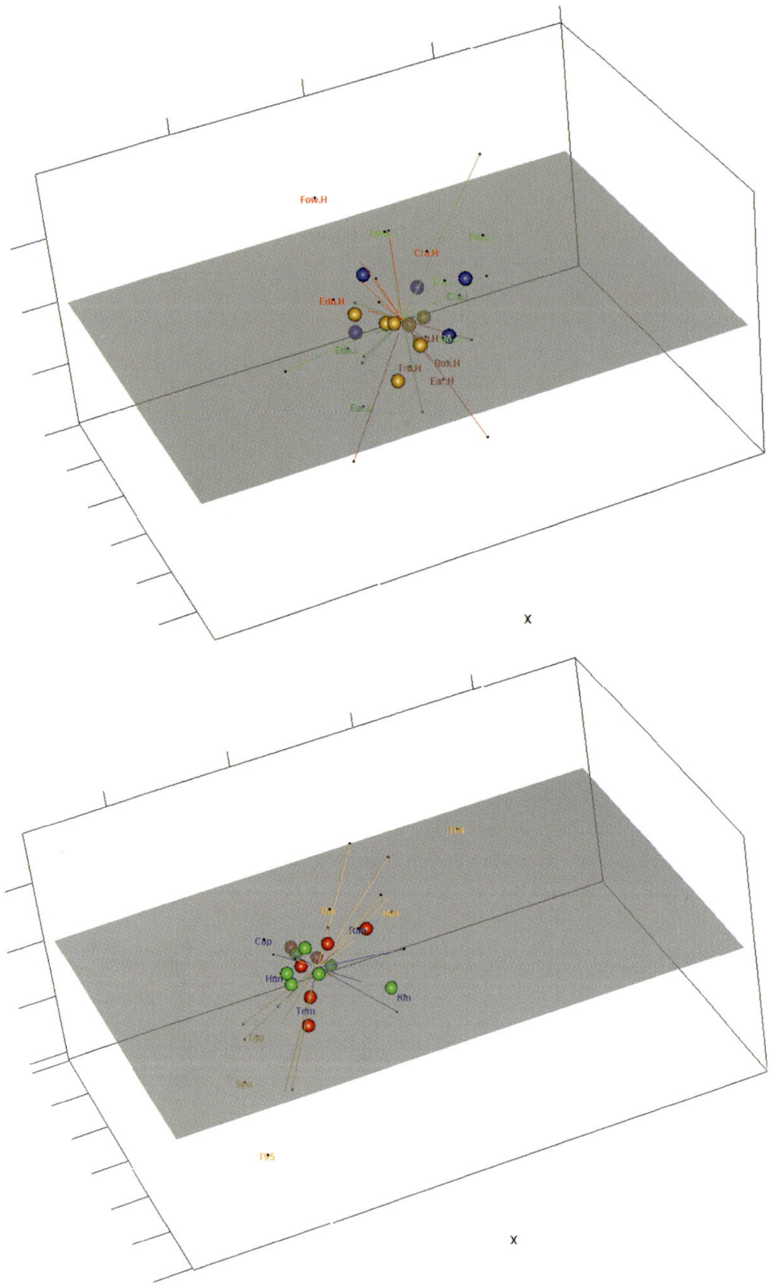

Figure 6.15 Screenshots from three-dimensional biplot of the interaction matrix of the wheat data after the SVD $\mathbf{Z} = \mathbf{U\Sigma V}'$. In the top panel varieties are plotted from the first three columns of $\mathbf{U\Sigma}^{1/2}$ and sites from the first three columns of $\mathbf{V\Sigma}^{1/2}$. In the bottom panel sites and varieties switch roles.

intermediate cases where the rank-2 structure is retained, the so-called row and column regression models which may be written as

$$\mathbf{X} = \mu\mathbf{1}\mathbf{1}' + \boldsymbol{\alpha}\mathbf{1}' + \mathbf{1}\boldsymbol{\beta}' + \lambda\boldsymbol{\gamma}\boldsymbol{\beta}' \quad \text{and} \quad \mathbf{X} = \mu\mathbf{1}\mathbf{1}' + \boldsymbol{\alpha}\mathbf{1}' + \mathbf{1}\boldsymbol{\beta}' + \lambda\boldsymbol{\alpha}\boldsymbol{\delta}',$$

in each of which one of the biadditive terms is proportional to a main effect parameter. The first of these may be rewritten

$$\mathbf{X} = (\mu\mathbf{1} + \boldsymbol{\alpha})\mathbf{1}' + (\mathbf{1} + \lambda\boldsymbol{\gamma})\boldsymbol{\beta}',$$

giving a biplot of coordinates $(\mu\mathbf{1} + \boldsymbol{\alpha}, \mathbf{1} + \lambda\boldsymbol{\gamma})$ and $(\mathbf{1}, \boldsymbol{\beta})$ with a straight line for the columns and general scatter for the rows; the row regression model does the opposite. When $\mathbf{X} = \mu\mathbf{1}\mathbf{1}' + \boldsymbol{\alpha}\mathbf{1}' + \mathbf{1}\boldsymbol{\beta}' + \lambda\boldsymbol{\alpha}\boldsymbol{\beta}'$ we get two nonorthogonal straight lines. In this way all these varieties of models with one biadditive term may be separated and identified from the SVD of $\mathbf{X}$.

Gower (1990) studied the three-dimensional geometry of these models, showing that the row points of the biplot lay in one plane and the column points in another plane. The projection of these two sets of points onto the intersection of the planes gives the biadditive parameters, and projections onto lines, one within each plane, give the main effects. Two-dimensional projections of this set-up give the diagnostic biplots discussed above. In principle, the row and column planes of the three-dimensional geometry can be extended to four dimensions including rank-2 interactions and could be used as a basis for additional types of biplot, but this possibility has not yet been explored.

7 Two-way tables: biplots associated with correspondence analysis

7.1 Introduction

Previous chapters have been concerned with biplots for a variety of forms of data matrix where, typically, the rows refer to n samples and the columns to p variables. As we have seen, samples and variables are very different concepts entailing different kinds of statistical treatment. In the previous chapter we looked at two-way tables with p rows and q columns which refer to similar entities. There, the body of the table consists of numerical values of a single variable playing the role of a dependent variable, while the two variables represented by the rows and columns have the role of independent variables. In this chapter, the body of the table is still regarded as a dependent variable with the rows and column classifiers treated as independent variables. The difference is that the dependent variable is no longer restricted to be a numerical variable measured on an interval or ratio scale but is available in the form of counts or frequencies, thus defining a contingency table. Correspondence analysis (CA) is concerned with the analysis and visualization of contingency tables, especially two-way contingency tables, and is discussed in this chapter. Our primary aim will be to understand the several variant forms of biplot related to CA without going into great detail about the methodology. Nevertheless, for a proper understanding, some methodological underpinning cannot be avoided. A CA biplot is very similar visually to the biplots discussed in Chapters 3 (PCA) and 6 (biadditive models). To get a preliminary feeling for the kind of visualizations involved, the reader may glance at the examples in Section 7.6 below. CA is unusual among the methods discussed in this book in having several variant forms, and

Understanding Biplots John Gower, Sugnet Lubbe and Niël le Roux

the reader is warned that it is not always easy to decide what variant is being offered in the wide range of available software. We hope that after reading this chapter readers will be in a position to decide the CA method and its biplot representation that best suit their purposes.

Before discussing CA biplots, we note that a data set may also consist of the different category levels of a single categorical variable. These category levels may also to be regarded as the values of a dependent or response variable depending on a single independent categorical variable. This situation calls for an optimal scoring procedure where the categories are replaced by optimal scores. Optimal scoring will be discussed in the latter part of Chapter 8. The latter situation can also be handled by multiple correspondence analysis (MCA) with three variables – that labelling the rows, that labelling the columns, and the categorical variable in the body of the table – but then the dependency relationship is ignored. MCA is discussed in the first part of Chapter 8.

7.2 The correspondence analysis biplot

Correspondence analysis (Benzécri, 1973; Greenacre, 2007) analyses the association in a $p \times q$ two-way contingency table $\mathbf{X}$. Just as the biadditive model is concerned with deviations from main effects, so CA is concerned with deviations from independence. Although CA is ideally performed on a contingency table, computationally, the elements of $\mathbf{X}$ may contain any nonnegative values; indeed only the row and column totals need be positive. Notice also that we are assuming a specific model for CA, namely, the *independence model*.

We begin with some notation. Let $\mathbf{R}$ and $\mathbf{C}$ denote the diagonal matrices containing the row and column sums $\mathbf{1}'\mathbf{X}$ and $\mathbf{X}\mathbf{1}$, respectively, of $\mathbf{X}$, treated as diagonal elements. The total sum of $\mathbf{X}$ is denoted by $n = \mathbf{1}'\mathbf{X}\mathbf{1} = \mathbf{1}'\mathbf{R}\mathbf{1} = \mathbf{1}'\mathbf{C}\mathbf{1}$. With this notation, the independence model, stating that the row classification of $\mathbf{X}$ is independent of its column classification, is given by $\mathbf{E} = \mathbf{R}\mathbf{1}\mathbf{1}'\mathbf{C}/n$. The CA procedure may be expressed in a variety of closely related variants. Greenacre (1984, 2007) and Le Roux and Rouanet (2004), in common with others, work in terms of frequencies and therefore initially divide $\mathbf{X}$ by n; this has no material effect and is ignored in the following. Here we cannot give an exhaustive treatment, but content ourselves with discussing some of the main variants of CA, especially with regard to biplot interpretation. Central to all these variants is the approximation to the deviations $\mathbf{X} - \mathbf{E}$ from the independence model:

$$\mathbf{X} - \mathbf{E} = \mathbf{X} - \mathbf{R}\mathbf{1}\mathbf{1}'\mathbf{C}/n. \tag{7.1}$$

7.2.1 Approximation to Pearson's chi-squared

A simple possibility is to base biplots directly on the singular value decomposition of the deviations $\mathbf{X} - \mathbf{E}$ from the independence model, but it turns out that the weighted deviations

$$\mathbf{R}^{-1/2}(\mathbf{X} - \mathbf{E})\mathbf{C}^{-1/2} \tag{7.2}$$

are of more interest. Indeed, writing $x_{i.}$ and $x_{.j}$ for row and column totals, the elements of $\mathbf{R}^{-1/2}(\mathbf{X}-\mathbf{E})\mathbf{C}^{-1/2}$ can be written as

$$\frac{x_{ij}-x_{i.}x_{.j}/n}{\sqrt{x_{i.}x_{.j}}}=\frac{1}{\sqrt{n}}\left(\frac{x_{ij}-x_{i.}x_{.j}/n}{\sqrt{x_{i.}x_{.j}/n}}\right). \tag{7.3}$$

The hypothesis that the independence model $\mathbf{E}=\mathbf{R}\mathbf{1}\mathbf{1}'\mathbf{C}/n$ describes the observed frequencies in the two-way contingency table $\mathbf{X}$ satisfactorily can be tested by the Pearson's chi-squared statistic

$$\begin{aligned}\chi^2&=\sum\frac{(\text{observed}-\text{expected})^2}{\text{expected}}\\&=\sum_i\sum_j\frac{(x_{ij}-x_{.j}x_{i.}/n)^2}{x_{.j}x_{i.}/n}.\end{aligned} \tag{7.4}$$

Comparing (7.3) with (7.4) shows that $n^{1/2}$ times the elements of $\mathbf{R}^{-1/2}(\mathbf{X}-\mathbf{E})\mathbf{C}^{-1/2}$ gives exactly the square roots of the contributions to Pearson's χ^2 for a contingency table; these are therefore sometimes termed the Pearson standardized residuals. Thus, we may seek to minimize

$$\|\mathbf{R}^{-1/2}(\mathbf{X}-\mathbf{E})\mathbf{C}^{-1/2}-\hat{\mathbf{X}}\|^2, \tag{7.5}$$

with the usual solution, based on the SVD

$$\mathbf{R}^{-1/2}(\mathbf{X}-\mathbf{E})\mathbf{C}^{-1/2}=\mathbf{U\Sigma V}', \tag{7.6}$$

of setting

$$\hat{\mathbf{X}}=\mathbf{U\Sigma JV}', \tag{7.7}$$

with $\mathbf{J}$ the diagonal matrix with units in its first r positions. This suggests that, for a k-dimensional approximation $\hat{\mathbf{X}}$ to the χ^2 contributions, we plot the first r columns of $\mathbf{U\Sigma}^{1/2}$ and $\mathbf{V\Sigma}^{1/2}$. Biplots of this inner product allow the identification of those elements of $\mathbf{X}$ which diverge from the independence assumption by contributing most, or least, to Pearson's χ^2. Alternative partitions such as $\mathbf{U\Sigma}, \mathbf{V}$ that preserve the inner product are also permissible.

7.2.2 Approximating the deviations from independence

Instead of approximating the Pearson residuals $\mathbf{U\Sigma}^{1/2}$ and $\mathbf{V\Sigma}^{1/2}$, an alternative possibility is that one wishes to approximate $\mathbf{X}-\mathbf{E}$, the deviations from independence, but weighted by the inverse square roots of the row and column totals. This requires the minimization of

$$\|\mathbf{R}^{-1/2}\{(\mathbf{X}-\mathbf{E})-\hat{\mathbf{X}}\}\mathbf{C}^{-1/2}\|^2, \tag{7.8}$$

which is given by $\mathbf{R}^{-1/2}\hat{\mathbf{X}}\mathbf{C}^{-1/2}=\mathbf{U\Sigma JV}'$. Now $\hat{\mathbf{X}}=\mathbf{R}^{1/2}\mathbf{U\Sigma JV}'\mathbf{C}^{1/2}$ and we may plot the first columns of $\mathbf{R}^{1/2}\mathbf{U\Sigma}^{1/2}$ and $\mathbf{C}^{1/2}\mathbf{V\Sigma}^{1/2}$. The difference between the two approaches demonstrates the subtleties engendered by weights appearing as part of the model and weights being used in the least-squares criterion itself. Other least-squares weights might be preferred, in which case the approximation would be based on a different SVD, but with no special difficulty.

7.2.3 Approximation to the contingency ratio

Surprisingly, plots based on interpretations of the Pearson residuals seem to be little used. Rather, redefining $\hat{\mathbf{X}}$, we may rewrite (7.8) as the minimization of

$$\|\mathbf{R}^{1/2}\{\mathbf{R}^{-1}(\mathbf{X}-\mathbf{E})\mathbf{C}^{-1}-\hat{\mathbf{X}}\}\mathbf{C}^{1/2}\|^2. \tag{7.9}$$

Equation (7.9) represents a least-squares problem with weights $\mathbf{R}^{1/2}$ and $\mathbf{C}^{1/2}$ where now the matrix $\hat{\mathbf{X}}$ approximates $\mathbf{R}^{-1}(\mathbf{X}-\mathbf{E})\mathbf{C}^{-1}$ whose elements

$$\frac{1}{n}\left(\frac{x_{ij}-e_{ij}}{e_{ij}}\right)=\frac{1}{n}\left(\frac{x_{ij}}{e_{ij}}-1\right)$$

are proportional to the deviations of x_{ij} from independence relative to the approximation e_{ij} under independence. Note that the ratio x_{ij}/e_{ij}, sometimes called the contingency ratio (see Greenacre, 2007) can be expanded as

$$\frac{x_{ij}}{e_{ij}}=\frac{x_{ij}/x_{.j}}{x_{i.}/n}=\frac{x_{ij}/x_{i.}}{x_{.j}/n}=n\frac{x_{ij}}{x_{i.}x_{.j}}.$$

We have thus shown that

$$\frac{1}{n}\left(\frac{x_{ij}-e_{ij}}{e_{ij}}\right)=\frac{1}{n}\left(\frac{nx_{ij}}{x_{i.}x_{.j}}-1\right)=\frac{1}{n}(\text{contigency ratio}-1). \tag{7.10}$$

From (7.9) we have that $\hat{\mathbf{X}}$ approximates $\mathbf{R}^{-1}(\mathbf{X}-\mathbf{E})\mathbf{C}^{-1}=\mathbf{R}^{-1/2}\mathbf{U}\mathbf{\Sigma}\mathbf{V}'\mathbf{C}^{-1/2}$, so we may plot the first r columns of $\mathbf{R}^{-1/2}\mathbf{U}\mathbf{\Sigma}^{1/2}$ and $\mathbf{C}^{-1/2}\mathbf{V}\mathbf{\Sigma}^{1/2}$ as coordinates to give visualizations of the departure from unity of the contingency ratios. Alternatively, we get the same inner product by plotting the first r columns of $\mathbf{R}^{-1/2}\mathbf{U}\mathbf{\Sigma}$ and $\mathbf{C}^{-1/2}\mathbf{V}$ as coordinates. Note that these biplots do not give the departure from unity directly but the departure divided by n. However, as we have seen previously, this scaling factor does not influence the shape of the biplot and can easily be provided for when constructing the calibrations of the biplot axes.

The biplot resulting from using $\mathbf{R}^{-1/2}\mathbf{U}\mathbf{\Sigma}$ and $\mathbf{C}^{-1/2}\mathbf{V}$ as coordinates has an interesting property, which we refer to as the *centroid property* and which follows from noting that the rows of $\mathbf{R}^{-1}\mathbf{X}$ give the proportions of each column category in each row. Weighting the column coordinates by these proportions gives

$$\begin{aligned}\mathbf{R}^{-1}\mathbf{X}(\mathbf{C}^{-1/2}\mathbf{V})&=\mathbf{R}^{-1/2}(\mathbf{R}^{-1/2}\mathbf{X}\mathbf{C}^{-1/2})\mathbf{V}\\&=\mathbf{R}^{-1/2}(\mathbf{U}\mathbf{\Sigma}\mathbf{V}'+\mathbf{R}\mathbf{1}\mathbf{1}'\mathbf{C}/n)\mathbf{V}\\&=\mathbf{R}^{-1/2}\mathbf{U}\mathbf{\Sigma},\end{aligned} \tag{7.11}$$

where the second term on the right-hand-side vanishes because of the orthogonality property $\mathbf{1}'\mathbf{CV} = \mathbf{0}'$. The meaning of (7.11) is that the row coordinates $\mathbf{R}^{-1/2}\mathbf{U\Sigma}$ are at the centroid of the column coordinates, weighted by the relative row frequencies $\mathbf{R}^{-1}\mathbf{X}$. Of course, given one set of coordinates, for the rows (say), it is simple to place column points at their centroids with or without weights given by the relative row frequencies. However, the centroid property is satisfied automatically for the CA scaling discussed in this section; it does not apply for other forms of CA scaling. The centroid property helps with interpreting the biplot because the column points for a row with high (low) weights will be tightly (loosely) clustered around the corresponding row point.

7.2.4 Approximation to chi-squared distance

A third and frequently used variant of CA is expressed in terms of chi-squared distance. This comes in two forms, (i) the chi-squared distance between the rows and (ii) the chi-squared distance between the columns of $\mathbf{X}$. The chi-squared distance $d_{ii'}$ between the ith and i'th rows of $\mathbf{X}$ is defined by

$$d_{ii'}^2 = \sum_{j=1}^{q} \frac{1}{x_{.j}} \left(\frac{x_{ij}}{x_{i.}} - \frac{x_{i'j}}{x_{i'.}} \right)^2 \tag{7.12}$$

or, in matrix terms,

$$d_{ii'}^2 = \left(\frac{\mathbf{x}_i}{x_{i.}} - \frac{\mathbf{x}_{i'}}{x_{i'.}} \right)' \mathbf{C}^{-1} \left(\frac{\mathbf{x}_i}{x_{i.}} - \frac{\mathbf{x}_{i'}}{x_{i'.}} \right). \tag{7.13}$$

Equation (7.13) is in the form of the square of a 'Mahalanobis distance', in the metric of the column totals, between points whose coordinates are the row proportions. Thus, when the ith and i'th rows have the same proportions, the chi-squared distance is zero, implying that we may amalgamate the rows without affecting the row chi-squared distances between any pairs of rows, including those that we have amalgamated. Rather less obvious is that amalgamating columns with equal proportions also has no affect on the row chi-squared distances. This statement may be verified by noting that proportionality of columns j and j' implies that $x_{ij'} = \tau x_{ij}, i = 1, \ldots, p$. Without amalgamation, these two columns contribute to $d_{ii'}^2$ the quantity

$$\frac{1}{x_{.j}} \left(\frac{x_{ij}}{x_{i.}} - \frac{x_{i'j}}{x_{i'.}} \right)^2 + \frac{\tau^2}{x_{.j'}} \left(\frac{x_{ij}}{x_{i.}} - \frac{x_{i'j}}{x_{i'.}} \right)^2. \tag{7.14}$$

After amalgamation into a single column the contribution is

$$\frac{1}{(1+\tau)x_{.j}} \left(\frac{(1+\tau)x_{ij}}{x_{i.}} - \frac{(1+\tau)x_{i'j}}{x_{i'.}} \right)^2. \tag{7.15}$$

But (7.14) can be written as

$$\left(\frac{1}{x_{.j}} + \frac{\tau^2}{x_{.j'}} \right) \left(\frac{x_{ij}}{x_{i.}} - \frac{x_{i'j}}{x_{i'.}} \right)^2 = \left(\frac{1}{x_{.j}} + \frac{\tau}{x_{.j}} \right) \left(\frac{x_{ij}}{x_{i.}} - \frac{x_{i'j}}{x_{i'.}} \right)^2 = (7.15).$$

This property of chi-squared distance, the so-called *principle of distributional equivalence*, is attractive because it suggests that chi-squared distance is not sensitive to small changes when row and column categorizations are amalgamated.

However, a less attractive property of chi-squared distance is that the division by $x_{\cdot j}$ gives higher weights to rarely occurring column categories, which is not always (some would say, rarely) desirable. Thus, we have to balance the principle of distributional equivalence with whether the weighting incorporated into chi-squared distance is appropriate in the first place.

The row chi-squared distances defined by (7.13) can also be regarded as weighted Euclidean distances between all rows of $\mathbf{X}$, and it is easy to see that the row chi-squared distances can be computed as ordinary Euclidean distances between pairs of rows of the matrix $\mathbf{R}^{-1}\mathbf{X}\mathbf{C}^{-1/2}$. Since translation does not affect distance, we are free to adjust by the translation term $\mathbf{11}'\mathbf{C}^{1/2}/n$, showing that the chi-squared distances are also generated by $\mathbf{R}^{-1}\mathbf{X}\mathbf{C}^{-1/2} - \mathbf{11}'\mathbf{C}^{1/2}/n$. We may write this as

$$\mathbf{R}^{-1}(\mathbf{X} - \mathbf{R11}'\mathbf{C}/n)\mathbf{C}^{-1/2} = \mathbf{R}^{-1}(\mathbf{X} - \mathbf{E})\mathbf{C}^{-1/2},$$

and therefore consider the approximation

$$\|\mathbf{R}^{1/2}\{\mathbf{R}^{-1}(\mathbf{X} - \mathbf{E})\mathbf{C}^{-1/2} - \hat{\mathbf{X}}\}\|^2 \qquad (7.16)$$

which is a similar weighted least-squares problem to (7.9) but now $\hat{\mathbf{X}}$ approximates the points that generate the row chi-squared distances; also, the criterion no longer carries column weights. In contrast to the high weights given to rarely occurring column categories in the definition of chi-squared distance, in the fitting criterion (7.16), the row weights $\mathbf{R}^{1/2}$ give lower weights to rarely occurring row categories. Equation (7.16) may be written

$$\|\mathbf{U\Sigma V}' - \mathbf{R}^{1/2}\hat{\mathbf{X}}\|^2. \qquad (7.17)$$

So the approximation $\hat{\mathbf{X}}$ is obtained from the inner product $\mathbf{R}^{-1/2}\mathbf{U\Sigma V}'$ and we may plot the first r columns of $\mathbf{R}^{-1/2}\mathbf{U\Sigma}$ for the rows and $\mathbf{V}$ for the columns. Note that $\mathbf{V}$, being an orthogonal matrix, does not affect the distances given by the row coordinates. This derivation is very close indeed to PCA with weights $\mathbf{R}^{1/2}$, the distances between pairs of row points now approximating chi-squared distance rather than the Pythagorean distance between the rows of $\mathbf{X}$. It follows from the orthogonality of the vector $\mathbf{1}'\mathbf{R}^{1/2}$ to the remaining $p-1$ singular vectors, that the weighted mean of the plotted points is $\mathbf{1}'\mathbf{R}^{1/2}(\mathbf{R}^{-1/2}\mathbf{U\Sigma}) = \mathbf{1}'\mathbf{U\Sigma} = \mathbf{0}'$, which is the usual centring for PCA. Furthermore, the vectors $\mathbf{V}$ are acting like axes and may be calibrated in the usual way (see Section 3.2).

Corresponding results apply to column chi-squared distance,

$$d_{jj'}^2 = \sum_{i=1}^{p} \frac{1}{x_{i\cdot}} \left(\frac{x_{ij}}{x_{\cdot j}} - \frac{x_{ij'}}{x_{\cdot j'}} \right)^2, \qquad (7.18)$$

generated by the columns of $\mathbf{R}^{-1/2}\mathbf{X}\mathbf{C}^{-1}$, leading to plotting the first r columns of $\mathbf{C}^{-1/2}\mathbf{V\Sigma}$ for the columns and $\mathbf{U}$ for the rows.

Very commonly the two chi-squared distance plots discussed above are amalgamated by plotting the columns of $\mathbf{R}^{-1/2}\mathbf{U\Sigma}$ and $\mathbf{C}^{-1/2}\mathbf{V\Sigma}$ simultaneously as two sets of points. This gives approximations to the row and column chi-squared distance, but the relationships between the two sets of points seem to have no simple interpretation. Furthermore, their inner product seems to be of little interest. Nevertheless, this seems to be the most commonly occurring form of correspondence analysis.

Before leaving chi-squared distance we establish a result that goes some way towards justifying the name. Let $w_i, i = 1, 2, \ldots, n$, be a set of nonnegative quantities, called weights. Define the weighted mean

$$\bar{x} = \frac{\sum_{i=1}^{n} w_i x_i}{\sum_{i=1}^{n} w_i}. \tag{7.19}$$

Then

$$\sum_{i=1}^{n} w_i (x_i - a)^2 = \sum_{i=1}^{n} w_i (x_i - \bar{x})^2 + (\bar{x} - a)^2 \sum_{i=1}^{n} w_i, \text{ for any } a. \tag{7.20}$$

Setting $a = x_{i'}$ in (7.20) leads to

$$\begin{aligned}\sum_{i'=1}^{n}\sum_{i=1}^{n} w_i w_{i'} (x_i - x_{i'})^2 &= \sum_{i'=1}^{n}\sum_{i=1}^{n} w_i w_{i'} (x_i - \bar{x})^2 + \sum_{i'=1}^{n} w_{i'} (\bar{x} - x_{i'})^2 \left(\sum_{i=1}^{n} w_i\right) \\ &= 2\sum_{i=1}^{n} w_i (x_i - \bar{x})^2 \left(\sum_{i=1}^{n} w_i\right),\end{aligned}$$

that is,

$$\sum_{i=1}^{n} w_i (x_i - \bar{x})^2 \left(\sum_{i=1}^{n} w_i\right) = \sum_{i<i'}^{n} w_i w_{i'} (x_i - x_{i'})^2. \tag{7.21}$$

Using weights $w_i = 1/n, i = 1, 2, \ldots, n$, in (7.21) gives the well-known identity

$$n \sum_{i=1}^{n} (x_i - \bar{x})^2 = \sum_{i<i'}^{n} (x_i - x_{i'})^2. \tag{7.22}$$

The total sum of squares of the Pearson residuals is given by

$$\chi^2 = \sum_{i=1}^{p}\sum_{j=1}^{q} \frac{(x_{ij} - x_{i.}x_{.j}/n)^2}{x_{i.}x_{.j}/n},$$

which may be written as

$$\chi^2 = \sum_{i=1}^{p}\sum_{j=1}^{q} \frac{x_{i.}(x_{ij}/x_{i.} - x_{.j}/n)^2}{x_{.j}/n} = n \sum_{j=1}^{q} \frac{1}{x_{.j}} \left[\sum_{i=1}^{p} x_{i.}(x_{ij}/x_{i.} - x_{.j}/n)^2\right]. \tag{7.23}$$

Replacing in (7.19) $w_i = x_{i.}$, $x_i = x_{ij}/x_{i.}$ and $n = p$, then

$$\bar{x} = \frac{\sum_{i=1}^{p} x_{i.}x_{ij}/x_{i.}}{\sum_{i=1}^{p} x_{i.}} = \frac{x_{.j}}{n},$$

where

$$n = \sum_{j=1}^{q} \sum_{i=1}^{p} x_{ij},$$

and, using the weighted identity on the expression in square brackets of (7.23), we have

$$\begin{aligned} \chi^2 &= n \sum_{j=1}^{q} \frac{1}{x_{.j}} \left[\sum_{i<i'}^{p} x_{i.} x_{i'.} (x_{ij}/x_{i.} - x_{i'j}/x_{i'.})^2 \right] \\ &= n \sum_{i<i'}^{p} x_{i.} x_{i'.} \left[\sum_{j=1}^{q} \frac{1}{x_{.j}} (x_{ij}/x_{i.} - x_{i'j}/x_{i'.})^2 \right]. \end{aligned} \tag{7.24}$$

The expression in the square brackets on the right-hand side of (7.24) is the chi-squared distance (7.12) between the ith and i'th rows of $\mathbf{X}$. Thus we have the simple result that

$$\chi^2 = n \sum_{i<i'}^{p} x_{i.} x_{i'.} d_{ii'}^2 = \frac{n}{2} \mathbf{1}'\mathbf{RDR1}, \tag{7.25}$$

where $\mathbf{D} = \{d_{ii'}^2\}$ is the $p \times p$ matrix of all the row chi-squared distances (7.12). Similarly, for the column chi-squared distances, we have

$$\chi^2 = n \sum_{j<j'}^{q} x_{.j} x_{.j'} d_{jj'}^2 = \frac{n}{2} \mathbf{1}'\mathbf{CDC1},$$

where now $\mathbf{D} = \{d_{jj'}^2\}$ is the $q \times q$ matrix of all the column chi-squared distances (7.18). These results link the chi-squared distances to the total Pearson's χ^2 for $\mathbf{X}$.

7.2.5 Canonical correlation approximation

Probably the oldest derivation of CA is due to Hirschfeld (1935) who asked what quantification of the categorical levels of the two variables classifying the contingency table, maximized their correlation. To express this idea algebraically, we define two indicator matrices, $\mathbf{G}_1$ and $\mathbf{G}_2$, of sizes $n \times p$ and $n \times q$ respectively, identifying row and column membership of the n cases.

In terms of our previous notation, we have

$$\mathbf{X} = \mathbf{G}_1'\mathbf{G}_2, \quad \mathbf{R} = \mathbf{G}_1'\mathbf{G}_1, \quad \mathbf{C} = \mathbf{G}_2'\mathbf{G}_2.$$

Next, we define quantification vectors $\mathbf{z}_1 : p \times 1$ and $\mathbf{z}_2 : q \times 1$, to be determined, which transform the categorical variables into quantitative variables $\mathbf{G}_1\mathbf{z}_1$ and $\mathbf{G}_2\mathbf{z}_2$. These two variables have squared (uncentred) correlations ρ^2 given by

$$\rho^2 = \frac{(\mathbf{z}_1'\mathbf{G}_1'\mathbf{G}_2\mathbf{z}_2)^2}{(\mathbf{z}_1'\mathbf{G}_1'\mathbf{G}_1\mathbf{z}_1)(\mathbf{z}_2'\mathbf{G}_2'\mathbf{G}_2\mathbf{z}_2)} = \frac{(\mathbf{z}_1'\mathbf{X}\mathbf{z}_2)^2}{(\mathbf{z}_1'\mathbf{R}\mathbf{z}_1)(\mathbf{z}_2'\mathbf{C}\mathbf{z}_2)}. \tag{7.26}$$

In order to maximize ρ^2 we first differentiate (7.26) with respect to $\mathbf{z}_1$ and $\mathbf{z}_2$ giving the derivatives

$$2(\mathbf{z}_1'\mathbf{X}\mathbf{z}_2)(\mathbf{X}\mathbf{z}_2)(\mathbf{z}_1'\mathbf{R}\mathbf{z}_1)^{-1}(\mathbf{z}_2'\mathbf{C}\mathbf{z}_2)^{-1} - 2(\mathbf{z}_1'\mathbf{R}\mathbf{z}_1)^{-2}(\mathbf{R}\mathbf{z}_1)(\mathbf{z}_1'\mathbf{X}\mathbf{z}_2)^2(\mathbf{z}_2'\mathbf{C}\mathbf{z}_2)^{-1}$$

and

$$2(\mathbf{z}_2'\mathbf{X}'\mathbf{z}_1)(\mathbf{X}\mathbf{z}_1)(\mathbf{z}_1'\mathbf{R}\mathbf{z}_1)^{-1}(\mathbf{z}_2'\mathbf{C}\mathbf{z}_2)^{-1} - 2(\mathbf{z}_2'\mathbf{C}\mathbf{z}_2)^{-2}(\mathbf{C}\mathbf{z}_2)(\mathbf{z}_2'\mathbf{X}'\mathbf{z}_1)^2(\mathbf{z}_1'\mathbf{R}\mathbf{z}_1)^{-1}.$$

Equating these derivatives to zero and rearranging, we obtain

$$(\mathbf{z}_1'\mathbf{R}\mathbf{z}_1)\mathbf{X}\mathbf{z}_2 = (\mathbf{z}_1'\mathbf{X}\mathbf{z}_2)\mathbf{R}\mathbf{z}_1 \tag{7.27}$$

and

$$(\mathbf{z}_2'\mathbf{C}\mathbf{z}_2)\mathbf{X}'\mathbf{z}_1 = (\mathbf{z}_1'\mathbf{X}\mathbf{z}_2)\mathbf{C}\mathbf{z}_2. \tag{7.28}$$

On normalizing $\mathbf{z}_1'\mathbf{R}\mathbf{z}_1 = 1$ and $\mathbf{z}_2'\mathbf{C}\mathbf{z}_2 = 1$, it follows from (7.27) and (7.28) that (7.26) is maximized when

$$\begin{aligned} \mathbf{X}\mathbf{z}_2 &= \rho\mathbf{R}\mathbf{z}_1, \\ \mathbf{X}'\mathbf{z}_1 &= \rho\mathbf{C}\mathbf{z}_2, \end{aligned} \tag{7.29}$$

the familiar equations for canonical correlation. These may be rewritten as

$$\begin{aligned} (\mathbf{R}^{-1/2}\mathbf{X}\mathbf{C}^{-1/2})\mathbf{C}^{1/2}\mathbf{z}_2 &= \rho\mathbf{R}^{1/2}\mathbf{z}_1, \\ (\mathbf{C}^{-1/2}\mathbf{X}'\mathbf{R}^{-1/2})\mathbf{R}^{1/2}\mathbf{z}_1 &= \rho\mathbf{C}^{1/2}\mathbf{z}_2. \end{aligned} \tag{7.30}$$

But

$$(\mathbf{R}^{-1/2}\mathbf{X}\mathbf{C}^{-1/2})\mathbf{C}^{1/2}\mathbf{1} = \mathbf{R}^{-1/2}\mathbf{X}\mathbf{1} = 1\mathbf{R}^{1/2}\mathbf{1}$$

and

$$(\mathbf{C}^{-1/2}\mathbf{X}'\mathbf{R}^{-1/2})\mathbf{R}^{1/2}\mathbf{1} = \mathbf{C}^{-1/2}\mathbf{X}'1 = 1\mathbf{C}^{1/2}\mathbf{1},$$

from which it is evident that $\mathbf{R}^{1/2}\mathbf{1}$ and $\mathbf{C}^{1/2}\mathbf{1}$ are a singular vector pair of $\mathbf{R}^{-1/2}\mathbf{X}\mathbf{C}^{-1/2}$ with unity as corresponding singular value. Since the elements of $\mathbf{R}^{1/2}\mathbf{1}, \mathbf{C}^{1/2}\mathbf{1}$ and $\mathbf{R}^{-1/2}\mathbf{X}\mathbf{C}^{-1/2}$ are all nonnegative, it follows from the Frobenius theorem (see Gower and Hand, 1996, Appendix A.11) that all other singular values of $\mathbf{R}^{-1/2}\mathbf{X}\mathbf{C}^{-1/2}$ must be smaller than unity. Therefore, it follows that

$$\mathbf{R}^{-1/2}\mathbf{X}\mathbf{C}^{-1/2} = \mathbf{R}^{1/2}\mathbf{1}\mathbf{1}'\mathbf{C}^{1/2}/n + \rho\mathbf{R}^{1/2}\mathbf{z}_1\mathbf{z}_2'\mathbf{C}^{1/2} + \dots \tag{7.31}$$

are the first terms in the SVD of $\mathbf{R}^{-1/2}\mathbf{X}\mathbf{C}^{-1/2}$, provided $\mathbf{R}^{1/2}\mathbf{z}_1$ and $\mathbf{C}^{1/2}\mathbf{z}_2$ are normalized so that $\mathbf{z}_1'\mathbf{R}\mathbf{z}_1 = 1$ and $\mathbf{z}_2'\mathbf{C}\mathbf{z}_2 = 1$.

The above development is in terms of uncentred variables, but correlation requires centring. The changes are straightforward, replacing $\mathbf{G}_1'\mathbf{G}_2, \mathbf{G}_1'\mathbf{G}_1, \mathbf{G}_2'\mathbf{G}_2$ by $\mathbf{G}_1'(\mathbf{I}-\mathbf{N})\mathbf{G}_2, \mathbf{G}_1'(\mathbf{I}-\mathbf{N})\mathbf{G}_1, \mathbf{G}_2'(\mathbf{I}-\mathbf{N})\mathbf{G}_2$, where $\mathbf{N}$ is the $n \times n$ centring matrix with $1/n$ in every position. Because $\mathbf{1}'\mathbf{G}_1 = \mathbf{1}'\mathbf{R}$ and $\mathbf{1}'\mathbf{G}_2 = \mathbf{1}'\mathbf{C}$, we have that

$$\begin{aligned} \mathbf{G}_1'(\mathbf{I}-\mathbf{N})\mathbf{G}_2 &= \mathbf{X} - \mathbf{R}\mathbf{1}\mathbf{1}'\mathbf{C}/n = \mathbf{X} - \mathbf{E}, \\ \mathbf{G}_1'(\mathbf{I}-\mathbf{N})\mathbf{G}_1 &= \mathbf{R} - \mathbf{R}\mathbf{1}\mathbf{1}'\mathbf{R}/n, \\ \mathbf{G}_2'(\mathbf{I}-\mathbf{N})\mathbf{G}_2 &= \mathbf{C} - \mathbf{C}\mathbf{1}\mathbf{1}'\mathbf{C}/n, \end{aligned} \tag{7.32}$$

so the basic equations now become

$$\begin{aligned}(\mathbf{X}-\mathbf{E})\mathbf{z}_2 &= \rho(\mathbf{R}-\mathbf{R}\mathbf{1}\mathbf{1}'\mathbf{R}/n)\mathbf{z}_1,\\ (\mathbf{X}-\mathbf{E})'\mathbf{z}_1 &= \rho(\mathbf{C}-\mathbf{C}\mathbf{1}\mathbf{1}'\mathbf{C}/n)\mathbf{z}_2,\end{aligned} \tag{7.33}$$

which may be rewritten as

$$\begin{aligned}[\mathbf{R}^{-1/2}(\mathbf{X}-\mathbf{E})\mathbf{C}^{-1/2}]\mathbf{C}^{1/2}\mathbf{z}_2 &= \rho\mathbf{R}^{1/2}\mathbf{z}_1-\rho\mathbf{R}^{1/2}\mathbf{1}(\mathbf{1}'\mathbf{R}\mathbf{z}_1)/n,\\ [\mathbf{C}^{-1/2}(\mathbf{X}-\mathbf{E})'\mathbf{R}^{-1/2}]\mathbf{R}^{1/2}\mathbf{z}_1 &= \rho\mathbf{C}^{1/2}\mathbf{z}_2-\rho\mathbf{C}^{1/2}\mathbf{1}(\mathbf{1}'\mathbf{C})\mathbf{z}_2/n.\end{aligned} \tag{7.34}$$

Now, from the usual orthogonality of the singular vectors of $\mathbf{R}^{-1/2}\mathbf{X}\mathbf{C}^{-1/2}$, we know that $\mathbf{1}'\mathbf{R}\mathbf{z}_1$ and $\mathbf{1}'\mathbf{C}\mathbf{z}_2$ are zero, but $\mathbf{R}^{-1/2}(\mathbf{X}-\mathbf{E})\mathbf{C}^{-1/2}]\mathbf{C}^{1/2}\mathbf{1}=\mathbf{0}$ and $\mathbf{C}^{-1/2}(\mathbf{X}-\mathbf{E})'\mathbf{R}^{-1/2}]\mathbf{R}^{1/2}\mathbf{1}=\mathbf{0}$, showing that $\mathbf{R}^{1/2}\mathbf{1}$ and $\mathbf{C}^{1/2}\mathbf{1}$ are also a singular vector pair of $\mathbf{R}^{-1/2}(\mathbf{X}-\mathbf{E})\mathbf{C}^{-1/2}$ with zero as corresponding singular value. Furthermore, $\mathbf{R}^{-1/2}\mathbf{X}\mathbf{C}^{-1/2}$ and $\mathbf{R}^{-1/2}(\mathbf{X}-\mathbf{E})\mathbf{C}^{-1/2}$ have the same set of singular vectors (apart from multiplication by -1) and, apart from the singular value associated with the singular vector pair $\mathbf{R}^{1/2}\mathbf{1}$ and $\mathbf{C}^{1/2}\mathbf{1}$, also the same singular values. It follows that to take care of centring, we only need to replace $\mathbf{X}$ by $\mathbf{X}-\mathbf{E}$ in (7.31). Thus, $\mathbf{z}_1$ and $\mathbf{z}_2$ are given by the columns of $\mathbf{R}^{-1/2}\mathbf{U}\boldsymbol{\Sigma}$ and $\mathbf{C}^{-1/2}\mathbf{V}\boldsymbol{\Sigma}$ respectively, as for chi-squared distance. However, there is nothing in the correlational criterion to suggest an interest in chi-squared distance; the aim is purely to derive the quantifications. If we plot the quantified coordinates of the n samples, $\mathbf{G}_1\mathbf{z}_1$ and $\mathbf{G}_2\mathbf{z}_2$, then these merely repeat (for as many times as in the corresponding values of the diagonal of $\mathbf{R}$) the p values of $\mathbf{z}_1$ and (for as many times as in the corresponding values of the diagonal of $\mathbf{C}$) the q values of $\mathbf{z}_2$ and nothing is gained. We note also that, by maximizing correlation, a one-dimensional solution is implied, in which case the scaling by the first singular value is immaterial and may be ignored altogether, so that it suffices to set $\mathbf{z}_1=\mathbf{R}^{-1/2}\mathbf{u}$ and $\mathbf{z}_2=\mathbf{C}^{-1/2}\mathbf{v}$. Multidimensional solutions need additional justification, for example by appealing to chi-squared distance. Perhaps the main interest in the correlational derivation is as an introduction to multiple correspondence analysis (Chapter 8).

7.2.6 Approximating the row profiles

Yet another variant comes from focusing on the row profiles, given by $\mathbf{R}^{-1}\mathbf{X}$, suggesting an interest in fitting

$$\|\mathbf{R}^{1/2}\{\mathbf{R}^{-1}(\mathbf{X}-\mathbf{E})-\hat{\mathbf{X}}\}\mathbf{C}^{-1/2}\|^2, \tag{7.35}$$

giving $\mathbf{U}\boldsymbol{\Sigma}\mathbf{V}'=\mathbf{R}^{1/2}\hat{\mathbf{X}}\mathbf{C}^{-1/2}$ and $\hat{\mathbf{X}}=\mathbf{R}^{-1/2}\mathbf{U}\boldsymbol{\Sigma}\mathbf{V}'\mathbf{C}^{1/2}$, with plots of $\mathbf{R}^{-1/2}\mathbf{U}\boldsymbol{\Sigma}$ (as for chi-squared distance) for the rows and $\mathbf{C}^{1/2}\mathbf{V}$ for the columns. The latter provides axes that may be calibrated for the row profiles. Actually, we do not get the pure row profiles but rather their deviations from the marginal row profile $\mathbf{1}'\mathbf{C}/n$. This follows from noting that

$$\mathbf{R}^{-1}\mathbf{E}=\mathbf{R}^{-1}(\mathbf{R}\mathbf{1}\mathbf{1}'\mathbf{C}/n)=\mathbf{1}\mathbf{1}'\mathbf{C}/n.$$

However, proper approximations to the row profiles may be obtained by approximating the row profiles $\mathbf{R}^{-1}\mathbf{X}$ directly, omitting the deviations from the independence model.

7.2.7 Analysis of variance and generalities

The basic weighted analysis of variance associated with a matrix $\mathbf{Y}$ is given by

$$\|\mathbf{W}_1\mathbf{Y}\mathbf{W}_2\|^2 = \|\mathbf{W}_1\hat{\mathbf{Y}}\mathbf{W}_2\|^2 + \|\mathbf{W}_1(\mathbf{Y} - \hat{\mathbf{Y}})\mathbf{W}_2\|^2 \tag{7.36}$$

or, in words, the 'total sum of squares' is the sum of the 'fitted sum of squares' and the 'residual sum of squares', where $\hat{\mathbf{Y}}$ is obtained by minimizing the residual sum of squares. In our context, all the variants of CA discussed above are special cases of:

$$\begin{aligned}&\|\mathbf{W}_1\{\mathbf{W}_1^{-1}\mathbf{R}^{-1/2}(\mathbf{X} - \mathbf{E})\mathbf{C}^{-1/2}\mathbf{W}_2^{-1}\}\mathbf{W}_2\|^2\\ &= \|\mathbf{W}_1\hat{\mathbf{X}}\mathbf{W}_2\|^2 + \|\mathbf{W}_1\{\mathbf{W}_1^{-1}\mathbf{R}^{-1/2}(\mathbf{X} - \mathbf{E})\mathbf{C}^{-1/2}\mathbf{W}_2^{-1} - \hat{\mathbf{X}}\}\mathbf{W}_2\|^2,\end{aligned} \tag{7.37}$$

which relates to (7.36) by setting $\mathbf{Y} = \{\mathbf{W}_1^{-1}\mathbf{R}^{-1/2}(\mathbf{X} - \mathbf{E})\mathbf{C}^{-1/2}\mathbf{W}_2^{-1}\}$, which is approximated by $\hat{\mathbf{Y}} = \hat{\mathbf{X}}$. Substituting $\mathbf{R}^{-1/2}(\mathbf{X} - \mathbf{E})\mathbf{C}^{-1/2} = \mathbf{U\Sigma V}'$ into (7.37) gives

$$\chi^2 = \|\mathbf{U\Sigma V}'\|^2 = \|\mathbf{U\Sigma JV}'\|^2 + \|\mathbf{U\Sigma}(\mathbf{I} - \mathbf{J})\mathbf{V}'\|^2,$$

where $\mathbf{J}$ is zero apart from r units in its first diagonal places, giving a simple way of expressing the r-dimensional approximation. Thus, when there are q nonzero singular values, all methods give the same analysis of variance irrespective of the particular weights $\mathbf{W}_1, \mathbf{W}_2$ and the choice of $\mathbf{Y}$:

$$\chi^2 = \sigma_1^2 + \sigma_2^2 + \cdots + \sigma_q^2 = (\sigma_1^2 + \sigma_2^2 + \cdots + \sigma_r^2) + (\sigma_{r+1}^2 + \sigma_{r+2}^2 + \cdots + \sigma_q^2).$$

In particular, we have

$$\begin{aligned}\text{row predictivities} &= \operatorname{diag}(\mathbf{U\Sigma J\Sigma U}')[\operatorname{diag}(\mathbf{U\Sigma\Sigma U}')]^{-1}\\ &= \operatorname{diag}(\mathbf{U\Sigma}^2\mathbf{JU}')[\operatorname{diag}(\mathbf{U\Sigma}^2\mathbf{U}')]^{-1}.\end{aligned} \tag{7.38}$$

Similarly,

$$\text{column predictivities} = \operatorname{diag}(\mathbf{V\Sigma}^2\mathbf{JV}')[\operatorname{diag}(\mathbf{V\Sigma}^2\mathbf{V}')]^{-1}. \tag{7.39}$$

The only things that change are the models being fitted,

$$\mathbf{W}_1^{-1}\mathbf{R}^{-1/2}(\mathbf{X} - \mathbf{E})\mathbf{C}^{-1/2}\mathbf{W}_2^{-1} = \mathbf{W}_1^{-1}\mathbf{U\Sigma V}'\mathbf{W}_2^{-1},$$

and the approximation $\hat{\mathbf{X}} = \mathbf{W}_1^{-1}\mathbf{U\Sigma JV}'\mathbf{W}_2^{-1}$. That the weights $\mathbf{W}_1$ and $\mathbf{W}_2$ occur both in the model being fitted and as the weights used in the least-squares criterion is confusing, though not unique. Thus, for example, in multidimensional scaling one may wish to approximate a set of distances by distances between points in a few dimensions, weighting by some function of the observed, or even the fitted, distances so as to reduce the importance of small or large distances (see Shepard and Carroll, 1966). What may be questioned in CA is why the particular weights are those chosen, apart from the convenience of always ensuring dependence on the SVD of $\mathbf{R}^{-1/2}(\mathbf{X} - \mathbf{E})\mathbf{C}^{-1/2}$.

Table 7.1 summarizes the main results discussed above. We now turn to the implications of Table 7.1 for biplot displays.

Table 7.1 Listing of the function being approximated (model), the weights used in the weighted least-squares criterion and the inner product approximation to the model used as a basis for plotting the variants of CA discussed in this chapter. Matrices giving coordinates for typical biplots are given in the final column. The variants are all special cases of the general result given in the final row of the table.

CA variant	Model	Weights $\mathbf{W}_1, \mathbf{W}_2$	Inner product Approximation	Typical biplot
Pearson residuals	$\mathbf{R}^{-1/2}(\mathbf{X}-\mathbf{E})\mathbf{C}^{-1/2}$	$\mathbf{I}, \mathbf{I}$	$\mathbf{U\Sigma V}'$	$\mathbf{U\Sigma}^{1/2}, \mathbf{V\Sigma}^{1/2}$ (case A) $\mathbf{U\Sigma}, \mathbf{V}$ (case B)
Independence deviations	$\mathbf{X}-\mathbf{E}$	$\mathbf{R}^{-1/2}, \mathbf{C}^{-1/2}$	$\mathbf{R}^{1/2}\mathbf{U\Sigma V}'\mathbf{C}^{1/2}$	$\mathbf{R}^{1/2}\mathbf{U\Sigma}^{1/2}, \mathbf{C}^{1/2}\mathbf{V\Sigma}^{1/2}$
Contingency ratio	$\mathbf{R}^{-1}(\mathbf{X}-\mathbf{E})\mathbf{C}^{-1}$	$\mathbf{R}^{1/2}, \mathbf{C}^{1/2}$	$\mathbf{R}^{-1/2}\mathbf{U\Sigma V}'\mathbf{C}^{-1/2}$	$\mathbf{R}^{-1/2}\mathbf{U\Sigma}^{1/2}, \mathbf{C}^{-1/2}\mathbf{V\Sigma}^{1/2}$ (case A) $\mathbf{R}^{-1/2}\mathbf{U\Sigma}, \mathbf{C}^{-1/2}\mathbf{V}$ (case B)
Chi-squared distance	$\mathbf{R}^{-1}(\mathbf{X}-\mathbf{E})\mathbf{C}^{-1/2}$ $\mathbf{R}^{-1/2}(\mathbf{X}-\mathbf{E})\mathbf{C}^{-1}$	$\mathbf{R}^{1/2}, \mathbf{I}$ $\mathbf{I}, \mathbf{C}^{1/2}$	$\mathbf{R}^{-1/2}\mathbf{U\Sigma V}'$ $\mathbf{U\Sigma V}'\mathbf{C}^{-1/2}$	$\mathbf{R}^{-1/2}\mathbf{U\Sigma}, \mathbf{V}$ (row χ^2) $\mathbf{U}, \mathbf{C}^{-1/2}\mathbf{V\Sigma}$ (col. χ^2)
Correlation	$\mathbf{R}^{-1/2}(\mathbf{X}-\mathbf{E})\mathbf{C}^{-1/2}$	$\mathbf{I}, \mathbf{I}$		$\mathbf{R}^{-1/2}\mathbf{U\Sigma}, \mathbf{C}^{-1/2}\mathbf{V\Sigma}$
Row profiles	$\mathbf{R}^{-1}(\mathbf{X}-\mathbf{E})$	$\mathbf{R}^{1/2}, \mathbf{C}^{-1/2}$	$\mathbf{R}^{-1/2}\mathbf{U\Sigma V}'\mathbf{C}^{1/2}$	$\mathbf{R}^{-1/2}\mathbf{U\Sigma}^{1/2}, \mathbf{C}^{1/2}\mathbf{V\Sigma}^{1/2}$ (case A) $\mathbf{R}^{-1/2}\mathbf{U\Sigma}, \mathbf{C}^{1/2}\mathbf{V}$ (case B)
General	$\mathbf{W}_1^{-1}\mathbf{R}^{-1/2}(\mathbf{X}-\mathbf{E})\mathbf{C}^{-1/2}\mathbf{W}_2^{-1}$	$\mathbf{W}_1, \mathbf{W}_2$	$\mathbf{W}_1^{-1}\mathbf{U\Sigma V}'\mathbf{W}_2^{-1}$	$\mathbf{W}_1^{-1}\mathbf{U\Sigma}^{1/2}, \mathbf{W}_2^{-1}\mathbf{V\Sigma}^{1/2}$

The column labelled 'model' in Table 7.1 relates to the function being minimized by the weighted least-squares criterion, and the next column gives the row and/or column weights used. As can be seen, these weights appear in inverted form in the inner product approximation $\hat{\mathbf{X}}$ to the model, given in the fourth column of the table. In the final column, we give typical matrices usually used to construct biplots for the rows and columns. We use the word 'typical' because the inner products may be split into two components in an infinite number of ways. For example, we have seen that when approximating the contingency ratio, and one wishes to use the centroid property discussed at the end of Section 7.2.2, then we have to plot $\mathbf{R}^{-1/2}\mathbf{U\Sigma}, \mathbf{C}^{-1/2}\mathbf{V}$(or $\mathbf{R}^{-1/2}\mathbf{U}, \mathbf{C}^{-1/2}\mathbf{V\Sigma}$) rather than the typical, symmetric values given in Table 7.1. Writing the SVD of $\mathbf{R}^{-1/2}(\mathbf{X}-\mathbf{E})\mathbf{C}^{-1/2} = (\mathbf{U\Sigma}^{\alpha})\ (\mathbf{V\Sigma}^{\beta})'$, the commonest choices of scales are $\alpha = \beta = 1/2, \alpha = \beta = 1, \alpha = 1$ and $\beta = 0$, or $\alpha = 0$ and $\beta = 1$. When $\alpha = \beta$, the plots are said to be symmetric, otherwise asymmetric. When $\alpha + \beta = 1$, the inner product is preserved and calibrated axes may be used (see Section 2.3). The inner product is not preserved when $\alpha = \beta = 1$, as happens (i) when simultaneously plotting points that approximate both row and column chi-squared distances and (ii) for the correlational approach, so care has to be taken not to use inner product interpretations for such diagrams. Having said this, Gabriel (2002) and Gower (2004) showed that whichever scaling is used has little effect on the biplot visualization, at least in the sense that they are all very highly correlated. This implies that they are all displaying similar information, whichever of the criteria discussed above is being used. One thing they all have in common is a concern with different aspects of departure from an assumption of independence. It seems to us that when the rows and columns of a contingency table are of equal status (see Section 7.1) there is little justification for using asymmetric biplots. However, sometimes, as with chi-squared distance, rows and columns do not have equal status and then it is not unreasonable to treat them differently. Of more importance is to use the appropriate weightings of $\mathbf{R}$ and $\mathbf{C}$ that are consistent with the choice of criterion.

Quite independently of the choice of α and β or, indeed, of any way of partitioning the inner product, we have an additional decision as how to represent the biplot. There are three possibilities:

(i) use points to represent both the row and column elements;

(ii) use lines (calibrated axes) to represent both the row and column elements;

(iii) use points for one of the classifications and axes for the other.

Possibility (i) is by far the most common usage for CA. Moreover, it is essential when the centroid property (Section 7.2.3) is central to interpretation. However, in this book we have emphasized the advantages of using calibrated axes as in (iii). This may appear to be introducing an asymmetric element into CA as we have to decide whether the rows of the columns are represented by axes. Fortunately, the choice is entirely arbitrary since it makes no difference to the predicted values. This is not so for the chi-squared distance variant of CA, where the row and column distances are essentially asymmetric. Although (ii) is a symmetric option, it is rarely, if ever, used. Both points and axes may be combined so providing calibrations on the axes but also highlighting the positions of row and/or column elements.

A fourth option, not discussed further in this book, rests on the observation that $ab\cos(\theta) = ab\sin(\theta + \pi/2)$. The left-hand side represents a simple inner product and the right-hand side an area. This implies that if we rotate one set of the scaffolding axes through 90 degrees we may present the biplot as two sets of points as in (i) but using area (including the origin) rather than inner products for interpretation. This introduces some new geometrical concepts that need to be assimilated (see Gower *et al.*, 2010).

As well as the partitioning of the inner product and the choices of points and axes for display, all the devices such as axis translation and λ-scaling of axes discussed in Chapter 2 remain available.

We conclude with a few words on terminology. The extensive CA literature has developed its own terminology, based on English translations from the seminal French text of Benzécri (1973), the fount of much subsequent research. Benzécri used terms drawn from basic concepts in mechanics, such as mass and inertia. We have used standard statistical terminology, referring to row (or column) sums rather than mass, and sums of squares rather than inertia. Greenacre (1984, 2007) also introduced some terminology, referring to standard coordinates, which do not include weights, and principal coordinates, which do. In the above, we ourselves have referred to symmetric and asymmetric representations in the Greenacre sense. This conflicts with the distinction made in Chapter 1 between symmetric and asymmetric classes of biplot, according to which all correspondence analyses of two-way contingency tables would be symmetric. The problem is that asymmetry (*sensu* Greenacre) arises when treating the two-way table in an asymmetric way, as if it were a data matrix, by focusing on inter-row or inter-column chi-squared distance. As there is some ambiguity in many of these terms, for an understanding of the biplots we prefer to state algebraically, as in the final column of Table 7.1 and in the examples, precisely what it is we plot in the different circumstances.

7.3 Interpolation of new (supplementary) points in CA biplots

We have seen that the model $\mathbf{Y} = \mathbf{W}_1^{-1}\mathbf{R}^{-1/2}(\mathbf{X} - \mathbf{E})\mathbf{C}^{-1/2}\mathbf{W}_2^{-1}$ is approximated by

$$\hat{\mathbf{Y}} = \mathbf{W}_1^{-1}\mathbf{U}\boldsymbol{\Sigma}\mathbf{J}\mathbf{V}'\mathbf{W}_2^{-1}$$

which, in turn, is written as an inner product

$$\hat{\mathbf{Y}} = \mathbf{A}\mathbf{B}',$$

giving a biplot where the rows of $\mathbf{A}$ give the coordinates in r dimensions for plotting row points and the rows of $\mathbf{B}$ give the coordinates for plotting column points. Immediately, we have

$$\begin{aligned} \mathbf{A} &= \hat{\mathbf{Y}}\mathbf{B}(\mathbf{B}'\mathbf{B})^{-1}, \\ \mathbf{B}' &= (\mathbf{A}'\mathbf{A})^{-1}\mathbf{A}'\hat{\mathbf{Y}}, \end{aligned} \tag{7.40}$$

which are known as *transition formulae* (see also Section 2.7) that relate the row and column coordinates. From the form of (7.40) it is clear that the transition formulae are a manifestation of the regression method discussed in Section 2.7.

To add a new row $\mathbf{x}' : 1 \times q$ to $\mathbf{X}$ we require a new row $\hat{\mathbf{y}}' : 1 \times q$ given by

$$\hat{\mathbf{y}}' = w_1^{-1} r^{-1/2}(\mathbf{x}' - r\mathbf{1}'\mathbf{C}/n)\mathbf{C}^{-1/2}\mathbf{W}_2^{-1}, \tag{7.41}$$

where $r = \mathbf{1}'\mathbf{x}$, the total of the new row, and w_1 is its weight. Similarly, to add a new column $\mathbf{x}$: $p \times 1$ requires

$$\hat{\mathbf{y}} : p \times 1 = \mathbf{W}_1^{-1}\mathbf{R}^{-1/2}(\mathbf{x} - c\mathbf{R1}/n)c^{-1/2}w_2^{-1} \tag{7.42}$$

where c denotes the total of the new column.

These settings may be inserted into (7.40) to give the new row and column coordinates $\mathbf{a}'$ and $\mathbf{b}$ to give:

$$\begin{aligned} \mathbf{a}' : 1 \times q &= \hat{\mathbf{y}}'\mathbf{B}(\mathbf{B}'\mathbf{B})^{-1}, \\ \mathbf{b} : p \times 1 &= (\mathbf{A}'\mathbf{A})^{-1}\mathbf{A}'\hat{\mathbf{y}}. \end{aligned} \tag{7.43}$$

This is all rather general and simplifies when we consider the actual values of $\mathbf{A}$ and $\mathbf{B}$ used in the various forms of correspondence analysis that we have considered. In every case we have

$$\mathbf{A} = \mathbf{W}_1^{-1}\mathbf{U}\boldsymbol{\Sigma}^{\alpha}\mathbf{J},$$

which may be written

$$\mathbf{A} = \mathbf{W}_1^{-1}(\mathbf{U}\boldsymbol{\Sigma}\mathbf{V}')\mathbf{V}\boldsymbol{\Sigma}^{-\beta}\mathbf{J} = \mathbf{W}_1^{-1}\mathbf{R}^{-1/2}(\mathbf{X} - \mathbf{E})\mathbf{C}^{-1/2}\mathbf{V}\boldsymbol{\Sigma}^{-\beta}\mathbf{J}.$$

Now, from the orthogonality relationships of the SVD we have that $\mathbf{E}\mathbf{C}^{-1/2}\mathbf{V} = \mathbf{0}$, so

$$\mathbf{A} = \mathbf{W}_1^{-1}(\mathbf{R}^{-1/2}\mathbf{X}\mathbf{C}^{-1/2})\mathbf{V}\boldsymbol{\Sigma}^{-\beta}\mathbf{J}.$$

Interpolating a new row $\mathbf{x}' : 1 \times q$ now gives

$$\mathbf{a}' = w_1^{-1}r^{-1/2}\mathbf{x}'\mathbf{C}^{-1/2}\mathbf{V}\boldsymbol{\Sigma}^{-\beta}\mathbf{J}. \tag{7.44}$$

Similarly, for interpolating a new column $\mathbf{x}$: $p \times 1$,

$$\mathbf{B} = \mathbf{W}_2^{-1}(\mathbf{C}^{-1/2}\mathbf{X}'\mathbf{R}^{-1/2})\mathbf{U}\boldsymbol{\Sigma}^{-\alpha}\mathbf{J},$$

so that

$$\mathbf{b}' = w_2^{-1}c^{-1/2}\mathbf{x}'\mathbf{R}^{-1/2}\mathbf{U}\boldsymbol{\Sigma}^{-\alpha}\mathbf{J} \tag{7.45}$$

provides the coordinates for interpolating the new column $\mathbf{x}$.

Thus, (7.44) and (7.45) are our basic formulae which cover all the variants of Table 7.1 where the settings of $\mathbf{W}_1$ and $\mathbf{W}_2$ are listed; the choices of α and β must satisfy $\alpha + \beta = 1$, but otherwise are free. These formulae can be derived from (7.40) by plugging in the settings of $\mathbf{A}$ and $\mathbf{B}$, but the direct approach is simpler. If λ-scaling is used then $\mathbf{a}$ and $\mathbf{b}$ will have to be scaled appropriately.

Measures of fit for an interpolated row (column) follow directly from (7.38) and (7.39).

7.4 Other CA related methods

It was pointed out in Section 7.1 that although CA was developed for two-way contingency tables, it remains computationally feasible whenever the margins are positive and,

Table 7.2 Number of people (out of 100) assessing four products as 'good' on six attributes.

Product	Attributes						Total
	Atr 1	Atr 2	Atr 3	Atr 4	Atr 5	Atr 6	
A	12	10	8	11	21	11	73
B	43	37	71	44	51	44	290
C	83	79	80	95	82	94	513
D	73	67	33	29	66	21	289

in particular, when **X** is a table of positive elements. Potentially, this opens the way for the CA analysis of two-way tables of several types. Of special importance are tables of preference data or other forms of rating data. These include the cases where a set of products are scored on a rating scale (say, five-point or ten-point) for several attributes or where the attributes are ranked in order of preference. Another possibility is where x_{ij} represents the number of judges out of N who rate product i $(i = 1, 2, \ldots, p)$ as *good* on attribute j $(j = 1, 2, \ldots, q)$; by implication $N - x_{ij}$ judges rate the product as *not good*. Table 7.2 illustrates this type of data where 100 people have rated four products, ***A***, ***B***, ***C***, ***D*** as *satisfactory* on six attributes.

Thus, the value 12 in the first cell of Table 7.2 means that 12 of the 100 persons assessed product ***A*** to be satisfactory on attribute ***1***. We notice that product ***A*** gets few *satisfactory* ratings on all attributes, while product ***C*** gets high ratings on all attributes. If the table were analysed by CA, then both ***A*** and ***C***, having similar row profiles, would be close together on the CA map. Clearly, chi-squared distance measures distance between proportions and is not a satisfactory measure if we wish to distinguish good ratings from bad ratings. Thus chi-squared distance would indicate that the proportions relative to the row margins for ***A*** and ***C*** were similar but not that the absolute values were very different.

To address this problem and remain within the context of CA, Greenacre (1984) suggested that we should include the implicit information on unsatisfactory assessments of the attributes. Then Table 7.2 becomes Table 7.3. The doubling of the data is apparent. Now the chi-squared distances between ***A*** and ***C*** have been adjusted but the constant row totals introduce some special effects that need elucidation. This is best seen by rearranging Table 7.3 as in Table 7.4.

Table 7.3 Similar to Table 7.2 but including the negative assessments.

Product	Attributes												Total
	Atr 1		Atr 2		Atr 3		Atr 4		Atr 5		Atr 6		
	+	−	+	−	+	−	+	−	+	−	+	−	
A	12	88	10	90	8	92	11	89	21	79	11	89	600
B	43	57	37	63	71	29	44	56	51	49	44	56	600
C	83	17	79	21	80	20	95	5	82	18	94	6	600
D	73	27	67	33	33	67	29	71	66	34	21	79	600

Table 7.4 Rearranged Table 7.3 separating the positive assessments from the negative assessments.

Product	Attributes Positive assessments						Attributes Negative assessments						Total
	Atr 1	Atr 2	Atr 3	Atr 4	Atr 5	Atr 6	Atr 1	Atr 2	Atr 3	Atr 4	Atr 5	Atr 6	
A	12	10	8	11	21	11	88	90	92	89	79	89	600
B	43	37	71	44	51	44	57	63	29	56	49	56	600
C	83	79	80	95	82	94	17	21	20	5	18	6	600
D	73	67	33	29	66	21	27	33	67	71	34	79	600

With the rearrangement shown in Table 7.4 the nonproportionate nature of the rows for ***A*** and ***C*** is clear. Tables 7.3 and 7.4 both give the same CA analysis. The constant row sums give weights that simplify the chi-squared distances. The precise nature of this simplification is best seen by writing the Table 7.4 data matrix in algebraic form:

$$(\mathbf{X}, N\mathbf{11}' - \mathbf{X}), \tag{7.46}$$

where N is the number of assessors. The row and column sums of (7.46), expressed as diagonal matrices, are $\mathbf{R}^* = Nq\mathbf{I}$ and $\mathbf{C}^* = (\mathbf{C}, Np\mathbf{I} - \mathbf{C})$, where $\mathbf{C}$ is the usual column-sum diagonal matrix of $\mathbf{X}$ itself. Note, that $\mathbf{R}$ has been eliminated from further consideration and that $n = Npq$ is the total sum of the extended data matrix. We may now find formulae for the chi-squared distances using the results of Section 7.2.4 (see also Table 7.1). Thus, the row chi-squared distances are generated by the rows of

$$\mathbf{R}^{*-1}(\mathbf{X}, N\mathbf{11}' - \mathbf{X})\mathbf{C}^{*-1/2}, \tag{7.47}$$

and the column chi-squared distances are generated by the columns of

$$\mathbf{R}^{*-1/2}(\mathbf{X}, N\mathbf{11}' - \mathbf{X})\mathbf{C}^{*-1}. \tag{7.48}$$

From (7.47) the row chi-squared distances are generated by the rows of

$$(Nq)^{-1}(\mathbf{X}, -\mathbf{X})(\mathbf{C}, Np\mathbf{I} - \mathbf{C})^{-1/2} \tag{7.49}$$

because $\mathbf{11}'\mathbf{C}^{*-1/2}$ represents a constant term added to every item of each column and so has no effect on distance and may be eliminated. Furthermore, the same columns of $\mathbf{X}$ occur in both parts of (7.47); only the column weights change. Thus the kth attribute contributes to the distance between products i and i' the quantity

$$(Nq)^{-2}(x_{ik} - x_{i'k})^2\{c_k^{-1} + (Np - c_k)^{-1}\},$$

which simplifies to

$$Np(Nq)^{-2}(x_{ik} - x_{i'k})^2\{c_k(Np - c_k)\}^{-1}. \tag{7.50}$$

The factor $\{c_k(Np - c_k)\}$ is minimum (so $\{c_k(Np - c_k)\}^{-1}$ has maximal weight in (7.50)) when c_k is close to zero or Np, and is maximum at $c_k = Np/2$. Greenacre (2007) refers to it as a measure of polarization of the kth attribute.

From (7.48) the column chi-squared distances are generated by the columns of

$$(Nq)^{-1/2}(\mathbf{X}, N\mathbf{1}\mathbf{1}' - \mathbf{X})(\mathbf{C}, Np\mathbf{I} - \mathbf{C})^{-1}. \tag{7.51}$$

Now the columns of $(Nq)^{-1/2}\mathbf{X}\mathbf{C}^{-1}$ give the coordinates of the positive assessments while the columns of $(Nq)^{-1/2}(N\mathbf{1}\mathbf{1}' - \mathbf{X})(Np\mathbf{I} - \mathbf{C})^{-1}$ give the coordinates of the negative assessments. Writing $\mathbf{x}_k$ for the kth column of $\mathbf{X}$, we see that the coordinates of the kth attribute in its positive form are given by $\mathbf{u}_k = \mathbf{x}_k/c_k$ and in its negative form by $\mathbf{v}_k = (N\mathbf{1} - \mathbf{x}_k)/(Np - c_k)$. It follows that any point on the line joining these two points has the form

$$\lambda\frac{\mathbf{x}_k}{c_k} + (1 - \lambda)\frac{N\mathbf{1} - \mathbf{x}_k}{Np - c_k}.$$

Setting $\lambda = c_k/Np$, this becomes $1/p$, showing that all attributes share this point, which, of course, is the mean. The points representing the positive and negative forms of an attribute are at the extremities of a line through the mean which divides its two ends in the ratio $c_k : Np - c_k$.

7.5 Functions for constructing CA biplots

7.5.1 Function `cabipl`

Our main function for constructing the one-, two- and three-dimensional CA biplots described in Sections 7.2–7.4 is the function `cabipl`, which we now describe in detail. The reader is encouraged to study and experiment with its arguments. As is the case with our other main functions such as `PCAbipl`, `CVAbipl` and `biadbipl`, several functions are called by `cabipl` for drawing the biplot, adding features to it and changing its appearance that are generally not directly called by the user. In addition to these functions, there are several closely related functions that users may like to call in order to add enhancements to an existing biplot or to obtain information associated with an existing biplot. These functions are briefly introduced in Sections 7.5.2–7.5.5.

Usage

`cabipl` uses the same calling conventions as `PCAbipl` and shares the following arguments with it:

`alpha.3d`	`constant`	`line.length`	`predictions.3d`
`aspect.3d`	`dim.biplot`	`markers`	`predictions.sample`
`ax`	`exp.factor`	`markers.size`	`reflect`
`ax.col.3d`	`e.vects`	`n.int`	`rotate.degrees`
`ax.name.col`	`factor.x`	`offset`	`side.label`
`ax.name.size`	`factor.y`	`offset.m`	`size.ax.3d`
`cex.3d`	`font.3d`	`ort.lty`	`size.points.3d`
`col.plane.3d`	`ID.labs`	`pos`	`Title`
`col.text.3d`	`ID.3d`	`pos.m`	`Titles.3d`

Arguments with a specific meaning in `cabipl`

`X`	Required argument. A two-way contingency table in the form of a matrix with p rows and q columns.
`X.new`	Optional argument specifying t new points to be interpolated into the biplot. Must be in the form of a matrix with t rows and q columns or, if $t = 1$, a q-component vector.
`ca.variant`	One of `"PearsonResA"`, `"PearsonResB"`, `"IndepDev"`, `"ConRatioA"`, `"ConRatioB"`, `"Chisq2A"`, `"Chisq2B"`, `"Corr"`, `"RowProfA"`, `"RowProfB"`, `"PCA"`. Defaults to `"PearsonResA"`.
`axis.col`	Digits or character values for specifying `col` parameter for printing biplot axes. Defaults to `"red"`. Vector input allows for different colours for the various biplot axes.
`col.points.col`	Specifies `col` parameter controlling the colour of character for printing column points. Default is `"red"`.
`col.points.size`	Specifies `cex` parameter controlling the size of character for printing column points. Defaults to 1.
`col.points.label.size`	Specifies `cex` parameter controlling the size of printing the labels of column points. Defaults to 0.6.
`col.points.text`	Logical TRUE or FALSE for turning on or off printing of the labels of column points.
`ConRatioMinOne`	Logical TRUE or FALSE for displaying contingency ratio (FALSE, the default) or contingency ratio minus one.
`lambda`	Logical TRUE or FALSE. If FALSE, lambda is taken as 1. Lambda controls stretching or shrinking of column and row distances. Default is FALSE.
`logCRat`	Logical TRUE or FALSE for specifying whether the biplot axes of contingency ratio approximation CA biplot must be calibrated on a log scale or not. Default is FALSE.
`marker.col`	Specifies the `col` argument for controlling the colour of markers on biplot axes.
`pch.row.points`	Plotting character for plotting of row points. Default is 16.
`pch.col.points`	Plotting character for plotting of column points. Default is 15.
`PearsonRes.scaled.markers`	Logical TRUE or FALSE. Default is FALSE, suppressing multiplication by sqrt(sum of elements of `X`).

`plot.col.points`	Logical TRUE or FALSE for plotting or suppressing the plot of column points. Default is TRUE.
`predictions.X.new`	Logical TRUE or FALSE for specifying whether predictions are required for new point to be interpolated. Default is FALSE.
`propshift`	Numeric value for shifting all one-dimensional biplot axes upwards (positive value) or downwards (negative value). The constant argument regulates the spacing between axes.
`row.points.col`	Specifies `col` parameter controlling the colour of character for printing row points. Defaults to `"green"`.
`row.points.size`	Specifies `cex` parameter controlling the size of character for printing row points.
`row.points.label.size`	Specifies `cex` parameter controlling the size of character for printing the labels of row points.
`RowProf.scaled.markers`	Logical TRUE or FALSE. Default of FALSE results in axis calibrations in terms of deviations from the marginal row profile. TRUE yields axis calibrations in terms of row profiles.
`samples.plot`	Logical TRUE or FALSE. Used to turn off printing of sample plotting characters in three-dimensional biplots. Default is TRUE.
`scaled.mat`	Used when `ca.variant = "PCA"` to scale matrix to unit column variances before performing the SVD. Default is FALSE.
`text.pos`	Integer-valued two-component vector. The first element controls the position of row point labels; the second element controls the position of column point labels. Default is `c(1,1)`. Values of 1, 2, 3 and 4 respectively indicate positions below, to the left of, above and to the right of the plotting character.
`tick.marker.col`	Specifies colour of tickmark label. Default is `"grey"`.
`output`	Integer-valued or named vector specifying which components of value list to print. Default is to print the entire value list.
`X.new.pch`	Plotting character indicating interpolated `X.new` points. Default of 1 indicates a circular plotting character.
`X.new.col`	Colour of plotting character indicating interpolated `X.new` points. Default colour is `"blue"`.
`X.new.pch.cex`	Numeric value controlling size of plotting character indicating interpolated `X.new` points. Default is 1.5.

`X.new.labels`	Character vector labelling interpolated `X.new` points. Defaults to `np1`, `np2`, ...
`X.new.labels.cex`	Numeric value controlling size of the label of plotting character indicating interpolated `X.new` points. Default is 0.6.
`zoomval`	Specifies zooming factor. Defaults to NULL for no zooming; specify a value less than unity for zooming in and a value larger than unity for zooming out

Value

The output of `cabipl` is a graph of the desired biplot together with an R list with the following components:

`out`	Dataframe containing predictions for all samples specified in `predictions.sample`.
`X.mat`	Input two-way contingency table matrix **X**.
`E.mat`	The $p \times q$ independence matrix of Section 7.2.
`R.mat`	Diagonal matrix formed from the row sums of input **X**.
`C.mat`	Diagonal matrix formed from the column sums of input **X**.
`dev.mat`	Matrix of deviations (7.1), `X.mat - E.mat`.
`weighted.dev.mat`	`dev.mat` multiplied from the left by `R.mat^(-1/2)` and from the right by `C.mat^(-1/2)`, i.e. matrix (7.2).
`Contingency.mat.prop`	The matrix `dev.mat` multiplied from the left by `R.mat^(-1)` and from the right by `C.mat^(-1)` as defined in Section 7.2.3.
`Contingency.mat`	The matrix formed by adding unity to `Contingency.mat.prop` times n.
`svd.weighted.dev.mat`	Singular value decompoisition of `weighted.dev.mat`.
`lambda`	Value of λ when performing λ-scaling.
`Quality`	Overall quality of the biplot display.
`predictions`	NULL for one- and two-dimensional biplots. In the case of a three-dimensional biplot with argument `predictions.3D = TRUE` the dataframe containing the three-dimensional predictions for all the rows of the `weighted.dev.mat`.
`weighted.dev.mat.new`	If `X.new` is non-NULL the `weighted.dev.mat` of `X.new`, else NULL.
`predictions.new`	The predictions specified by the argument `predictions.X.new` in the form of a dataframe.
`column.plot.coords`	A matrix containing the coordinates for plotting column points.
`calibrations plot`	A list with jth element a vector containing the calibrations of the jth biplot axis.

The actual plotting of the CA biplot is done by either the function `drawbipl.ca` or the function `drawbipl.3dim.ca`. These functions are called by `cabipl` and are usually not called by a user. Two functions that can be called by a user, are the functions `ca.predictivities` (Section 7.5.2) and `ca.predictions.mat` (Section 7.5.3).

7.5.2 Function `ca.predictivities`

The function `ca.predictivities` is for calculating overall quality, axis predictivities and sample predictivities associated with a CA.

Arguments

`data`	Contingency table in the form of a $p \times q$ matrix.
`out`	Integer-valued vector specifying the output. Default is that entire output list is printed.

Value

The output of `ca.predictivities` is an R list with the following components:

`Quality`	The overall quality of the biplot display.
`Weights`	The weights for calculating the quality from the axis predictivities.
`Adequacies`	Adequacies associated with the individual biplot axes.
`Axis.predictivities`	Axis predictivities associated with the individual biplot axes.
`Sample.predictivities`	Sample predictivities associated with the rows of the input matrix.
`X.hat`	Predicted matrix (7.7).

7.5.3 Function `ca.predictions.mat`

The function `ca.predictions.mat` is an alternative function for calculating row profile predictions of the input matrix. It gives identical predictions to function `cabipl` with arguments `ca.variant = "RowProfA"` or `"RowProfB"` and `RowProf.scaled.markers = TRUE`, but provides also for predictions in more than three dimensions.

Arguments

`Pmat`	$p \times q$ matrix with positive elements. Preferably a correspondence matrix, i.e. all elements of `Pmat` are nonnegative and sum to 1.
`e.vects`	Integer-valued vector specifying the eigenvectors used in calculating the predictions.

Value

The output of `ca.predictions.mat` is a matrix of similar size than `Pmat` containing the row profile predictions.

Details

If the input of `ca.predictions.mat` is not in the form of a correspondence matrix but instead a matrix with nonnegative elements, the relationship between the output of `ca.predictions.mat` and the output of the function `cabipl` can be determined from the following examples:

- `ca.predictions.mat(Pmat, e.vects = 1:2)` outputs the same predictions as `cabipl with e.vects = 1;`
- `ca.predictions.mat(Pmat, e.vects = 1:3)` outputs the same predictions as `cabipl with e.vects = 1:2;`
- `ca.predictions.mat(Pmat, e.vects = 1:4)` outputs the same predictions as `cabipl with e.vects = 1:3;`
- `ca.predictions.mat(Pmat, e.vects = c(1, 2, 4))` outputs the same predictions as `cabipl with e.vects = c(1,3).`

7.5.4 Functions `indicatormat`, `construct.df`, `Chisq.dist`

The functions `indicatormat`, `construct.df` and `Chisq.dist` are useful when performing a CA.

The function `indicatormat` calculates the indicator matrix associated with a two-way table of frequencies. It takes only the one argument, `table`, a two-way table of frequencies in the form of a $p \times q$ matrix. The output of `indicatormat` is the indicator matrix associated with the input $p \times q$ matrix of frequencies. This function must not be confused with the function `indmat` that returns the indicator matrix associated with its argument, a grouping vector.

The function `construct.df` is useful when a data set is in a similar form to our data set `RSACrime.data`, that is, a list of matrices each representing a contingency table. Given such a data set as its argument, `construct.df` extracts any required element of the list in the form of a dataframe. It takes the following arguments:

`dat.list`	A list with elements forming a complete data set.
`year`	Elements of the list to extract in the form of dataframes.

The output of `construct.df` is a dataframe corresponding to a specified element of a list containing an entire data set.

The function `Chisq.dist` calculates the chi-squared distances as well as ordinary Euclidean distances between the rows of a matrix. It takes as its only argument a $p \times q$ matrix of frequencies. The output of `Chisq.dist` is a two-component list with elements:

`Sq.Chis2.dis`	A symmetric $p \times p$ matrix containing the squared chi-squared distances between the rows of its matrix argument $\mathbf{X}$: $p \times q$.
`Sq.Euclid.dis`	A symmetric $p \times p$ matrix containing the squared Euclidean distances between the rows of its matrix argument $\mathbf{X}$: $p \times q$.

7.5.5 Function `cabipl.doubling`

This is a function for implementing the doubling procedure as described in Section 7.4.

Arguments

`X`	A $p \times q$ matrix with elements the number of favourable ratings for each of q attributes of p products by N judges.
`N`	The number of judges asked to rate each of q attributes of p different products as satisfactory or not.
`...`	Any set of arguments to pass to `cabipl`.

Value

The matrix of positive and negative assessments of the form of Table 7.4 is returned together with two biplots: a CA biplot of the doubled matrix of the number of positive and negative assessments of the q attributes of p products; and a biplot with the positive and negative extremities of each attribute connected with lines that divide its two ends in the ratio $c_k : Np - c_k$ as explained in Section 7.4.

7.6 Examples

7.6.1 The RSA crime data set

The RSA crime data set is used extensively in this section to illustrate the different biplots that can be constructed to approximate various aspects of model (7.1). It consists of the annual number of 14 serious crimes reported to the South African police in each of the nine provinces of South Africa for the period 2001–08. It is tabulated according to the nature of the crime, with province boundaries standardized at 2007 boundaries. The data have been extracted from the official website of the South African police (http://www.saps.gov.za/) and can be represented in the form of 9×14 contingency tables. The R object `RSACrime.data` is a list with seven matrices as elements. Each of these matrices represents the 10×14 contingency table for one of the periods 2001/02, 2002/03, ..., 2007/08. The extra row is the aggregate over all provinces. Here is a brief summary of the crimes considered, together with the abbreviations used in the biplots and tables discussed in this chapter:

Arsn	Annual number of reported *arson* cases.
AGBH	Annual number of reported *assault with the intention of inflicting grievous bodily harm* cases.
AtMr	Annual number of reported *attempted murder* cases.
BNRs	Annual number of reported *burglary: nonresidential premises* cases.
BRs	Annual number of reported *burglary: residential premises* cases.
CrJk	Annual number of reported *carjacking* cases, a subcategory of *aggravated robbery* cases.
CmAs	Annual number of reported *common assault* cases.
CmRb	Annual number of reported *common robbery* cases.
DrgR	Annual number of reported *drug related crime* cases.
InAs	Annual number of reported *indecent assault* cases.
Mrd	Annual number of reported *murder* cases.
PubV	Annual number of reported *public violence* cases.
Rape	Annual number of reported *rape* cases.
RAC	Annual number of reported *robbery with aggravating circumstances* cases.

Each contingency table has as columns the 14 serious crimes listed above and as its rows the nine provinces:

ECpe	Eastern Cape
FrSt	Free State
Gaut	Gauteng
KZN	KwaZulu Natal
Limp	Limpopo
Mpml	Mpumalanga
NCpe	Northern Cape
NWst	North West
WCpe	Western Cape

As an example of a contingency table we consider the crime data for 2007/08. This will be our data matrix **X**, to be subjected to a CA. Note that the symbols **X**, **R**, **C** and **E** have dual meanings in this chapter, to be distinguished in context:

(i) **X**, a general $p \times q$ contingency table in the form of a matrix; **R**, the $p \times p$ diagonal matrix containing the row sums of **X**; **C**, the $q \times q$ diagonal matrix containing the column sums of **X**; and **E**, the independence matrix $\mathbf{R11}'\mathbf{C}/n$;

(ii) shorthand for the 9×14 crime data contingency table for 2007/08 and its associated row-sum and column-sum diagonal matrices as well as its calculated independence matrix.

Our data matrix (contingency table) is given in its transposed form in Table 7.5.

It follows from Tables 7.5 and 7.6 that the row sums – that is, the diagonal elements of the matrix **R** – are 127 627, 75 580, ..., 192 050. Symbolically, we write these sums as $x_{1.}, x_{2.}, \ldots, x_{9.}$ respectively, where a dot indicates that summation is carried out over the

Table 7.5 The matrix **X**: 9 × 14 representing the 2007/08 crime data two-way contingency table.

	Arsn	AGBH	AtMr	BNRs	BRs	CrJk	CmAs	CmRb	DrgR	InAs	Mrd	PubV	Rape	RAC
Ecpe	1235	34479	2160	5946	29508	604	19875	7086	7929	764	3515	88	5499	8939
FrSt	432	16833	939	4418	15705	156	19885	4193	4525	427	879	59	2628	4501
Gaut	1824	46993	5257	15117	62703	7466	57153	22152	12348	1501	3674	174	8073	50970
KZN	1322	30606	4946	10258	37203	3889	29410	9264	24174	1197	4713	76	6502	24290
Limp	573	13670	722	5401	11857	203	11024	3760	3198	215	696	31	2816	2447
Mpml	588	16849	1271	4273	18855	664	12202	4752	1770	251	835	61	2635	5907
NWst	624	15861	881	4987	14722	291	10406	3863	7004	345	917	117	3017	5528
NCpe	169	9898	775	1956	4924	5	5431	1337	2201	213	422	32	1020	1175
WCpe	629	24915	1844	10639	42376	923	32663	8578	45985	1850	2836	257	4000	14555

Table 7.6 The transposed version of **X**: 9 × 14 with the row and column sums of **X** added.

	ECpe	FrSt	Gaut	KZN	Limp	Mpml	NWst	NCpe	WCpe	Total
Arsn	1235	432	1824	1322	573	588	624	169	629	7396
AGBH	34479	16833	46993	30606	13670	16849	15861	9898	24915	210104
AtMr	2160	939	5257	4946	722	1271	881	775	1844	18795
BNRs	5946	4418	15117	10258	5401	4273	4987	1956	10639	62995
BRs	29508	15705	62703	37203	11857	18855	14722	4924	42376	237853
CrJk	604	156	7466	3889	203	664	291	5	923	14201
CmAs	19875	19885	57153	29410	11024	12202	10406	5431	32663	198049
CmRb	7086	4193	22152	9264	3760	4752	3863	1337	8578	64985
DrgR	7929	4525	12348	24174	3198	1770	7004	2201	45985	109134
InAs	764	427	1501	1197	215	251	345	213	1850	6763
Mrd	3515	879	3674	4713	696	835	917	422	2836	18487
PubV	88	59	174	76	31	61	117	32	257	895
Rape	5499	2628	8073	6502	2816	2635	3017	1020	4000	36190
RAC	8939	4501	50970	24290	2447	5907	5528	1175	14555	118312
Total	127627	75580	295405	187850	56613	70913	68563	29558	192050	1104159

subscript that has been replaced by the dot. Similarly, the column sums are the diagonal elements of the matrix **C**, namely 7396, 210 104, ..., 118 312, written symbolically as $x_{.1}, x_{.2}, \ldots, x_{.14}$, respectively. The sum of the elements in **X**, $n = x_{..} = 1\ 104\ 159$, is the total number of the 14 categories of crime reported in the country as a whole.

The meaning of the independence model for the contingency table given in Table 7.5 is that the type of crime reported is independent of the province in which it occurs. The matrix **E**: 9 × 14 of the independence model is given in Table 7.7, while the matrix of deviations from independence, (7.1), is given in Table 7.8.

The hypothesis that the independence model $\mathbf{E} = \mathbf{R11'C}/n$ describes the observed frequencies in Table 7.5 satisfactorily is most implausible. Unsurprisingly, a formal Pearson's chi-squared test (7.4) gives 123 183 with 8 × 13 degrees of freedom. The natural question that arises is whether a biplot can be constructed that provides information on how the different cells in **X** contribute to the lack of fit of the independence model. Table 7.9 shows the data in weighted deviation form $\mathbf{R}^{-1/2}(\mathbf{X} - \mathbf{E})\mathbf{C}^{-1/2}$.

Table 7.7 The calculated independence matrix $\mathbf{E} = \mathbf{R11}'\mathbf{C}/n$ for the 2007/08 crime data.

	Arsn	AGBH	AtMr	BNRs	BRs	CrJk	CmAs	CmRb	DrgR	InAs	Mrd	PubV	Rape	RAC
ECpe	854.9	24285.4	2172.5	7281.4	27492.8	1641.5	22892.0	7511.5	12614.5	781.7	2136.9	103.5	4183.1	13675.4
FrSt	506.3	14381.7	1286.5	4312.0	16281.1	972.1	13556.5	4448.2	7470.3	462.9	1265.4	61.3	2477.2	8098.5
Gaut	1978.7	56210.9	5028.4	16853.6	63634.8	3799.3	52985.7	17386.0	29197.5	1809.4	4946.0	239.4	9682.2	31653.0
KZN	1258.3	35744.9	3197.6	10717.3	40465.8	2416.0	33694.0	11055.9	18566.9	1150.6	3145.2	152.3	6157.0	20128.4
Limp	379.2	10772.6	963.7	3229.9	12195.3	728.1	10154.5	3331.9	5595.6	346.8	947.9	45.9	1855.6	6066.2
Mpml	475.0	13493.6	1207.1	4045.8	15275.8	912.0	12719.4	4173.6	7009.0	434.3	1187.3	57.5	2324.2	7598.4
NWst	459.3	13046.5	1167.1	3911.7	14769.5	881.8	12297.9	4035.3	6776.7	419.9	1148.0	55.6	2247.2	7346.6
NCpe	198.0	5624.4	503.1	1686.4	6367.3	380.2	5301.7	1739.6	2921.5	181.0	494.9	24.0	968.8	3167.2
WCpe	1286.4	36544.1	3269.1	10956.9	41370.6	2470.0	34447.3	11303.1	18982.0	1176.3	3215.5	155.7	6294.6	20578.4

Table 7.8 The calculated matrix of deviations from independence $\mathbf{X} - \mathbf{E}$ for the 2007/08 crime data.

	Arsn	AGBH	AtMr	BNRs	BRs	CrJk	CmAs	CmRb	DrgR	InAs	Mrd	PubV	Rape	RAC
Ecpe	380.1	10193.6	−12.5	−1335.4	2015.2	−1037.5	−3017.0	−425.5	−4685.5	−17.7	1378.1	−15.5	1315.9	−4736.4
FrSt	−74.3	2451.3	−347.5	106.0	−576.1	−816.1	6328.5	−255.2	−2945.3	−35.9	−386.4	−2.3	150.8	−3597.5
Gaut	−154.7	−9217.9	228.6	−1736.6	−931.8	3666.7	4167.3	4766.0	−16849.5	−308.4	−1272.0	−65.4	−1609.2	19317.0
KZN	63.7	−5138.9	1748.4	−459.3	−3262.8	1473.0	−4284.0	−1791.9	5607.1	46.4	1567.8	−76.3	345.0	4161.6
Limp	193.8	2897.4	−241.7	2171.1	−338.3	−525.1	869.5	428.1	−2397.6	−131.8	−251.9	−14.9	960.4	−3619.2
Mpml	113.0	3355.4	63.9	227.2	3579.2	−248.0	−517.4	578.4	−5239.0	−183.3	−352.3	3.5	310.8	−1691.4
NWst	164.7	2814.5	−286.1	1075.3	−47.5	−590.8	−1891.9	−172.3	227.3	−74.9	−231.0	61.4	769.8	−1818.6
NCpe	−29.0	4273.6	271.9	269.6	−1443.3	−375.2	129.3	−402.6	−720.5	32.0	−72.9	8.0	51.2	−1992.2
WCpe	−657.4	−11629.1	−1425.1	−317.9	1005.4	−1547.0	−1784.3	−2725.1	27003.0	673.7	−379.5	101.3	−2294.6	−6023.4

Table 7.9 The 2007/2008 crime contingency table in weighted deviation form $\mathbf{R}^{-1/2}(\mathbf{X} - \mathbf{E})\mathbf{C}^{-1/2}$.

	Arsn	AGBH	AtMr	BNRs	BRs	CrJk	CmAs	CmRb	DrgR	InAs	Mrd	PubV	Rape	RAC
Ecpe	0.0124	0.0622	−0.0003	−0.0149	0.0116	−0.0244	−0.0190	−0.0047	−0.0397	−0.0006	0.0284	−0.0014	0.0194	−0.0385
FrSt	−0.0031	0.0195	−0.0092	0.0015	−0.0043	−0.0249	0.0517	−0.0036	−0.0324	−0.0016	−0.0103	−0.0003	0.0029	−0.0380
Gaut	−0.0033	−0.0370	0.0031	−0.0127	−0.0035	0.0566	0.0172	0.0344	−0.0938	−0.0069	−0.0172	−0.0040	−0.0156	0.1033
KZN	0.0017	−0.0259	0.0294	−0.0042	−0.0154	0.0285	−0.0222	−0.0162	0.0392	0.0013	0.0266	−0.0059	0.0042	0.0279
Limp	0.0095	0.0266	−0.0074	0.0364	−0.0029	−0.0185	0.0082	0.0071	−0.0305	−0.0067	−0.0078	−0.0021	0.0212	−0.0442
Mpml	0.0049	0.0275	0.0018	0.0034	0.0276	−0.0078	−0.0044	0.0085	−0.0596	−0.0084	−0.0097	0.0004	0.0061	−0.0185
NWst	0.0073	0.0235	−0.0080	0.0164	−0.0004	−0.0189	−0.0162	−0.0026	0.0026	−0.0035	−0.0065	0.0078	0.0155	−0.0202
NCpe	−0.0020	0.0542	0.0115	0.0062	−0.0172	−0.0183	0.0017	−0.0092	−0.0127	0.0023	−0.0031	0.0016	0.0016	−0.0337
WCpe	−0.0174	−0.0579	−0.0237	−0.0029	0.0047	−0.0296	−0.0091	−0.0244	0.1865	0.0187	−0.0064	0.0077	−0.0275	−0.0400

The observed χ^2-value of 123 183 can be verified by multiplying the sum of the squared elements of Table 7.9 by $n = 1\ 104\ 159$.

The SVD $\mathbf{R}^{-1/2}(\mathbf{X} - \mathbf{E})\mathbf{C}^{-1/2} = \mathbf{U\Sigma V}'$ for Table 7.9 gives

$$\mathbf{U} =$$

$$\begin{pmatrix}
-0.1496 & 0.4188 & -0.4404 & 0.1598 & 0.4907 & -0.1346 & 0.3941 & -0.2301 & -0.3400 \\
-0.1003 & 0.2729 & 0.5702 & -0.4899 & 0.2588 & -0.3156 & 0.0482 & 0.3361 & -0.2616 \\
-0.4558 & -0.6061 & 0.2151 & 0.0741 & 0.0544 & 0.2339 & 0.1607 & -0.1487 & -0.5172 \\
0.1342 & -0.2595 & -0.5625 & -0.4055 & -0.2319 & -0.3595 & -0.2298 & 0.1669 & -0.4125 \\
-0.1015 & 0.3381 & 0.2075 & 0.0459 & -0.6847 & -0.2489 & 0.1029 & -0.4874 & -0.2264 \\
-0.2312 & 0.2059 & 0.0504 & 0.4958 & 0.0954 & -0.1041 & -0.7431 & 0.1391 & -0.2534 \\
0.0184 & 0.2122 & -0.0803 & 0.2580 & -0.3717 & 0.3134 & 0.3410 & 0.6833 & -0.2492 \\
-0.0503 & 0.3299 & -0.1028 & -0.4635 & -0.0092 & 0.7189 & -0.2904 & -0.1876 & -0.1636 \\
0.8217 & -0.0691 & 0.2359 & 0.1877 & 0.1391 & 0.1024 & -0.0175 & -0.1585 & -0.4171
\end{pmatrix}$$

$$\mathbf{\Sigma} =$$

$$\begin{pmatrix}
0.2514 & 0 & 0 & 0 & 0 & 0 & 0 & 0 & 0 & 0 & 0 & 0 & 0 & 0 \\
0 & 0.1865 & 0 & 0 & 0 & 0 & 0 & 0 & 0 & 0 & 0 & 0 & 0 & 0 \\
0 & 0 & 0.0845 & 0 & 0 & 0 & 0 & 0 & 0 & 0 & 0 & 0 & 0 & 0 \\
0 & 0 & 0 & 0.0501 & 0 & 0 & 0 & 0 & 0 & 0 & 0 & 0 & 0 & 0 \\
0 & 0 & 0 & 0 & 0.0472 & 0 & 0 & 0 & 0 & 0 & 0 & 0 & 0 & 0 \\
0 & 0 & 0 & 0 & 0 & 0.0325 & 0 & 0 & 0 & 0 & 0 & 0 & 0 & 0 \\
0 & 0 & 0 & 0 & 0 & 0 & 0.0223 & 0 & 0 & 0 & 0 & 0 & 0 & 0 \\
0 & 0 & 0 & 0 & 0 & 0 & 0 & 0.0116 & 0 & 0 & 0 & 0 & 0 & 0 \\
0 & 0 & 0 & 0 & 0 & 0 & 0 & 0 & 0 & 0 & 0 & 0 & 0 & 0
\end{pmatrix}$$

and

$$\mathbf{V} =$$

$$\begin{pmatrix}
-0.064 & 0.066 & -0.133 & 0.099 & -0.137 & -0.180 & 0.201 & 0.093 & -0.385 & 0.152 & -0.232 & -0.468 & -0.340 & -0.552 \\
-0.226 & 0.547 & -0.284 & -0.138 & 0.143 & 0.525 & 0.047 & -0.061 & 0.110 & -0.000 & 0.346 & -0.284 & 0.137 & -0.132 \\
-0.065 & -0.056 & -0.339 & -0.370 & -0.093 & -0.058 & -0.653 & 0.122 & -0.414 & 0.090 & -0.124 & 0.002 & 0.304 & 0.017 \\
0.002 & 0.116 & 0.144 & 0.035 & -0.799 & -0.001 & -0.083 & -0.142 & -0.115 & 0.154 & 0.423 & -0.042 & -0.180 & 0.221 \\
-0.012 & 0.045 & 0.048 & 0.643 & 0.287 & -0.296 & -0.389 & 0.114 & 0.018 & 0.181 & 0.387 & -0.224 & 0.121 & 0.004 \\
-0.143 & -0.400 & -0.179 & -0.115 & -0.147 & -0.079 & -0.224 & -0.283 & 0.531 & -0.009 & 0.186 & 0.047 & -0.039 & -0.544 \\
-0.083 & 0.006 & 0.645 & -0.530 & 0.188 & -0.251 & 0.050 & 0.079 & -0.069 & 0.072 & 0.297 & -0.204 & 0.158 & -0.155 \\
-0.156 & -0.093 & 0.163 & 0.274 & -0.084 & 0.095 & 0.173 & -0.550 & -0.358 & -0.110 & -0.070 & 0.068 & 0.576 & -0.186 \\
0.907 & -0.096 & -0.088 & -0.053 & -0.038 & 0.127 & 0.052 & -0.007 & -0.009 & -0.049 & 0.154 & -0.213 & 0.214 & -0.142 \\
0.085 & -0.011 & -0.003 & -0.065 & 0.133 & 0.108 & 0.074 & -0.241 & 0.070 & 0.925 & -0.137 & 0.110 & 0.012 & 0.059 \\
0.024 & 0.032 & -0.471 & -0.181 & 0.214 & -0.496 & 0.316 & -0.292 & -0.219 & -0.045 & 0.390 & 0.181 & -0.103 & 0.143 \\
0.031 & 0.023 & 0.042 & 0.097 & -0.000 & 0.194 & 0.075 & 0.411 & -0.255 & 0.095 & 0.296 & 0.659 & -0.026 & -0.423 \\
-0.086 & 0.168 & -0.187 & 0.019 & -0.320 & -0.352 & 0.325 & 0.401 & 0.320 & 0.131 & -0.109 & -0.011 & 0.553 & -0.015 \\
-0.225 & -0.685 & -0.150 & 0.011 & 0.025 & 0.296 & 0.267 & 0.278 & -0.136 & 0.090 & 0.247 & -0.276 & 0.074 & 0.227
\end{pmatrix}$$

A one-dimensional biplot constructed from the first columns of $\mathbf{U\Sigma}^{1/2}$ and $\mathbf{V\Sigma}^{1/2}$ is given in Figure 7.1. In this one-dimensional biplot all axes representing the columns are actually on top of each other but have been vertically shifted, allowing their scalings to be used for determining the values for each of the provinces. This can be done by dropping a line vertically from a point representing any of the provinces onto the axis and reading off the value as is done with an ordinary scatterplot. We will provide more details when discussing the two-dimensional biplot constructed using the first two columns of $\mathbf{U\Sigma}^{1/2}$ and $\mathbf{V\Sigma}^{1/2}$. We construct the one-dimensional biplot in Figure 7.1 by using the R instruction

```
cabipl(X = as.matrix(SACrime08.data), constant = 0.055,
   dim.biplot = 1, n.int = rep(5,14), offset = c(0, 2, 0, 0.5),
   ort.lty = 2, plot.col.points = FALSE, predictions.sample = 9,
   propshift = -0.24)
```

leaving all other arguments at their defaults.

A screenshot of a three-dimensional biplot constructed from the first three columns of $\mathbf{U}\boldsymbol{\Sigma}^{1/2}$ and $\mathbf{V}\boldsymbol{\Sigma}^{1/2}$ is given in Figure 7.2. This biplot is an example of the output of our R function `cabipl` with argument `dim.bipl = 3`. The full call to `cabipl` is:

```
cabipl(X = as.matrix(SACrime08.data), dim.biplot = 3,
   samples.plot = TRUE, size.points.3d = 0.025, ID.labs = TRUE,
   adjust.3d = c(1.5, 0.5),  ID.3d = rownames(SACrime08.data))
```

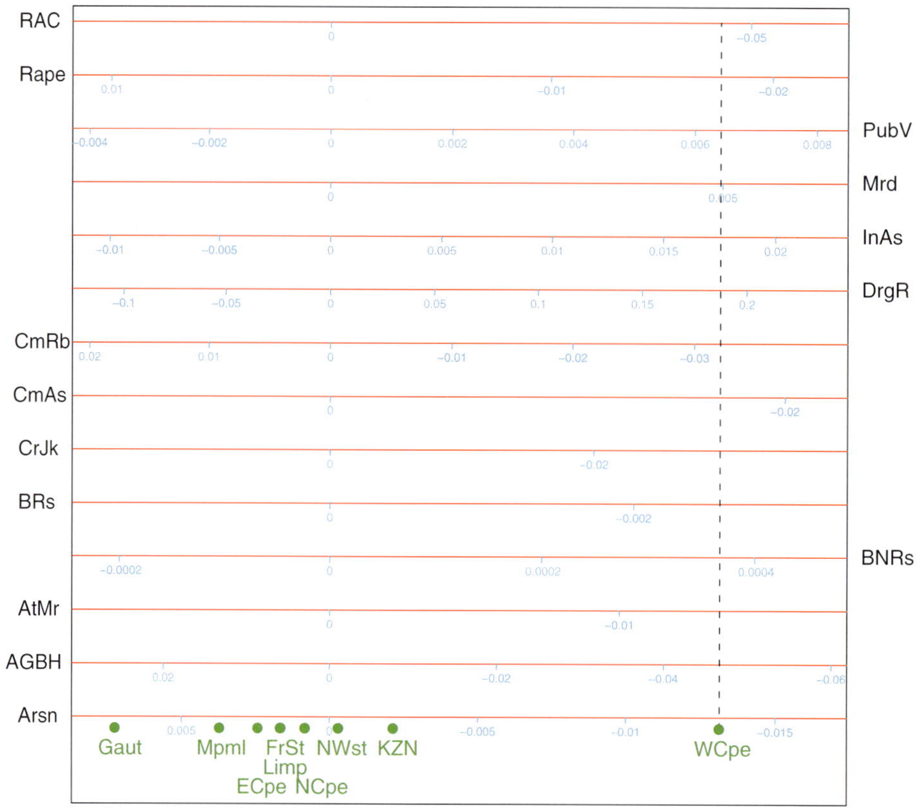

Figure 7.1 One-dimensional CA biplot for the 2007/08 crime contingency table. Proportional to Pearson residuals in deviation form $\mathbf{R}^{-1/2}(\mathbf{X} - \mathbf{E})\mathbf{C}^{-1/2}$; constructed from $\mathbf{U}\boldsymbol{\Sigma}^{1/2}$ and $\mathbf{V}\boldsymbol{\Sigma}^{1/2}$.

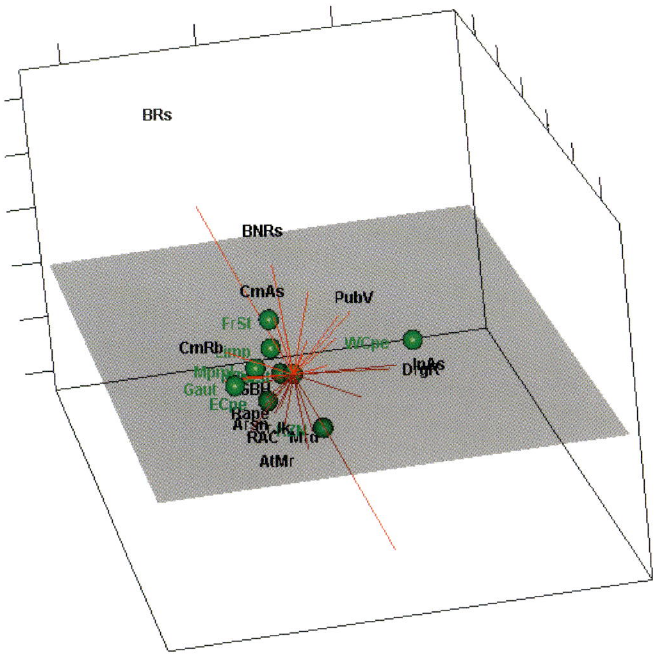

Figure 7.2 Three-dimensional CA biplot for the 2007/08 crime contingency table. Proportional to Pearson residuals in deviation form $\mathbf{R}^{-1/2}(\mathbf{X} - \mathbf{E})\mathbf{C}^{-1/2}$; constructed from $\mathbf{U\Sigma}^{1/2}$ and $\mathbf{V\Sigma}^{1/2}$.

keeping all other arguments at their defaults. In order to appreciate this three-dimensional biplot we suggest the reader construct it using function `cabipl` and then interactively rotate the biplot using the left mouse button and zooming in or out using the right mouse button. Here we will discuss the two-dimensional biplot (given in Figure 7.3) in more detail, referring to the one- and three-dimensional CA biplots by way of comparison only.

The two-dimensional biplot constructed from the first two columns of $\mathbf{U\Sigma}^{1/2}$ and $\mathbf{V\Sigma}^{1/2}$ is given in Figure 7.3 with both rows and columns in the form of points. Figure 7.4 is the same biplot but now, according to our aim of constructing biplots to be used analogously to ordinary scatterplots, with columns represented by calibrated axes. In Figure 7.4 we have retained the points representing the columns to illustrate that these points (in red) appear exactly on the corresponding axes. This biplot is the result of the R call

```
cabipl(X = as.matrix(SACrime08.data), axis.col = "grey",
   marker.col = "black", offset = c(2, 2, 0.5, 0.5),
   offset.m = rep(-0.2,14))
```

Figure 7.3 Two-dimensional CA biplot for the 2007/08 crime contingency table. Proportional to Pearson residuals in deviation form $\mathbf{R}^{-1/2}(\mathbf{X}-\mathbf{E})\mathbf{C}^{-1/2}$; constructed from $\mathbf{U\Sigma}^{1/2}$ and $\mathbf{V\Sigma}^{1/2}$. Both rows and columns are represented as points: points representing rows are in green and points representing columns are in red.

Changing the function call to

```
cabipl (X = as.matrix(SACrime08.data), axis.col = "grey",
   marker.col = "black", offset = c(2, 2, 0.5, 0.5), ort.lty = 2,
   predictions.sample = 9, offset.m = rep(-0.2, 14))
```

produces Figure 7.5. Here we show how to obtain predictions for the Western Cape province. Predictions for all the provinces can be obtained simultaneously by changing argument `predictions.sample` to `predictions.sample = 1:9`. In Table 7.10 we give the full set of predictions made from the two-dimensional biplot in Figure 7.5.

Remember that $n^{1/2}$ times the elements of $\mathbf{R}^{-1/2}(\mathbf{X}-\mathbf{E})\mathbf{C}^{-1/2}$ gives exactly the square roots of the contributions to Pearson's χ^2 for a contingency table. Therefore, the values of $n^{1/2}$ times the calibrations on the biplot axes of Figures 7.4 and 7.5 can be interpreted in terms of Pearson residuals associated with the row points (the provinces in our data set). However, as we saw in Chapter 2, it is easy to calibrate the axes

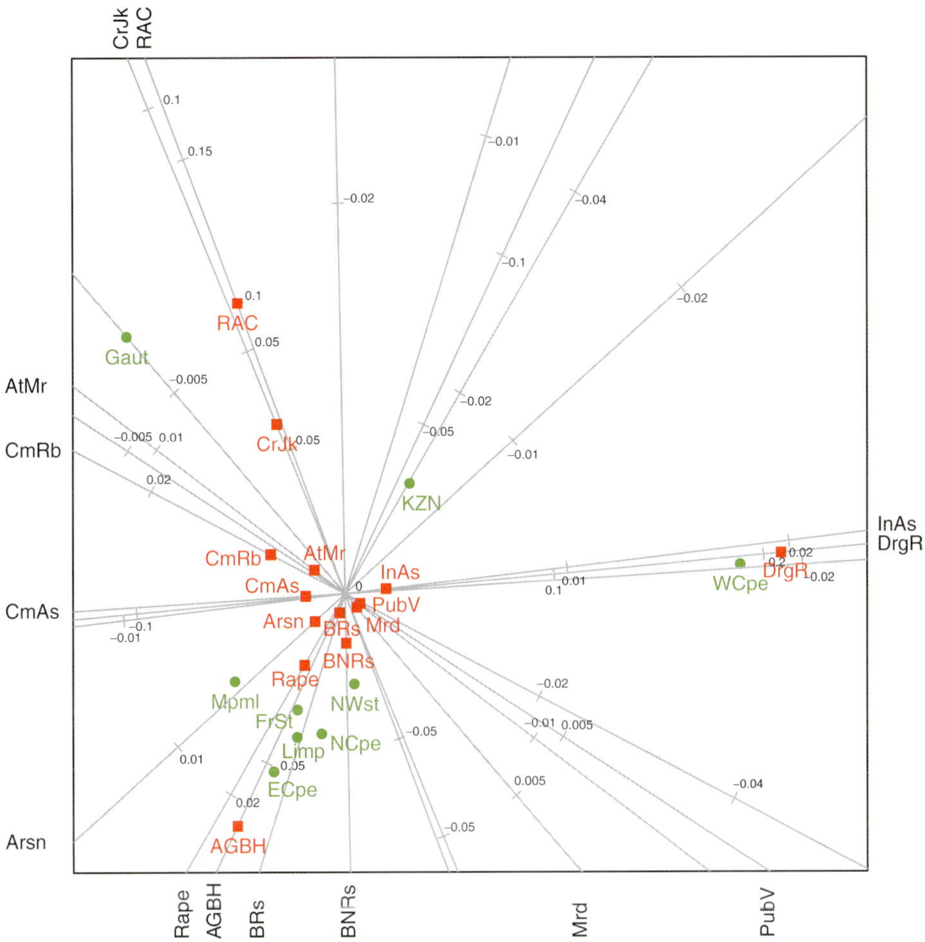

Figure 7.4 Two-dimensional CA biplot for the 2007/08 crime contingency table, constructed from the first two columns of $\mathbf{U\Sigma}^{1/2}$ and $\mathbf{V\Sigma}^{1/2}$. Columns are represented by axes. Calibrations on axes are in terms of $\mathbf{R}^{-1/2}(\mathbf{X}-\mathbf{E})\mathbf{C}^{-1/2}$, that is, proportional to Pearson residuals.

directly in terms of Pearson residuals by incorporating the factor $n^{1/2}$ directly in the calibrations. Setting in `cabipl` the argument `PearsonRes.scaled.markers = TRUE` provides axes calibrated in terms of Pearson residuals. The resulting biplot is shown in Figure 7.6, and its associated overall quality is a satisfactory 87.84% (see Table 7.11).

The sample predictivities and axis predictivities in Tables 7.12 and 7.13 are generally referred to as the row qualities and column qualities, respectively (see Greenacre 2007). The two-dimensional axis predictivities of ***AtMr***, ***BNRs***, ***BRs***, ***CmAs*** and ***Mrd*** show that these axes should be used with caution in a two-dimensional biplot. Increasing the biplot dimension to 3 leads to a considerable increase in the axis predictivities of ***AtMr***, ***CmAs*** and ***Mrd***. In two dimensions the sample predictivities are high apart from moderate values for ***FrSt*** and ***KZN***.

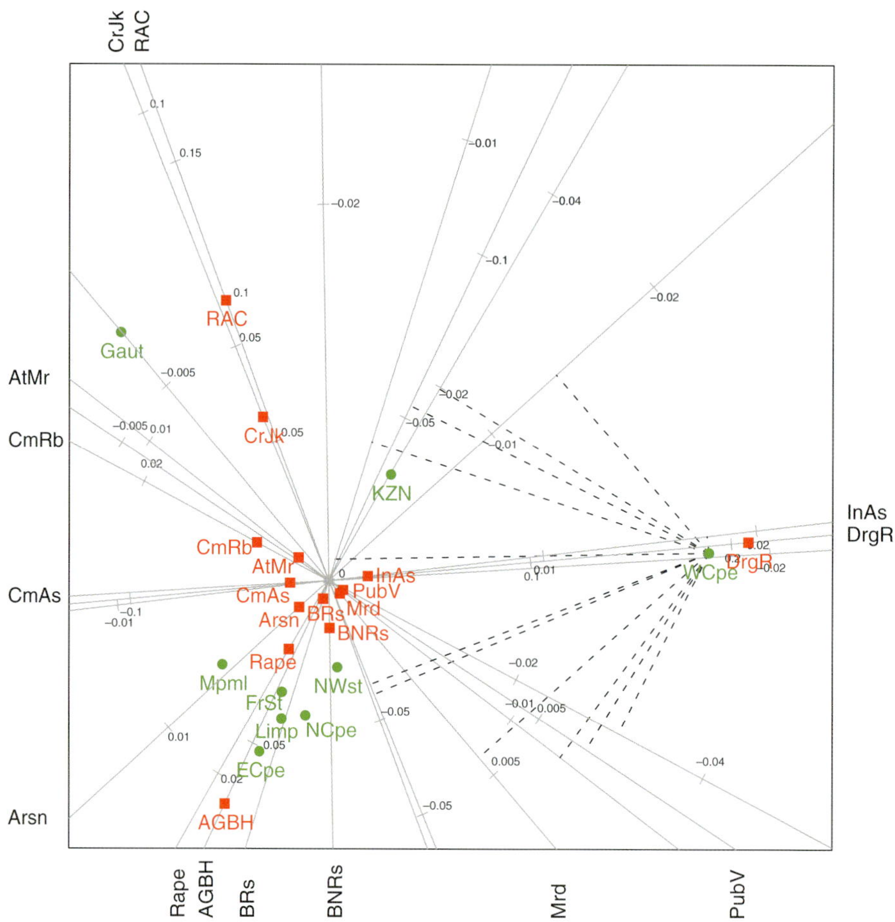

Figure 7.5 Two-dimensional CA biplot for the 2007/08 crime contingency table, constructed from the first two columns of $\mathbf{U\Sigma}^{1/2}$ and $\mathbf{V\Sigma}^{1/2}$. Columns are represented by axes. Calibrations on axes are in terms of $\mathbf{R}^{-1/2}(\mathbf{X}-\mathbf{E})\mathbf{C}^{-1/2}$, that is, proportional by a factor of $n^{1/2}$ to Pearson residuals. It is shown how to obtain these predictions for the Western Cape province.

We now illustrate the CA biplot resulting from the SVD $\mathbf{R}^{-1/2}(\mathbf{X}-\mathbf{E})\mathbf{C}^{-1/2} = \mathbf{U\Sigma V}'$ but plotting $\mathbf{U\Sigma}$ and $\mathbf{V}$ instead of $\mathbf{U\Sigma}^{1/2}$ and $\mathbf{V\Sigma}^{1/2}$. The biplot in Figure 7.7 is obtained by specifying the arguments `ca.variant = "PearsonResB"` and `PearsonRes.scaled.markers = FALSE` in our function `cabipl`.

Although the predictions made from Figure 7.7 are exactly those obtained from Figure 7.5, the biplot in Figure 7.7 differs in a very obvious way from that in Figure 7.5: the row points are squeezed towards each other, making graphical prediction difficult. There is an easy way to address this problem: setting argument `lambda = TRUE` utilizes the lambda tool described in Section 2.3 by requiring lambda to satisfy

$$\lambda = \frac{p\,\mathrm{tr}(\mathbf{VV}')}{q\,\mathrm{tr}(\mathbf{U\Sigma}^2\mathbf{U}')}. \tag{7.52}$$

Table 7.10 Predictions from the two-dimensional biplot for the 2007/2008 crime contingency table in weighted deviation form $\mathbf{R}^{-1/2}(\mathbf{X}-\mathbf{E})\mathbf{C}^{-1/2}$.

	Arsn	AGBH	AtMr	BNRs	BRs	CrJk	CmAs	CmRb	DrgR	InAs	Mrd	PubV	Rape	RAC
Ecpe	0.0075	0.0512	−0.0020	0.0090	0.0040	−0.0259	0.0036	−0.0014	−0.0416	−0.0041	0.0016	0.0006	0.0164	−0.0451
FrSt	0.0049	0.0335	−0.0012	0.0059	0.0026	−0.0168	0.0024	−0.0008	−0.0277	−0.0027	0.0010	0.0004	0.0107	−0.0292
Gaut	−0.0001	−0.0360	0.0138	−0.0134	−0.0037	0.0616	0.0089	0.0283	−0.0931	−0.0084	−0.0064	−0.0062	−0.0092	0.1032
KZN	−0.0053	−0.0341	0.0005	−0.0056	−0.0026	0.0145	−0.0031	−0.0007	0.0352	0.0034	−0.0007	−0.0001	−0.0110	0.0256
Limp	0.0058	0.0403	−0.0019	0.0073	0.0032	−0.0216	0.0025	−0.0019	−0.0292	−0.0029	0.0014	0.0007	0.0128	−0.0375
Mpml	0.0062	0.0341	0.0016	0.0044	0.0025	−0.0071	0.0051	0.0055	−0.0564	−0.0054	−0.0002	−0.0009	0.0114	−0.0132
NWst	0.0023	0.0206	−0.0025	0.0046	0.0017	−0.0165	−0.0001	−0.0044	0.0004	−0.0001	0.0014	0.0011	0.0063	−0.0282
Ncpe	0.0048	0.0365	−0.0026	0.0071	0.0029	−0.0228	0.0014	−0.0038	−0.0174	−0.0018	0.0017	0.0010	0.0114	−0.0393
Wcpe	−0.0140	−0.0537	−0.0127	−0.0011	−0.0031	−0.0244	−0.0173	−0.0309	0.1886	0.0177	0.0046	0.0061	−0.0198	−0.0376

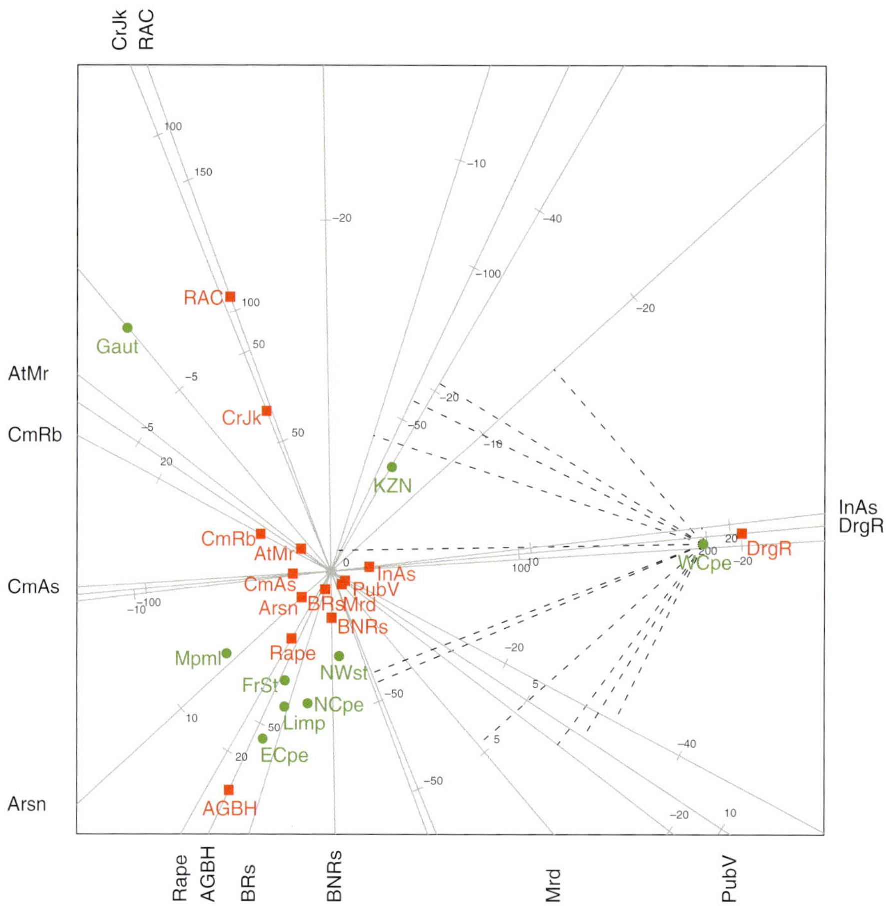

Figure 7.6 Two-dimensional CA biplot for the 2007/08 crime contingency table. Similar to the biplot in Figure 7.5, but calibrations scaled by a factor of $n^{1/2}$ using methods described in Chapter 2. Thus, calibrations are in terms of Pearson residuals.

Table 7.11 Quality expressed as percentages for the 2007/08 crime contingency table in weighted deviation form $\mathbf{R}^{-1/2}(\mathbf{X}-\mathbf{E})\mathbf{C}^{-1/2}$.

Dim 1	Dim 2	Dim 3	Dim 4	Dim 5	Dim 6	Dim 7	Dim 8	Dim 9
56.67	87.84	94.24	96.49	98.49	99.44	99.88	100.00	100.00

It turns out from the output of `cabipl` that in this case $\lambda = 1.9032$, where λ is defined by (7.52).

Figure 7.8 demonstrates the usefulness of the lambda tool and also that the biplot resulting from plotting $\mathbf{U\Sigma}$ and $\mathbf{V}$ conveys the same information as the one resulting from plotting $\mathbf{R}^{1/2}\mathbf{U\Sigma}^{1/2}$ and $\mathbf{C}^{1/2}\mathbf{V\Sigma}^{1/2}$. The above characteristics are also true for one- and three-dimensional biplots. The reader can verify this by setting `dim.biplot = 1` (or `3`) in the above calls to `cabipl`.

Table 7.12 Axis predictivities for the 2007/2008 crime contingency table in weighted deviation form $\mathbf{R}^{-1/2}(\mathbf{X}-\mathbf{E})\mathbf{C}^{-1/2}$.

	Arsn	AGBH	AtMr	BNRs	BRs	CrJk	CmAs	CmRb	DrgR	InAs	Mrd	PubV	Rape	RAC
Dim_1	0.392	0.221	0.151	0.000	0.007	0.179	0.103	0.668	0.992	0.855	0.017	0.337	0.217	0.161
Dim_2	0.621	0.934	0.213	0.230	0.054	0.951	0.103	0.799	0.998	0.863	0.034	0.440	0.680	0.985
Dim_3	0.813	0.974	0.674	0.302	0.065	0.983	0.800	0.882	1.000	0.863	0.767	0.509	0.796	0.993
Dim_4	0.851	0.977	0.867	0.303	0.763	0.988	0.965	0.965	1.000	0.883	0.805	0.640	0.797	0.993
Dim_5	0.915	0.980	0.878	0.997	0.886	0.994	0.984	0.972	1.000	0.957	0.852	0.640	0.904	0.993
Dim_6	0.968	1.000	0.880	0.997	0.949	0.995	1.000	0.976	1.000	0.980	0.972	0.859	0.965	0.998
Dim_7	0.998	1.000	0.999	0.999	0.999	0.999	1.000	0.982	1.000	0.985	0.995	0.874	0.990	1.000
Dim_8	1.000	1.000	1.000	1.000	1.000	1.000	1.000	1.000	1.000	1.000	1.000	1.000	1.000	1.000
Dim_9	1.000	1.000	1.000	1.000	1.000	1.000	1.000	1.000	1.000	1.000	1.000	1.000	1.000	1.000

Table 7.13 Sample predictivities for the 2007/2008 crime contingency table in weighted deviation form $\mathbf{R}^{-1/2}(\mathbf{X}-\mathbf{E})\mathbf{C}^{-1/2}$.

	ECpe	FrSt	Gaut	KZN	Limp	Mpml	NWst	NCpe	WCpe
Dim_1	0.147	0.099	0.499	0.177	0.107	0.583	0.009	0.031	0.984
Dim_2	0.783	0.503	0.984	0.541	0.760	0.837	0.680	0.766	0.988
Dim_3	0.927	0.864	0.996	0.891	0.811	0.840	0.700	0.780	0.997
Dim_4	0.933	0.958	0.997	0.956	0.811	0.947	0.772	0.885	0.999
Dim_5	0.989	0.981	0.997	0.974	0.983	0.950	0.904	0.885	1.000
Dim_6	0.991	0.997	0.999	0.995	0.994	0.952	0.948	0.991	1.000
Dim_7	0.999	0.998	1.000	0.999	0.995	1.000	0.973	0.999	1.000
Dim_8	1.000	1.000	1.000	1.000	1.000	1.000	1.000	1.000	1.000
Dim_9	1.000	1.000	1.000	1.000	1.000	1.000	1.000	1.000	1.000

Specifying in `cabipl` the argument `ca.variant = IndepDev` leads to what might be called an independence deviations CA biplot (see (7.8)). We illustrate this type of biplot with the two-dimensional biplot given in Figure 7.9 for the 2007/08 crime data set.

For approximating the contingency ratio our function `cabipl` provides for constructing biplots using $\mathbf{R}^{-1/2}\mathbf{U}\boldsymbol{\Sigma}^{1/2}$ and $\mathbf{C}^{-1/2}\mathbf{V}\boldsymbol{\Sigma}^{1/2}$ as well as $\mathbf{R}^{-1/2}\mathbf{U}\boldsymbol{\Sigma}$ and $\mathbf{C}^{-1/2}\mathbf{V}$. Furthermore, users have the option to calibrate axes in terms of the contingency ratios or in terms of contingency ratio minus 1. The computed contingency ratio matrix of the 2007/08 crime data is given in Table 7.14.

The two-dimensional contingency approximating biplot can be obtained either by specifying in the function `cabipl` the argument `ca.variant = "ConRatioA"` as in Figure 7.10, or `ca.variant = "ConRatioB"` resulting in Figure 7.11. Both these biplots give exactly the same predictions as are given in Table 7.15. The user can set argument `ConRatioMinOne = TRUE` to obtain the calibrations of the axes (as well as the predictions) in terms of the deviations of x_{ij} from independence relative to the approximation e_{ij} under independence. It is clear that Figures 7.9 and 7.10 are similar but that the provinces (the row points) are squeezed in towards the origin of the biplot axes, while the column points (the red squares) are relatively more spread out. Again, setting argument `lambda = TRUE` will result in a biplot in which the row points appear more spread out. This is illustrated with the biplot given in Figure 7.12. By specifying the

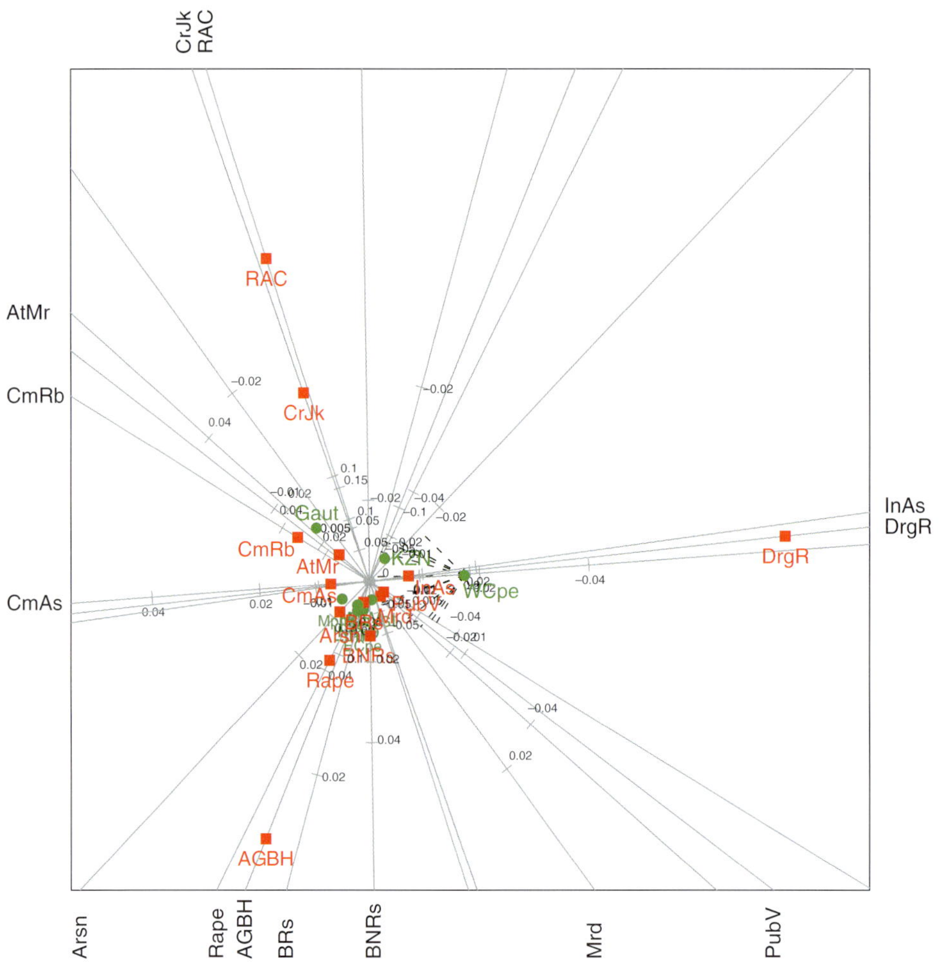

Figure 7.7 Two-dimensional CA biplot of the 2007/08 crime data set resulting from the SVD $\mathbf{R}^{-1/2}(\mathbf{X}-\mathbf{E})\mathbf{C}^{-1/2} = \mathbf{U\Sigma V}'$ and plotting $\mathbf{U\Sigma}$ and $\mathbf{V}$. Predictions for the Western Cape are illustrated.

argument `lambda = TRUE` a biplot is obtained in which it is much easier to use the biplot axes for prediction purposes. Note that low-dimensional approximations to e_{ij} can become negative for poorly approximated elements, where the original element is close to zero (see Table 7.15).

Recall from Section 7.2.3 that, as a result of using $\mathbf{R}^{-1/2}\mathbf{U\Sigma}$ and $\mathbf{C}^{-1/2}\mathbf{V}$ as coordinates, the biplot shown in Figure 7.11 satisfies the centroid property that the row points are given by the mean of the column points, weighted by the row means $\mathbf{R}^{-1}\mathbf{X}$. This results in the clustering of the row points as they must all lie within the weighted centroids of the column points. The origin represents the average weighting, given by the elements of $\mathbf{C1}/n$. If a row point coincides with a column point then all the weight for that row must be associated with that column, the remainder being zero. More generally, row points are attracted to column points with high weights. Column points collinear

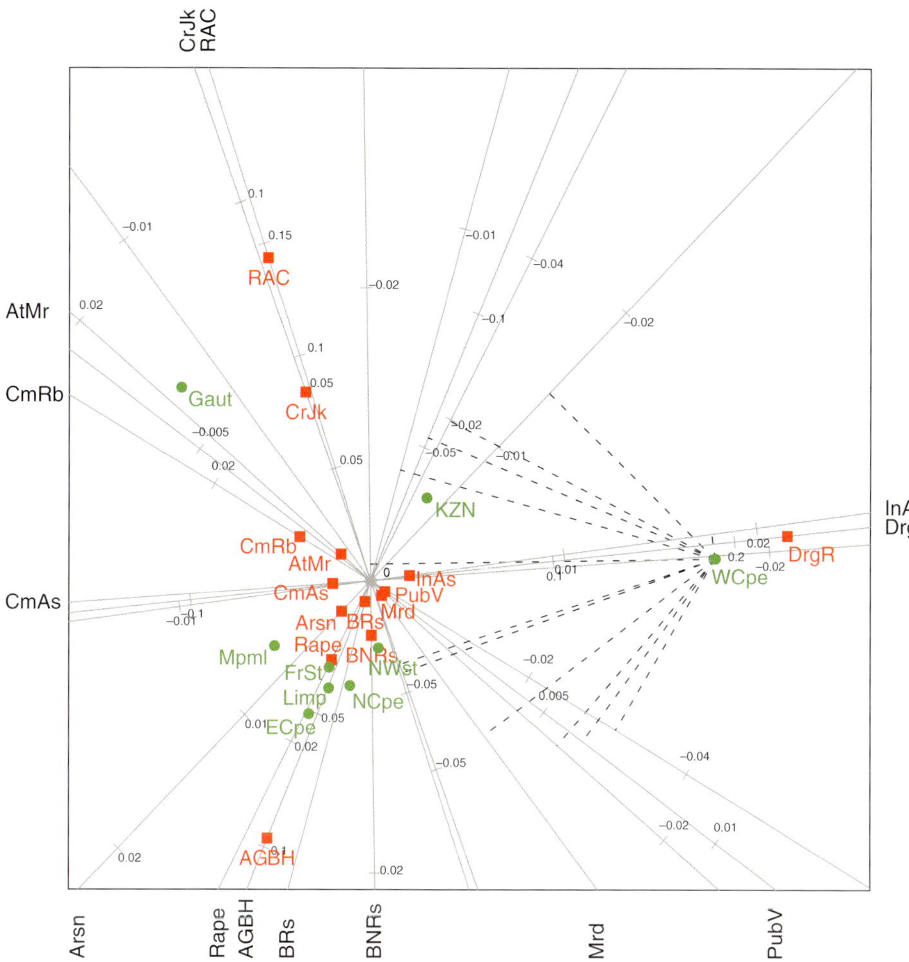

Figure 7.8 Two-dimensional CA biplot of the 2007/08 crime data set resulting from the SVD $\mathbf{R}^{-1/2}(\mathbf{X}-\mathbf{E})\mathbf{C}^{-1/2} = \mathbf{U\Sigma V}'$ and plotting $\mathbf{U\Sigma}$ and $\mathbf{V}$ with argument `lambda = TRUE`. Determination of predictions for the Western Cape is illustrated.

with the origin suggest that any associated row points will be behaving averagely except for the weights for the collinear column points. Thus, it can be seen that the Western Cape diverges from the average in respect of ***DrgR***, ***InAs*** and ***CmAs***. Similarly, ***Gaut*** is unusual for ***CrJk***, ***RAC***, ***CmRb*** and ***AtMr***.

It is interesting to compare the actual values of the lambda parameter obtained when specifying `ca.variant = "ConRatioA"` and `ca.variant = "ConRatioB"`, respectively. When `"ConRatioA"` is specified a lambda value of 1.0648 is obtained. This shows that the specifying argument `lambda = TRUE` should not lead to a biplot that differs much from the biplot obtained with the default `lambda = FALSE` setting. On the other hand, a lambda value of 2.3072 is obtained when `ca.variant = "ConRatioB"` is specified together with `lambda = TRUE`. The effect of this value is obvious when Figure 7.11 is compared with Figure 7.12.

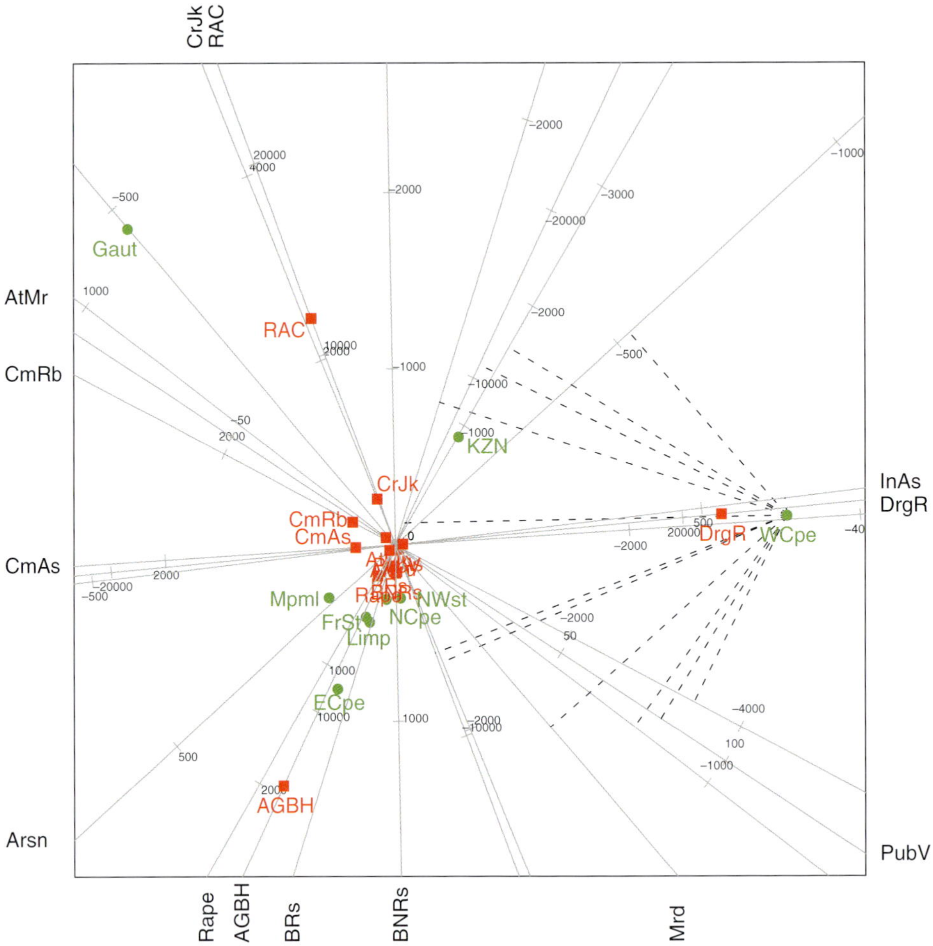

Figure 7.9 Two-dimensional CA biplot of the 2007/08 crime data set. Deviations from independence are approximated by plotting $\mathbf{R}^{1/2}\mathbf{U}\mathbf{\Sigma}^{1/2}$ and $\mathbf{C}^{1/2}\mathbf{V}\mathbf{\Sigma}^{1/2}$.

Since the deviance for a log-linear model is defined as

$$2\sum \text{observed} \times \log\left(\frac{\text{observed}}{\text{expected}}\right), \tag{7.53}$$

it might be of interest to calibrate axes in terms of the logarithm of the contingency ratio. Notice that this is an example of a linear axis with unequal spaced intervals. Setting the argument `logCRat = TRUE` in the call to `cabipl` allows this functionality by utilizing the calibration tool discussed in Section 2.3. Figure 7.13 provides an example of approximating the contingency ratio in terms of a logarithmic scale. We set `ax = 9` to suppress plotting of all the axes except the ***DrgR***-axis in order to draw attention to the calibrations on this axis.

In Table 7.16 we show $\sqrt{n}$ times the chi-squared distance matrix between the rows of our 2007/08 crime data set. The corresponding biplots (setting the arguments

Table 7.14 Contingency ratio matrix of the 2007/08 crime data, that is, the matrix $n\mathbf{R}^{-1}(\mathbf{X}-\mathbf{E})\mathbf{C}^{-1}+1$. Subtracting 1 from this matrix gives the deviations of the x_{ij} from independence relative to the approximation $e_{ij}=x_{i.}x_{.j}/n$.

	Arsn	AGBH	AtMr	BNRs	BRs	CrJk	CmAs	CmRb	DrgR	InAs	Mrd	PubV	Rape	RAC
ECpe	1.4446	1.4197	0.9943	0.8166	1.0733	0.3680	0.8682	0.9434	0.6286	0.9773	1.6449	0.8506	1.3146	0.6537
FrSt	0.8533	1.1704	0.7299	1.0246	0.9646	0.1605	1.4668	0.9426	0.6057	0.9224	0.6946	0.9631	1.0609	0.5558
Gaut	0.9218	0.8360	1.0455	0.8970	0.9854	1.9651	1.0786	1.2741	0.4229	0.8296	0.7428	0.7267	0.8338	1.6103
KZN	1.0506	0.8562	1.5468	0.9571	0.9194	1.6097	0.8729	0.8379	1.3020	1.0403	1.4985	0.4991	1.0560	1.2068
Limp	1.5110	1.2690	0.7492	1.6722	0.9723	0.2788	1.0856	1.1285	0.5715	0.6200	0.7343	0.6755	1.5176	0.4034
Mpml	1.2379	1.2487	1.0530	1.0562	1.2343	0.7280	0.9593	1.1386	0.2525	0.5779	0.7033	1.0612	1.1337	0.7774
NWst	1.3587	1.2157	0.7549	1.2749	0.9968	0.3300	0.8462	0.9573	1.0335	0.8215	0.7988	2.1053	1.3425	0.7525
NCpe	0.8536	1.7598	1.5403	1.1599	0.7733	0.0132	1.0244	0.7686	0.7534	1.1765	0.8527	1.3356	1.0529	0.3710
WCpe	0.4890	0.6818	0.5641	0.9710	1.0243	0.3737	0.9482	0.7589	2.4226	1.5727	0.8820	1.6509	0.6355	0.7073

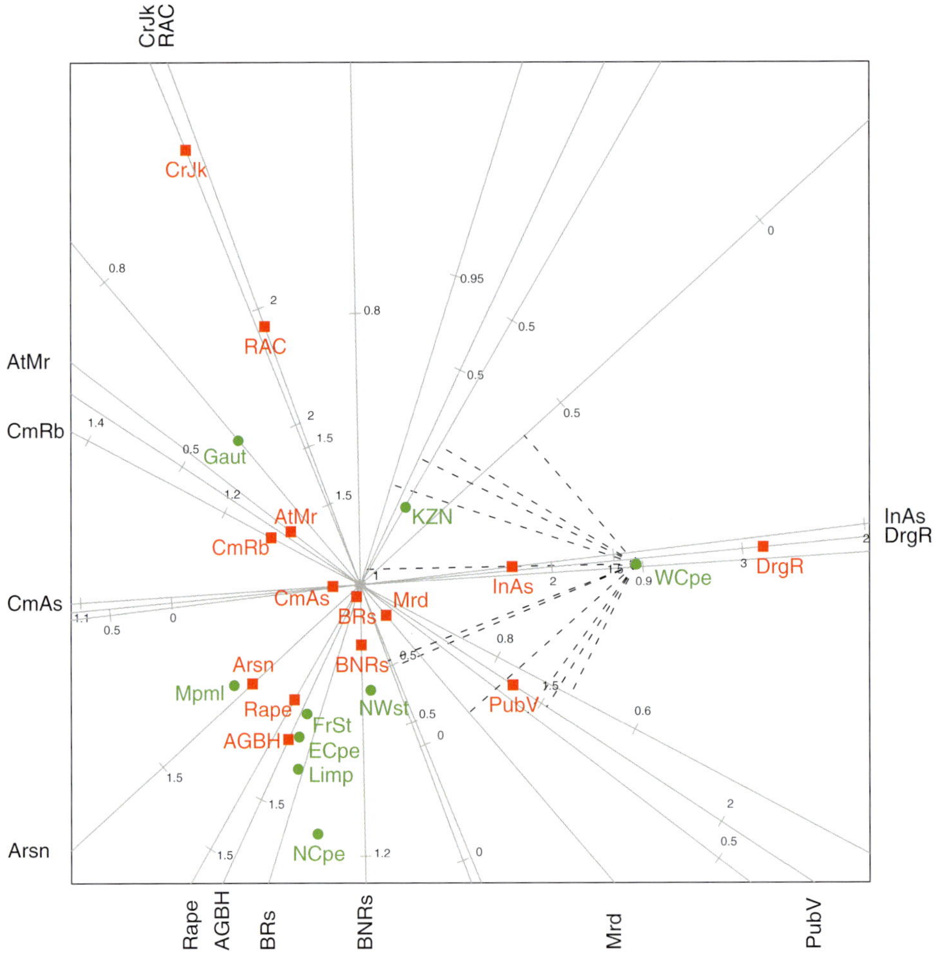

Figure 7.10 Two-dimensional CA biplot of the 2007/08 crime data set. The contingency ratio is approximated by plotting $\mathbf{R}^{-1/2}\mathbf{U}\boldsymbol{\Sigma}^{1/2}$ and $\mathbf{C}^{-1/2}\mathbf{V}\boldsymbol{\Sigma}^{1/2}$ with correctly adjusted scales.

`ca.variant = Chisq2Rows` and `ca.variant = Chisq2Cols`, respectively) are given in Figures 7.14 and 7.15. These biplots were used to obtain predictions for the chi-squared distance matrix. The two-dimensional approximation of $\sqrt{n}$ times the chi-squared distance matrix is given in Table 7.17.

The distances in Table 7.17 were calculated as follows. The biplot in Figure 7.14 provides the approximation to the matrix $\mathbf{R}^{-1}(\mathbf{X}-\mathbf{E})\mathbf{C}^{-1/2}$. This approximation was used for calculating the between-row chi-squared distances precisely in the same way as the actual distances were calculated from $\mathbf{R}^{-1}(\mathbf{X}-\mathbf{E})\mathbf{C}^{-1/2}$ using the function `Chisq.dist`.

Likewise, we calculated $\sqrt{n}$ times the chi-squared distance matrix between the columns of our 2007/08 crime data set. This matrix is shown in Table 7.18. The approximation based upon the two-dimensional biplot in Figure 7.15 is given in Table 7.19.

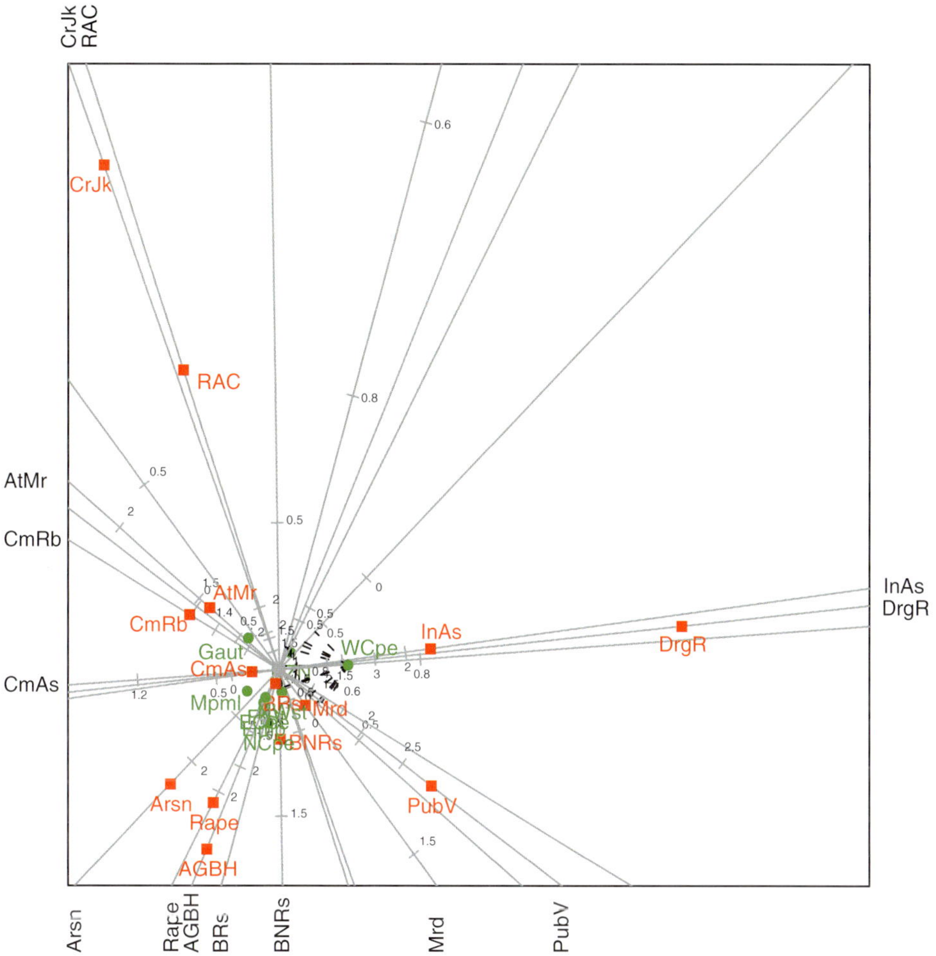

Figure 7.11 Two-dimensional CA biplot of the 2007/08 crime data set. The contingency ratio is approximated by plotting $\mathbf{R}^{-1/2}\mathbf{U}\boldsymbol{\Sigma}$ and $\mathbf{C}^{-1/2}\mathbf{V}$ with correctly adjusted scales and with the argument `lambda = FALSE` (the default).

The distances in Table 7.19 were calculated as follows. The biplot in Figure 7.15 provides the approximation to the matrix $\mathbf{R}^{-1/2}(\mathbf{X}-\mathbf{E})\mathbf{C}^{-1}$. This approximation was used for calculating the between-column chi-squared distances precisely in the same way as the actual distances were calculated from $\mathbf{R}^{-1/2}(\mathbf{X}-\mathbf{E})\mathbf{C}^{-1}$ using the function `Chisq.dist`.

In Figure 7.16 we have switched the role of the columns and the rows by providing as input to `cabipl` the transposed matrix $\mathbf{X}$, while the correlation approach discussed in Section 7.2.5 is illustrated in Figure 7.17.

In Table 7.20 we show the first 10 rows of the matrix $\mathbf{G}$: 1 104 159 × 23 of the 2007/08 crime data set. This matrix was constructed from the contingency table given in Table 7.5 by using our function `indicatormat` for converting a contingency table into indicator matrix form. The first nine columns of Table 7.20 are the indicator matrix for the row categories and the last 14 columns that for the column categories.

Table 7.15 The predicted contingency ratio matrix of the 2007/08 crime data obtained from the two-dimensional biplot shown in Figure 7.9. Subtracting 1 from the elements of this matrix matrix gives the predicted deviations of the x_{ij} from independence relative to the approximation $e_{ij} = x_{i.}x_{.j}/n$.

	Arsn	AGBH	AtMr	BNRs	BRs	CrJk	CmAs	CmRb	DrgR	InAs	Mrd	PubV	Rape	RAC
ECpe	1.2699	1.3454	0.9559	1.1112	1.0253	0.3290	1.0250	0.9829	0.6108	0.8463	1.0364	1.0659	1.2658	0.5951
FrSt	1.2306	1.2939	0.9640	1.0941	1.0215	0.4351	1.0217	0.9872	0.6628	0.8670	1.0304	1.0529	1.2264	0.6589
Gaut	0.9973	0.8406	1.2049	0.8918	0.9847	2.0499	1.0404	1.2258	0.4274	0.7913	0.9046	0.5805	0.9016	1.6096
KZN	0.8425	0.8105	1.0099	0.9434	0.9864	1.3109	0.9823	0.9925	1.2718	1.1060	0.9861	0.9940	0.8523	1.1894
Limp	1.3105	1.4076	0.9359	1.1349	1.0301	0.1596	1.0261	0.9655	0.5901	0.8370	1.0482	1.1033	1.3121	0.4944
Mpml	1.2996	1.3088	1.0489	1.0722	1.0209	0.7544	1.0472	1.0890	0.2922	0.7288	0.9950	0.8727	1.2492	0.8403
NWst	1.1126	1.1895	0.9221	1.0776	1.0149	0.4165	0.9986	0.9272	1.0053	0.9971	1.0429	1.1493	1.1388	0.6548
NCpe	1.3610	1.5116	0.8761	1.1827	1.0386	−0.2289	1.0205	0.9054	0.6625	0.8612	1.0789	1.2211	1.3860	0.2660
WCpe	0.5901	0.7047	0.7663	0.9886	0.9837	0.4851	0.9023	0.6943	2.4385	1.5425	1.0845	1.5158	0.7371	0.7245

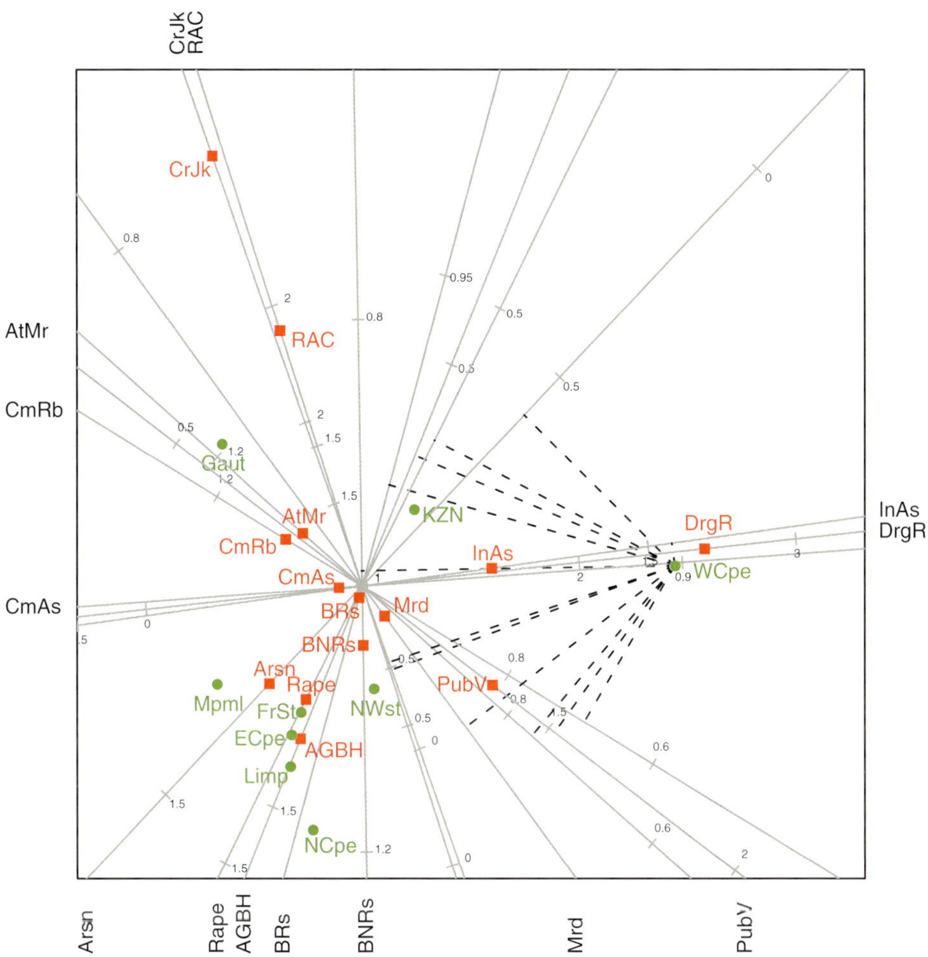

Figure 7.12 Two-dimensional CA biplot of the 2007/08 crime data set. The contingency ratio is approximated by plotting $\mathbf{R}^{-1/2}\mathbf{U}\boldsymbol{\Sigma}$ and $\mathbf{C}^{-1/2}\mathbf{V}$ with correctly adjusted scales and with the argument `lambda = TRUE`.

As an illustration, the row profiles of the 2007/08 crime data contingency table are shown, in deviation form, in Table 7.21. Instead of approximating the row profiles directly, we can construct a biplot from $\mathbf{R}^{-1/2}\mathbf{U}\boldsymbol{\Sigma}$ for the rows and $\mathbf{C}^{1/2}\mathbf{V}$ for the columns as described above, or from $\mathbf{R}^{-1/2}\mathbf{U}\boldsymbol{\Sigma}^{1/2}$ for the rows and $\mathbf{C}^{1/2}\mathbf{V}\boldsymbol{\Sigma}^{1/2}$ for the columns. Then, using the calibration tool, we can calibrate the axes in terms of the actual row profiles. These different biplots are shown in Figures 7.18–7.20.

Figure 7.16 showed that in order to plot the rows of a contingency table as biplot axes with the columns as points, all that is necessary is to transpose the data matrix as input to the argument `X` of the function `cabipl`. As a further example of this we provide Figure 7.21 as a CA biplot of the 2007/08 crime data set approximating the column profiles $(\mathbf{X}-\mathbf{E})\mathbf{C}^{-1}$. The biplot in Figure 7.21 can be considered as an approximation of the profiles of the types of crime with the provinces as calibrated axes.

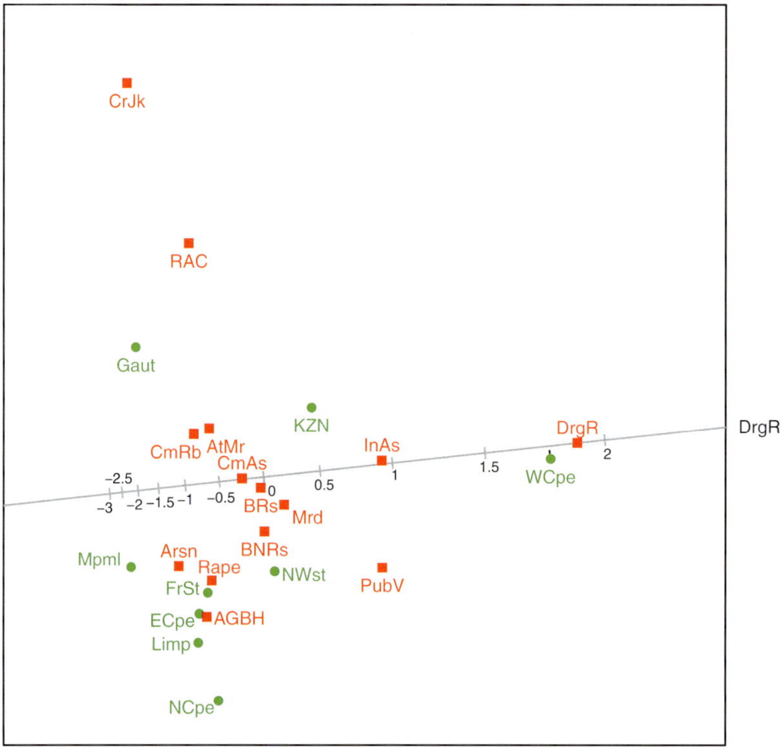

Figure 7.13 Two-dimensional CA biplot of the 2007/08 crime data set. The contingency ratio is approximated by plotting $\mathbf{R}^{-1/2}\mathbf{U}\boldsymbol{\Sigma}$ and $\mathbf{C}^{-1/2}\mathbf{V}$ with correctly adjusted scales and with the argument `lambda = TRUE`. Only biplot axis ***DrgR*** is shown with calibrations in terms of $2\times$ log(contingency ratio).

Table 7.16 $\sqrt{n}$ times the chi-squared distance matrix between the rows of the 2007/08 crime data set. The chi-squared distance is defined as the square root of $d^2_{ii'}$ as given in (7.12).

	ECpe	FrSt	Gaut	KZN	Limp	Mpml	NWst	NCpe	WCpe
ECpe	0.0000	0.3218	0.4894	0.4167	0.2881	0.2296	0.2302	0.2943	0.6811
FrSt	0.3218	0.0000	0.4730	0.4759	0.2590	0.3013	0.3186	0.3607	0.6609
Gaut	0.4894	0.4730	0.0000	0.3624	0.5343	0.3897	0.4627	0.6504	0.7415
KZN	0.4167	0.4759	0.3624	0.0000	0.5001	0.4618	0.3279	0.5586	0.4621
Limp	0.2881	0.2590	0.5343	0.5001	0.0000	0.2722	0.2429	0.3241	0.7029
Mpml	0.2296	0.3013	0.3897	0.4618	0.2722	0.0000	0.2935	0.4023	0.7550
NWst	0.2302	0.3186	0.4627	0.3279	0.2429	0.2935	0.0000	0.3414	0.5297
NCpe	0.2943	0.3607	0.6504	0.5586	0.3241	0.4023	0.3414	0.0000	0.7422
WCpe	0.6811	0.6609	0.7415	0.4621	0.7029	0.7550	0.5297	0.7422	0.0000

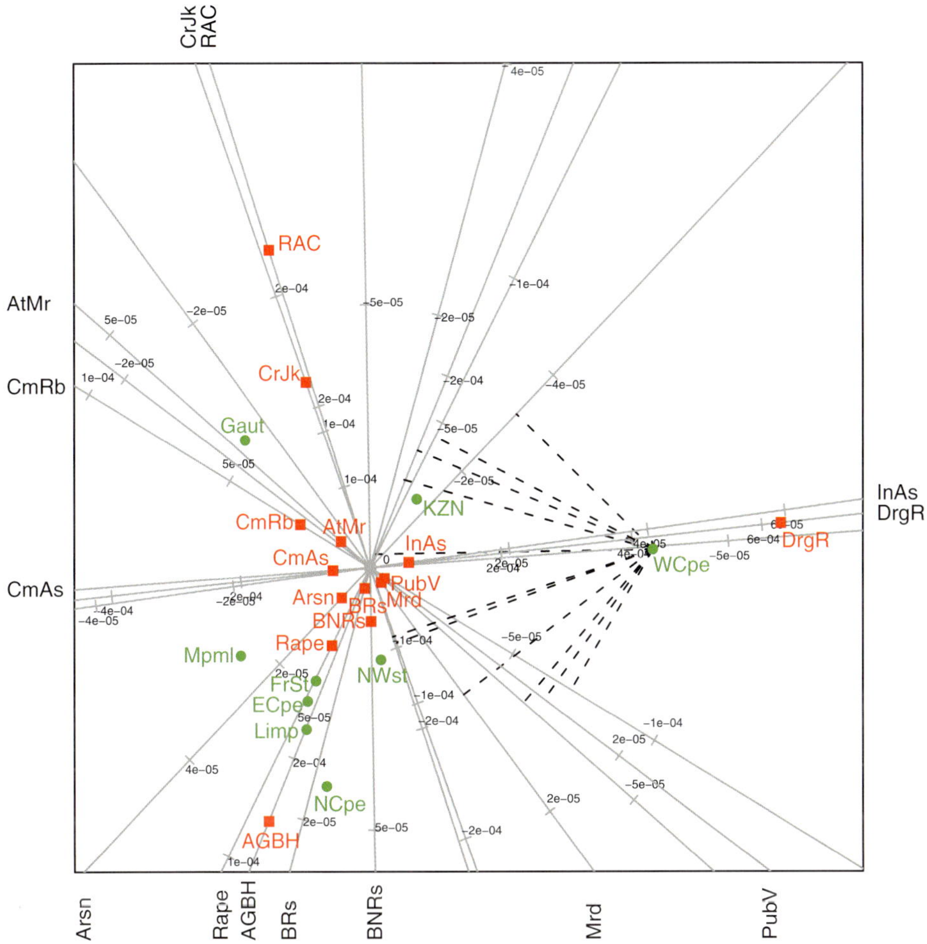

Figure 7.14 Two-dimensional CA biplot of the 2007/08 crime data set. Chi-squared distance approximation plotting $\mathbf{R}^{-1/2}\mathbf{U}\mathbf{\Sigma}$ and $\mathbf{V}$ (row chi-squared distance) with the argument `lambda = TRUE`. (Lambda evaluates to 36.33).

We now give some examples of interpolating new points in CA biplots. We first form a set of points to interpolate by taking the crime data for the Western Cape province for each of the periods 2001/02, 2002/03, ..., 2007/08 and arrange them as the rows of an R data matrix called `SACrimeWcpe.data`. A call to `cabipl` with arguments `X.new = SACrimeWcpe.data` and `X.new.labels = c("02","03","04","05","06","07","08")` leads to the biplot given in Figure 7.22.

Notice that in Figure 7.22 the point representing the row for the 2007/08 data in `X.new` interpolates into its correct place. Of importance are the interpolated positions of the other points: there is a clear progression of the interpolated points away from the 2007/08 point with the earlier data. Interpolating values for Gauteng province does not show this tendency: see Figure 7.23.

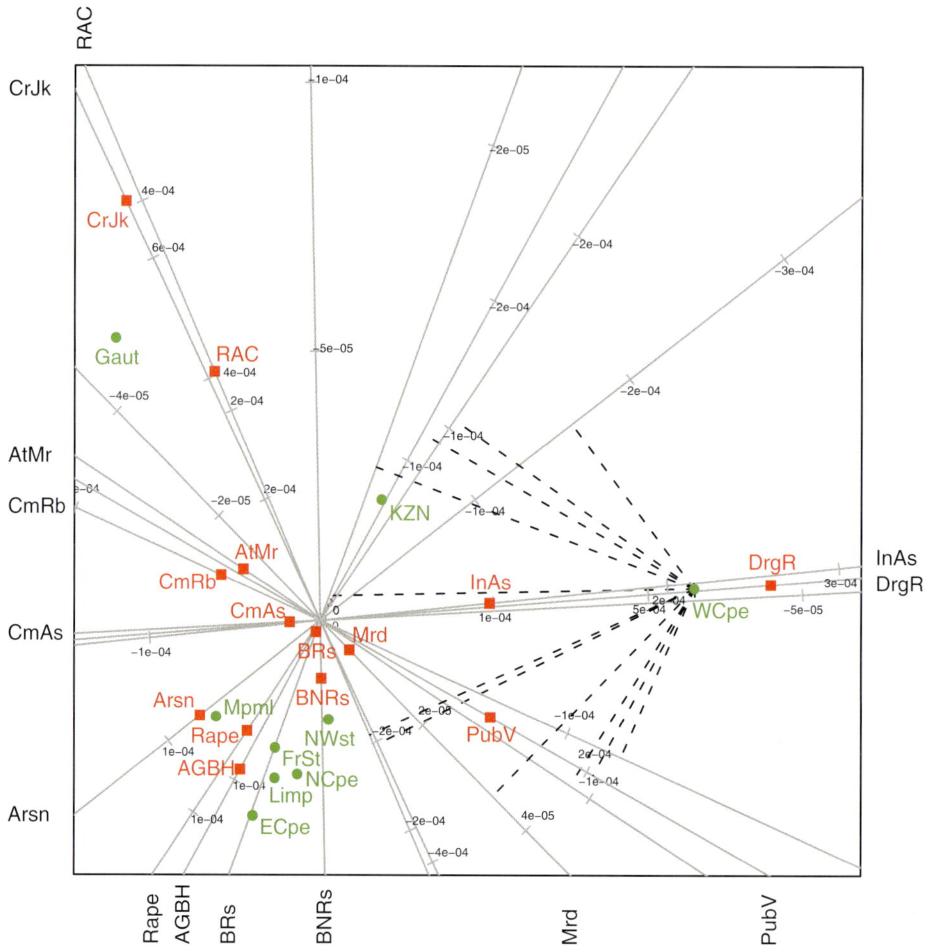

Figure 7.15 Two-dimensional CA biplot of the 2007/08 crime data set. Chi-squared distance approximation plotting **U** and $\mathbf{C}^{-1/2}\mathbf{V}\boldsymbol{\Sigma}$ (column chi-squared distance) with the argument `lambda = TRUE`. (Lambda evaluates to 0.0264).

Before we can draw final conclusions on our comparison of the interpolated values for the two provinces shown in Figures 7.22 and 7.23, we would like to know how well the interpolated points are represented in the respective graphs. Although the interpolated points do not contribute to the position of the axes, we can still answer this question by considering how well the axes predict the actual values of the interpolated points. Therefore, we first obtain the predictions for the interpolated values just as we have done in the case of the original points. Setting the argument `predictions.X.new = TRUE` allows us to find these predictions, as is illustrated in Figure 7.24 for the Western Cape province.

The actual values of the weighted Pearson residual matrix (see (7.2))

$$\mathbf{R}_{\text{new}}^{-1/2}(\mathbf{X}_{\text{new}} - \mathbf{E}_{\text{new}})\mathbf{C}^{-1/2} \tag{7.54}$$

Table 7.17 Approximation based upon two-dimensional biplot given in Figure 7.14 of $\sqrt{n}$ times the chi-squared distances between the rows of the 2007/08 crime data set.

	ECpe	FrSt	Gaut	KZN	Limp	Mpml	NWst	NCpe	WCpe
ECpe	0.0000	0.0374	0.4611	0.3962	0.0499	0.1421	0.1470	0.1507	0.6589
FrSt	0.0374	0.0000	0.4311	0.3591	0.0862	0.1395	0.1209	0.1829	0.6331
Gaut	0.4611	0.4311	0.0000	0.3191	0.5090	0.3702	0.4478	0.6118	0.7406
KZN	0.3962	0.3591	0.3191	0.0000	0.4416	0.4113	0.2839	0.5185	0.4227
Limp	0.0499	0.0862	0.5090	0.4416	0.0000	0.1726	0.1781	0.1036	0.6824
Mpml	0.1421	0.1395	0.3702	0.4113	0.1726	0.0000	0.2487	0.2713	0.7473
NWst	0.1470	0.1208	0.4478	0.2839	0.1781	0.2487	0.0000	0.2371	0.5127
NCpe	0.1507	0.1829	0.6118	0.5185	0.1036	0.2713	0.2371	0.0000	0.7021
WCpe	0.6589	0.6331	0.7406	0.4227	0.6824	0.7473	0.5127	0.7021	0.0000

Table 7.18 $\sqrt{n}$ times the chi-squared distance matrix between the columns of the 2007/08 crime data set. The chi-squared distance is defined as the square root of $d_{ii'}^2$ as given in (7.12).

	Arsn	AGBH	AtMr	BNRs	BRs	CrJk	CmAs	CmRb	DrgR	InAs	Mrd	PubV	Rape	RAC
Arsn	0.000	0.219	0.372	0.310	0.306	0.832	0.380	0.319	0.961	0.567	0.380	0.646	0.112	0.569
AGBH	0.219	0.000	0.400	0.285	0.278	0.922	0.315	0.338	0.905	0.490	0.409	0.550	0.166	0.627
AtMr	0.372	0.400	0.000	0.410	0.365	0.627	0.395	0.369	0.892	0.505	0.335	0.735	0.363	0.414
BNRs	0.310	0.285	0.410	0.000	0.216	0.843	0.232	0.277	0.758	0.392	0.463	0.468	0.231	0.546
BRs	0.306	0.278	0.365	0.216	0.000	0.773	0.186	0.199	0.743	0.321	0.378	0.462	0.262	0.454
CrJk	0.832	0.922	0.627	0.843	0.773	0.000	0.766	0.650	1.212	0.890	0.835	1.110	0.865	0.341
CmAs	0.380	0.315	0.395	0.232	0.186	0.766	0.000	0.202	0.795	0.363	0.469	0.520	0.316	0.453
CmRb	0.319	0.338	0.369	0.277	0.199	0.650	0.202	0.000	0.893	0.463	0.485	0.583	0.314	0.341
DrgR	0.961	0.905	0.892	0.758	0.743	1.212	0.795	0.893	0.000	0.466	0.765	0.613	0.885	0.958
InAs	0.567	0.490	0.505	0.392	0.321	0.890	0.363	0.463	0.466	0.000	0.425	0.417	0.496	0.585
Mrd	0.380	0.409	0.335	0.463	0.378	0.835	0.469	0.485	0.765	0.425	0.000	0.687	0.361	0.590
PubV	0.646	0.550	0.735	0.468	0.462	1.110	0.520	0.583	0.613	0.417	0.687	0.000	0.579	0.782
Rape	0.112	0.166	0.363	0.231	0.262	0.865	0.316	0.314	0.885	0.496	0.361	0.579	0.000	0.583
RAC	0.569	0.627	0.414	0.546	0.454	0.341	0.453	0.341	0.958	0.585	0.590	0.782	0.583	0.000

as well as the predicted values of (7.54) as obtained from the output list of the (two-dimensional) biplots that include the interpolated points are given in Tables 7.22–7.25. The element `out` of the output list of `cabipl` returns the predicted values and as element `weighted.dev.mat.new` the matrix (7.54). The positions of the interpolated points in Figure 7.22 draw our attention to the remarkable increase in the reporting of drug related crimes (highlighted in Table 7.22) in the Western Cape since the 2003/04 period.

The contents of Tables 7.22–7.25 can be used to calculate the sample predictivities for the interpolated points in Figures 7.22 and 7.23. The row predictivities for the Western Cape province, for example, are calculated according to (7.38) as follows. First, regarding Table 7.24 as a matrix of predicted values, calculate the diagonal of the product of this matrix and its transpose. Do exactly the same with Table 7.22 considered as the matrix

Table 7.19 Approximation based upon two-dimensional biplot given in Figure 7.14 of $\sqrt{n}$ times the chi-squared distances between the columns of the 2007/08 crime data set.

	Arsn	AGBH	AtMr	BNRs	BRs	CrJk	CmAs	CmRb	DrgR	InAs	Mrd	PubV	Rape	RAC
Arsn	0.000	0.354	0.078	0.045	0.038	0.292	0.037	0.103	0.646	0.117	0.064	0.072	0.072	0.469
AGBH	0.354	0.000	0.429	0.347	0.385	0.630	0.385	0.429	0.918	0.456	0.406	0.414	0.283	0.799
AtMr	0.078	0.429	0.000	0.108	0.064	0.215	0.044	0.056	0.618	0.090	0.068	0.067	0.149	0.393
BNRs	0.045	0.347	0.108	0.000	0.045	0.322	0.077	0.145	0.621	0.109	0.060	0.068	0.071	0.500
BRs	0.038	0.385	0.064	0.045	0.000	0.277	0.045	0.108	0.608	0.078	0.026	0.033	0.103	0.455
CrJk	0.292	0.630	0.215	0.322	0.277	0.000	0.256	0.202	0.633	0.254	0.271	0.266	0.360	0.178
CmAs	0.037	0.385	0.044	0.077	0.045	0.256	0.000	0.068	0.642	0.109	0.064	0.069	0.105	0.433
CmRb	0.103	0.429	0.056	0.145	0.108	0.202	0.068	0.000	0.668	0.146	0.120	0.121	0.163	0.375
DrgR	0.646	0.918	0.618	0.621	0.608	0.633	0.642	0.668	0.000	0.534	0.583	0.576	0.690	0.707
InAs	0.117	0.456	0.090	0.109	0.078	0.254	0.109	0.146	0.534	0.000	0.053	0.045	0.178	0.427
Mrd	0.064	0.406	0.068	0.060	0.026	0.271	0.064	0.120	0.583	0.053	0.000	0.008	0.126	0.448
PubV	0.072	0.414	0.067	0.068	0.033	0.266	0.069	0.121	0.576	0.045	0.008	0.000	0.134	0.443
Rape	0.072	0.283	0.149	0.071	0.103	0.360	0.105	0.163	0.690	0.178	0.126	0.134	0.000	0.536
RAC	0.469	0.799	0.393	0.500	0.455	0.178	0.433	0.375	0.707	0.427	0.448	0.443	0.536	0.000

of actual values. Dividing the first set of diagonal values by their counterparts in the second set gives the sample predictivities for the interpolated points. Table 7.26 contains the required predictivities. As expected, we see that the 2007/08 values are identical to those displayed in Table 7.13.

The interpolation of selected row points in the CA biplots displayed in Figures 7.22 and 7.23 has drawn our attention to striking changes in the reported cases of drug related crimes in the Western Cape province. Instead of relying on the positions of interpolated points, we can investigate the crime data for the Western Cape over the entire study period by constructing a CA biplot for the corresponding data. Table 7.27 contains the data for the Western Cape province over the entire period from 2001/02 to 2007/08.

The Pearson residuals case A biplot for the Table 7.27 data is given in Figure 7.25. We have set the argument `markers = FALSE` in order to display only the two markers at the ends of each biplot axis. This biplot has a very high overall quality of 98.22%. The column and row predictivities are given in Tables 7.28 and 7.29, respectively.

The positions occupied by the row points in the Figure 7.25 biplot suggest that even a one-dimensional CA biplot should adequately represent the distances among them. This conclusion is supported by a one-dimensional overall quality of 92.84%.

Does the exceptionally high two-dimensional quality of 98.22% necessarily ensure that all rows and columns are well represented in the biplot? For guidance we consider the column and row predictivities: it is clear that the column predictivity of ***Rape*** is a very low 0.1783, with that of ***PubV*** a moderate 0.4636. In two dimensions all row points have very high row predictivity values, indicating that their true weighted residual values for all the different crimes can be accurately determined from the biplot. We also notice that in a one-dimensional biplot it is only row point ***04/05*** that has a very low row predictivity but that with the addition of only one further dimension it jumps to 0.9513.

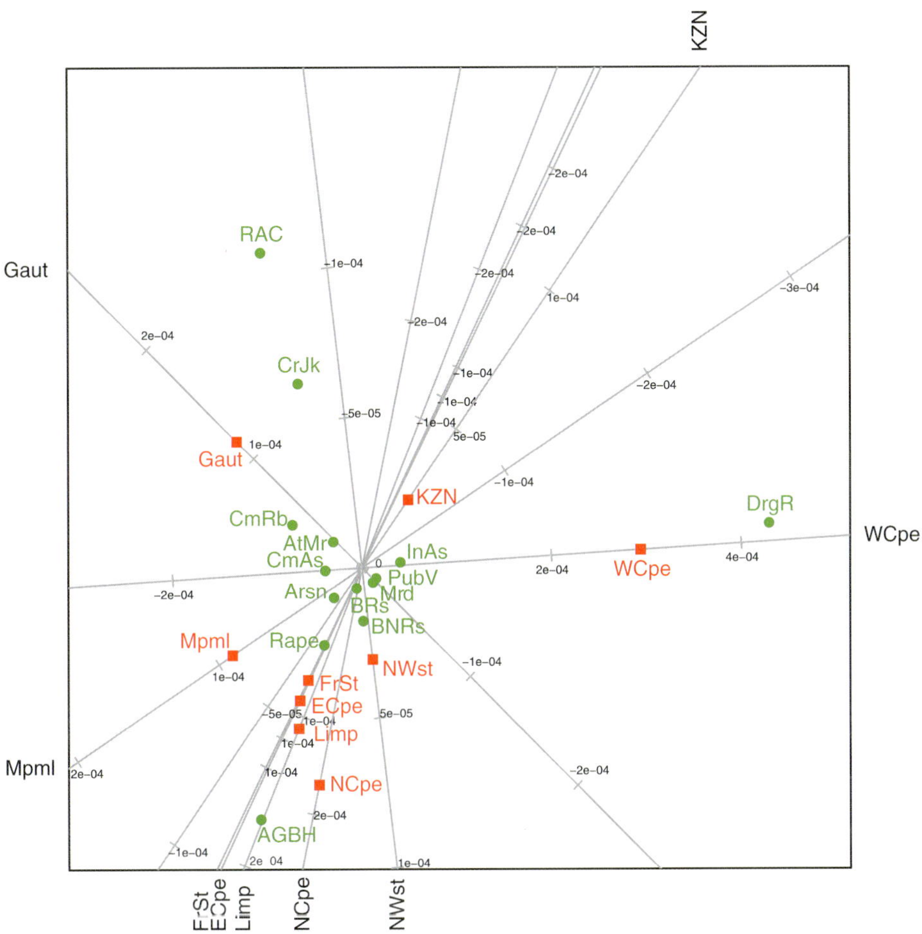

Figure 7.16 Similar to Figure 7.15 but plotting the provinces as calibrated axes with the different crimes as points. Two-dimensional CA biplot of the 2007/08 crime data set. Chi-squared distance approximation plotting $\mathbf{U}$ and $\mathbf{C}^{-1/2}\mathbf{V}\boldsymbol{\Sigma}$ (column chi-squared distance) with the argument `lambda = TRUE`. (Lambda evaluates to 0.0275).

The previous example raises the question of how to include the time aspect in a CA biplot. Indeed, the full crime data set can be considered to be a three-way contingency table. An obvious way is to concatenate the seven two-way contingency tables row-wise. As an example of this we show in Figure 7.26 the Pearson residuals case A CA biplot for the concatenated 2001/02 and 2007/08 crime data. This biplot shows some evidence of a reduction in several of the crimes for most of the provinces: the main exceptions are ***KZN*** and ***WCpe***, where the position of ***KZN*** indicates a large increase in murder and that of ***WCpe*** a large increase in drug related issues. In order to understand this biplot, as already stated, we have to take into account the calculated measures of fit: the overall quality of the display is a satisfactory 84.65% (see Table 7.30); the column predictivities and row predictivities are given in Tables 7.31 and 7.32.

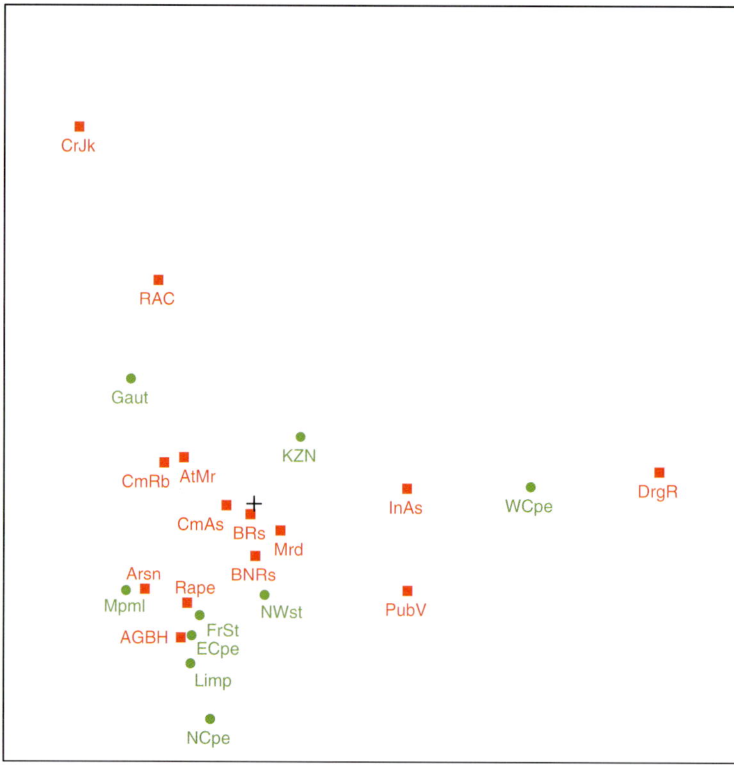

Figure 7.17 Two-dimensional CA biplot of the 2007/08 crime data set. Correlation approximation plotting $\mathbf{R}^{-1/2}\mathbf{U}\mathbf{\Sigma}$ and $\mathbf{C}^{-1/2}\mathbf{V}\mathbf{\Sigma}$ with the argument `lambda = FALSE`. (Lambda evaluates to 1.0704, indicating that setting `lambda = TRUE` would have had a small effect on the appearance of the biplot.) The origin is indicated by a black cross.

Table 7.20 The first 10 rows of the concatenated two-indicator matrices identifying row and column membership of the 2007/08 crime data set.

ECpe	FrSt	Gaut	KZN	Limp	Mpml	NWst	NCpe	WCpe	Arsn	AGBH	AtMr	BNRs	BRs	CrJk	CmAs	CmRb	DrgR	InAs	Mrd	PubV	RaPe	RAC
1	0	0	0	0	0	0	0	0	1	0	0	0	0	0	0	0	0	0	0	0	0	0
1	0	0	0	0	0	0	0	0	1	0	0	0	0	0	0	0	0	0	0	0	0	0
1	0	0	0	0	0	0	0	0	1	0	0	0	0	0	0	0	0	0	0	0	0	0
1	0	0	0	0	0	0	0	0	1	0	0	0	0	0	0	0	0	0	0	0	0	0
1	0	0	0	0	0	0	0	0	1	0	0	0	0	0	0	0	0	0	0	0	0	0
1	0	0	0	0	0	0	0	0	1	0	0	0	0	0	0	0	0	0	0	0	0	0
1	0	0	0	0	0	0	0	0	1	0	0	0	0	0	0	0	0	0	0	0	0	0
1	0	0	0	0	0	0	0	0	1	0	0	0	0	0	0	0	0	0	0	0	0	0
1	0	0	0	0	0	0	0	0	1	0	0	0	0	0	0	0	0	0	0	0	0	0
1	0	0	0	0	0	0	0	0	1	0	0	0	0	0	0	0	0	0	0	0	0	0

Table 7.21 Row profiles of the 2007/08 crime data as deviations from the marginal row profile. The marginal row profile is included as the last row of the table.

	Arsn	AGBH	AtMr	BNRs	BRs	CrJk	CmAs	CmRb	DrgR	InAs	Mrd	PubV	Rape	RAC
ECpe	0.0097	0.2702	0.0169	0.0466	0.2312	0.0047	0.1557	0.0555	0.0621	0.0060	0.0275	0.0007	0.0431	0.0700
FrSt	0.0057	0.2227	0.0124	0.0585	0.2078	0.0021	0.2631	0.0555	0.0599	0.0056	0.0116	0.0008	0.0348	0.0596
Gaut	0.0062	0.1591	0.0178	0.0512	0.2123	0.0253	0.1935	0.0750	0.0418	0.0051	0.0124	0.0006	0.0273	0.1725
KZN	0.0070	0.1629	0.0263	0.0546	0.1980	0.0207	0.1566	0.0493	0.1287	0.0064	0.0251	0.0004	0.0346	0.1293
Limp	0.0101	0.2415	0.0128	0.0954	0.2094	0.0036	0.1947	0.0664	0.0565	0.0038	0.0123	0.0005	0.0497	0.0432
Mpml	0.0083	0.2376	0.0179	0.0603	0.2659	0.0094	0.1721	0.0670	0.0250	0.0035	0.0118	0.0009	0.0372	0.0833
NWst	0.0091	0.2313	0.0128	0.0727	0.2147	0.0042	0.1518	0.0563	0.1022	0.0050	0.0134	0.0017	0.0440	0.0806
NCpe	0.0057	0.3349	0.0262	0.0662	0.1666	0.0002	0.1837	0.0452	0.0745	0.0072	0.0143	0.0011	0.0345	0.0398
WCpe	0.0033	0.1297	0.0096	0.0554	0.2207	0.0048	0.1701	0.0447	0.2394	0.0096	0.0148	0.0013	0.0208	0.0758
$\mathbf{1}'\mathbf{C}/n$	0.0067	0.1903	0.0170	0.0571	0.2154	0.0129	0.1794	0.0589	0.0988	0.0061	0.0167	0.0008	0.0328	0.1072

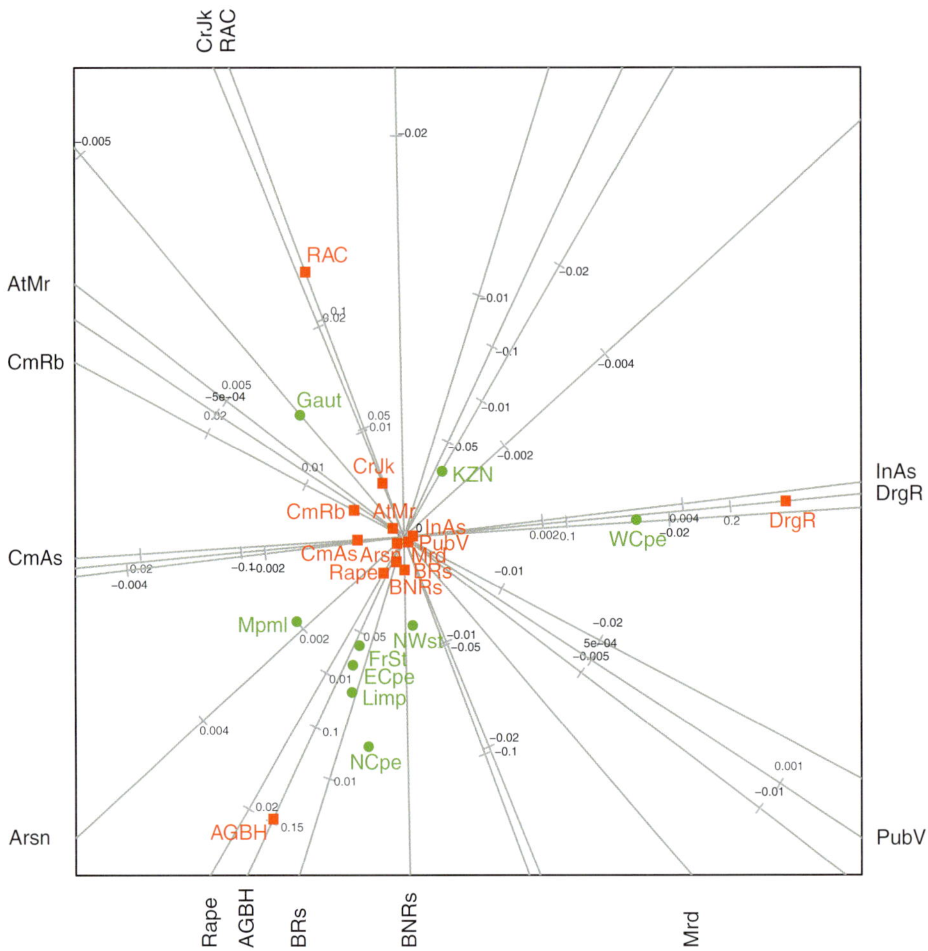

Figure 7.18 Two-dimensional CA biplot of the 2007/08 crime data set. Approximating the row profiles $\mathbf{R}^{-1}(\mathbf{X}-\mathbf{E})$ by plotting $\mathbf{R}^{-1/2}\mathbf{U}\mathbf{\Sigma}^{1/2}$ and $\mathbf{C}^{1/2}\mathbf{V}\mathbf{\Sigma}^{1/2}$ (case A) with arguments `ca.variant = RowProfA`; `RowProf.scaled.markers = FALSE` (the default); `lambda = TRUE`. (Lambda evaluates to 311.9243, indicating that setting `lambda = FALSE` would result in a biplot in which all the row points are squeezed into one another with the column points more spread out.) Calibrations on axes are in the form of deviations from the marginal row profile.

From Table 7.31 it follows that while ***DrgR*** has a high two-dimensional axis predictivity, those for ***Arsn***, ***AtMr***, ***BRs***, ***CmAs*** and ***Mrd*** are all very low. In Table 7.32 we see that both ***KZN*** and ***WCpe*** have low row predictivities in 2002 but high values in 2008. This raises the question of what the positions of ***KZN*** and ***WCpe*** would have been if the 2002 data had been considered on their own. The Pearson residuals case A CA biplot for 2001/02 given in Figure 7.27, with an overall quality of 85.20%, provides the answer.

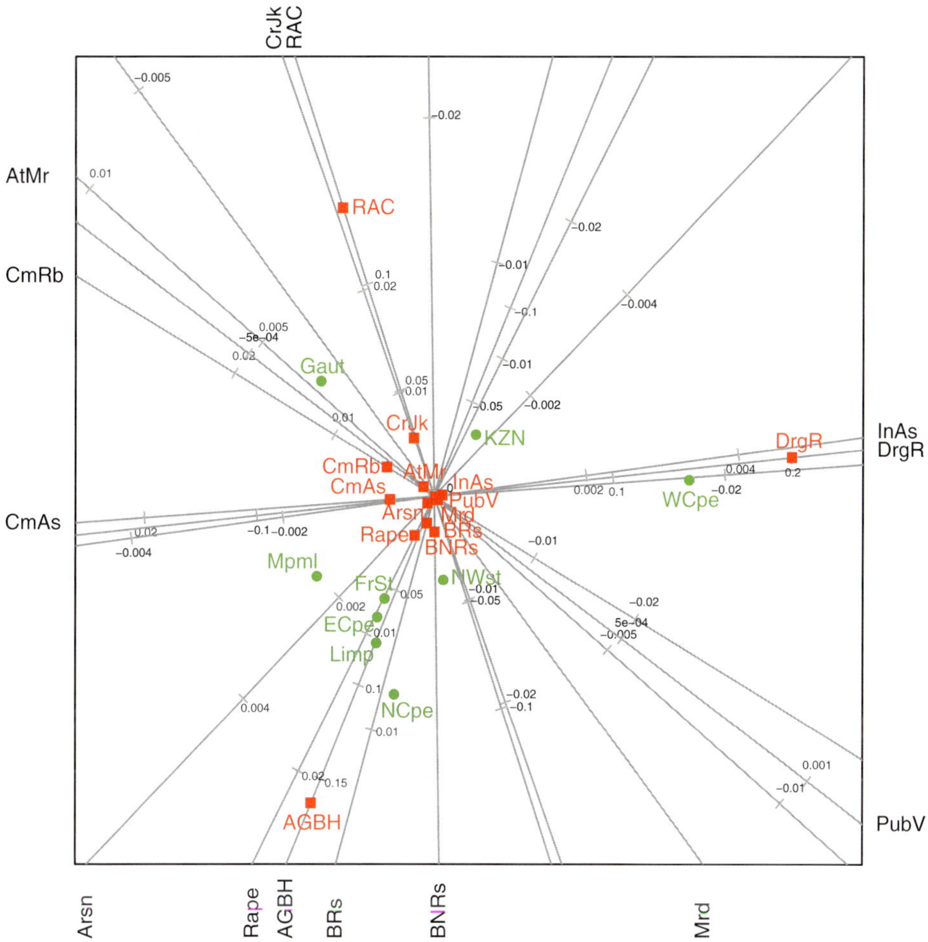

Figure 7.19 Two-dimensional CA biplot of the 2007/08 crime data set. Approximating the row profiles $\mathbf{R}^{-1}(\mathbf{X}-\mathbf{E})$ by plotting $\mathbf{R}^{-1/2}\mathbf{U\Sigma}$ and $\mathbf{C}^{1/2}\mathbf{V}$ (case B) with arguments `ca.variant = RowProfB`; `RowProf.scaled.markers = FALSE` (the default); `lambda = TRUE`. (Lambda evaluates to 672.66, indicating that setting `lambda = FALSE` would result in a biplot in which all the row points are squeezed into one another with the column points more spread out.) Calibrations on axes are as in Figure 7.18.

Figure 7.27 is an example where a reflection and/or a rotation of the biplot might be considered for comparison purposes. In Figure 7.27 we have made the following changes in the arguments of `cabipl: reflect = "y", rotate = 70`. In the resulting biplot shown in Figure 7.28 all the distance properties of the biplot have been retained but the outcome is a biplot that is visually much easier to compare with previous biplots such as Figure 7.5.

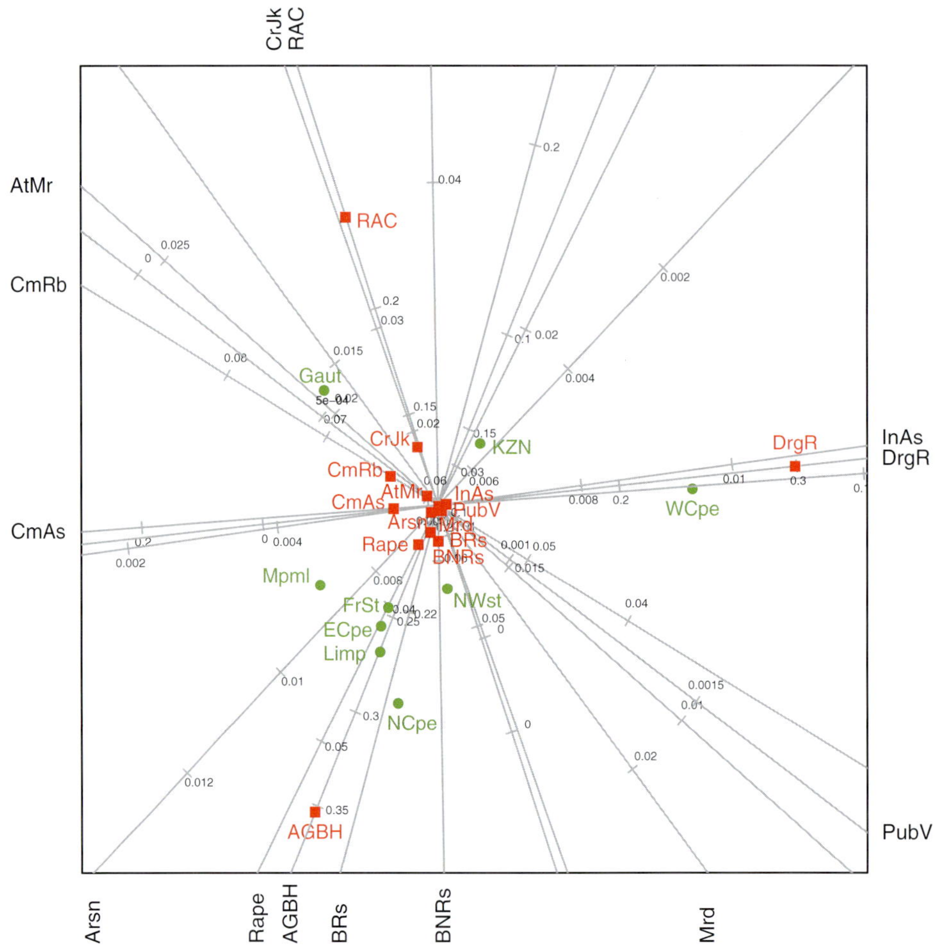

Figure 7.20 Two-dimensional CA biplot of 2007/08 crime data set. Approximating the row profiles $\mathbf{R}^{-1}(\mathbf{X}-\mathbf{E})$ by plotting $\mathbf{R}^{-1/2}\mathbf{U}\boldsymbol{\Sigma}$ and $\mathbf{C}^{1/2}\mathbf{V}$ (case B) with calibrations scaled to provide predictions of actual profile values by setting the argument `RowProf.scaled.markers = TRUE` in the function `cabipl`. All other arguments are kept at their Figure 7.19 values.

The reader can check with our function `ca.predictivities` that ***Gaut***, ***KZN*** and ***WCpe*** have row predictivities of 0.9735, 0.5366 and 0.9499, respectively. The three-dimensional row predictivity of ***KZN*** is 0.8479. The reader can also verify that the two-dimensional axis predictivities of ***DrgR*** and ***Mrd*** are 0.8871 and 0.1482, respectively. If a third dimension is added the three-dimensional column predictivity of ***Mrd*** increases sharply to 0.8469. This suggests that looking at a biplot with the second and third eigenvectors as scaffolding might be worthwhile. So changing, in the argument list of `cabipl`, `e.vects = 2:3`, we obtain the CA biplot given in Figure 7.29.

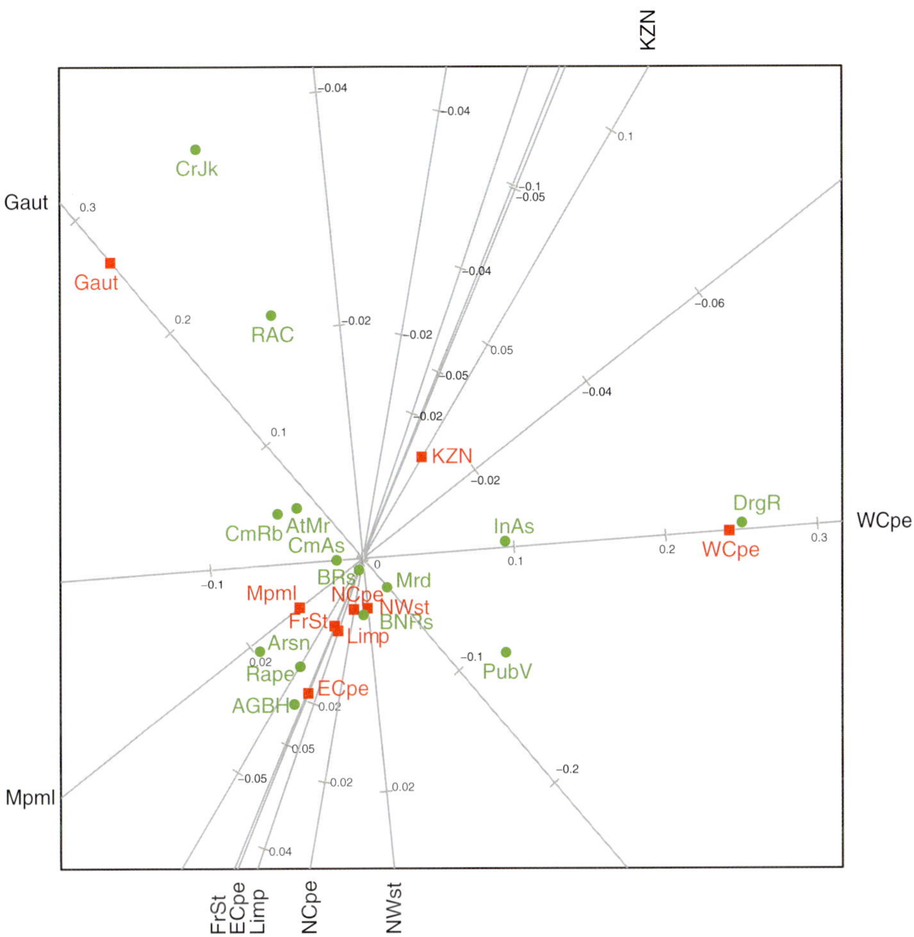

Figure 7.21 Two-dimensional CA biplot of the 2007/08 crime data set but with the transpose of the data matrix used as input. Approximating the row profiles $\mathbf{C}^{-1}(\mathbf{X}-\mathbf{E})'$ by plotting $\mathbf{R}^{-1/2}\mathbf{U}\boldsymbol{\Sigma}^{1/2}$ and $\mathbf{C}^{1/2}\mathbf{V}\boldsymbol{\Sigma}^{1/2}$ (case A) of the *transposed matrix* $\mathbf{X}$ with argument `lambda = TRUE` ($\lambda = 368.29$, indicating that setting `lambda = FALSE` would result in a biplot where all the row points are sqeezed into one another with the column points more spread out).

7.6.2 Ordinary PCA biplot of the weighted deviations matrix

Specifying in `cabipl` the argument `ca.variant = PCA` results in an ordinary PCA biplot of the weighted deviations matrix $\mathbf{R}^{-1/2}(\mathbf{X}-\mathbf{E})\mathbf{C}^{-1/2}$. The user can specify, if required, the argument `scaled.mat = TRUE` for first scaling the columns of $\mathbf{R}^{-1/2}(\mathbf{X}-\mathbf{E})\mathbf{C}^{-1/2}$ to have unit variances. As an example we again consider the 2007/08 crime data set. The PCA biplot of $\mathbf{R}^{-1/2}(\mathbf{X}-\mathbf{E})\mathbf{C}^{-1/2}$ without any further scaling is given in

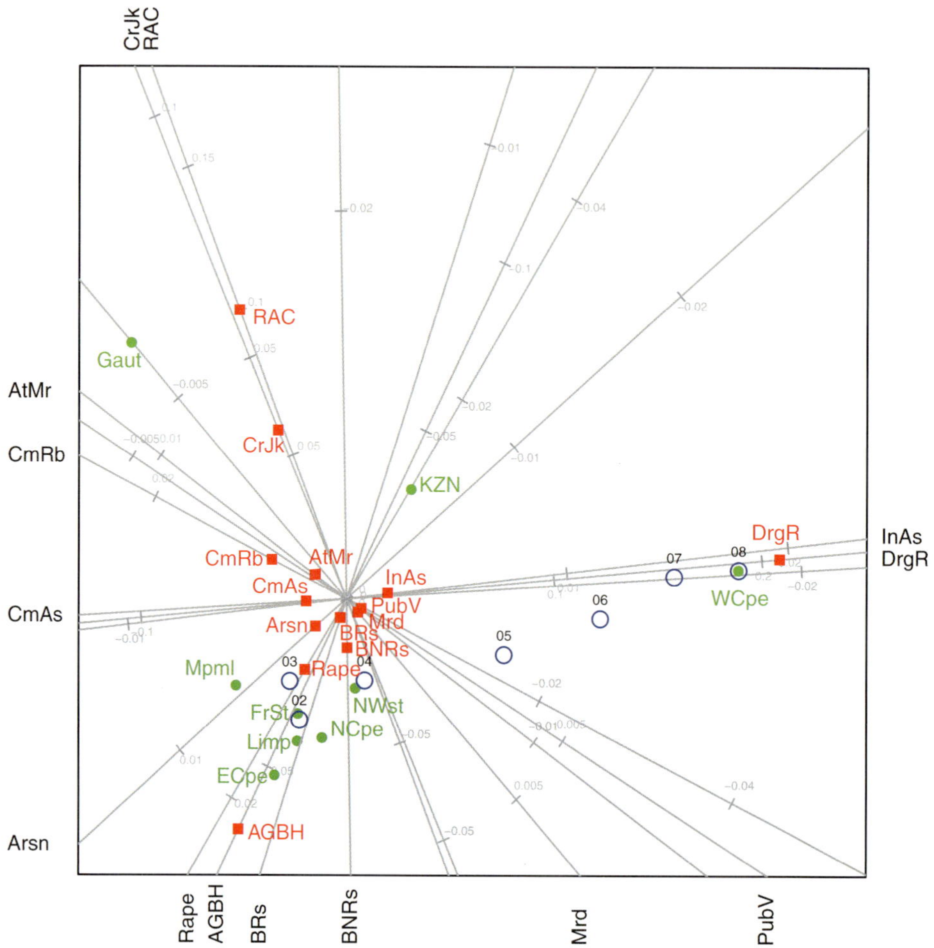

Figure 7.22 Pearson residuals case A CA biplot of the 2007/08 crime data with the interpolated points for the Western Cape province for the periods 2001/02, 2002/03, ..., 2007/08.

Figure 7.30. This figure looks remarkably similar to those obtained by specifying one of the other alternatives for the argument `ca.variant`. A quality of display of 87.56% is obtained. We leave as an exercise to the reader to find the column predictivities and row predictivities with our function `ca.predictivities`.

7.6.3 Doubling in a CA biplot

The effect of doubling on chi-squared distance in CA was discussed in Section 7.4. We use the artificial example of Table 7.2 to illustrate the method. Figure 7.31 shows a

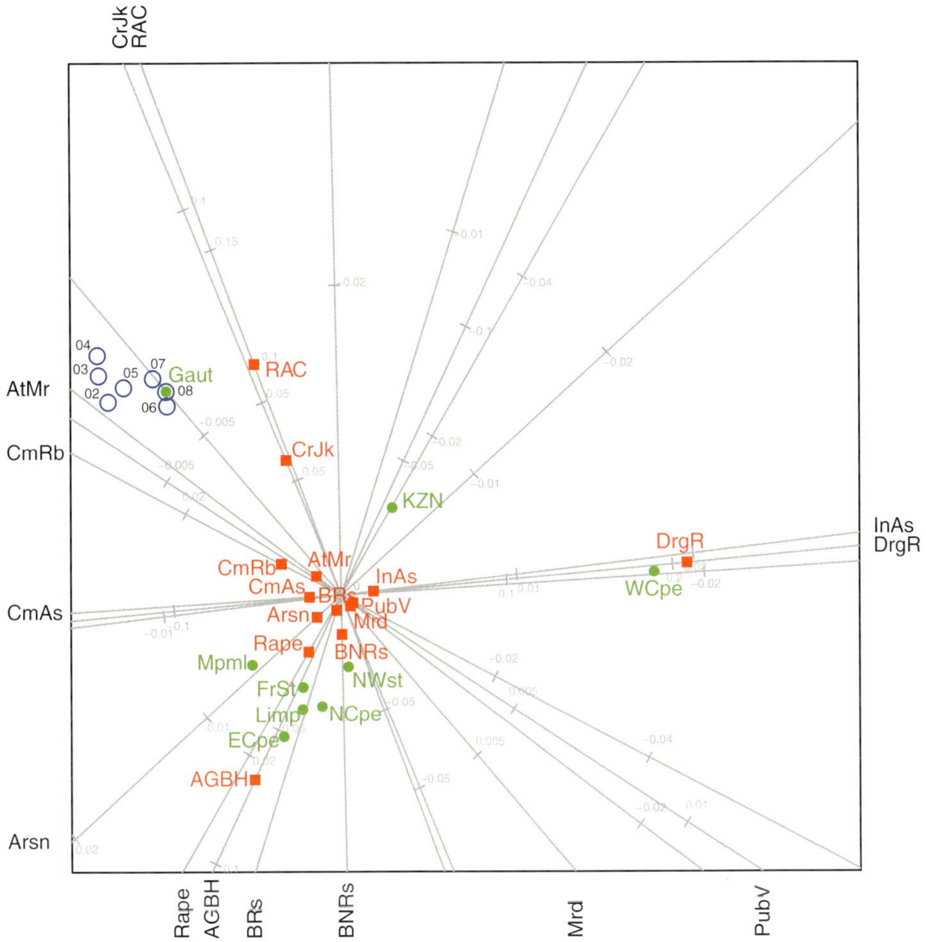

Figure 7.23 Pearson residuals case A CA biplot of the 2007/08 crime data with the interpolated points for the Gauteng province for the periods 2001/02, 2002/03, ..., 2007/08. For this biplot it was necessary to set in `cabipl` the argument `exp.factor = 1.5` to allow the appearance of all interpolated data points in the display.

two-dimensional approximation to the row chi-squared distances. From Figure 7.31 we see that products $\boldsymbol{A}$ and $\boldsymbol{C}$ are close together, as is to be expected from the similarity of their row profiles. But this does not reflect the commercially important information that $\boldsymbol{A}$ has much lower ratings than does $\boldsymbol{C}$. The doubling process is designed to rectify this. The result is shown in Figure 7.32, where $\boldsymbol{A}$ and $\boldsymbol{C}$ are now the most distant pair of products, reflecting the actual ratings. In these figures, we have included axes representing the attributes; these have no particular interest, but we note a pattern of clustering which we discuss further when considering the column chi-squared distances.

Figures 7.31 and 7.32 are constructed with the following calls to `cabipl` and `cabipl.doubling`, respectively:

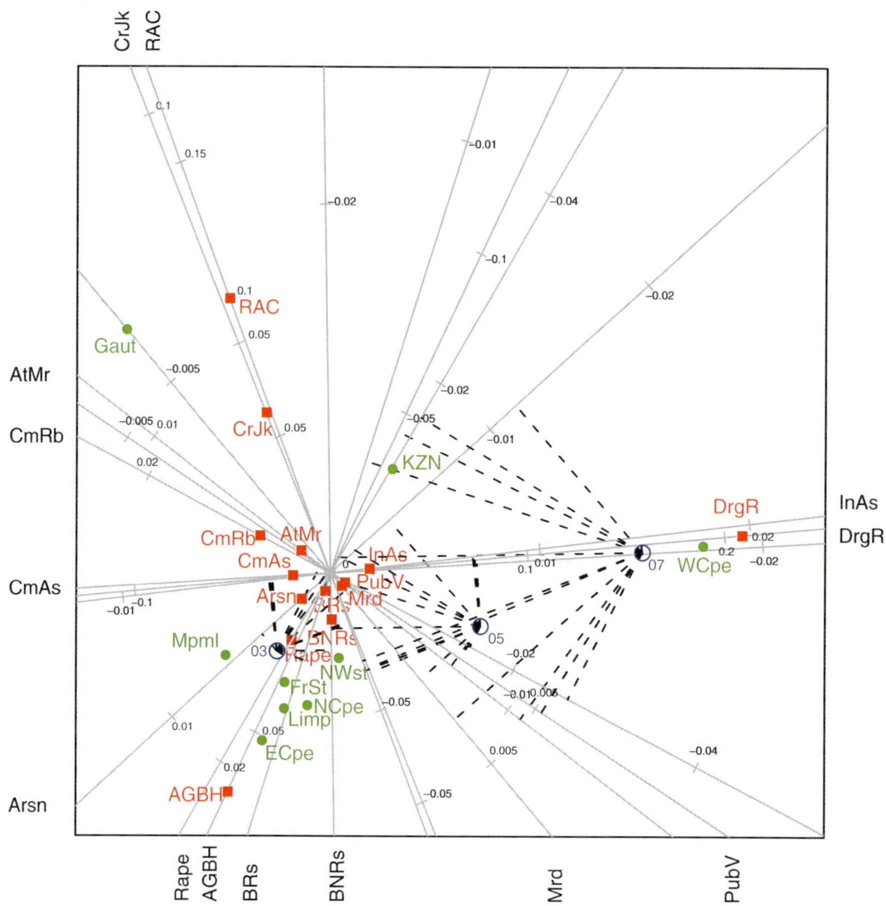

Figure 7.24 Pearson residuals case A CA biplot of the 2007/08 crime data with the interpolated points together with their predictions for the Western Cape province for the periods 2002/03, 2004/05 and 2006/07.

```
cabipl (X = Table.7.2.data, ca.variant = "Chisq2Rows",
   markers = F, offset = c(-0.1, 0.2, 0, 0.25),
   plot.col.points = FALSE, pos = "Hor", show.origin = TRUE,
   zoomval = 0.025)
```

```
cabipl.doubling(X = Table.7.2.data, N = 100, ca.variant =
   "Chisq2Rows", markers = F, offset = c(3, 2.75, 0.21, 0.21),
   offset.m = c(-0.1, -0.1, -0.3, -0.1, -0.1, -0.1),
   plot.col.points = FALSE, show.origin = TRUE, zoomval = 0.05)
```

Figure 7.33 gives the approximation to the column chi-squared distances of Table 7.2 before doubling, and Figure 7.34 after doubling. These figures are obtained by making

Table 7.22 Actual values $\mathbf{R}_{new}^{-1/2}(\mathbf{X}_{new} - \mathbf{E}_{new})\mathbf{C}^{-1/2}$ for the Western Cape province.

	Arsn	AGBH	AtMr	BNRs	BRs	CrJk	CmAs	CmRb	DrgR	InAs	Mrd	PubV	Rape	RAC
01/02	−0.0124	−0.0170	0.0063	0.0336	0.0387	−0.0350	0.0489	0.0172	−0.0487	0.0130	−0.0013	0.0056	−0.0205	−0.0650
02/03	−0.0129	−0.0275	0.0168	0.0050	0.0428	−0.0335	0.0576	0.0324	−0.0517	0.0163	−0.0005	0.0064	−0.0271	−0.0579
03/04	−0.0125	−0.0224	−0.0015	−0.0062	0.0299	−0.0323	0.0624	0.0171	−0.0112	0.0179	−0.0131	0.0064	−0.0283	−0.0598
04/05	−0.0174	−0.0289	−0.0172	−0.0263	0.0079	−0.0329	0.0544	0.0080	0.0640	0.0268	−0.0134	0.0084	−0.0201	−0.0593
05/06	−0.0176	−0.0350	−0.0222	−0.0247	0.0044	−0.0278	0.0253	−0.0142	0.1151	0.0309	−0.0062	0.0122	−0.0184	−0.0471
06/07	−0.0176	−0.0532	−0.0204	−0.0078	0.0080	−0.0299	0.0030	−0.0235	0.1522	0.0243	−0.0057	0.0191	−0.0250	−0.0357
07/08	−0.0174	−0.0579	−0.0237	−0.0029	0.0047	−0.0296	−0.0091	−0.0244	0.1865	0.0187	−0.0064	0.0077	−0.0275	−0.0400

Table 7.23 Actual values $\mathbf{R}_{new}^{-1/2}(\mathbf{X}_{new} - \mathbf{E}_{new})\mathbf{C}^{-1/2}$ for the Gauteng province.

	Arsn	AGBH	AtMr	BNRs	BRs	CrJk	CmAs	CmRb	DrgR	InAs	Mrd	PubV	Rape	RAC
01/02	−0.0111	−0.0369	0.0260	−0.0058	0.0272	0.0744	−0.0039	0.0586	−0.1324	−0.0191	−0.0126	−0.0096	−0.0129	0.0878
02/03	−0.0099	−0.0466	0.0393	−0.0302	0.0186	0.0659	0.0209	0.0690	−0.1366	−0.0184	−0.0149	−0.0059	−0.0218	0.0977
03/04	−0.0094	−0.0520	0.0266	−0.0416	0.0122	0.0571	0.0373	0.0659	−0.1362	−0.0161	−0.0222	−0.0076	−0.0221	0.1135
04/05	−0.0068	−0.0450	0.0097	−0.0456	0.0083	0.0402	0.0393	0.0673	−0.1209	−0.0120	−0.0250	−0.0064	−0.0158	0.1030
05/06	−0.0047	−0.0403	0.0033	−0.0364	0.0184	0.0463	0.0191	0.0429	−0.0957	−0.0104	−0.0235	−0.0043	−0.0117	0.0920
06/07	−0.0007	−0.0391	0.0054	−0.0222	−0.0001	0.0503	0.0096	0.0464	−0.0998	−0.0102	−0.0181	−0.0044	−0.0114	0.1144
07/08	−0.0033	−0.0370	0.0031	−0.0127	−0.0035	0.0566	0.0172	0.0344	−0.0938	−0.0069	−0.0172	−0.0040	−0.0156	0.1033

Table 7.24 Two-dimensional predicted values of $\mathbf{R}_{new}^{-1/2}(\mathbf{X}_{new} - \mathbf{E}_{new})\mathbf{C}^{-1/2}$ for the Western Cape province.

	Arsn	AGBH	AtMr	BNRs	BRs	CrJk	CmAs	CmRb	DrgR	InAs	Mrd	PubV	Rape	RAC
01/02	0.0051	0.0349	−0.0014	0.0062	0.0027	−0.0179	0.0024	−0.0012	−0.0273	−0.0027	0.0011	0.0005	0.0111	−0.0312
02/03	0.0043	0.0266	−0.0001	0.0042	0.0020	−0.0104	0.0027	0.0012	−0.0303	−0.0029	0.0005	−0.0001	0.0087	−0.0183
03/04	0.0018	0.0177	−0.0027	0.0042	0.0015	−0.0158	−0.0006	−0.0048	0.0051	0.0004	0.0014	0.0011	0.0053	−0.0269
04/05	−0.0037	−0.0052	−0.0068	0.0030	0.0001	−0.0217	−0.0067	−0.0152	0.0728	0.0068	0.0028	0.0031	−0.0029	−0.0355
05/06	−0.0079	−0.0254	−0.0092	0.0013	−0.0013	−0.0226	−0.0111	−0.0216	0.1205	0.0113	0.0035	0.0044	−0.0100	−0.0361
06/07	−0.0116	−0.0444	−0.0107	−0.0008	−0.0026	−0.0207	−0.0144	−0.0260	0.1576	0.0148	0.0038	0.0051	−0.0164	−0.0321
07/08	−0.0140	−0.0537	−0.0127	−0.0011	−0.0031	−0.0244	−0.0173	−0.0309	0.1886	0.0177	0.0046	0.0061	−0.0198	−0.0376

Table 7.25 Two-dimensional predicted values of $\mathbf{R}_{new}^{-1/2}(\mathbf{X}_{new} - \mathbf{E}_{new})\mathbf{C}^{-1/2}$ for the Gauteng province.

	Arsn	AGBH	AtMr	BNRs	BRs	CrJk	CmAs	CmRb	DrgR	InAs	Mrd	PubV	Rape	RAC
01/02	0.0027	−0.0240	0.0160	−0.0127	−0.0029	0.0648	0.0121	0.0338	−0.1288	−0.0118	−0.0071	−0.0072	−0.0049	0.1079
02/03	0.0022	−0.0305	0.0172	−0.0145	−0.0035	0.0715	0.0126	0.0362	−0.1332	−0.0122	−0.0078	−0.0078	−0.0068	0.1193
03/04	0.0015	−0.0364	0.0179	−0.0158	−0.0040	0.0760	0.0125	0.0373	−0.1327	−0.0121	−0.0081	−0.0081	−0.0086	0.1270
04/05	0.0016	−0.0307	0.0158	−0.0137	−0.0034	0.0665	0.0112	0.0329	−0.1186	−0.0108	−0.0071	−0.0071	−0.0071	0.1110
05/06	0.0004	−0.0317	0.0133	−0.0124	−0.0033	0.0583	0.0089	0.0275	−0.0933	−0.0085	−0.0061	−0.0060	−0.0079	0.0975
06/07	0.0000	−0.0378	0.0148	−0.0142	−0.0039	0.0657	0.0095	0.0304	−0.1004	−0.0091	−0.0068	−0.0066	−0.0096	0.1100
07/08	−0.0001	−0.0360	0.0138	−0.0134	−0.0037	0.0616	0.0089	0.0283	−0.0931	−0.0084	−0.0064	−0.0062	−0.0092	0.1032

Table 7.26 Interpolated row predictivities for the Gauteng and Western Cape provinces.

Period	2001/02	2002/03	2003/04	2004/05	2005/06	2006/07	2007/08
Gaut	0.9224	0.9188	0.9198	0.8796	0.9193	0.9733	0.9844
WCpe	0.2437	0.1410	0.1179	0.4892	0.8473	0.9599	0.9875

Table 7.27 Crimes reported in the Western Cape province for period from 2001/02 to 2007/08.

	Arsn	AGBH	AtMr	BNRs	BRs	CrJk	CmAs	CmRb	DrgR	InAs	Mrd	PubV	Rape	RAC
01/02	921	36482	3979	15886	54021	793	47752	14400	13429	1782	3447	248	5115	12300
02/03	960	36122	4843	13197	57399	965	51677	16889	13813	1985	3664	269	4819	14311
03/04	967	36912	3633	11784	54069	1015	52339	14943	19940	2032	2839	268	4664	13855
04/05	720	33869	2490	8950	46977	901	48739	13283	30432	2296	2680	285	5126	13143
05/06	595	28479	1856	7944	41000	965	38226	9387	34788	2236	2750	308	4589	12945
06/07	625	25905	2046	10118	43142	911	35083	8697	41067	2054	2881	406	4217	15226
07/08	629	24915	1844	10639	42376	923	32663	8578	45985	1850	2836	257	4000	14555

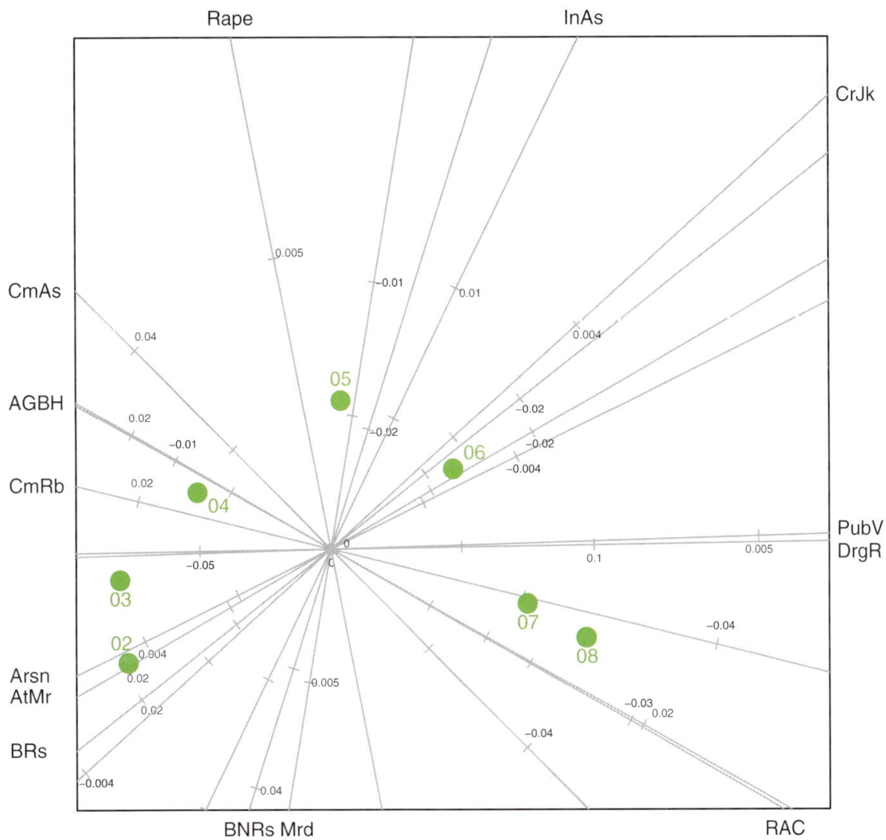

Figure 7.25 Pearson residuals case A CA biplot of the crime data for the Western Cape province for the period from 2001/02 to 2007/08.

Table 7.28 Column predictivities calculated from the weighted deviation form $\mathbf{R}^{-1/2}(\mathbf{X}-\mathbf{E})\mathbf{C}^{-1/2}$ of the crime data for the Western Cape province for the period from 2001/02 to 2007/08.

	Arsn	AGBH	AtMr	BNRs	BRs	CrJk	CmAs	CmRb	DrgR	InAs	Mrd	PubV	Rape	RAC
Dim_1	0.8644	0.8624	0.8710	0.2914	0.8514	0.4878	0.8062	0.9426	0.9997	0.4149	0.0674	0.4635	0.0251	0.7640
Dim_2	0.9129	0.9277	0.9376	0.9472	0.9733	0.5797	0.9958	0.9559	0.9997	0.8162	0.6247	0.4636	0.1783	0.8199
Dim_3	0.9189	0.9950	0.9966	0.9963	0.9960	0.6960	0.9966	0.9773	0.9997	0.8199	0.6440	0.4779	0.6226	0.9566
Dim_4	0.9416	0.9965	0.9967	0.9989	0.9960	0.7438	0.9979	0.9894	1.0000	0.9893	0.8264	0.8211	0.8166	0.9790
Dim_5	0.9877	0.9971	0.9998	0.9990	0.9978	0.7786	0.9990	0.9994	1.0000	1.0000	0.9929	0.8836	0.9846	0.9974
Dim_6	1.0000	1.0000	1.0000	1.0000	1.0000	1.0000	1.0000	1.0000	1.0000	1.0000	1.0000	1.0000	1.0000	1.0000

Table 7.29 Row predictivities calculated from the weighted deviation form $\mathbf{R}^{-1/2}(\mathbf{X}-\mathbf{E})\mathbf{C}^{-1/2}$ of the crime data for the Western Cape province for the period from 2001/02 to 2007/08.

	01/02	02/03	03/04	04/05	05/06	06/07	07/08	01/02
Dim_1	0.9096	0.9657	0.9411	0.0177	0.8879	0.9736	0.9688	0.9096
Dim_2	0.9746	0.9704	0.9797	0.9513	0.9746	0.9908	0.9955	0.9746
Dim_3	0.9997	0.9982	0.9806	0.9689	0.9799	0.9934	0.9956	0.9997
Dim_4	0.9999	0.9984	0.9843	0.9820	0.9960	0.9967	0.9995	0.9999
Dim_5	1.0000	1.0000	0.9993	0.9917	0.9967	0.9986	0.9998	1.0000
Dim_6	1.0000	1.0000	1.0000	1.0000	1.0000	1.0000	1.0000	1.0000

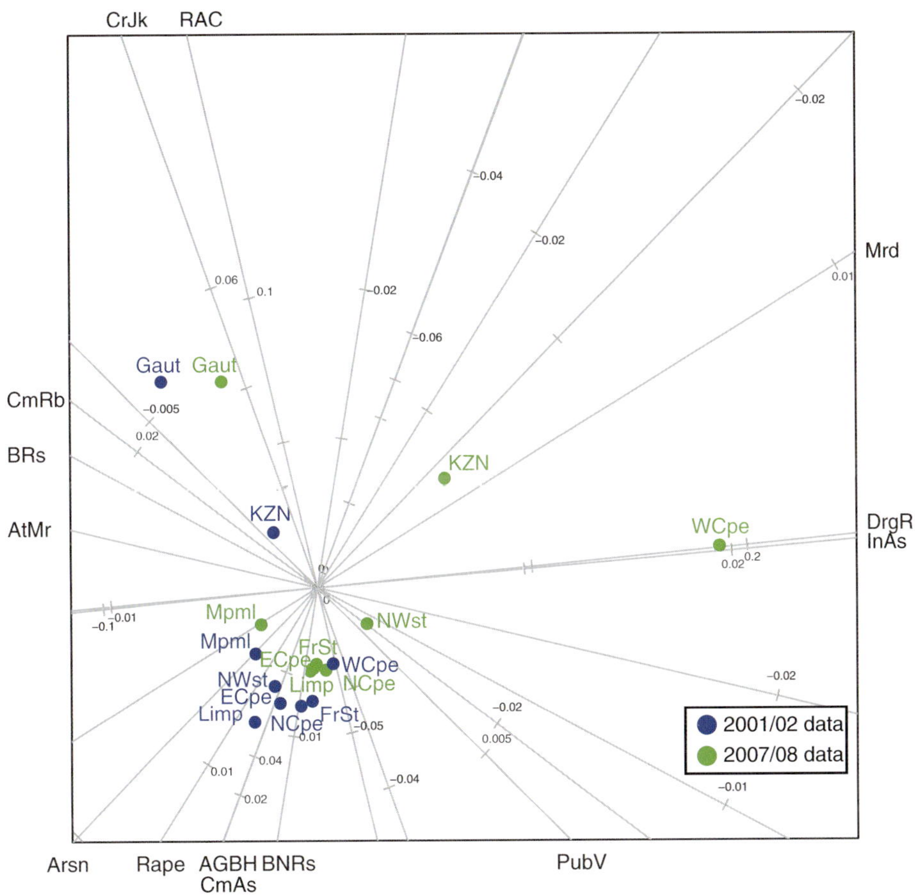

Figure 7.26 Pearson residuals case A CA biplot of the concatenated 2001/02 and 2007/08 crime data.

Table 7.30 Quality of CA biplot display of concatenated 2001/02 and 2007/08 crime data.

	Dim1	Dim2	Dim3	Dim4	Dim5	Dim6	Dim7	Dim8	Dim9	Dim10	Dim11	Dim12
Quality	48.6	84.7	91.4	94.7	96.8	98.0	98.9	99.4	99.7	99.9	99.9	100.0

Table 7.31 Column predictivities of the CA biplot display of concatenated 2001/02 and 2007/08 crime data.

	Arsn	AGBH	AtMr	BNRs	BRs	CrJk	CmAs	CmRb	DrgR	InAs	Mrd	PubV	Rape	RAC
Dim_1	0.146	0.116	0.175	0.015	0.038	0.123	0.042	0.494	0.990	0.640	0.024	0.232	0.113	0.060
Dim_2	0.271	0.784	0.183	0.455	0.047	0.949	0.282	0.729	0.998	0.644	0.031	0.423	0.350	0.982
Dim_3	0.661	0.947	0.270	0.545	0.081	0.952	0.877	0.767	0.999	0.684	0.343	0.423	0.611	0.987
Dim_4	0.661	0.987	0.369	0.581	0.730	0.970	0.948	0.802	1.000	0.699	0.585	0.435	0.629	0.997
Dim_5	0.693	0.993	0.888	0.608	0.956	0.971	0.979	0.812	1.000	0.701	0.799	0.517	0.632	0.997
Dim_6	0.723	0.995	0.889	0.977	0.965	0.972	0.994	0.818	1.000	0.753	0.898	0.533	0.724	0.997
Dim_7	0.879	1.000	0.979	0.978	0.965	0.976	0.999	0.852	1.000	0.757	0.972	0.663	0.968	0.997
Dim_8	0.886	1.000	0.991	0.979	1.000	0.983	1.000	0.969	1.000	0.821	0.987	0.684	0.968	0.997
Dim_9	0.890	1.000	0.992	0.989	1.000	0.994	1.000	0.997	1.000	0.891	0.989	0.783	0.975	1.000
Dim_10	0.953	1.000	0.995	0.995	1.000	0.998	1.000	0.999	1.000	0.991	0.994	0.820	0.977	1.000
Dim_11	0.983	1.000	1.000	0.999	1.000	1.000	1.000	1.000	1.000	0.992	0.998	0.863	0.995	1.000
Dim_12	0.987	1.000	1.000	0.999	1.000	1.000	1.000	1.000	1.000	0.992	0.998	1.000	1.000	1.000
Dim_13	1.000	1.000	1.000	1.000	1.000	1.000	1.000	1.000	1.000	1.000	1.000	1.000	1.000	1.000

the change `ca.variant="Chisq2Cols"` in the above calls. Both figures show two clusters of attributes, (***1***,***2***,***5***) and (***3***,***4***,***6***), indicating similar column profiles within these clusters. Thus, the doubling process has preserved this information, but Figure 7.34 clearly shows the bipolar structure of the variables. Recall that the points representing the positive and negative forms of an attribute are at the extremities of a line through the mean which divides its two ends in the ratio $c_k : Np - c_k$. It can be seen that this ratio is close to 0.5 for all attributes, so there is little polarization with this data set. Had some of the values been closer to zero or unity, then the column chi-squared analysis before and after doubling would have been more different. Note also that the ***plus*** pole is closer to the point of intersection of the lines in the case of attributes ***1*** and ***5***, whereas the ***min*** pole is closer to the intersection point in all other cases. This shows that if votes are aggregated for the four products only attributes ***1*** and ***5*** receive a majority of positive votes.

Doubling is an ingenious idea but a final comment might be that it underlines that chi-squared distance is an inadequate measure when absolute values rather than ratios are important.

7.7 Conclusion

The previous section exemplifies the versatility of the R code. In practice the main uses of correspondence analysis are concerned with approximating chi-squared distance and, to a lesser extent, with approximating the Pearson residuals. We recommend the latter but admit that in not being overenthusiastic about chi-squared distance we are in a minority.

Table 7.32 Row predictivities of the CA biplot display of concatenated 2001/02 and 2007/08 crime data.

	2001/02									2007/08								
Dim	Ecpe	FrSt	Gaut	KZN	Limp	Mpml	NWst	NCpe	WCpe	ECpe	FrSt	Gaut	KZN	Limp	Mpml	NWst	NCpe	WCpe
1	0.097	0.002	0.400	0.166	0.186	0.357	0.131	0.020	0.014	0.002	0.000	0.190	0.514	0.008	0.518	0.375	0.007	0.986
2	0.822	0.578	0.981	0.381	0.867	0.684	0.933	0.778	0.309	0.372	0.588	0.922	0.823	0.717	0.695	0.545	0.661	0.994
3	0.924	0.938	0.983	0.406	0.877	0.754	0.934	0.793	0.766	0.841	0.799	0.934	0.915	0.718	0.736	0.749	0.821	0.996
4	0.936	0.949	0.983	0.719	0.968	0.798	0.938	0.829	0.967	0.846	0.915	0.988	0.921	0.720	0.781	0.760	0.921	0.996
5	0.936	0.963	0.988	0.966	0.979	0.884	0.938	0.895	0.969	0.874	0.915	0.990	0.984	0.726	0.971	0.839	0.924	0.999
6	0.965	0.968	0.988	0.972	0.988	0.894	0.969	0.895	0.972	0.952	0.964	0.992	0.984	0.944	0.971	0.949	0.926	0.999
7	0.973	0.980	0.994	0.981	0.994	0.897	0.976	0.996	0.974	0.990	0.992	0.992	0.988	0.974	0.971	0.961	0.977	1.000
8	0.977	0.995	0.994	0.982	0.996	0.987	0.986	0.996	0.981	0.995	0.992	0.994	0.994	0.991	0.986	0.970	0.978	1.000
9	0.984	0.998	1.000	0.986	0.996	0.998	0.992	0.998	0.988	0.996	0.993	0.999	0.994	0.996	0.986	0.985	0.989	1.000
10	0.991	0.998	1.000	0.997	0.996	0.998	0.994	0.998	0.998	0.999	0.994	1.000	1.000	0.999	0.991	0.989	0.996	1.000
11	0.999	0.999	1.000	0.999	0.998	0.999	0.997	0.998	1.000	1.000	1.000	1.000	1.000	1.000	0.995	0.990	0.998	1.000
12	1.000	1.000	1.000	1.000	0.999	0.999	1.000	0.999	1.000	1.000	1.000	1.000	1.000	1.000	0.995	1.000	0.999	1.000
13	1.000	1.000	1.000	1.000	1.000	1.000	1.000	1.000	1.000	1.000	1.000	1.000	1.000	1.000	1.000	1.000	1.000	1.000

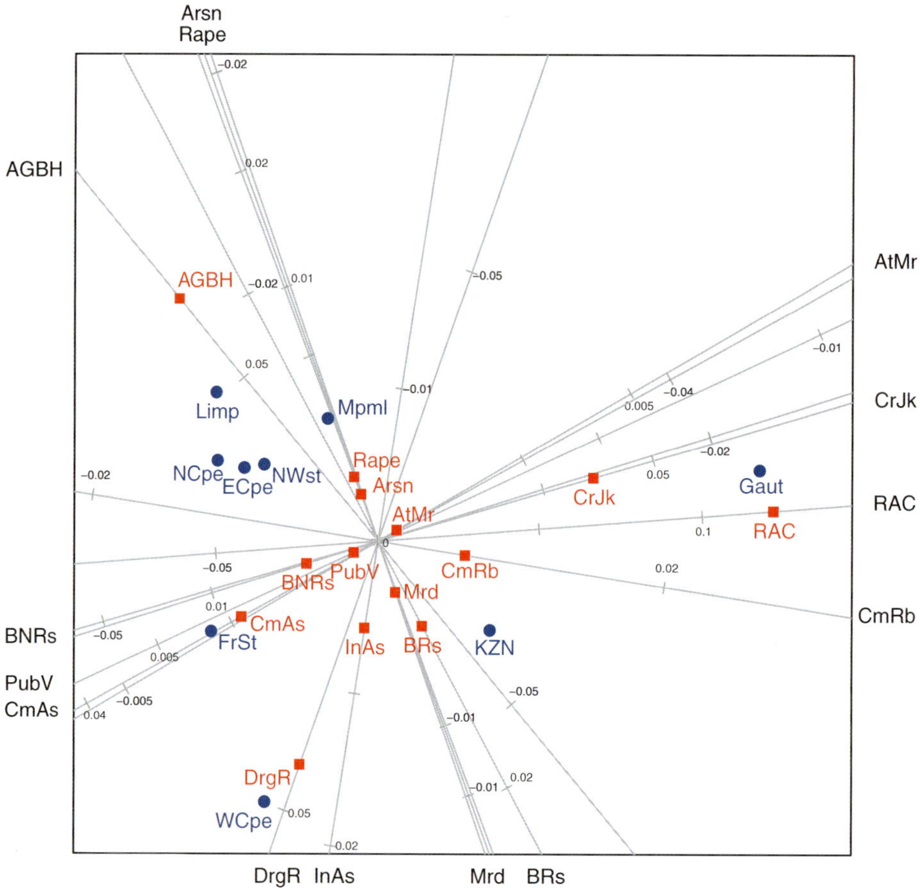

Figure 7.27 Pearson residuals case A CA biplot of the 2001/02 crime data.

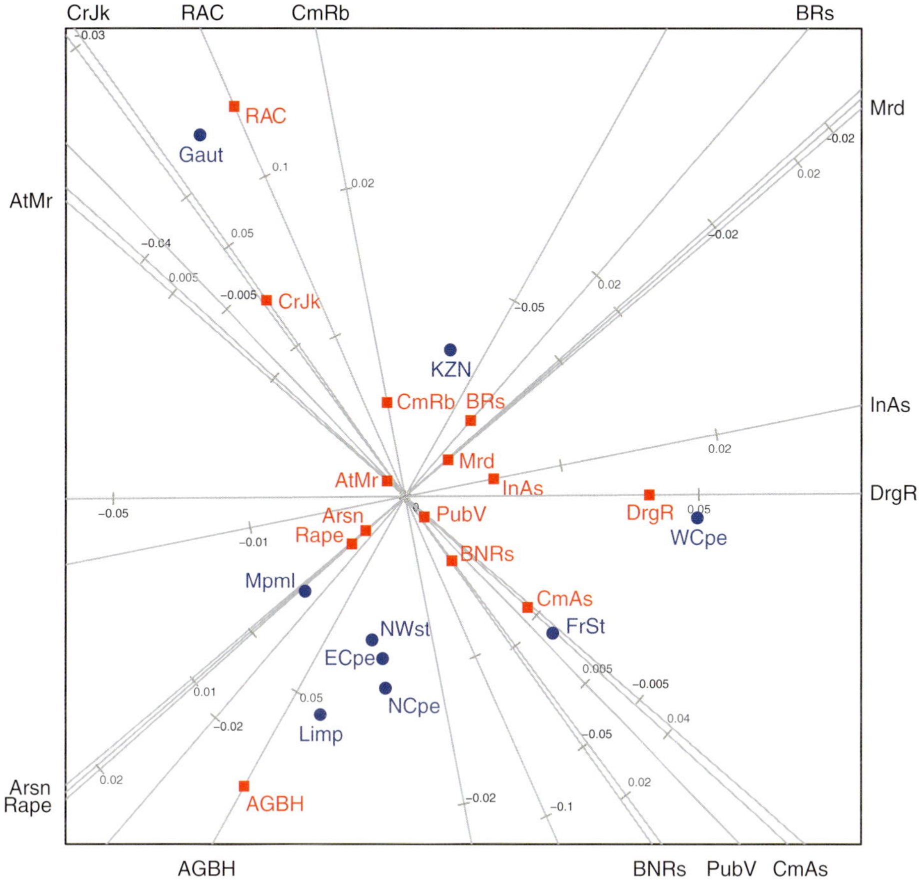

Figure 7.28 The Pearson residuals case A CA biplot of the 2001/02 crime data shown in Figure 7.27, but reflected about the "y" axis and rotated 70 degrees clockwise.

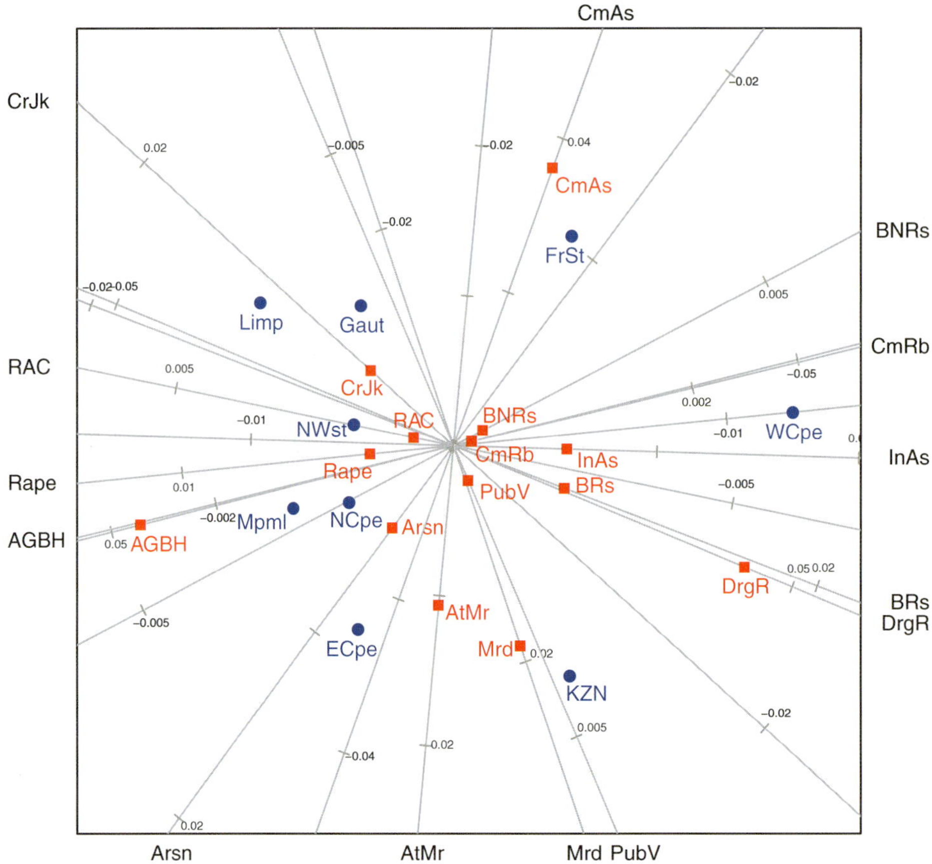

Figure 7.29 Pearson residuals case A CA biplot of the 2001/02 crime data using the second and third eigenvectors as scaffolding.

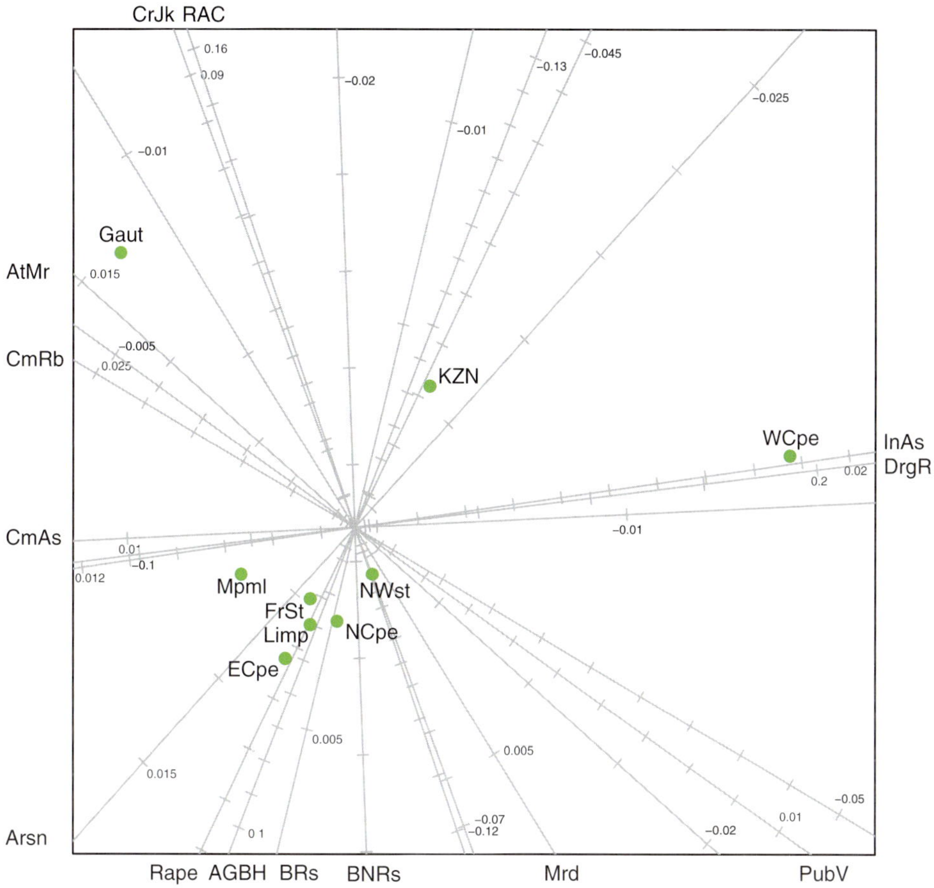

Figure 7.30 Ordinary PCA biplot of $\mathbf{R}^{-1/2}(\mathbf{X}-\mathbf{E})\mathbf{C}^{-1/2}$ for the 2007/08 crime data.

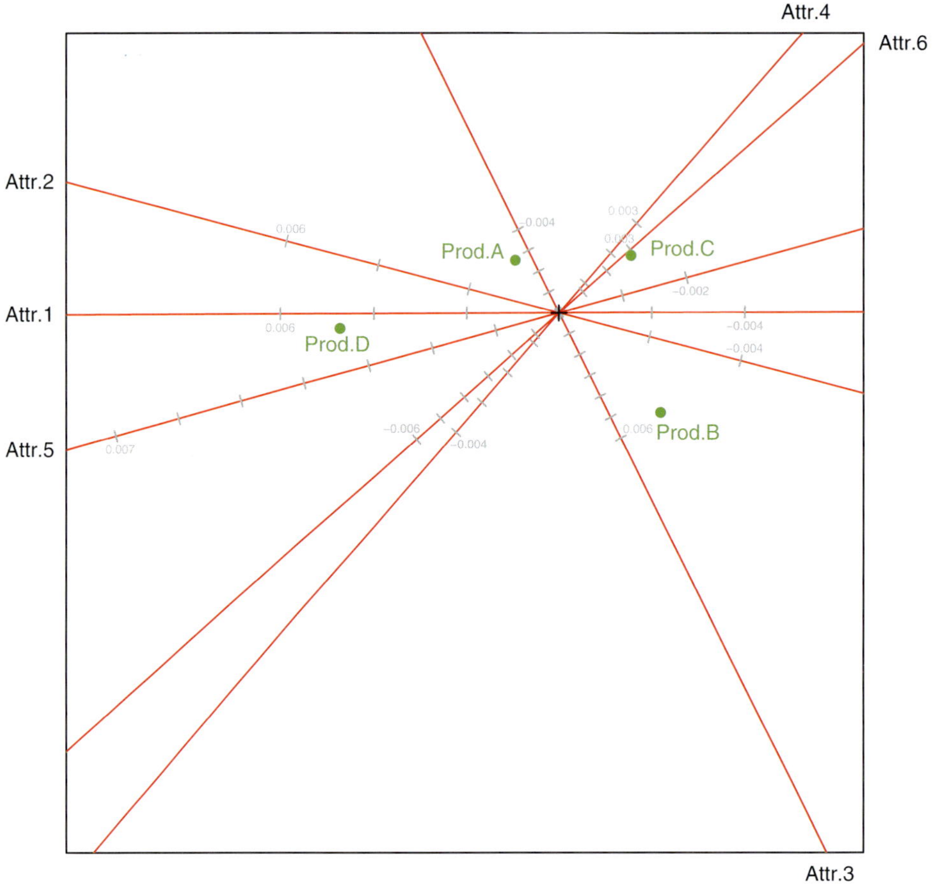

Figure 7.31 Plot of $\mathbf{R}^{-1/2}\mathbf{U}\mathbf{\Sigma}$ generating row chi-squared distances for Table 7.2 before doubling. No lambda-scaling, but using the zoom utility of `cabipl` to zoom into positions occupied by row points.

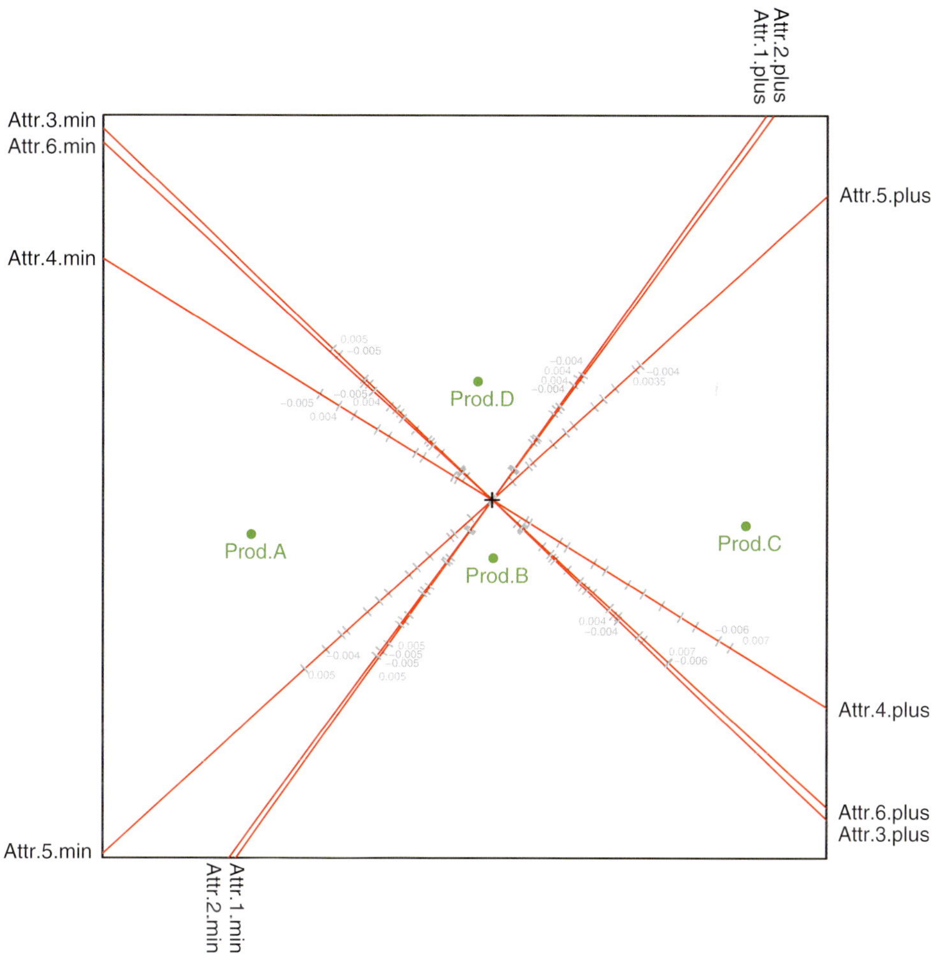

Figure 7.32 Similar to Figure 7.31 but after doubling.

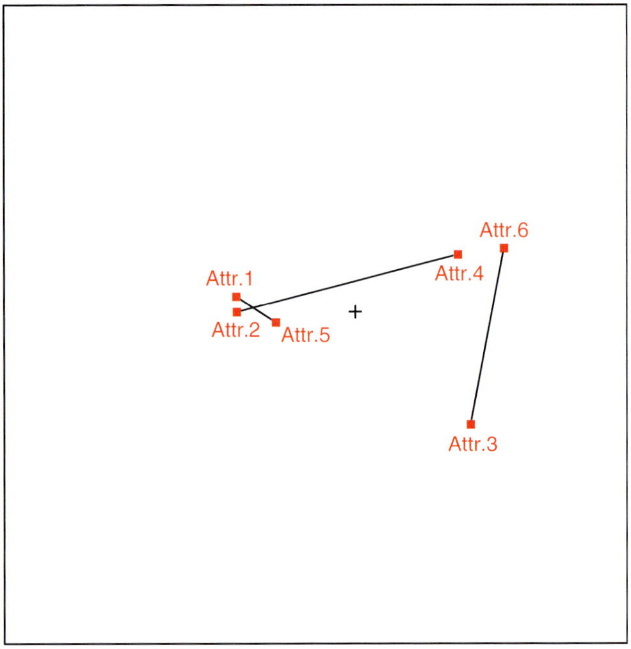

Figure 7.33 Plot of $\mathbf{C}^{-1/2}\mathbf{V}\boldsymbol{\Sigma}$ generating column chi-squared distances for the data of Table 7.2 before doubling. No lambda-scaling but using zoom utility to zoom into positions occupied by column points.

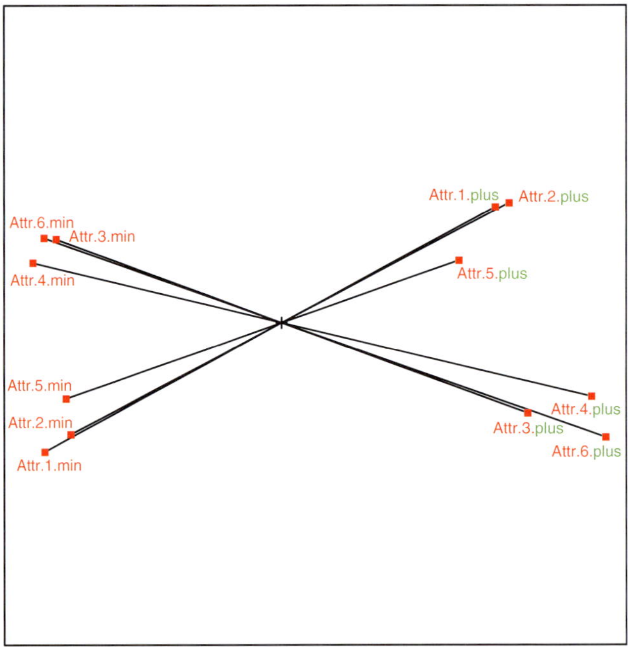

Figure 7.34 Similar to Figure 7.33 but after doubling and with connecting lines added to show the bipolar structure.

Chi-squared distance is not a symmetric approach, requiring separate maps for the row and column approximations. These maps are often superimposed by taking one set of points from each map, to give a single map, but because neither row-column distances nor inner products can be readily interpreted, the justification for this practice is not clear. The Pearson residuals have a simple interpretation in terms of illustrating the extent of departures from the independence model that is analogous to examining the interaction term in biadditive models. We also differ from much of the correspondence analysis literature in our attitude to underlying models. It is often stressed that correspondence analysis is not model based, but we believe that it is only by examining underlying models that it is possible to appreciate the different variants and how to interpret them. As we have mentioned, and is evident from the above examples, there is considerable agreement between the maps produced by the different variants, so at some level, all may be regarded as giving similar information. Nevertheless, it is good to understand what are the underlying approximations.

Whichever method is used, there remains the question of its graphical presentation, which we have seen may be in terms of two sets of points, two sets of axes or one of each. Distances are best appreciated between pairs of points, inner products are best appreciated by projection onto calibrated axes. However, two sets of calibrated axes are confusing and, therefore, usually unacceptable. We may also show points on calibrated axes, but these too should be used with circumspection. In general we prefer one set of points and one set of axes. This allows distance interpretations for the points together with inner product interpretations. There is an element of asymmetry when used with symmetric methods, such as the approximation of Pearson residuals, because the axes have to be chosen to represent either the rows or the columns. Note that this is an asymmetry in presentation, not in the analysis. Sometimes two sets of points are acceptable, as when representing the contingency ratio and using the centroid interpretation; centroids require points.

We have presented λ-scaling as a method for improving presentation when one set of points is concentrated near the origin, and the other set is more scattered. This becomes unnecessary when the concentrated set is replaced by calibrated axes. The situation is analogous to presenting a set of scales in millimetres when the axes are better calibrated in metres. Nevertheless, λ-scaling can be useful but it cannot be used with centroid interpretations and in its general form, which allows different scaling for different dimensions, it will destroy distance interpretations. The simple form of λ-scaling used in the above examples gives harmless isotropic scaling.

8 Multiple correspondence analysis

8.1 Introduction

Biplots for the correspondence analysis of a two-way contingency table were discussed in Chapter 7. Here we shall examine biplots for more than two categorical variables. There are two areas of confusion. The first is analogous to the difficulty we have already met (Chapter 3) of whether, in principal component analysis of a data matrix $\mathbf{X}$, we are concerned with the analysis of $\mathbf{X}$ itself or with $\mathbf{X}'\mathbf{X}$; the second is the relationship with the correspondence analysis of a two-way contingency table. All the techniques we shall discuss share much common mathematics, which may be regarded as a valuable unifying feature, although Gower (2006) referred to them as 'divided by a common language' because important statistical differences tend to become blurred.

The analogy with PCA is very close because, in place of an $n \times p$ data matrix $\mathbf{X}$ we have an $n \times p$ matrix whose columns give the category levels taken by p categorical variables for each of the n samples. Thus a categorical variable ***Hair Colour*** may have category levels *Dark, Grey, Fair* and *Brown*. Table 8.1 is an example of a small data matrix of categorical variables.

Such nominal data is usually coded in pseudo-numerical form where the kth variable is recorded in an $n \times L_k$ matrix $\mathbf{G}_k$ and L_k is the number of category levels ($L_k = 4$ in the case of hair colour). The ith row of $\mathbf{G}_k$ is zero, apart from a single unit in the column pertaining to the actual category level taken by the ith sample. The column sums of $\mathbf{G}_k$ give the frequencies of each category level in the n samples and will be denoted by $\mathbf{L}_k$, which will be considered as a diagonal matrix. Thus $\mathbf{G}_k\mathbf{1} = \mathbf{1}$ and $\mathbf{1}'\mathbf{G}_k = \mathbf{1}'\mathbf{L}_k$; also $\mathbf{1}'\mathbf{L}_k\mathbf{1} = n$ as every sample must take some category level; $\mathbf{G}_k$ is termed an indicator matrix. The indicator matrix $\mathbf{G}$ for the complete data is obtained by combining the

Understanding Biplots John Gower, Sugnet Lubbe and Niël le Roux

Table 8.1 A simple data matrix for information on five categorical variables for seven individuals.

Case	Sex	Hair Colour	Region	Work	Education
1 George	M	Brown	England	Manual	School
2 Alisdair	M	Dark	Scotland	Clerical	University
3 Jane	F	Brown	Scotland	Professional	University
4 Ivor	M	Grey	Wales	Professional	University
5 Myfanwy	F	Fair	Wales	Clerical	School
6 Harriet	F	Brown	England	Manual	School
7 Jeremy	M	Grey	England	Professional	Postgrad

Table 8.2 Recoding of Table 8.1 as an indicator matrix **G**. Here $\mathbf{G}_1$ has two levels (M, F), $\mathbf{G}_2$ has four levels (B, D, F, G), $\mathbf{G}_3$ has three levels (E, S, W), $\mathbf{G}_4$ has three levels (M, C, P) and $\mathbf{G}_5$ has three levels (S, U, P). The frequencies $\mathbf{1}'\mathbf{L}_1, \mathbf{1}'\mathbf{L}_2, \mathbf{1}'\mathbf{L}_3, \mathbf{1}'\mathbf{L}_4$ and $\mathbf{1}'\mathbf{L}_5$ are given in the final row.

Case	Sex		Hair Colour				Region			Work			Education		
	M	F	B	D	F	G	E	S	W	M	C	P	S	U	P
1 George	1	0	1	0	0	0	1	0	0	1	0	0	1	0	0
2 Alisdair	1	0	0	1	0	0	0	1	0	0	1	0	0	1	0
3 Jane	0	1	1	0	0	0	0	1	0	0	0	1	0	1	0
4 Ivor	1	0	0	0	0	1	0	0	1	0	0	1	0	1	0
5 Myfanwy	0	1	0	0	1	0	0	0	1	0	1	0	1	0	0
6 Harriet	0	1	1	0	0	0	1	0	0	1	0	0	1	0	0
7 Jeremy	1	0	0	0	0	1	1	0	0	0	0	1	0	0	1
Frequencies	4	3	3	1	1	2	3	2	2	2	2	3	3	3	1

indicator matrices for all categorical variables to give

$$\mathbf{G} = [\mathbf{G}_1, \mathbf{G}_2, \mathbf{G}_3, \ldots, \mathbf{G}_p]\colon n \times L, \quad \text{where } L = L_1 + L_2 + \ldots + L_p.$$

Table 8.2 shows Table 8.1 coded as an indicator matrix. Thus **G**, consisting entirely of 0s and 1s, is the categorical equivalent of the quantitative data matrix **X** of PCA. Because every categorical variable has one level for every sample, we have that the rows of **G** all sum to p. Further, the column sums give the frequencies of all the category levels assumed to be held in an $L \times L$ diagonal matrix $\mathbf{L} = \text{diag}(\text{diag}(\mathbf{L}_1), \text{diag}(\mathbf{L}_2), \ldots, \text{diag}(\mathbf{L}_p))$. Hence, $\mathbf{G1} = p\mathbf{1}, \mathbf{1}'\mathbf{G} = \mathbf{1}'\mathbf{L}$ and $\mathbf{1}'\mathbf{L1} = np$.

8.2 Multiple correspondence analysis of the indicator matrix

One way of generalizing CA is to treat the categorical data matrix **G** as if it were a two-way contingency table. This compares with the CA of chi-squared distance where we saw in Chapter 7 that the two-way contingency table is sometimes treated as if it were a data matrix where either the rows or the columns are treated as if they were variables.

Corresponding to $\mathbf{R}^{-1/2}\mathbf{X}\mathbf{C}^{-1/2}$ for the correspondence analysis of a contingency table $\mathbf{X}$, for the analysis of $\mathbf{G}$ we have $\mathbf{R} = p\mathbf{I}$ and $\mathbf{C} = \mathbf{L}$, giving

$$p^{-1/2}\mathbf{G}\mathbf{L}^{-1/2} = \mathbf{U}\boldsymbol{\Sigma}\mathbf{V}'. \tag{8.1}$$

The factor $p^{-1/2}$ is superfluous, but we retain it below to maintain the link with simple CA. As in Chapter 7, the first singular vectors, corresponding to a unit singular value, may be ignored, being equivalent to working in deviations from the column means. These vectors are $\mathbf{1}$ and $\mathbf{L}^{1/2}\mathbf{1}$ which, through the orthogonality properties of singular vectors, imply that the remaining singular vectors satisfy

$$\mathbf{1}'\mathbf{U} = \mathbf{0}' \quad \text{and} \quad \mathbf{1}'\mathbf{L}^{1/2}\mathbf{V} = \mathbf{0}'. \tag{8.2}$$

As in Chapter 7 there are several variants that might be plotted, as summarized in Table 7.1. These remain available, but the usual choice for MCA is that for chi-squared distance. We saw in Chapter 7 that this is equivalent to a weighted PCA but the row weights are now all equal ($\mathbf{R} = p\mathbf{I}_n$), leading to simplifications, where for approximating row chi-squared distance we plot

$$\mathbf{Z}_0 = \mathbf{U}\boldsymbol{\Sigma}. \tag{8.3}$$

In CA itself we would normally include in the same diagram, approximations to the column chi-squared distances but, for reasons discussed below, here we represent the columns by the projected category-level points (CLPs; see Section 8.3):

$$\mathbf{Z} = p^{-1/2}\mathbf{L}^{-1/2}\mathbf{V}. \tag{8.4}$$

In these expressions, the singular value of unity together with the first singular vectors (8.2) are excluded, so $\mathbf{U}$ and $\mathbf{V}$ refer to the second and subsequent columns of the SVD of (8.1). As a simple example of an MCA plot let us construct the above row chi-squared distance MCA plot for the data in Table 8.2 by issuing the following function call:

```
MCAbipl(X = MCA.Table.1.data[,-1], e.vects = 1:2, mca.variant =
   "indicator", column.points.size = 1.2, row.points.size = 1.2,
   pch.col.points = 15, pch.row.points = 16, offset.m =
   list(rep(0.4, 20),rep(0.4, 20)), pos.m=list(c(4,4,2,4,4,4,4),
   rep(4,15)), row.points.col = "green", column.points.col =
   c("red","red","blue","blue","blue","blue","pink","pink",
   "pink","brown","brown","brown","black","black","black"))
```

This call provides the MCA representations in Figures 8.1–8.3.

Inspection of Figure 8.1 shows that the weighted centroid of each set of category points is at the centroid of the row points. Furthermore, each row point is at the vector-sum of the CLPs referring to its categories. Thus, ***George*** is at the vector-sum of *Male, Brown, England, Manual*, and *School* (see Figure 8.2). These results are generally true and may be summarized as follows:

$$\begin{aligned} \mathbf{Z}_0 &= \mathbf{G}\mathbf{Z}, \\ \mathbf{1}'\mathbf{L}_k^*\mathbf{Z}_k &= 0, \\ \mathbf{1}'\mathbf{Z}_0 &= 0. \end{aligned} \tag{8.5}$$

Figure 8.1 Row chi-squared MCA biplot of the data in Table 8.2. First the SVD of $p^{-1/2}\mathbf{GL}^{-1/2}$ is performed. The row points (the green filled circles) are then plotted using the first two columns (after discarding the column associated with the singular value of unity) of $\mathbf{Z}_0 = \mathbf{U\Sigma}$ and the column points (the filled squares) are plotted as the projected positions of the CLPs i.e. the first two columns of $\mathbf{Z} = p^{-1/2}\mathbf{L}^{-1/2}\mathbf{V}$. The column points are colour-coded such that the categories of any particular categorical variable appear in the same colour. The quality of the display is 56.79%, which on adding a third dimension increases to 77.50%, suggesting that a three-dimensional plot may be worthwhile.

Here, $\mathbf{L}_k^*$ is of size $L \times L$ with only the $L_k \times L_k$ diagonal block $\mathbf{L}_k$ nonzero; for computational purposes the zero parts of $\mathbf{L}_k^*$ would be ignored. The first of the expressions (8.5) follows immediately from the definition of $\mathbf{Z}$, giving

$$\mathbf{GZ} = (p^{-1/2}\mathbf{GL}^{-1/2})\mathbf{V} = (\mathbf{U\Sigma V}')\mathbf{V} = \mathbf{U\Sigma} = \mathbf{Z}_0.$$

The interpretation is that every row point is given by the vector-sum of the relevant CLPs, as is consistent with the vector-sum method of interpolation.

The second expression in (8.5) is an extension of the orthogonality results. We have

$$(\mathbf{U\Sigma V}')\mathbf{L}^{1/2}\mathbf{1}_k^* = (p^{-1/2}\mathbf{GL}^{-1/2})\mathbf{L}^{1/2}\mathbf{1}_k^* = p^{-1/2}\mathbf{G1}_k^* = p^{-1/2}\mathbf{1},$$

where $\mathbf{1}_k^*$ denotes a vector of full size (L) with units in those parts pertaining to the kth categorical variable, else zero. Premultiplication by $\mathbf{\Sigma}^{-1}\mathbf{U}'$ and using the orthogonality

Figure 8.2 Demonstrating the vector-sum method for ***George*** by using the function `vectorsum.interp`.

relationship (8.2) gives $\mathbf{V}'(\mathbf{L}_k^*)^{1/2}\mathbf{1}_k^* = \mathbf{0}$. Because $\mathbf{Z}_k = p^{-1/2}(\mathbf{L}_k^*)^{-1/2}\mathbf{V}$, it follows that $\mathbf{1}'\mathbf{L}_k^*\mathbf{Z}_k = p^{-1/2}\mathbf{1}'(\mathbf{L}_k^*)^{1/2}\mathbf{V} = \mathbf{0}'$ as required.

The third expression of (8.5) follows from $\mathbf{1}'\mathbf{Z}_0 = \mathbf{1}'\mathbf{GZ} = \mathbf{1}'\mathbf{LZ}$ which, on summing the second expression over all variables, is verified to be zero.

In Figure 8.1 the row points tend to occupy the peripheral positions, so we might like to find a more balanced presentation. The graphical representation of $\mathbf{Z}_0$ gives a visual approximation of the row chi-squared distances (see below). We have not provided for inner product interpretations which would require a plot of $\mathbf{V}$ and offer approximations to $\mathbf{G}$ or $p^{-1/2}\mathbf{GL}^{-1/2}$, both of which are uninteresting. For these reasons, using lambda-scaling to improve the figure is not an acceptable option. Furthermore, lambda-scaling would destroy the vector-sum properties of (8.5). The remaining possibility is to scale the row points isotropically relative to the column points. This is acceptable because the relative chi-squared distances are unchanged. A bonus is that if we scale the row points by a factor p the vector-sums become more easily interpreted centroids (see Figure 8.4). There is a further minor advantage as follows. Suppose we ask where the centroids are of all row points with the same categories. For example, where is the centroid of all the

Figure 8.3 The row chi-squared MCA biplot of Figure 8.1 but with row point coordinates reduced to $p^{-1}\mathbf{U\Sigma}$. Columns are plotted as in Figure 8.1 by using $\mathbf{Z} = p^{-1/2}\mathbf{L}^{-1/2}\mathbf{V}$.

English people in Figure 8.1? The coordinates of all such points are given by the rows of

$$\mathbf{L}^{-1}\mathbf{G}'\mathbf{Z}_0, \tag{8.6}$$

which we term the *category centroids*. Here, $\mathbf{G}'\mathbf{Z}_0$ sums the relevant coordinates, while $\mathbf{L}^{-1}$ ensures that each sum is divided by the correct frequency. From (8.3)–(8.5) we have that

$$\begin{aligned}\mathbf{L}^{-1}\mathbf{G}'\mathbf{Z}_0 &= \mathbf{L}^{-1}\mathbf{G}'(p^{-1/2}\mathbf{G}\mathbf{L}^{-1/2}\mathbf{V}) = p^{-1/2}\mathbf{L}^{-1/2}(\mathbf{L}^{-1/2}\mathbf{G}'\mathbf{G}\mathbf{L}^{-1/2})\mathbf{V} \\ &= p^{1/2}\mathbf{L}^{-1/2}(\mathbf{V\Sigma}^2\mathbf{V}')\mathbf{V} = p\mathbf{Z\Sigma}^2. \end{aligned} \tag{8.7}$$

If it were not for the factor $\mathbf{\Sigma}^2$ the category centroids would be at p times the actual category-level points; rescaling would make them coincide. Although the singular values interfere with this nice property, (8.7) encourages us to rescale as is shown in Figure 8.3. It also encourages us to consider replacing the projected CLPs by the category centroids as is shown in Figures 8.5 and 8.6.

If chi-squared distance can be controversial for a contingency table (see Chapter 7), it is even more so for a categorical data matrix $\mathbf{G}$. This is because the only contribution to

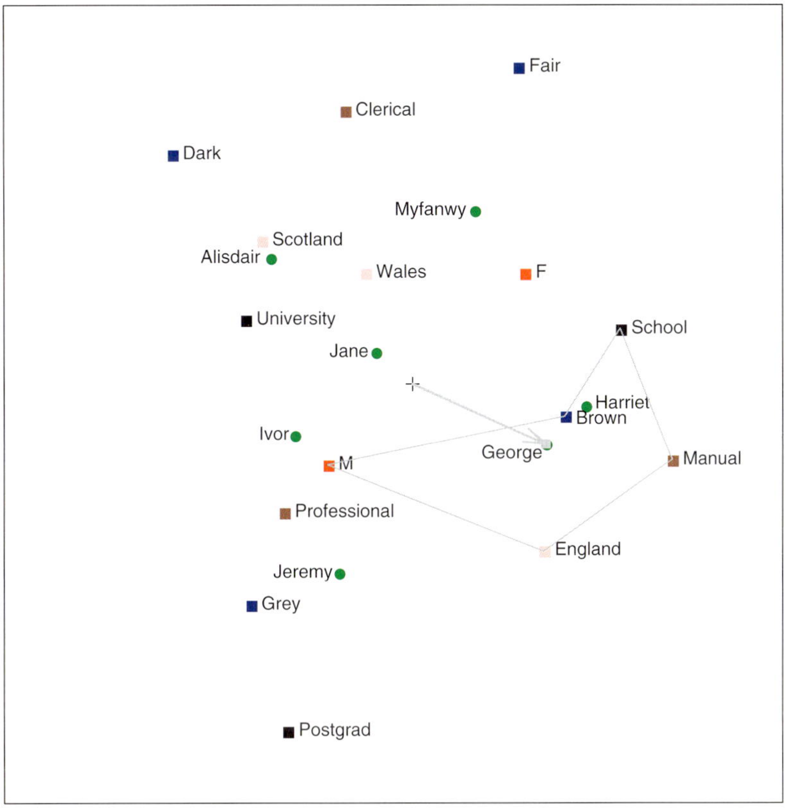

Figure 8.4 Demonstrating the vector-sum method when row point coordinates are reduced to $p^{-1}\mathbf{U\Sigma}$ and columns plotted by using $\mathbf{Z} = p^{-1/2}\mathbf{L}^{-1/2}\mathbf{V}$ as in Figure 8.3.

chi-squared distance between two samples i and i' occurs for those variables that differ. Thus, if the levels of the kth variable differ and the two differing levels have frequencies l_i and $l_{i'}$ then the contribution is

$$\frac{1}{p^2}\left(\frac{1}{l_i} + \frac{1}{l_{i'}}\right), \tag{8.8}$$

depending only on the inverse frequencies of the categories. This is indeed a somewhat strange distance and, at the very least, bears examination in every instance of its use. We discuss other choices of distance below. The interpretation of column chi-squared distance between two category levels j and j' is even harder to justify. This is because we have to reconcile the meaning of distances between levels that refer to the same categorical variable and those that do not (e.g. distance between *grey* and *brown*, interpretable as a mismatch, and distance between *grey* and *university*, which is meaningless). For this reason we do not explore the column chi-squared distance version of MCA. Rather, as we have seen, the columns (variables) are shown as projected CLPs with their useful centroid and interpolation properties. This gives an asymmetric representation as is appropriate for a data matrix.

Figure 8.5 The row chi-squared MCA biplot of Figure 8.1. The row points (the green filled circles) are plotted using the first two columns (after discarding the column associated with the singular value of unity) of $\mathbf{Z}_0 = \mathbf{U\Sigma}$ and the column points (the filled squares) are plotted as the category centroids $\mathbf{L}^{-1}\mathbf{G}'\mathbf{Z}_0$.

8.3 The Burt matrix

Just as PCA may find the SVD of $\mathbf{X}$ by evaluating the spectral decomposition of $\mathbf{X}'\mathbf{X}$, so may the SVD of the adjusted indicator matrix be found by evaluating the spectral decomposition of $p^{-1}\mathbf{L}^{-1/2}\mathbf{G}'\mathbf{GL}^{-1/2}$. Now, $\mathbf{G}'\mathbf{G}$, known as the Burt matrix, is a block matrix each of whose blocks $\{\mathbf{G}_j'\mathbf{G}_{j'}\}$ is the two-way contingency table for the jth and j'th categorical variables. Table 8.3 shows the Burt matrix derived from Tables 8.1 and 8.2. This is a trivial example, but it suffices to show that a Burt matrix is symmetric, has diagonal blocks giving the frequencies of the different categorical variables and has off-diagonal blocks giving the pairwise contingency tables.

Because of the contingency table structure of the Burt matrix, we may wish to approximate $\mathbf{G}'\mathbf{G}$ itself, rather than just use it as a stepping-stone to calculating an SVD. In this way we can have a multivariate extension of CA in which all the $\frac{1}{2}p(p-1)$ contingency tables are simultaneously approximated. Even better, the spectral decomposition of the normalized Burt matrix $\mathbf{L}^{-1/2}\mathbf{G}'\mathbf{GL}^{-1/2}$, given by

$$\mathbf{L}^{-1/2}\mathbf{G}'\mathbf{GL}^{-1/2} = p\mathbf{V\Sigma}^2\mathbf{V}', \tag{8.9}$$

Figure 8.6 Demonstrating the centroid property of Figure 8.3 for the English people.

Table 8.3 The Burt matrix derived from Tables 8.1 and 8.2.

		Sex		Hair Colour				Region			Work			Education		
		M	F	B	D	G	F	E	S	W	M	C	P	S	U	P
Sex	M	4	0	1	1	2	0	2	1	1	1	1	2	1	2	1
	F	0	3	2	0	0	1	1	1	1	1	1	1	2	1	0
Hair Colour	B	1	2	3	0	0	0	2	1	0	2	0	1	2	1	0
	D	1	0	0	1	0	0	0	1	0	0	1	0	0	1	0
	G	2	0	0	0	2	0	1	0	1	0	0	2	0	1	1
	F	0	1	0	0	0	1	0	0	1	0	1	0	1	0	0
Region	E	2	1	2	0	1	0	3	0	0	2	0	1	2	0	1
	S	1	1	1	1	0	0	0	2	0	0	1	1	0	2	0
	W	1	1	0	0	1	1	0	0	2	0	1	1	1	1	0
Work	M	1	1	2	0	0	0	2	0	0	2	0	0	2	0	0
	C	1	1	0	1	0	1	0	1	1	0	2	0	1	1	0
	P	2	1	1	0	2	0	1	1	1	0	0	3	0	2	1
Education	S	1	2	2	0	0	1	2	0	1	2	1	0	3	0	0
	U	2	1	1	1	1	0	0	2	1	0	1	2	0	3	0
	P	1	0	0	0	1	0	1	0	0	0	0	1	0	0	1

gives approximations to the contingency tables scaled, as in CA itself, by the square roots of the inverse row and column frequencies. For this reason, this type of MCA may be deemed more acceptable than the version based on $p^{-1/2}\mathbf{G}\mathbf{L}^{-1/2}$. Note that the diagonal blocks of the normalized Burt matrix are units, the off-diagonal blocks being the scaled two-way contingency tables (and their transposes) of CA.

The least-squares approximation to the symmetric normalized Burt matrix is given by $\mathbf{V}\mathbf{\Sigma}^2\mathbf{J}\mathbf{V}'$, which would indicate a plot of $\mathbf{V}\mathbf{\Sigma}\mathbf{J}$. This is a monoplot (Chapter 10) in which the pairwise inner products of the L plotted points give approximations to the normalized Burt matrix. Alternatively, we may regard $\mathbf{G}'\mathbf{G}$ as a giant two-way contingency table and use any of the approximations discussed in Chapter 7. Focusing on approximating chi-squared distance, we note that, corresponding to the argument leading to (7.17),

$$\mathbf{L}^{-1}\mathbf{G}'\mathbf{G}\mathbf{L}^{-1/2} \tag{8.10}$$

generates all the row chi-squared distances arising from the Burt matrix. This bland statement merits closer attention. Firstly, we note that (8.10) includes contributions from the unit diagonal matrices (see below). Secondly, each contingency table enters twice, first as $\mathbf{L}_i^{-1}\mathbf{G}_i'\mathbf{G}_j\mathbf{L}_j^{-1/2}$ and then in transposed form $\mathbf{L}_j^{-1}\mathbf{G}_j'\mathbf{G}_i\mathbf{L}_i^{-1/2}$, the second of which generates row chi-squared distances that are column chi-squared distances of the first. Thus (8.10) involves not only all the row chi-squared distances but also all the column chi-squared distances of the two-way contingency tables. Apart from the diagonal block, calculating chi-squared distances between different levels of the same variable generated by (8.10) gives the sum of the $p-1$ different estimates (one from each of the contingency tables involving that variable as a row classifier) and is an acceptable estimate. However, evaluating chi-squared distances between levels of different variables generated by (8.10) is as hard to justify as is calculating chi-squared distances between rows and columns of a two-way contingency table. With this background, as for chi-squared distance CA (Section 7.2.4), we plot

$$\mathbf{Z} = \mathbf{L}^{-1/2}\mathbf{V}\mathbf{\Sigma}^2\mathbf{J}. \tag{8.11}$$

Equation (8.11) gives p sets of CLPs but no representation of the n units, as is in accord with the common practice of CA. We may plot the units, as above, at the centroids of the their category points:

$$\mathbf{Z}_0 = \mathbf{G}\mathbf{Z}/p. \tag{8.12}$$

Although the chi-squared distances based on the Burt matrix are functions of those of the correspondence analysis of a contingency table, they differ from the row chi-squared distances discussed above (8.10) used in the analysis of $\mathbf{G}$. They differ yet again from the column chi-squared distances for $\mathbf{G}$, which, if they were used, would measure distances between as well as within the levels of the p categorical variables (see Section 8.2).

An MCA chi-squared distance biplot based upon the Burt matrix associated with the data of Table 8.1 is given in Figure 8.7 as a result of setting `mca.variant = "Burt"` in the call to `MCAbipl`.

When $p = 2$, our notations for CA and MCA are related by $\mathbf{R} = \mathbf{L}_1$ and $\mathbf{C} = \mathbf{L}_2$. Then, the normalized Burt matrix is

$$\begin{pmatrix} \mathbf{I} & \mathbf{R}^{-1/2}\mathbf{G}_1'\mathbf{G}_2\mathbf{C}^{-1/2} \\ \mathbf{C}^{-1/2}\mathbf{G}_2'\mathbf{G}_1\mathbf{R}^{-1/2} & \mathbf{I} \end{pmatrix}. \tag{8.13}$$

Figure 8.7 MCA biplot of the data in Table 8.1 based on the associated normalized Burt matrix $\mathbf{L}^{-1/2}\mathbf{G}'\mathbf{G}\mathbf{L}^{-1/2} = p\mathbf{V}\mathbf{\Sigma}^2\mathbf{V}'$. Category-level points plotted as $\mathbf{Z} = \mathbf{L}^{-1/2}\mathbf{V}\mathbf{\Sigma}^2\mathbf{J}_2$ and samples as $\mathbf{Z}_0 = \mathbf{G}\mathbf{Z}/p$.

Including the unit diagonal blocks implies that when $p = 2$, CA and MCA are not precisely the same, though the two are closely related; see Gower and Hand (1996) for the precise relationship. In general, the approximation of the unit diagonal blocks is of no interest and is detrimental to the good approximation of the contingency tables themselves. Greenacre (1988, 2007) omits these superfluous blocks from the approximation, terming this form of MCA 'joint correspondence analysis' (JCA). Note that, when $p = 2$, JCA retains only the two-way contingency table (twice), so then JCA and CA coincide, giving the kind of compatibility desirable in any generalization. JCA requires an iterative algorithm, at each stage of which the diagonal blocks are updated in a similar way to that in which simple factor analysis algorithms of a correlation matrix iteratively replace the unit diagonals by communalities.

Different estimates of goodness of fit accompany all these variants of MCA. Basically, there are four main types: (i) fits based on the least-squares approximation of the whole Burt matrix; (ii) fits based on the least-squares approximation of the Burt matrix but excluding the diagonal blocks; (iii) fits based on the least-squares approximation of the Burt matrix excluding the diagonal blocks but with a simple adjustment to reduce the residual sum of squares; and (iv) JCA which is as for (iii) but applying the adjustment iteratively. Each of these successively reduces the residual sum of squares until in (iv) it is globally minimized. However, (ii) and (iii) do not give an orthogonal analysis of variance (total = fit + residual) so 'variance accounted for' should be treated with caution. Gower and Hand (1996) give a detailed discussion of these issues, while Gower (2006) tabulates several measures of fit.

8.4 Similarity matrices and the extended matching coefficient

We have expressed doubts concerning the usefulness of approximating $p^{-1/2}\mathbf{G}\mathbf{L}^{-1/2}$ via chi-squared distance. Because $\mathbf{G}$ is a binary matrix, another possibility is to use any dissimilarity coefficient for binary variables to determine an $n \times n$ dissimilarity matrix and analyse it by any form of MDS. A problem with this is that, because of its underlying p-variable structure, $\mathbf{G}$ is not a conventional binary matrix. What is needed is a dissimilarity coefficient that respects this structure. Gower and Hand (1996) suggested the extended matching coefficient (EMC) which expresses the number of matches for every pair of samples as a ratio of the number p of variables. Thus the proportion of matches is given by the nondiagonal values of

$$\mathbf{G}\mathbf{G}'/p \tag{8.14}$$

and the corresponding proportion of dissimilarities by

$$\mathbf{1}\mathbf{1}' - \mathbf{G}\mathbf{G}'/p. \tag{8.15}$$

When every variable has two levels, the EMC coincides with the simple matching coefficient.

It is of some interest to compare the simple properties of the EMC with what happens if $\mathbf{G}$ is treated as a binary matrix, ignoring the structure of the underlying categorical variables. To be a little more explicit about the EMC, for some pair of rows of $\mathbf{G}$, let a be the number of 1–1 matches, b the number of 1–0 matches, c the number of 0–1 matches and d the number of 0–0 matches. Thus $a + b + c + d = L$. The number of positive matches is a out of p variables, giving an EMC of $E = a/p$. The relationship with the calculation of a conventional similarity coefficient is easy to determine. For, if there are a 1–1 matches then there must be $p-a$ 1–0 and 0–1 mismatches, leaving a remainder of $L - 2p + a$ 0–0 matches. Thus, if we use a similarity coefficient of the Jaccard family $S = a/(a + \theta\{b + c\})$, it takes the special form $S = a/(a + 2\theta\{p - a\})$, depending solely on a and hence equivalently on E. We have that

$$E/S = (a + 2\theta\{p - a\})/p = (1 - 2\theta)E + 2\theta,$$

showing that E and S are monotonically related. A similarity coefficient from the simple matching family, $S = (a + d)/(a + d + \theta\{b + c\})$ takes the special form $S = (L - 2\{p - a\})/(L - 2\{1 - \theta\}\{p - a\})$. Now, we have

$$1/S = \{1 - \theta\} + \{\theta L\}/(L - 2p\{1 - E\}),$$

again showing a monotonic relationship between E and S. Nonmetric methods of analysis are invariant to monotonically related data, so then it makes no difference what measure

Figure 8.8 Biplot based on the EMC plotting $\mathbf{U\Sigma J}$ for the samples and $(\mathbf{I} - \mathbf{11}'\mathbf{L}/n)\mathbf{VJ}$ for the CLPs, where $\mathbf{G} - \mathbf{11}'\mathbf{L}/n = \mathbf{U\Sigma V}'$. The quality of the two-dimensional display is 59.77%, which on adding a third dimension increases to 77.49%, suggesting that a three-dimensional plot may be worthwhile.

we may adopt from the Jaccard and simple matching families. In other cases there will be slight differences but it makes sense to use the EMC, which is the simplest choice.

The EMC matrix may be analysed by any method of multidimensional scaling but most simply by the principal component analysis of $\mathbf{G}$. Then, the CLPs for the EMC are given by the $L \times L$ unit matrix $\mathbf{I}$. Unlike with MCA, where the first eigenvector of the SVD and spectral decompositions adjusts for deviations from the mean, special attention has to be given to allow for this in handling the CLPs. Expressing $\mathbf{G}$ in deviations from the column means gives $(\mathbf{I} - \mathbf{11}'/n)\mathbf{G} = \mathbf{G} - \mathbf{11}'\mathbf{L}/n$. The same adjustment must be made to the CLPs to give $\mathbf{I} - \mathbf{11}'\mathbf{L}/n$. Thus if $\mathbf{G} - \mathbf{11}'\mathbf{L}/n$ has SVD $\mathbf{U\Sigma V}'$, we plot $\mathbf{U\Sigma J} = (\mathbf{G} - \mathbf{11}'\mathbf{L}/n)\mathbf{VJ}$ for the samples and $(\mathbf{I} - \mathbf{11}'\mathbf{L}/n)\mathbf{VJ}$ for the CLPs. This biplot is shown in Figure 8.8 and is a result of setting `mca.variant = "EMC"` in the call to `MCAbipl`.

8.5 Category-level points

Category-level points have been used extensively above. Here, we supply an amplified discussion and an introduction to the concept of prediction regions. Every dummy variable of the indicator matrix takes either zero or unit values, so the usual idea of a coordinate axis requires modification. Quantitative variables define a continuum of values represented by continuous, often linear, axes. Categorical variables take a finite number of levels L_k represented by L_k points, termed the category-level points. Just as a sample with a value x_k is closer to the marker for x_k on the kth axis, so must a sample with a particular category level be closest to the CLP for that category level. For the EMC it works out that the CLPs for all L category levels are given by the rows of the unit matrix $\mathbf{I}$. For MCA the CLPs are at the points given by the rows of the matrix $p^{-1/2}\mathbf{L}^{-1/2}$; for CVA the CLPs are the coordinates of the canonical means.

Thus, the CLPs act like coordinated axes, in that every sample is placed at the vector-sum of its associated CLPs and is nearest the p correct CLPs, one from each set, for its particular categories. Thus every CLP defines a set of nearest-neighbour regions. This causes some problems for representing categories in low-dimensional representations. For continuous variables we have seen that the concept of back-projection (Section 5.4.2.3) allows markers to be placed in the approximation that are nearest the true markers that inhabit some high-dimensional space, which in turn justifies constructing biplot axes. For CLPs the position is similar but there is a whole convex region in the approximation space that is nearer each CLP than the others. These regions are called *prediction regions* because any point in a prediction region will be predicted to have that same associated category level. The kth categorical variable will have L_k such prediction regions. This is shown in Figures 8.9 (MCA) and 8.10 (EMC).for the four variables of Table 8.1 The prediction regions for the two methods are similar, but this is not easy to see because the maps are differently oriented. One problem with this representation is that each categorical variable occupies the whole of the space of the display, not just a single biplot axis. This makes it impracticable, though not impossible, to represent the prediction regions for more than one variable on a single diagram. Gower and Hand (1996) and Gower (1993) sketch how a sophisticated algorithm might be developed for calculating the prediction regions. A simple alternative is to cover the whole approximation with pixels and associate a different colour with each pixel, to indicate the nearest CLP; this is what we have done in our examples. The pixel colouring procedure is implemented in our function `pred.regions` that is called by `MCAbipl` on setting argument `pred.regions = TRUE`. Note that the CLPs are not at the centroids of their nearest-neighbour regions, so neither are their projections. Indeed, the projections need not even lie within their predictions regions and some category levels may not have a prediction region in the approximation space, because it is hidden behind other prediction regions.

8.6 Homogeneity analysis

The methods discussed in this section are often described as optimal scores methods. Their aim is to replace the nominal category levels by numerical optimal scores. In homogeneity analysis (see Gifi, 1990) we seek scores $\mathbf{z} = (\mathbf{z}_1, \mathbf{z}_2, \ldots, \mathbf{z}_p)$, often termed *quantifications*, that replace $\mathbf{G}$ by $\mathbf{Gz}$. The criterion chosen is to minimize the dispersion within the rows; hence the reference to homogeneity. This is the same as maximizing dispersion between rows. If the scaling of $\mathbf{z}$ is left uncontrolled there is a trivial solution

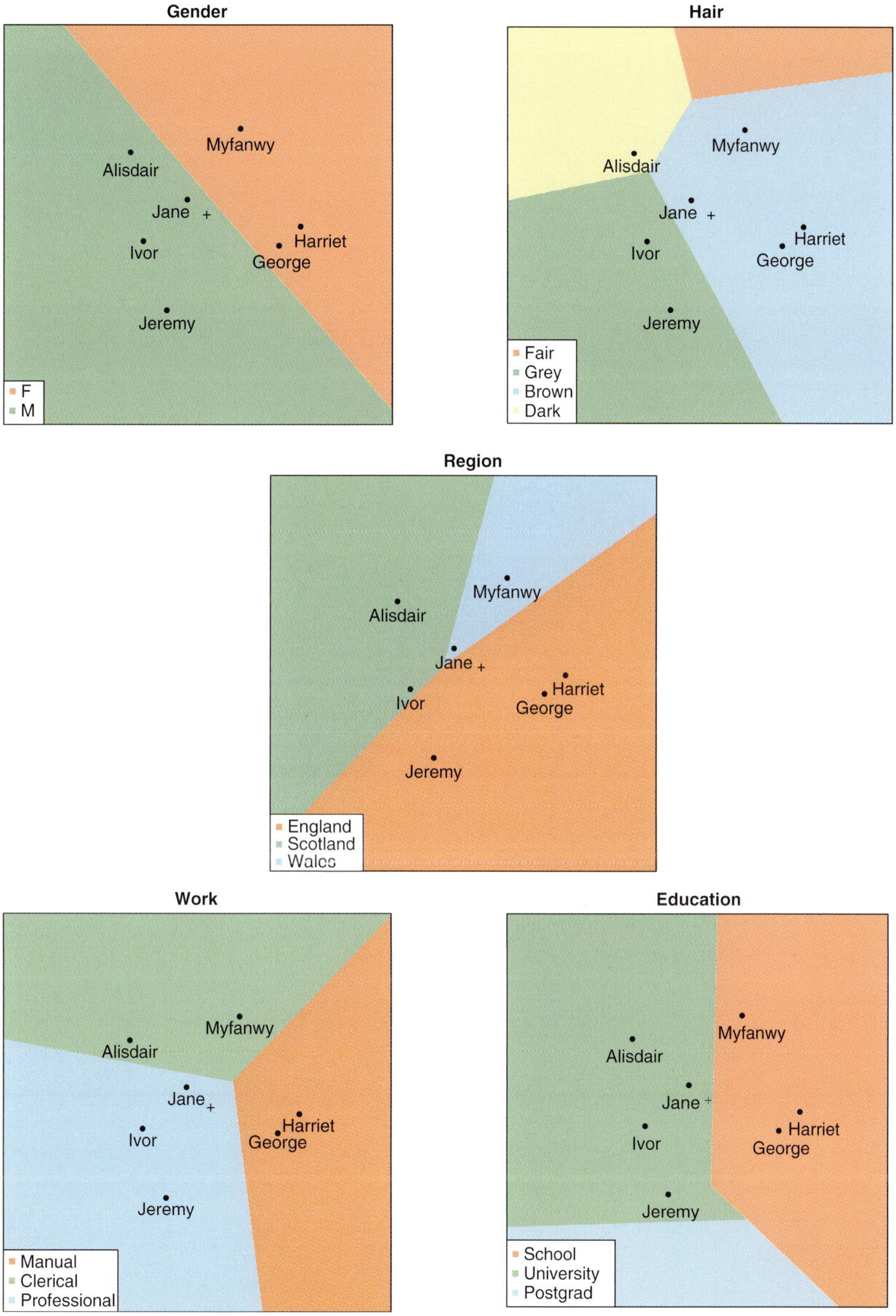

Figure 8.9 MCA biplot based on the normalized Burt matrix as given in Figure 8.7 with prediction regions added.

Figure 8.10 MCA biplot based on the EMC as given in Figure 8.8 with prediction regions added.

with $\mathbf{z} = \mathbf{0}$. To avoid this, we either fix the total sum of squares to 1, say, or express the criterion as a ratio,

$$\frac{\mathbf{z}'\mathbf{G}'\mathbf{G}\mathbf{z}}{\mathbf{z}'\mathbf{L}\mathbf{z}},$$

in which case the scaling is arbitrary (though for convenience usually fixed to $\mathbf{z}'\mathbf{L}\mathbf{z} = 1$). In this ratio, $\mathbf{z}'\mathbf{L}\mathbf{z}$ represents the total sum of squares, while the numerator is the sum of squares between the row totals $\mathbf{Gz}$. The maximization of the ratio requires the solution to the two-sided eigenvalue problem

$$\mathbf{G}'\mathbf{G}\mathbf{z} = \lambda\mathbf{L}\mathbf{z},$$

which may be written as

$$(\mathbf{L}^{-1/2}\mathbf{G}'\mathbf{G}\mathbf{L}^{-1/2})\mathbf{L}^{1/2}\mathbf{z} = \lambda\mathbf{L}^{1/2}\mathbf{z},$$

where, apart from the irrelevant factor p^{-1}, the matrix in parentheses is the normalized Burt matrix. Thus, we arrive back at our previous MCA eigenvalue problem. The eigenvector $\mathbf{L}^{1/2}\mathbf{z}$ may be normalized in the usual way so that $\mathbf{z}'\mathbf{L}\mathbf{z} = 1$, in which case the sum of squares of the row scores $\mathbf{z}'\mathbf{G}'\mathbf{G}\mathbf{z} = \lambda$, which is maximized by taking the largest nonunit eigenvalue. The deviations from the mean have been ignored in this derivation, but we have already seen when discussing CA that these are accounted for by the vectors associated with the rejected unit eigenvalue.

Just as with the correlational approach to CA (Section 7.2.5), the homogeneity analysis criterion seeks only a one-dimensional solution but the multidimensional solution may be justified as above.

8.7 Correlational approach

We have seen that the CA of a two-way table may be developed as seeking quantifications that maximize the correlation between the two categorical variables. This extends to MCA. One way of developing canonical correlation (CCA) for quantitative variables in data matrices $\mathbf{X}_1, \mathbf{X}_2$ is to find linear transformations $\mathbf{Z}_1$, $\mathbf{Z}_2$ that minimize

$$\|\mathbf{X}_1\mathbf{Z}_1 - \mathbf{X}_2\mathbf{Z}_2\|^2$$

normalized so that $\operatorname{diag}(\mathbf{Z}_1'\mathbf{X}_1'\mathbf{X}_1\mathbf{Z}_1 + \mathbf{Z}_2'\mathbf{X}_2'\mathbf{X}_2\mathbf{Z}_2) = 2\mathbf{I}$ (see Gower and Dijksterhuis, 2004, for a fuller discussion of appropriate constraints and the links with other formulations of CCA). Apart from trivial factors of proportionality, this criterion may be written as

$$\sum_{k=1}^{2} \|\mathbf{X}_k\mathbf{Z}_k - \mathbf{M}\|^2, \quad \text{where } \mathbf{M} = \frac{1}{2}\sum_{k=1}^{2}\mathbf{X}_k\mathbf{Z}_k,$$

to be minimized with the constraint $\operatorname{diag}(\mathbf{Z}_1'\mathbf{X}_1'\mathbf{X}_1\mathbf{Z}_1 + \mathbf{Z}_2'\mathbf{X}_2'\mathbf{X}_2\mathbf{Z}_2) = 2\mathbf{I}$. For indicator matrices $\mathbf{G}_1$, $\mathbf{G}_2$ this becomes the minimization of

$$\sum_{k=1}^{2} \|\mathbf{G}_k\mathbf{Z}_k - \mathbf{M}\|^2, \quad \text{where } \mathbf{M} = \frac{1}{2}\sum_{k=1}^{2}\mathbf{G}_k\mathbf{Z}_k,$$

where now, rather than as linear combinations, $\mathbf{Z}_1$, $\mathbf{Z}_2$ may be interpreted as giving quantifications. The constraint becomes $\text{diag}(\mathbf{Z}_1'\mathbf{G}_1'\mathbf{G}_1\mathbf{Z}_1 + \mathbf{Z}_2'\mathbf{G}_2'\mathbf{G}_2\mathbf{Z}_2) = 2\mathbf{I}$. The generalization to p categorical variables is to minimize

$$\sum_{k=1}^{p} \|\mathbf{G}_k\mathbf{Z}_k - \mathbf{M}\|^2, \quad \text{where } \mathbf{M} = \frac{1}{p}\sum_{k=1}^{p} \mathbf{G}_k\mathbf{Z}_k$$

subject to the constraint

$$\text{diag} \sum_{k=1}^{p} (\mathbf{Z}_k'\mathbf{G}_k'\mathbf{G}_k\mathbf{Z}_k) = p\mathbf{I}.$$

In terms of our notation for MCA, the criterion may be written

$$\text{trace}(\mathbf{Z}'\mathbf{L}\mathbf{Z}) - \frac{1}{p}\text{trace}(\mathbf{Z}'\mathbf{G}'\mathbf{G}\mathbf{Z})$$

and the constraint $\text{diag}(\mathbf{Z}'\mathbf{L}\mathbf{Z}) = p\mathbf{I}$. This constraint says that the sum of squares of each of the r columns of $\mathbf{GZ}$ is p. Writing $\mathbf{\Lambda}$ as a diagonal matrix of r Lagrange multipliers, minimization gives

$$\mathbf{LZ} - \frac{1}{p}(\mathbf{G}'\mathbf{G})\mathbf{Z} = \mathbf{LZ\Lambda}$$

or

$$(\mathbf{G}'\mathbf{G})\mathbf{Z} = p\mathbf{LZ}(\mathbf{I} - \mathbf{\Lambda}) = \mathbf{LZ\Phi}.$$

This is a two-sided eigenvalue problem and hence we have the usual orthogonality conditions that $\mathbf{Z}'\mathbf{L}\mathbf{Z}$ and $\mathbf{Z}'\mathbf{G}'\mathbf{G}\mathbf{Z}$ are both diagonal. To comply with the constraint, we normalize so that $\mathbf{Z}'\mathbf{L}\mathbf{Z} = p\mathbf{I}$. Note that not only $\text{diag}(\mathbf{Z}'\mathbf{L}\mathbf{Z}) = p\mathbf{I}$, as required by the constraint, but also the off-diagonal values are zero. This has not been imposed as a constraint but is a natural consequence of the solution. When $p = 2$, it turns out that $\mathbf{Z}'\mathbf{L}\mathbf{Z} = 2\mathbf{I}$ implies the separate condition $\mathbf{Z}_1'\mathbf{L}_1\mathbf{Z}_1 = \mathbf{Z}_2'\mathbf{L}_2\mathbf{Z}_2 = \mathbf{I}$ which is a usual normalization in CCA, but this separability does not extend to the case when $p > 2$.

We may rewrite the two-sided eigenvalue equation as

$$\{\mathbf{L}^{-1/2}(\mathbf{G}'\mathbf{G})\mathbf{L}^{-1/2}\}\mathbf{L}^{1/2}\mathbf{Z} = \mathbf{L}^{1/2}\mathbf{Z\Phi}$$

so that $\mathbf{L}^{1/2}\mathbf{Z}$ are eigenvectors of the normalized Burt matrix $\mathbf{L}^{-1/2}(\mathbf{G}'\mathbf{G})\mathbf{L}^{-1/2}$, just as is required for MCA. The interpretation of $\mathbf{Z}$ in terms of scores supports interest in the quantified variables $\mathbf{G}_1\mathbf{Z}_1, \mathbf{G}_2\mathbf{Z}_2, \ldots, \mathbf{G}_p\mathbf{Z}_p$ and their mean $\mathbf{M}$. Each of these gives n points that may be plotted as individual points on the MCA to give a proper biplot of individuals/samples and variables. However, because n is usually large in MCA, often we would prefer some form of density plot for the samples and, rather than p such plots, would be content just with a density plot of $\mathbf{M}$. For any eigenvalue ϕ with corresponding eigenvector $\mathbf{z}$, we have

$$\mathbf{1}'(\mathbf{G}'\mathbf{G})\mathbf{z} = \phi\mathbf{1}'\mathbf{Lz},$$

that is,

$$\mathbf{1}'\mathbf{Lz} = \phi\mathbf{1}'\mathbf{Lz}.$$

Thus, **1** is an eigenvector corresponding to $\phi = 1$, and it follows from the orthogonality relationships that $\mathbf{1}'\mathbf{L}\mathbf{z} = 0$ for any other eigenvector $\mathbf{z}$. This shows that $\mathbf{1}'\mathbf{M}\mathbf{z} = 0$ so that the means are centred, the usual result concerning the 'uninteresting' unit eigenvalue in CA and MCA.

Finally, let us examine the correlation between the quantifications of the sth dimension kth variable with the sth dimension of $\mathbf{M}$, or equivalently of $\mathbf{GZ}$. For simplicity, and without loss of generality, we take $s = 1$ and $k = 1$. Thus, we are interested in the correlation ρ_1 between $\mathbf{G}_1\mathbf{z}_1$ and $\mathbf{g} = \mathbf{G}_1\mathbf{z}_1 + \mathbf{G}_2\mathbf{z}_2 + \ldots + \mathbf{G}_p\mathbf{z}_p$ where $\mathbf{z}_k\,(k = 1, 2, \ldots, p)$ is the *first* column of $\mathbf{Z}_k$. Then, using the eigendecomposition, we have

$$\mathbf{g}'\mathbf{G}_1\mathbf{z}_1 = \phi_1\mathbf{z}_1'\mathbf{L}_1\mathbf{z}_1,$$
$$\mathbf{g}'\mathbf{g} = \sum_{k=1}^{p} \mathbf{z}_k'\mathbf{L}_k\mathbf{z}_k = p\phi_1,$$

so that

$$\rho_1^2 = \frac{(\mathbf{g}'\mathbf{G}_1\mathbf{z}_1)^2}{(\mathbf{z}_1'\mathbf{G}_1'\mathbf{G}_1\mathbf{z}_1)(\mathbf{g}'\mathbf{g})} = \frac{(\phi_1\mathbf{z}_1'\mathbf{L}_1\mathbf{z}_1)^2}{(\mathbf{z}_1'\mathbf{L}_1\mathbf{z}_1)(p\phi_1)} = \frac{\phi_1\mathbf{z}_1'\mathbf{L}_1\mathbf{z}_1}{p}.$$

Similar results follow for the correlations $\rho_2, \ldots, \rho_p$ of $\mathbf{G}_2\mathbf{z}_2, \ldots, \mathbf{G}_p\mathbf{z}_p$ with $\mathbf{g}$. Summing gives

$$\sum_{k=1}^{p} \rho_k^2 = \frac{\phi_1}{p}\sum_{k=1}^{p}(\mathbf{z}_k'\mathbf{L}_k\mathbf{z}_k) = \phi_1.$$

Thus, the sum of squares of the correlations of the first columns of $\mathbf{G}_1\mathbf{z}_1, \ldots, \mathbf{G}_p\mathbf{z}_p$ is maximized and equal to the maximum eigenvalue. Subsequent columns have sums of squares equal to the successively decreasing eigenvalues $\phi_2, \ldots, \phi_r$. We started by generalizing the two-variable CCA concept and have finished by showing that the generalization has a nice correlational interpretation in its own right.

8.8 Categorical (nonlinear) principal component analysis

Homogeneity analysis and MCA are essentially the same thing, justified by slightly different, but equivalent, criteria. A rather different approach is given by nonlinear PCA, increasingly and more appropriately referred to as categorical PCA. Like homogeneity analysis, this gives scores to the category levels but now we focus on the individual categorical variables, to give a data matrix

$$\mathbf{H} = [\mathbf{G}_1\mathbf{z}_1, \mathbf{G}_2\mathbf{z}_2, \ldots, \mathbf{G}_p\mathbf{z}_p].$$

In homogeneity analysis we are interested only in the total and row totals of $\mathbf{Gz}$ and their sums of squares. Here we are interested in the individual values of the table and ask what values of $\mathbf{z}$ give an optimal PCA in some specified number, r, of dimensions. By 'optimal' here we mean the choice of $\mathbf{z}$ that maximizes the sum of the first r eigenvalues. We assume that $\mathbf{z}$ is scaled so that the centred columns of $\mathbf{H}$ are normalized, that is, $\mathbf{z}_k'\mathbf{G}_k'(\mathbf{I} - \frac{1}{n}\mathbf{1}\mathbf{1}')\mathbf{G}_k\mathbf{z}_k = 1,\ k = 1, \ldots, p$. With this normalization, $\mathbf{H}'\mathbf{H}$ is a correlation matrix.

Thus, the computational problem for categorical PCA is

$$\min \|\mathbf{H} - \mathbf{Y}\|^2. \tag{8.16}$$

Here, rank($\mathbf{Y}$) is a given value r and the minimization is over $\mathbf{Y}$ and the quantifications $\mathbf{z}$, normalized as above. When $\mathbf{z}$ is known, $\mathbf{Y}$ is given by the usual r-dimensional Eckart-Young PCA solution (see Chapter 2) to approximating $\mathbf{H}$. When $\mathbf{Y}$ is known, the criterion (8.16) may be written

$$\min \sum_{k=1}^{p} \|\mathbf{G}_k \mathbf{z}_k - \mathbf{y}_k\|^2 \tag{8.17}$$

so, writing $\mathbf{y}_k$ for the kth column of $\mathbf{Y}$, we may find the quantifications independently for each variable by solving

$$\min \|\mathbf{G}_k \mathbf{z}_k - \mathbf{y}_k\|^2, \tag{8.18}$$

subject to the constraints on $\mathbf{z}_k$. We may arrange that $\mathbf{Y}$ always has zero column sums so only the constraint $\mathbf{z}_k'\mathbf{G}_k'\mathbf{G}_k\mathbf{z}_k = \mathbf{z}_k'\mathbf{L}_k\mathbf{z}_k = 1$ needs attention. This minimization is a constrained regression problem. To find the solution to (8.18) we introduce the Lagrange multiplier λ and consider

$$\mathbf{z}_k'\mathbf{G}_k'\mathbf{G}_k\mathbf{z}_k - 2\mathbf{z}_k'\mathbf{G}_k'\mathbf{y}_k + \mathbf{y}_k'\mathbf{y}_k + \lambda(\mathbf{z}_k'\mathbf{L}_k\mathbf{z}_k - 1). \tag{8.19}$$

Taking the derivatives of (8.19) with respect to λ and to $\mathbf{z}_k$ and setting to zero, we obtain

$$\mathbf{G}_k'\mathbf{G}_k\mathbf{z}_k - \mathbf{G}_k'\mathbf{y}_k + \lambda\mathbf{L}_k\mathbf{z}_k = \mathbf{0}. \tag{8.20}$$

From (8.20), noticing that $\mathbf{z}_k'\mathbf{G}_k'\mathbf{G}_k\mathbf{z}_k = \mathbf{z}_k'\mathbf{L}_k\mathbf{z}_k = 1$, it follows that $\lambda = \mathbf{z}_k'\mathbf{G}_k'\mathbf{y}_k - 1$. Therefore,

$$\mathbf{G}_k'\mathbf{G}_k\mathbf{z}_k - \mathbf{G}_k'\mathbf{y}_k + (\mathbf{z}_k'\mathbf{G}_k'\mathbf{y}_k - 1)\mathbf{L}_k\mathbf{z}_k = \mathbf{0}$$

that is,

$$\mathbf{z}_k - \mathbf{L}_k^{-1}\mathbf{G}_k'\mathbf{y}_k + (\mathbf{z}_k'\mathbf{G}_k'\mathbf{y}_k - 1)\mathbf{z}_k = \mathbf{0}$$

or

$$\mathbf{z}_k = \frac{1}{\mathbf{z}_k'\mathbf{G}_k'\mathbf{y}_k}\mathbf{L}_k^{-1}\mathbf{G}_k'\mathbf{y}_k.$$

But $\mathbf{z}_k'\mathbf{L}_k\mathbf{z}_k = 1$ implies that

$$\mathbf{z}_k'\mathbf{G}_k'\mathbf{y}_k = \pm\sqrt{\mathbf{y}_k'\mathbf{G}_k\mathbf{L}_k^{-1}\mathbf{G}_k'\mathbf{y}_k},$$

so that a solution to (8.18) is given by

$$\mathbf{z}_k = \frac{\mathbf{L}_k^{-1}\mathbf{G}_k'\mathbf{y}_k}{\sqrt{\mathbf{y}_k'\mathbf{G}_k'\mathbf{L}_k^{-1}\mathbf{G}_k'\mathbf{y}_k}}. \tag{8.21}$$

It is easy to check that $\mathbf{z}_k'\mathbf{L}_k\mathbf{z}_k = 1$ and $\mathbf{1}'\mathbf{L}_k\mathbf{z}_k = \mathbf{1}'\mathbf{y}_k = 0$. Note that $\mathbf{z}_k = \mathbf{L}_k^{-1}\mathbf{G}_k\mathbf{y}_k$ merely estimates the quantifications by the average values of $\mathbf{y}_i$ obtained for the different levels of the kth variable. Thus, a least-squares algorithm alternating between an Eckart-Young step to estimate $\mathbf{Y}$ and a quantification estimation step, should converge to the categorical PCA solution. Unlike PCA, categorical PCA solutions are not nested in decreasing numbers of dimensions. Thus, the categorical PCA solution for $r = 3$ dimensions does not include the solution for $r = 2$ dimensions, which has to be computed *ab initio*.

At convergence, it is clear that $\mathbf{Y}$ gives the conventional PCA r-dimensional approximation to $\mathbf{H}$. Thus, once quantifications have been evaluated, $\mathbf{H}$ becomes a numerical data matrix whose PCA and associated biplots remain as in Chapter 3. Note that although biplot axes could be calibrated as usual, the L_k quantifications for the kth categorical variable have only nominal validity, and do not support interpretation in terms of a numerical continuum. For this reason, calibrations should be marked only for the computed quantifications and should be labelled by their category names. Because predicted categories are given by the nearest category point, a useful device is to colour, or similarly demarcate, regions along a biplot axis that give the same predictions. To be precise, if z_a, z_b and z_c are three adjacent category points on an axis, then we could use a colour denoting level b of the variable to 'paint' the segment joining half way between z_a, and z_b to half way between z_b and z_c.

Sometimes, categorical variables are not purely nominal but have an ordinal nature. An example would be a variable ***Wealth*** with ordered levels *poor, adequate, comfortable, wealthy, millionaire*. Then it is natural to require that the quantifications be constrained to be ordered in the same way. We would recommend that axes should continue to be calibrated with the nominal value but now there is more validity in believing that the quantifications are part of a continuum. For example, quantifications for an ordinal categorical variable ***Wealth*** might be regarded as a surrogate for annual income. Computationally, it is not difficult to impose the order constraint on the regression estimates for the quantifications, though care is needed to ensure that the conditions $\mathbf{z}_k'\mathbf{L}_k\mathbf{z}_k = 1$ and $\mathbf{1}'\mathbf{L}_k\mathbf{z}_k = 0$ are maintained. When quantifications are not naturally ordered, so the ordering constraint has to be imposed, then sometimes this can be done only by introducing ties (see monotone regression: Figures 8.12 and 10.2) which means that z_a, z_b and z_c will have a pair of equal values. This causes difficulties in labelling the category levels but not in painting axis segments, unless three or more levels tie. In the latter case, it seems unwise to insist on ordering constraints.

Figure 8.11 shows a categorical PCA for Table 8.1 where all variables are treated as nominal. Each axis is linear but broken into coloured sections. Thus brown hair colour is indicated by a light blue section, consisting partly of all points that are nearer *Brown* than *Fair* and partly by all points that are nearer *Brown* than *Dark*. The other segments are derived similarly. Prediction is as usual by orthogonal projection, as is shown for ***Jane***. The quantifications are given in Figure 8.12, where it can be seen that only the variables ***Gender*** and ***Work*** have ordinal quantifications. When ordinal constraints are imposed on the variables ***Hair, Education*** and ***Work*** we get Figure 8.13 with quantifications given in Figure 8.15. It can be seen that ordinality can only be satisfied by allowing ties. In the plot, ordinality is indicated by thickening the axes, as shown, and ties are indicated by a special notation. Since prediction is by orthogonal projection, axes shifts are permissible, as is shown in Figure 8.14. This has shifted the axes away from most of the points and given a clearer presentation.

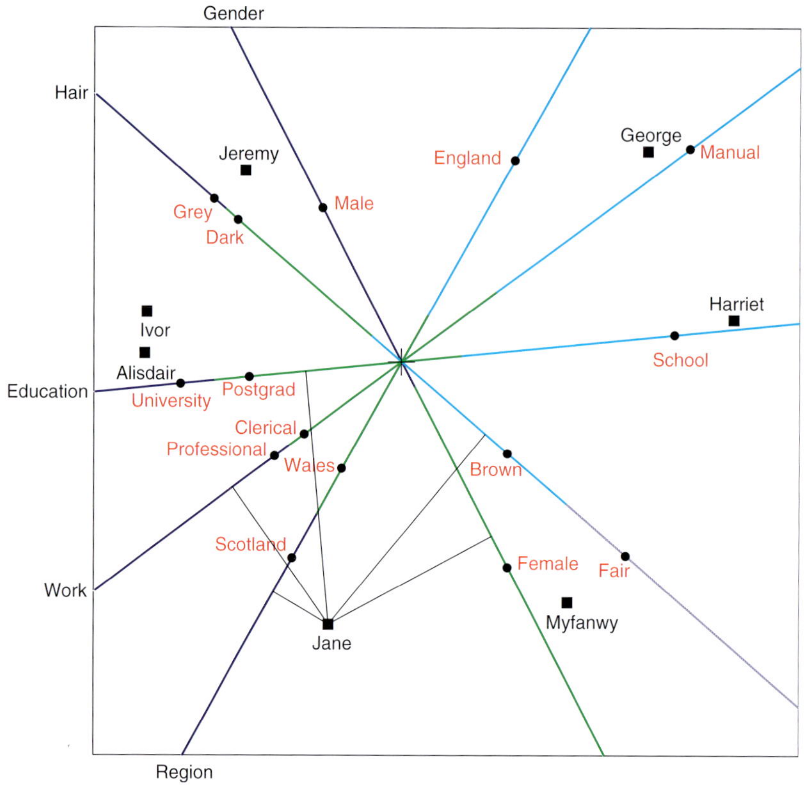

Figure 8.11 Categorical PCA. Scales on axes marked with solid black circles are labelled in red according to category. Each axis is in different colours, indicating that the scales are not continuous. Predictions are shown for ***Jane***.

8.9 Functions for constructing MCA related biplots

8.9.1 Function `cabipl`

In Section 8.2 we remarked that MCA may be regarded as a CA of the categorical data matrix **G** as if it were a two-way contingency table. Although we also expressed our reservations about this practice, it does open up the use of `cabipl` for constructing MCA biplots. Therefore, all the facilities of `cabipl` as discussed in Section 7.6.1 are available when the indicator matrix is used as input. However, some notational precautions must be kept in mind. These are summarized in Table 8.4.

8.9.2 Function `MCAbipl`

This is our main function for constructing the two-dimensional MCA biplots based on the Burt matrix and the EMC discussed in Sections 8.3 and 8.4, respectively.

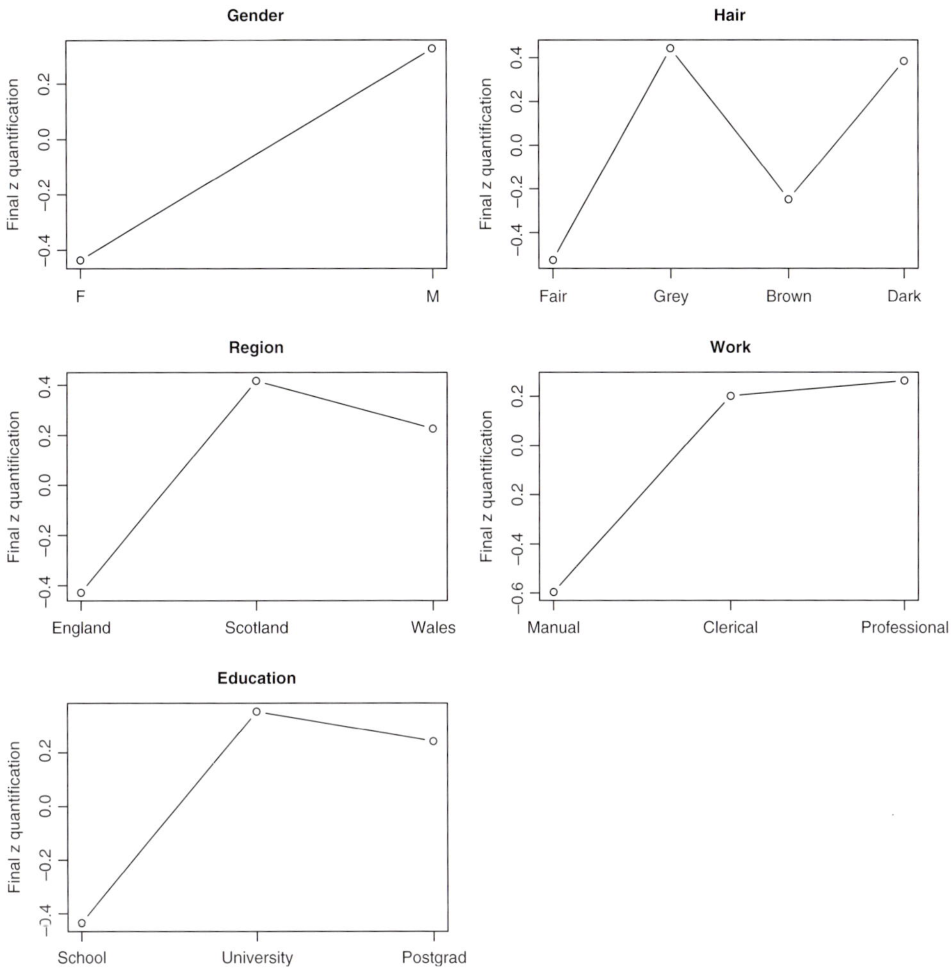

Figure 8.12 The final z-scores in the categorical PCA. All variables are treated as nominal.

Usage

`MCAbipl` uses the same calling conventions as `cabipl` and shares the following arguments with it:

```
category.labels
column.points.col
column.points.size
column.points.label.size,
col.points.text
exp.factor
e.vects
offset.m
parplotmar
pch.row.points
pch.col.points
pos.m
reflect
rotate.degrees
row.points.col
row.points.size
row.points.label.size
Title
zoomval
```

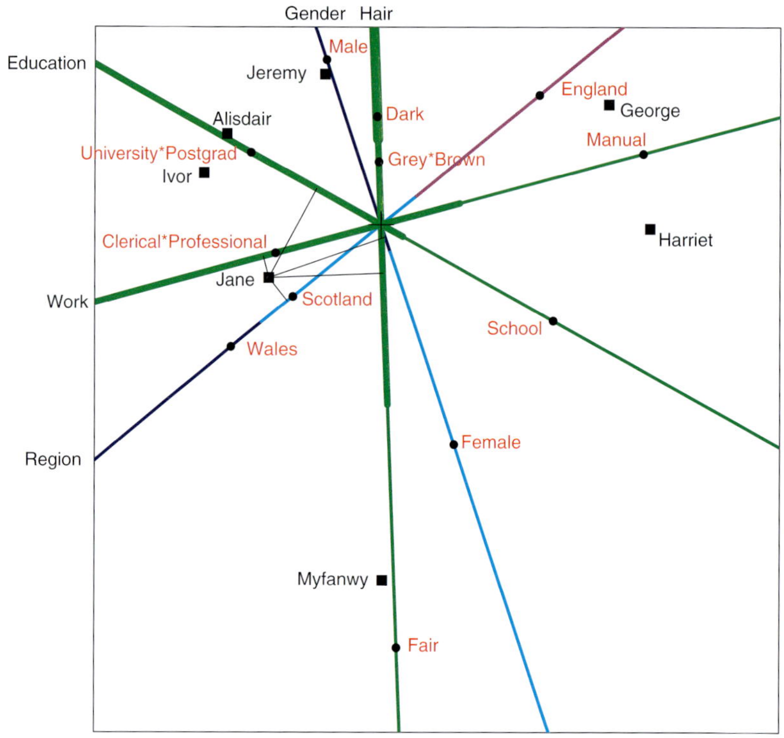

Figure 8.13 Categorical PCA. The variables ***Hair***, ***Education*** and ***Work*** are treated as ordinal and the variables ***Gender*** and ***Region*** are treated as nominal. The axes for the nominal variables are colour-calibrated as in Figure 8.11. Axes for the ordinal variables are calibrated in the same colour (here green) but with thickness increasing with order. Note the way ties are written, e.g. *University*Postgrad, Clerical*Professional*.

Arguments with a specific meaning in `MCAbipl`

`X`	Required argument. An $n \times p$ data matrix where each column represents a categorical variable, e.g. Table 8.1.
`mca.variant`	One of "indicator", "Burt", "EMC". Defaults to "indicator".
`main`	The titles to be used for each set of prediction regions. Default is to use the column names of `X`.
`prediction.regions`	If TRUE, the prediction regions associated with each categorical variable are constructed in separate graph windows. The actual construction of the prediction regions is performed by the functions `pred.regions` and `Colour.Pixel` that are called by `MCAbipl`. See also Section 4.7.1. Default is FALSE.
`pred.region.symbol`	Symbol used in pixel colouring. Default is 15 (solid square).

`pred.region.symbol.size`	Size of the symbol used in pixel colouring Default is 0.30.
`region.colours`	Vector for specifying the colours of the different prediction regions.
`sample.labels`	Logical TRUE or FALSE. If TRUE the sample points (rows of `X`) are labelled according to row names of `X`.
`text.size`	A two-component vector. The first element specifies the character size of sample labels and the second element the character size of the category levels. Default is `c(0.65, 0.65)`.
`x.grid`	Determines the x-coordinates of a grid overlaying the biplot space. Defaults to 0.05.
`y.grid`	Determines the y-coordinates of a grid overlaying the biplot space. Defaults to 0.05.

Value

The output of `MCAbipl` is a graph of the desired biplot: when `mca.variant = "indicator"` the three biplots discussed in Section 8.2 are produced, while `mca.variant = "Burt"` or `"MCA"` leads to the biplot discussed in Section 8.3 or 8.4 respectively. In addition, setting `prediction.regions = TRUE` together with `mca.variant = "Burt"` or `"MCA"` produces the prediction regions associated

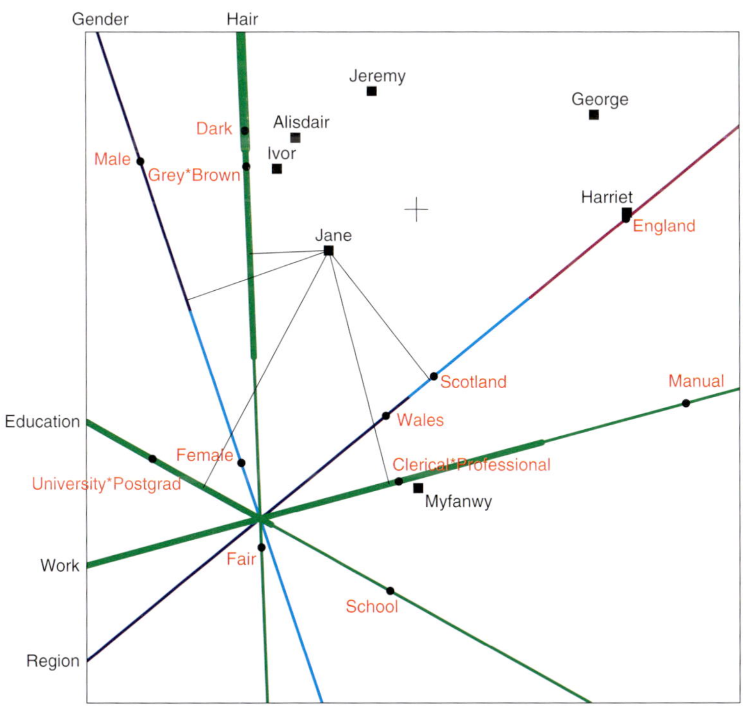

Figure 8.14 Similar to Figure 8.13 but with shift of origin.

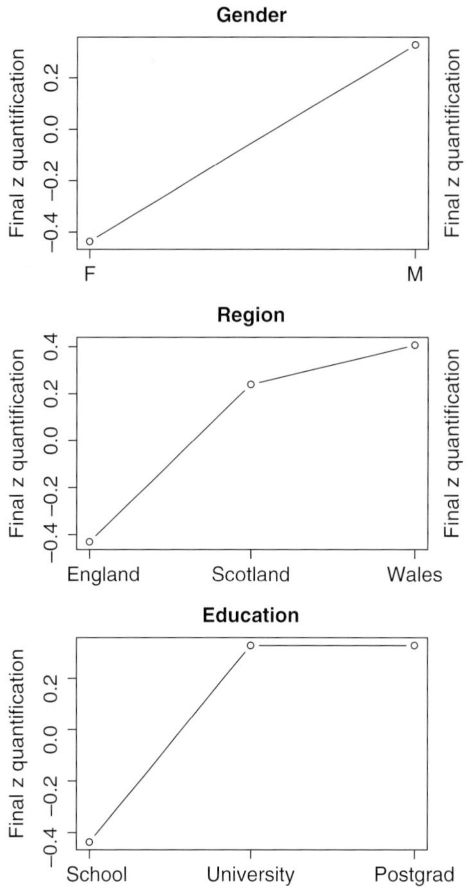

Figure 8.15 Final z-scores where ***Hair, Work*** and ***Education*** are treated as ordinal. No ties with ***Hair***, but ties occur in ***Work*** and ***Education***.

with each categorical variable together with an appropriate legend. Regardless of the `mca.variant` specification, a list with the following components is returned:

`Gmat`	The indicator matrix associated with the input `X`.
`BurtMat`	The Burt matrix associated with the input `X`.
`EMC.prop.match`	Matrix of the EMC matching proportions.
`Z`	$L \times 2$ matrix containing the coordinates of the plotted CLPs.
`Z.0`	$n \times 2$ matrix containing the coordinates of the plotted samples.
`CLPlist`	A list with kth element the $L_k \times L$ matrix of the L-dimensional coordinates of the CLPs associated with the kth categorical variable.
`e.values`	Vector containing the eigenvalues associated with the specified choice of `mca.variant` excluding the unity eigenvalue associated with the `"indicator"` or `"Burt"` options.

Table 8.4 MCA equivalents for the CA notation used in Chapter 7.

CA	MCA
$\mathbf{X}$: Two-way contingency table with p rows and q columns representing two categorical variables with p and q categories, respectively.	$\mathbf{G}$: Indicator matrix with n rows and L columns. $\mathbf{G} = [\mathbf{G}_1, \mathbf{G}_2, \ldots, \mathbf{G}_p]$ $\mathbf{G}_j$: $n \times L_j$. The indicator matrix for the jth ($j = 1, \ldots, p$) categorical variable having L_j categories.
n: Total number of samples, i.e. $n = \mathbf{1}'\mathbf{X}\mathbf{1}$	n: Total number of samples (number of rows of $\mathbf{G}$), i.e. $n = \mathbf{1}'\mathbf{G}\mathbf{1}/p$
$\mathbf{R}$: $p \times p = \text{diag}(\mathbf{X}\mathbf{1})$	$\text{diag}(\mathbf{G}\mathbf{1}) = \text{diag}(p\mathbf{1}) = p\mathbf{I}$: $n \times n$
$\mathbf{C}$: $q \times q = \text{diag}(\mathbf{1}'\mathbf{X})$	$\text{diag}(\mathbf{1}'\mathbf{G}) = \text{diag}(\mathbf{1}'\mathbf{L}) = \mathbf{L}$: $L \times L$ Also, $\mathbf{L} = \text{diag}(\text{diag}(\mathbf{L}_1), \text{diag}(\mathbf{L}_2), \ldots, \text{diag}(\mathbf{L}_p))$, where $\mathbf{L}_k$ is a diagonal matrix giving the frequencies $(l_{k1}, l_{k2}, \ldots, l_{kL_k})$ of the kth variable.
$\mathbf{E}$: $p \times q = \mathbf{R}\mathbf{1}\mathbf{1}'\mathbf{C}/n$	$p\mathbf{1}\mathbf{1}'\mathbf{L}/np = \mathbf{1}\mathbf{1}'\mathbf{L}/n$ (A subtle point to note is that the sum of all elements of the input matrix is needed here, not the number of samples.)
Column-centred $\mathbf{X} = (\mathbf{I} - \mathbf{1}\mathbf{1}'/p)\mathbf{X}$	Column-centred $\mathbf{G} = (\mathbf{I} - \mathbf{1}\mathbf{1}'/n)\mathbf{G}$
$\mathbf{W}_1^{-1}\mathbf{R}^{-1/2}(\mathbf{X} - \mathbf{E})\mathbf{C}^{-1/2}\mathbf{W}_2^{-1}$	$\mathbf{W}_1^{-1}(p^{-1/2}\mathbf{I})(\mathbf{G} - \mathbf{1}\mathbf{1}'\mathbf{L}/n)\mathbf{L}^{-1/2}\mathbf{W}_2^{-1}$ $= \mathbf{W}_1^{-1}(p^{-1/2}\mathbf{I})(\mathbf{G} - \mathbf{1}\mathbf{1}'\mathbf{G}/n)\mathbf{L}^{-1/2}\mathbf{W}_2^{-1}$ $= \mathbf{W}_1^{-1}\{(\mathbf{I} - \mathbf{1}\mathbf{1}'/n)(p^{-1/2}\mathbf{G}\mathbf{L}^{-1/2})\}\mathbf{W}_2^{-1}$ where the expression within the braces denotes the column-centred matrix $p^{-1/2}\mathbf{GL}\ ^{1/2}$

8.9.3 Function `CATPCAbipl`

`CATPCAbipl` is our main function for constructing the two-dimensional categorical PCA biplot discussed in Section 8.8.

Usage

`CATPCAbipl` uses the same calling conventions as `PCAbipl` and shares the following arguments with it:

```
alpha            line.width        ort.lty
ax               max.num           parplotmar
ax.name.size     offset            pos
c.hull.n         orthog.transx     predict.sample
exp.factor       orthog.transy     select.origin
```

Arguments with a specific meaning in `CATPCAbipl`

`Xcat`	Required argument. An $n \times p$ data matrix where each column represents a categorical variable, e.g. Table 8.1.
`factor.type`	p-vector with elements either `"nom"` or `"ord"` specifying whether categorical variable is nominal or ordinal. If `factor.type` is specified as `"ord"` then it is assumed that the levels are ordered as 1, 2, 3, ... and that the corresponding column of `Xcat` has been specified to be ordered.
`drawbagplots`	NULL or character vector specifying the groups for which α-bags are to be drawn.
`calibration.pch`	Integer specifying plotting symbol for CLPs. Defaults to 16.
`calibration.col`	Colour to use for CLPs.
`calibration.size`	Size of plotting symbols representing the CLPs. Defaults to 1.
`calibration.label.col`	Colour to use for CLP label.
`calibration.label.offset`	Used together with arguments `calibration.label.pos` to optimize placement of CLP labels. A numerical value controlling the offset from the position specified by `calibration.label.pos`. Default is no offset.
`class.vec`	Character vector defining the groups.
`class.cols`	Vector of same size as `class.vec` specifying the colour of the bag drawn for each class.
`class.pch`	Vector of same size as `class.vec` specifying the plotting character to be used when a convex hull is to be drawn.
`calibration.label.pos`	An integer value controlling placement of a CLP label: 1 = below, 2 = to left, 3 = above and 4 = to right of specified coordinates.
`calibration.label.size`	Numerical value specifying size of a CLP label. Defaults to 1.
`epsilon`	Convergence criterion in quantifications step (8.18). Defaults to 0.000001.
`ID`	Determines the y-coordinates of a grid overlaying the biplot space. Defualts to 0.05.
`label.samples`	Logical TRUE or FALSE for specifying if sample points are to be labelled. Default is FALSE.
`line.type.bags`	Vector of same size as `class.vec` specifying the line type to be used in drawing the bag for each class.
`nom.col`	Controlling colouring of CLPs associated with nominal categories.

`r`	The required number of dimensions in which to calculate the **z** that gives the optimal PCA as described in Section 8.8.
`ord.col`	Controlling colouring of CLPs associated with ordinal categories.
`plot.samples`	NULL (the default) or integer-valued vector specifying which sample points are to be drawn.
`reverse`	If set to TRUE, the ordering of an ordered categorical variable is reversed. Default is FALSE.
`samples.col`	Specifies colour to use for plotting the sample points.
`samples.pch`	Specifies plotting symbol for representing sample points. Default is 15.
`samples.size`	Specifies size of plotting character representing sample points. Default is 1.
`samples.label.pos`	An integer value controlling placement of label of a sample point: 1 = below, 2 = to left, 3 = above and 4 = to right of specified coordinates. Default is 1.
`samples.label.col`	Specifies colouring of labels of sample points.
`samples.label.offset`	A numerical value controlling the offset from the position specified by `samples.label.pos`. Default is no offset.
`samples.labcl.size`	A numerical value for controlling the size of the plotting symbol representing sample points. Defaults to 0.75.
`specify.bags`	Character vector specifying the classes for which α-bags are to be drawn.
`w.factor`	Factor controlling width of connecting lines for ordered and nominal categorical variables. Defaults to 2.

Value

`CATPCAbipl` constructs the categorical PCA biplot described in Section 8.8 as well as a graphical representation of the optimal scores. In addition, a list is returned with the following components:

`Hmat`	The **H** matrix as defined in Section 8.8.
`Hmat.cent`	The centred **H** matrix of Section 8.8.
`z.axes`	A list with each component a matrix containing the details of a biplot axis to be plotted.
`Yr`	The r-dimensional Eckart–Young PCA solution of the catogorical PCA criterion (8.16).
`zlist`	The optimal z-scores described in Section 8.8.

8.9.4 Function `CATPCAbipl.predregions`

`CATPCAbipl.predregions` is used for constructing prediction regions for the categorical variables used in a categorical PCA. It shares the following arguments with `CATPCAbipl`:

```
Xcat            orthog.transx     select.origin
factor.type     orthog.transy     w.factor
ord.col         parplotmar        predict.sample
nom.col         exp.factor
epsilon         reverse
```

It has the additional argument `prediction.regions` (with default `"all"`) for specifying the categorical variable(s) for which prediction regions are to be constructed.

8.9.5 Function `PCAbipl.cat`

This function is to be used only with `CATPCAbipl` where it calls `drawbipl.catPCA`.

8.10 Revisiting the remuneration data: examples of MCA and categorical PCA biplots

The remuneration data were introduced in Section 4.9.1 where all the variables used in constructing the CVA biplots were treated as continuous except the grouping variable ***Gender***. The categorical variable ***Faclty*** was excluded from the analysis. As an example of MCA we use here the 2002 data with all variables treated as nominal (unordered categorical) while as an illustration of categorical PCA some of the variables are treated as ordered categorical. The variables are categorized as follows:

Remun	in decile categories coded as *R1*, *R2*, ..., *R10* (highest)
Resrch	categories *Res0, Res1*, ..., *Res7* (highest)
Rank	categories *Rnk1, Rnk2*, ..., *Rnk5* (highest)
Age	categories *A1*, *A2*, ..., *A5* (oldest)
Gender	*male, female*
AQual	categories *A1*, *A2*, ..., *A5* (highest)
Faclty	categories *F1*, *F2*, ..., *F9* (unordered – nominal scale)

Figures 8.16 and 8.17 contain MCA biplots based on the indicator matrix of the dataframe `Remuneration.cat.data.2002`, while the biplot in Figure 8.18 results from

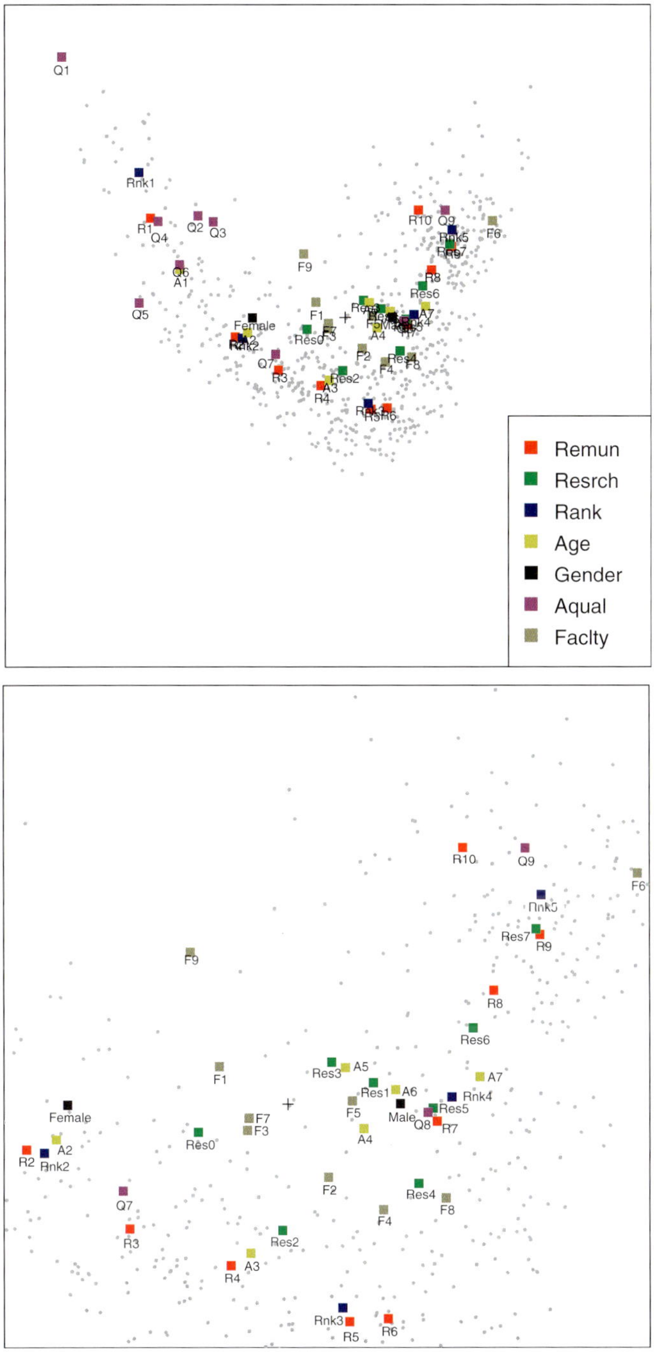

Figure 8.16 MCA indicator matrix based biplot of the dataframe `Remuneration.cat.data.2002`: (top) without zoom; (bottom) with `zoomval = 0.42`. The first two dimensions are used for scaffolding axes. Coloured solid squares are the category centroids defined by (8.6); grey solid circles indicate the sample points.

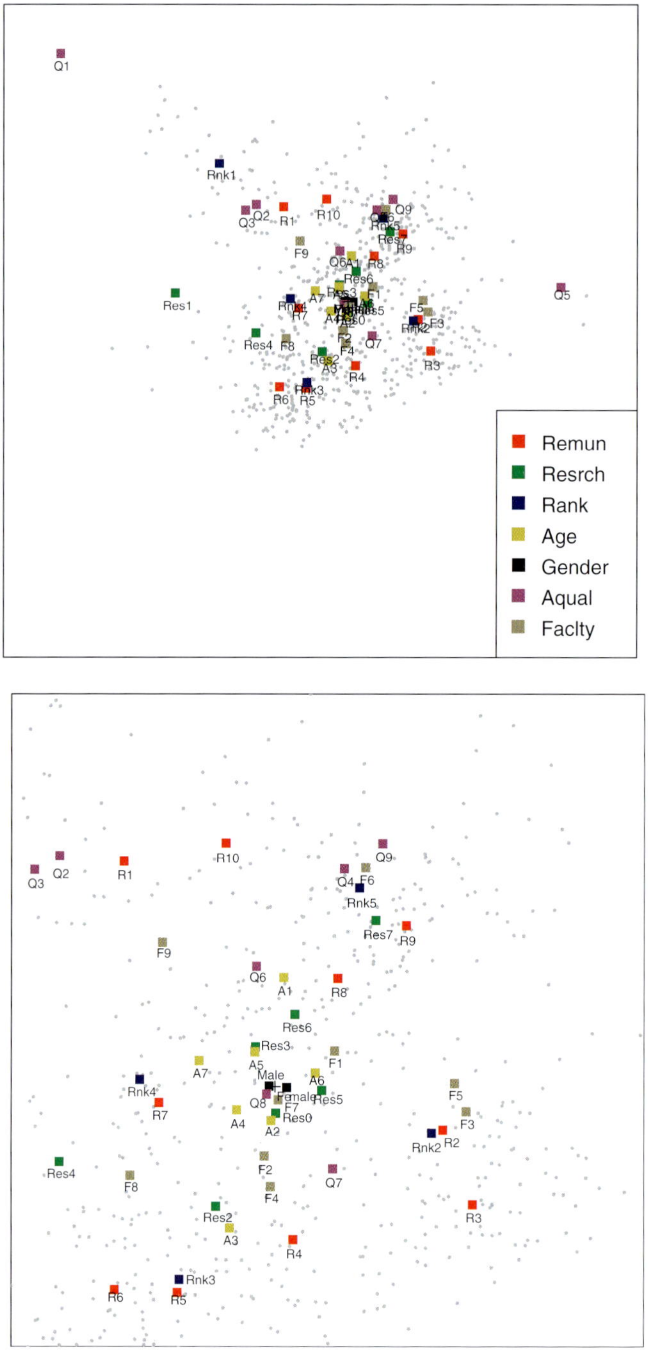

Figure 8.17 MCA biplot of the indicator matrix as in Figure 8.16 (without zoom in the top panel and with `zoomval = 0.42` in the bottom panel) but with dimension 3 replacing dimension 1 as the horizontal scaffolding axis (dimension 2 remains as the vertical scaffolding axis). Note that differences between the males and females have almost vanished.

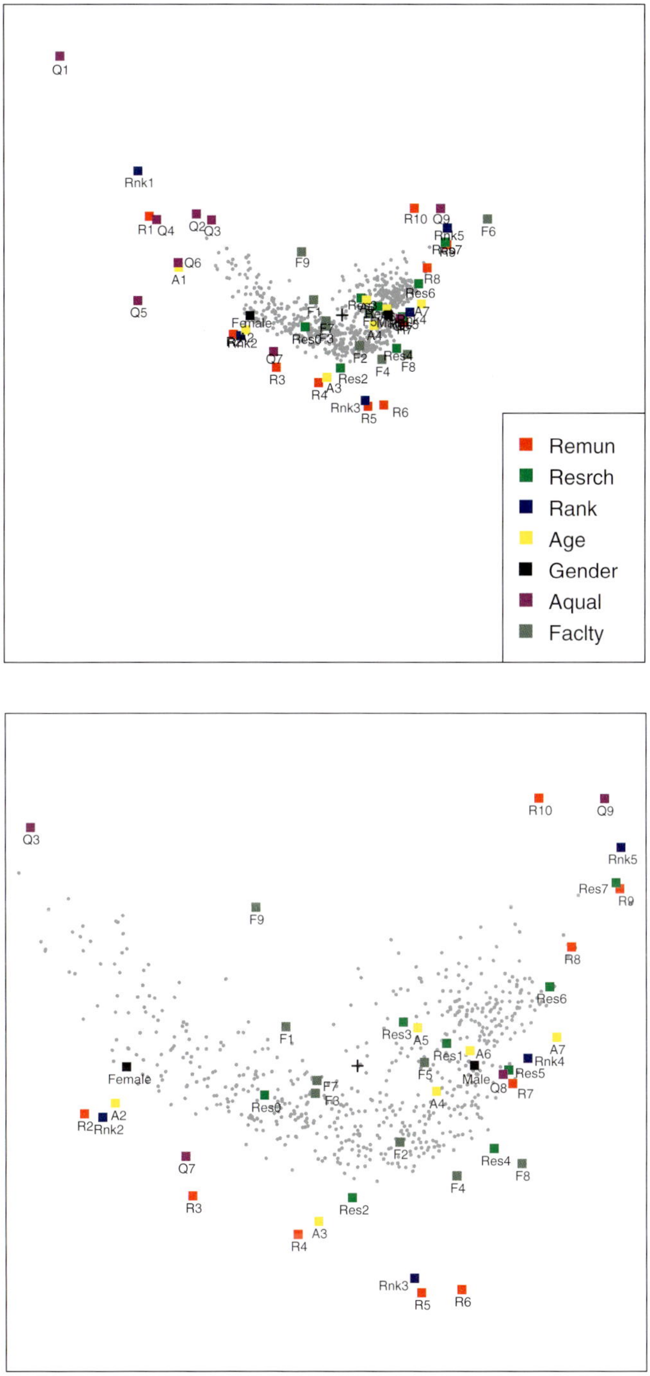

Figure 8.18 MCA Burt matrix based biplot of the dataframe `Remuneration.cat.data.2002`: (top) without zoom; (bottom) with `zoomval = 0.4`.

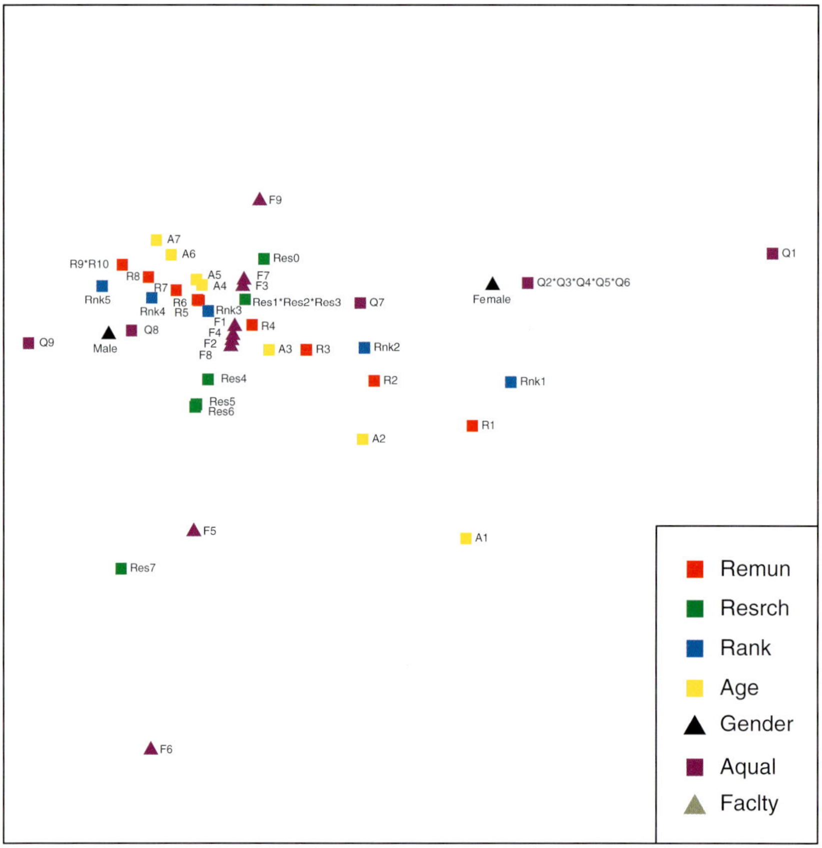

Figure 8.19 Categorical PCA biplot of the dataframe `Remuneration.cat.data.2002`. All variables are ordered categorical except ***Gender*** and ***Faclty***; CLPs are shown. Plotting of sample points is suppressed.

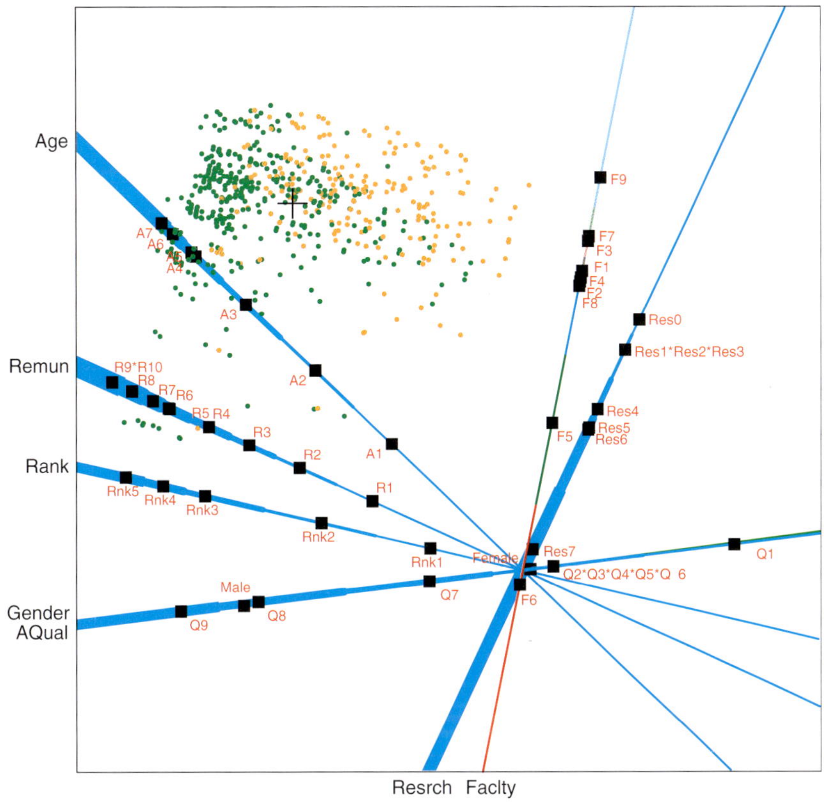

Figure 8.20 Categorical PCA biplot similar to Figure 8.19 but with lines connecting the CLPs associated with each categorical variable. All variables except ***Faclty*** and ***Gender*** are taken as ordered. Sample points are coloured according to gender (green = *male* and orange = *female*). 'Axes' shifted to avoid interference with sample points. CLPs for each variable are connected with lines. Connecting lines for ordered variables have width varying according to ordering; different colours are used in connecting line segments for nominal CLPs. Sample points can be represented as densities, concentration ellipses, α-bags treating all as one group or separated according to faculty, income decile, gender, etc.

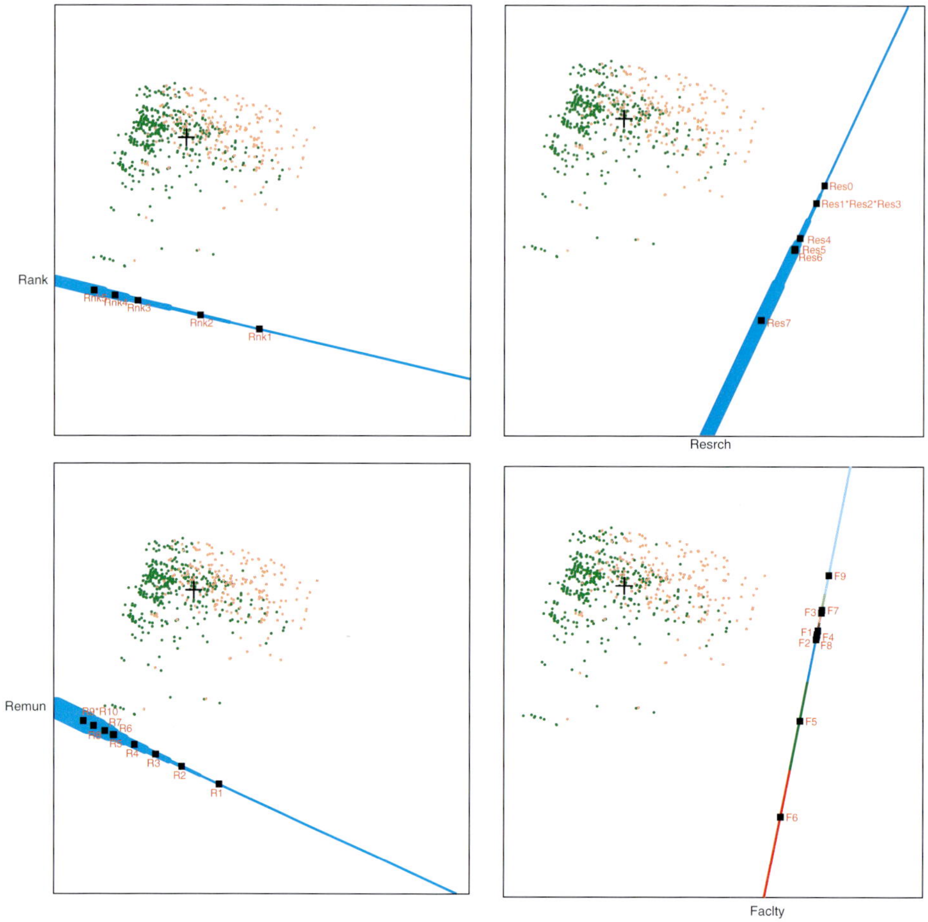

Figure 8.21 Same as Figure 8.20 but with only one axis shown in each panel. All variables, except ***Faclty*** and ***Gender***, are taken as ordered. Sample points are coloured according to gender (green = *male* and orange = *female*).

the Burt matrix. The category centroid of the females in Figure 8.16 is well separated from that of the males. Furthermore, the category centroid of the females is closer to the category centroids of staff in the lower income deciles, the younger staff, the more junior staff members, and those staff members with the smallest research output. This stands in sharp contrast to the category centroid of the males. When the first dimension (horizontal scaffolding axis) in Figure 8.16 is replaced by the third dimension, the positions of the category centroids of the males and females almost coincide. We note also large differences in the positions of the category centroids of the various faculties, with $F6$ quite distant from the rest (in both Figures 8.16 and 8.17). The relative positions of the CLPs in the Burt matrix based biplot of Figure 8.18 are similar to those in Figure 8.16 but they occupy more peripheral positions relative to the individual sample points.

In Figures 8.19–8.21 we give categorical PCA biplots of the `Remuneration.cat.data.2002` data with all variables taken as ordinal categorical except ***Faclty*** and ***Gender***. In these biplots several devices are illustrated for enhancing

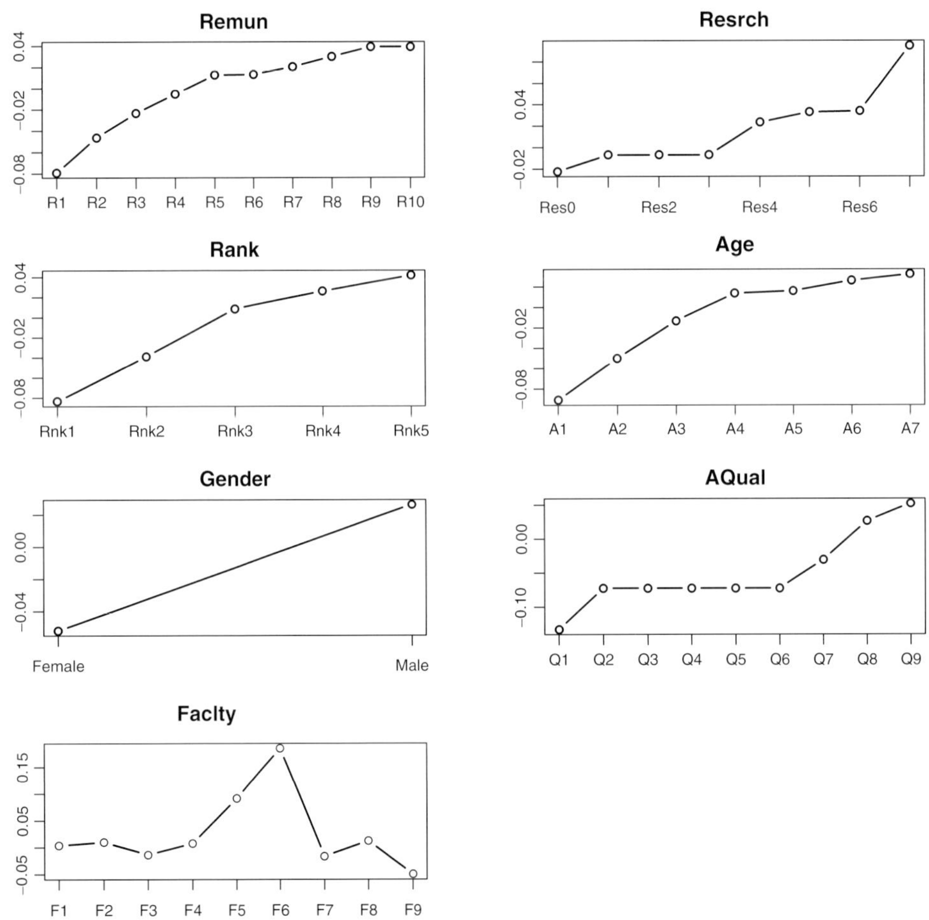

Figure 8.22 Final optimal z-scores for each variable, suggesting that some categories should be lumped together.

the readability of the biplot. We provide the optimal z-scores for the variables in Figure 8.22. This figure shows that some ordinal categories are tied (*R9* and *R10*; *Res1, Res2* and *Res3*; *Q2, Q3, Q4, Q5*, and *Q6*) while others are almost similar (*R5* and *R6*; *Res5* and *Res6*; *A4* and *A5*). A final categorical PCA biplot of the remuneration data is given in Figure 8.23, in which we show the biplot with all axes translated to peripheral positions so as not to interfere with the sample points.

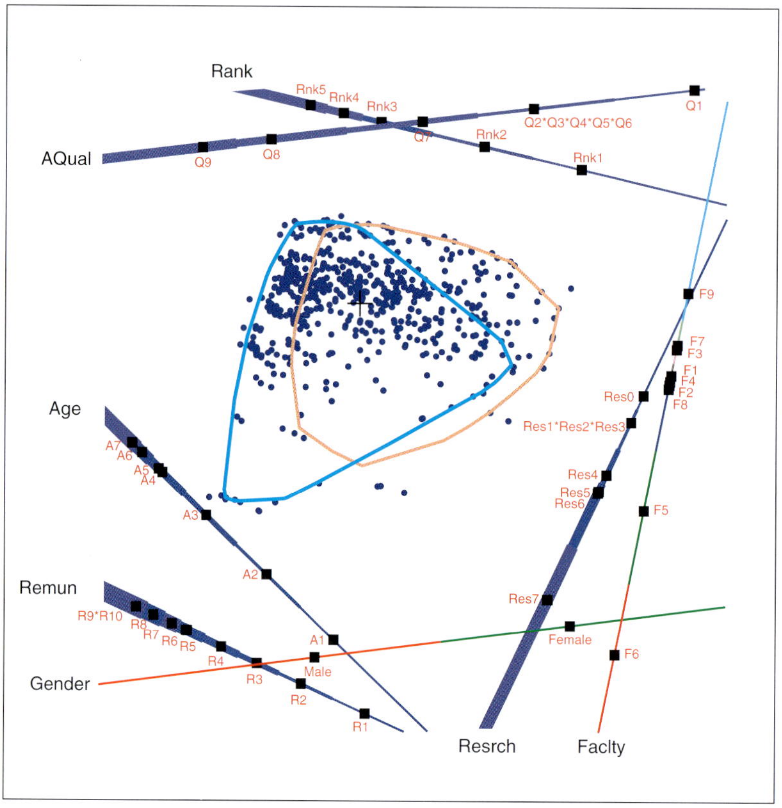

Figure 8.23 Categorical PCA biplot similar to Figure 8.20. In drawing this figure, we have used many of the devices discussed in earlier chapters to improve the display. We have shifted the axes into peripheral positions and translated the axes into familiar positions that are roughly horizontal and vertical. Rather than calibrate each nominal variable according to its quantifications, we have used colour to demarcate the different levels of ***Gender*** and ***Faclty***. For the remaining variables – all ordinal – we increase the width of the axis to demarcate the different levels of the ordinal categories. With large samples, it is impracticable to label individual samples or even to plot them all. If we are interested in possible gender differences we can enclose the sample points with 0.95-bags for each level of ***Gender***. The cyan-coloured bag denotes the males and the orange coloured bag the females. Similarly, if we are interested in, for example, the differences between the various faculties we can use the levels of ***Faclty*** for constructing the bags. Optionally density plots, convex hulls or concentration ellipses (see Section 2.9) are also available to the user.

The complete function call used to construct Figure 8.23 is:

```
CATPCAbipl (Xcat = Remuneration.cat.data.2002[,c(2,3,4,5,
   6,8,9)], ax = 1:7, calibration.size = 1.2, calibration.pch =
   15, calibration.label.size = 0.7, calibration.label.offset =
   rep(0.3,7), calibration.label.pos = c(2,1,1,2,1,1,4),
   c.hull.n = 10, class.cols = c("orange","cyan"),
   class.pch = rep(2,nrow(Remuneration.cat.data.2002)),
   class.vec = Remuneration.cat.data.2002[,6], exp.factor = 1.75,
   drawbagplots = TRUE, factor.type = c(rep("ord",4),
   "nom","ord","nom"), line.type.bags = rep(2,2),
   nom.col = UBcolours[1:9], offset = c(0, 0.3, 0.15, 0),
   ord.col = rep(4,12), orthog.transy = c(-0.35, -0.60,
   0.16, -0.31, -0.3, 0.15, -1.4), plot.samples =
   1:nrow(Remuneration.cat.data.2002), pos = "Hor",
   reverse = TRUE, samples.col = rep("blue",
   nrow(Remuneration.cat.data.2002)), samples.size = 0.75,
   select.origin = FALSE, specify.bags = levels(Remuneration.
   cat.data.2002[,6]), w.factor = 1.25)
```

The biplot in Figure 8.23 provides us with a display that mimics an ordinary scatterplot: all data points are shown together with information on all variables using biplot axes that do not distract attention from the sample points. Calibrations on axes are given in terms of CLPs. Axes for nominal variables are 'calibrated' in different colours, indicating no natural ordering of the categories; axes for ordinal variables are 'calibrated' in the same colour but with increasing width indicating ordered category levels. Blasius *et al*. (2009) give a further example of this type of categorical PCA biplot.

Although the 0.95-bags of the males and females in Figure 8.23 show a large degree of overlap, it is also clear that the males tend to occupy a position indicating higher-ranked positions than the females, higher qualifications, more advanced age, higher remuneration and more research output. Overall it can be concluded that the biplot shows evidence of a gender gap. The changes in the dynamics of this gap can be studied by constructing a similar biplot of the 2005 data. We leave this as an exercise to the reader.

In the next chapter we discuss how to construct biplots with some continuous and some categorical variables.

9 *Generalized biplots*

9.1 Introduction

Nonlinear biplots (Chapter 5) generalize the PCA biplot by providing for distance measures for quantitative variables other than Pythagorean distance; generalized biplots (Gower, 1992) offer a further generalization that allows categorical variables to be included. The map of the samples is achieved by allowing for the calculation of distances between samples with categorical measurements and plugging this distance matrix into the PCO method. Biplot axes for any quantitative variables are handled in a similar way to that discussed for nonlinear biplots.

Because a categorical variable can only assume a finite number of distinct values, called category levels, the concept of a continuous trajectory becomes invalid. Now, each categorical variable will be represented by a set of category-level points – one point for each level – in such a way that samples nearest to a certain CLP will be associated with that level of the categorical variable (see Chapter 8). The nearest-neighbour region for a CLP is the convex subspace that contains all points that are nearer to this CLP than to any of the other CLPs for this variable. The combination of nonlinear trajectories for continuous variables with CLPs for categorical variables defines a generalized coordinate system, termed a *reference system*. In low-dimensional approximations, continuous variables are represented by nonlinear biplot axes, while the contributions from the nearest-neighbour regions define convex regions termed prediction regions.

Table 9.1 shows a simple data set consisting of two variables, the first the height of each subject (a continuous variable in centimetres) and the second the colour of their eyes (a categorical variable with three levels: *blue*, *brown* and *green*). One possibility for plotting these data is to represent ***height*** on a horizontal axis and use three parallel bars for the categories, one below the other, as is shown in Figure 9.1. There is, however, no natural ordering of the categories and any other order would be equally valid. In the following, we show how categorical variables can be represented directly in a scatterplot.

Table 9.1 Example data set for both continuous and categorical data.

Subject	Height	Eye colour
a	172	blue
b	171	brown
c	175	green
d	168	brown

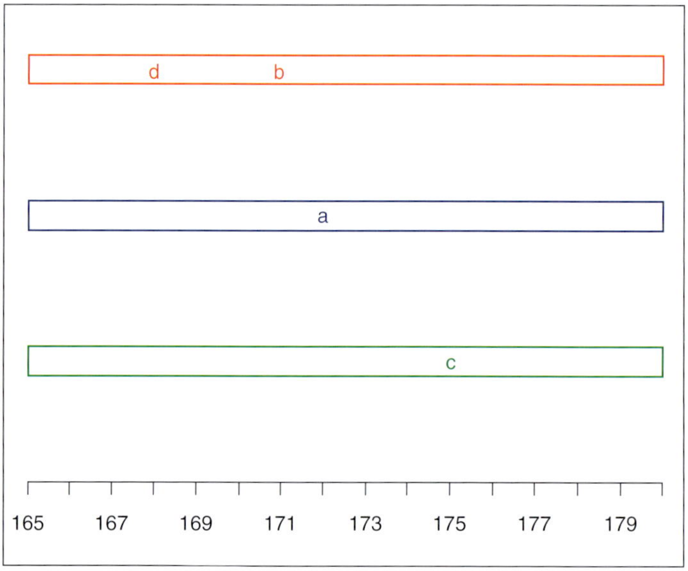

Figure 9.1 Graphical representation of height and eye colour data of Table 9.1.

9.2 Calculating inter-sample distances

In the nonlinear biplot described in Chapter 5 the matrix $\mathbf{X}$: $n \times p$ represents n observations on p variables and d_{ij} denotes the distance between observations $\mathbf{x}_i$ and $\mathbf{x}_j$. The matrix $\mathbf{D} = \{-\frac{1}{2}d_{ij}^2\}$ of *ddistances* forms the basis of the nonlinear biplot.

As in Chapter 5, it will be assumed that the matrix $\mathbf{D}$ is calculated with additive Euclidean embeddable distance measures. The *ddistances* can therefore be expressed in the form

$$-\frac{1}{2}d_{ij}^2 = \sum_{k=1}^{p} f_k(x_{ik}, x_{jk}). \tag{9.1}$$

For continuous variables, any additive Euclidean embeddable distance measure can be used. For example, in Chapter 5 examples of the function $f_k(x_{ik}, x_{jk})$ for the kth variable are defined as follows: for Pythagorean distance,

$$f_k(x_{ik}, x_{jk}) = -\frac{1}{2}(x_{ik} - x_{jk})^2; \tag{9.2}$$

for the square root of the Manhattan distance,

$$f_k(x_{ik}, x_{jk}) = -\frac{1}{2}\left|x_{ik} - x_{jk}\right|; \tag{9.3}$$

and for Clark's distance,

$$f_k(x_{ik}, x_{jk}) = -\frac{1}{2}\left(\frac{x_{ik} - x_{jk}}{x_{ik} + x_{jk}}\right)^2. \tag{9.4}$$

All these distances are defined in terms of differences between two values of the same variable, so there is no need to work in terms of deviations from the mean. However, (9.2) and (9.3) depend on the scaling used, which, as explained in Sections 2.5 and 3.6, makes it vital to use some form of normalization when combining measurements from variables measured on incommensurable scales. Gower (1992) and Gower and Hand (1996) consider scaling each variable to have unit sum of squares or unit range. In the following, we assume that all quantitative variables have been prescaled to correct for incommensurability. In our first example, we have scaled to unit range with the scaled value $\tilde{x}_{ik}$ for $i = 1, 2, \ldots, n$ and $k = 1, 2, \ldots, p$ given by

$$\tilde{x}_{ik} = \frac{x_{ik}}{\max_i(x_{ik}) - \min_i(x_{ik})}. \tag{9.5}$$

Then we have used Pythagorean distance.

Due to the assumption of additive distance it can be assumed without loss of generality that the variables are ordered such that the first $p_{(1)}$ are continuous and the remaining $p_{(2)}$ are categorical, with $p_{(1)} + p_{(2)} = p$ and

$$\mathbf{D} = \sum_{k=1}^{p_{(1)}} \mathbf{D}_k + \sum_{k=p_{(1)}+1}^{p} \mathbf{D}_k = \mathbf{D}_{(1)} + \mathbf{D}_{(2)}, \tag{9.6}$$

where $\mathbf{D}_k = \{f_k(x_{ik}, x_{jk})\}$, the *ddistance* matrix derived solely for the kth variable.

The matrix $\mathbf{D}_{(1)}$ is calculated as before, using an additive Euclidean embeddable distance measure on the $p_{(1)}$ continuous variables. If the kth variable is categorical, an indicator matrix (see Chapter 8) $\mathbf{G}_k$: $n \times L_k$ is formed with L_k the number of category levels for this variable. Each row of $\mathbf{G}_k$ represents a sample such that

$$g_{ih} = \begin{cases} 1 & \text{if the } i\text{th observation on variable } k \text{ falls into category level } h \\ 0 & \text{otherwise.} \end{cases}$$

To calculate $\mathbf{D}_{(2)}$ the matrix

$$\underset{n \times L}{\mathbf{G}} = \begin{bmatrix} \mathbf{G}_{p(1)+1} & \cdots & \mathbf{G}_p \end{bmatrix}$$

is formed, where $L = L_{p(1)+1} + \ldots + L_p$. If the L columns of $\mathbf{G}$ are, or are viewed as, dichotomous variables, an obvious approach would be to derive $\mathbf{D}_{(2)}$ as a matrix of dissimilarities. However, coding multilevel categories as a series of dichotomous variables leads to a situation where the number of negative matches (0–0) dominates the number of matches (1–1), which occurs only when both samples have the same category level. Furthermore, the number of matches (1–1) is the same as the number of agreements between the variables but the number of mismatches (1–0 or 0–1) is twice the number of disagreements between the category levels of the variables. To overcome this deficiency, Gower (1992) suggests an extension of the Jaccard coefficient, the *extended matching coefficient*, which is Euclidean embeddable, defined as

$$f_k(x_{ik}, x_{jk}) = -\frac{1}{2}\sum_{h=1}^{L_k} I\left(g_{ih}^{(k)} = g_{jh}^{(k)}\right), \tag{9.7}$$

where $I(a = b)$ is the indicator function which is unity when $a = b = 1$, and zero otherwise. Since the ijth element of $\mathbf{D}_k$ is $f_k(x_{ik}, x_{jk})$, it follows that

$$\mathbf{D}_k = -\frac{1}{2}(\mathbf{1}_n\mathbf{1}'_n - \mathbf{G}_k\mathbf{G}'_k)$$

and from $\mathbf{D}_{(2)} = \sum_{k=p_{(1)}+1}^{p} \mathbf{D}_k$ the matrix $\mathbf{D}_{(2)}$ is obtained as

$$\mathbf{D}_{(2)} = -\frac{1}{2}(p_{(2)}\mathbf{1}_n\mathbf{1}'_n - \mathbf{G}\mathbf{G}'). \tag{9.8}$$

Since the EMC contributes zero to the overall similarity of the two samples if the kth categorical variable mismatches and unity if it matches, $f_k(x_{ik}, x_{jk})$ can assume a maximum value of 1. This ensures that the contributions from the EMC can be combined commensurably with the contributions from the quantitative variables, for example as in (9.5).

The matrix $\mathbf{D}$ in (9.6) can be plugged into any multidimensional scaling method to obtain a map of the sample points. However, biplot axes and prediction regions are most readily constructed when using principal coordinates/classical scaling as we did for nonlinear biplots.

9.3 Constructing a generalized biplot

A PCO of an Euclidean embeddable *ddistance* matrix $\mathbf{D}$ gives a map $\mathbf{Y}$ in $\mathscr{R}$. In Figure 9.2 the map is based on Pythagorean distance for the heights and the EMC for the eye colours. This representation is exact and approximation is not relevant since four samples can be represented exactly in three dimensions.

9.4 Reference system

As explained above, the reference system in generalized biplots consists of the usual biplot trajectories representing the continuous variables, and a set of CLPs with accompanying nearest-neighbour and prediction regions defined for each categorical variable.

If all the variables were continuous, the biplot trajectories could be obtained similarly to the nonlinear biplot trajectories by interpolating 'new samples'

$$\phi\mathbf{e}_k = \frac{\tau}{\max\limits_{h=1,\ldots,n}(x_{hk}) - \min\limits_{h=1,\ldots,n}(x_{hk})}\mathbf{e}_k, k = 1, 2, \ldots, p \text{ and } -\infty < \tau < \infty,$$

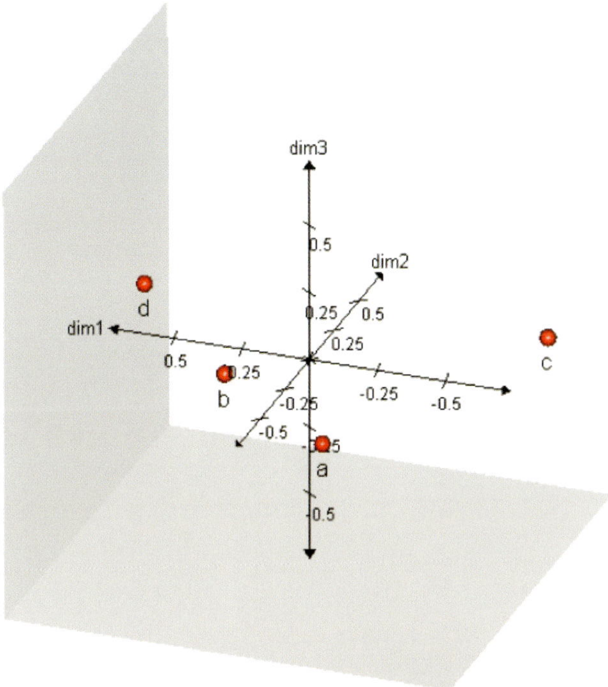

Figure 9.2 Three-dimensional map of the one continuous and the single categorical variable given in Table 9.1.

with the interpolation formula. The value $\phi\mathbf{e}_k$ is a pseudo-sample having zero values, representing the means, for all variables except the kth. As τ varies, the interpolated values trace out the trajectory for the kth variable. For a categorical variable the concept of a centred data matrix with zero representing the mean value of the variable is invalid. To make progress, for each admissible value τ of the kth variable we need a set of n pseudo-samples that do not require the zero mean. Thus, we consider

$$(x_{11}, x_{12}, \ldots, x_{1;k-1}, \tau, x_{1;k+1}, \ldots, x_{1p})$$
$$\vdots$$
$$(x_{i1}, x_{i2}, \ldots, x_{i;k-1}, \tau, x_{i;k+1}, \ldots, x_{ip})$$
$$\vdots$$
$$(x_{n1}, x_{n2}, \ldots, x_{n;k-1}, \tau, x_{n;k+1}, \ldots, x_{np}).$$

It follows that the pseudo-samples differ from the original $\mathbf{X}$ only in the common value τ assigned to the kth variable. If the kth variable is continuous then $-\infty < \tau < \infty$. and if it is categorical then τ will take the L_k distinct values representing the category levels for that variable.

We may interpolate the n pseudo-samples and find their centroid. It is the locus of this centroid as τ varies that defines a nonlinear trajectory for a continuous variable

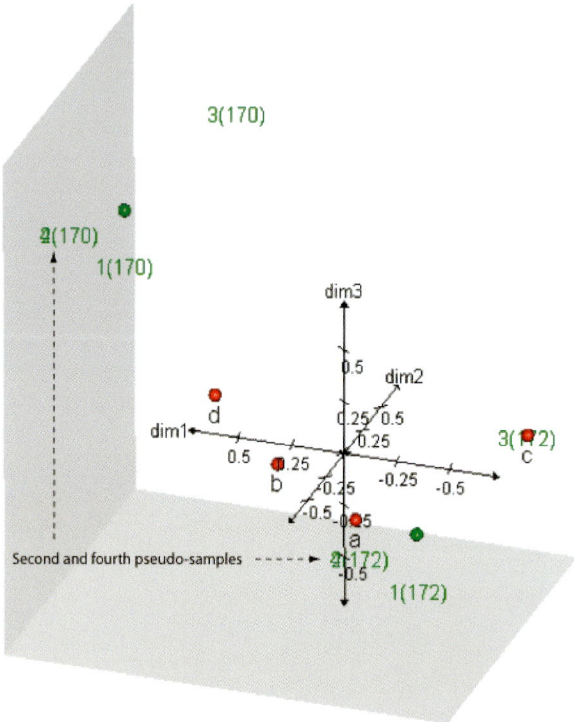

Figure 9.3 Pseudo-sample interpolation for variable ***height*** where $\tau = 170$ and $\tau = 172$.

and a set of L_k CLPs for a categorical variable. Gower (1992) shows that for a PCO for continuous variables, with a wide class of distance measures in common use, this choice of pseudo-samples effectively leads to the nonlinear biplot. The only difference is that the trajectory is parallel to the nonlinear biplot trajectory (see orthogonal parallel translation: Section 2.4). We note that, unlike nonlinear biplots, $\tau = 0$ is not common to all the continuous trajectories. In Figure 9.3 the pseudo-samples for variable $k = 1$, ***height***, are interpolated where $\tau = 170$ and $\tau = 172$. As an example, the pseudo-samples for $\tau = 170$ are given by: (170, blue); (170, brown); (170, green); (170, brown).

Since the second and fourth pseudo-samples are identical, their interpolated representations are at exactly the same point. The centroid of the four interpolated pseudo-samples is shown as a green sphere.

As explained above, when the assumption of additive distance is valid the biplot trajectory can be calculated by using the distance between the centroid of the pseudo-samples and the original samples. Define the matrix

$$\tilde{\mathbf{D}} = \begin{bmatrix} \mathbf{D}_{11} & \mathbf{D}_{12} \\ \mathbf{D}'_{12} & \mathbf{D}_{22} \end{bmatrix} \tag{9.9}$$

with $\mathbf{D}_{11}$ containing *ddistances* between the samples, $\mathbf{D}_{22}$ containing *ddistances* between the pseudo-samples and $\mathbf{D}_{12}$ containing *ddistances* between the ith sample point and the jth pseudo-sample.

Gower and Hand (1996) show that the squared distances between the centroid of the pseudo-samples and the original samples are given by the elements of the vector

$$\mathbf{a} = \frac{\mathbf{1}'\mathbf{D}_{22}\mathbf{1}}{n^2}\mathbf{1} - \frac{2}{n}\mathbf{D}_{12}\mathbf{1},$$

where n is the number of observations in the 'second' set forming the matrix $\mathbf{D}_{22}$. The centroid of the pseudo-samples can be interpolated onto $\mathscr{R}^+$ with the vector $\mathbf{d}_{n+1} = -\frac{1}{2}\mathbf{a}$ using the usual formula (5.9), giving

$$\mathbf{y} = \mathbf{\Lambda}^{-1}\mathbf{Y}'(\{f(x_{ik}, \tau)\} - \mathbf{D}_k\mathbf{1}/n). \tag{9.10}$$

This is the fundamental equation for generalized biplots, because for different choices of k and $f(x_{ik}, \tau)$ it gives the coordinates $\mathbf{y}$ for all points of the reference system. In particular, we see that when $\tau = x_{ik}$ then $f(x_{ik}, \tau) = 0$, showing that every sample point is nearer its marker on the kth trajectory than it is to any other marker on the same trajectory or CLP, thus satisfying the fundamental 'nearness' property of a coordinate system.

If we allow τ to vary in a sequence between 168 and 175, the embedded biplot trajectory of ***height*** is found as shown in Figure 9.4.

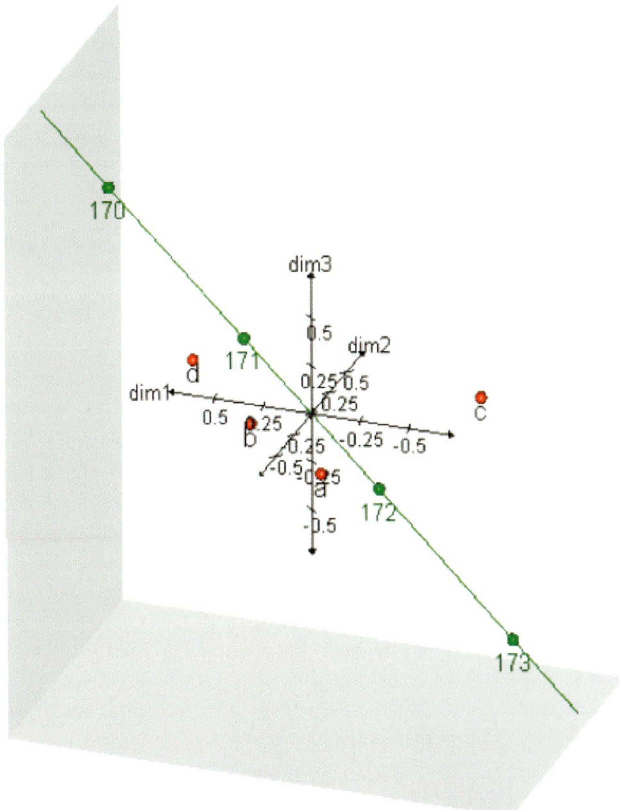

Figure 9.4 Biplot trajectory for ***height*** in $\mathscr{R}$.

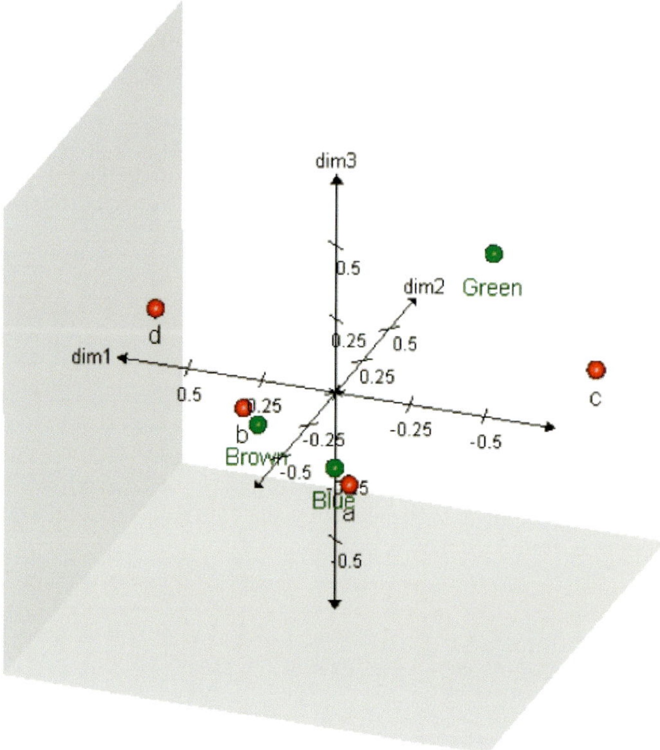

Figure 9.5 Generalized biplot ordination with CLPs (green spheres) in $\mathscr{R}$.

The CLPs are found in a similar manner by using the pseudo-samples $(172, \tau)$, $(171, \tau)$, $(175, \tau)$, $(168, \tau)$ where τ takes the 'values' *blue*, *brown* and *green*. The CLPs are represented in $\mathscr{R}$ in Figure 9.5.

Since the PCO is already referred to its principal axes, the biplot representation in, say, $r = 2$ dimensions is simply obtained from the first two dimensions of Figure 9.5, as shown in Figures 9.6 and 9.7. Figure 9.7 shows the interpolation biplot axis for the continuous variable ***height*** and the associated CLPs for the categorical variable ***eye colour***.

9.5 The basic points

From (9.10) we may compute the markers for all values, quantitative and categorical, occurring in $\mathbf{X}$. These are not of great interest for quantitative variables, but they give all the CLPs for categorical variables. From (9.10) we see that the basic points for the kth variable are given by (since $\mathbf{1}'\mathbf{Y} = \mathbf{0}'$)

$$\mathbf{Z}'_k = \mathbf{\Lambda}^{-1}\mathbf{Y}'\mathbf{D}_k(\mathbf{I} - \mathbf{1}\mathbf{1}'/n) = \mathbf{\Lambda}^{-1}\mathbf{Y}'(\mathbf{I} - \mathbf{1}\mathbf{1}'/n)\mathbf{D}_k(\mathbf{I} - \mathbf{1}\mathbf{1}'/n),$$

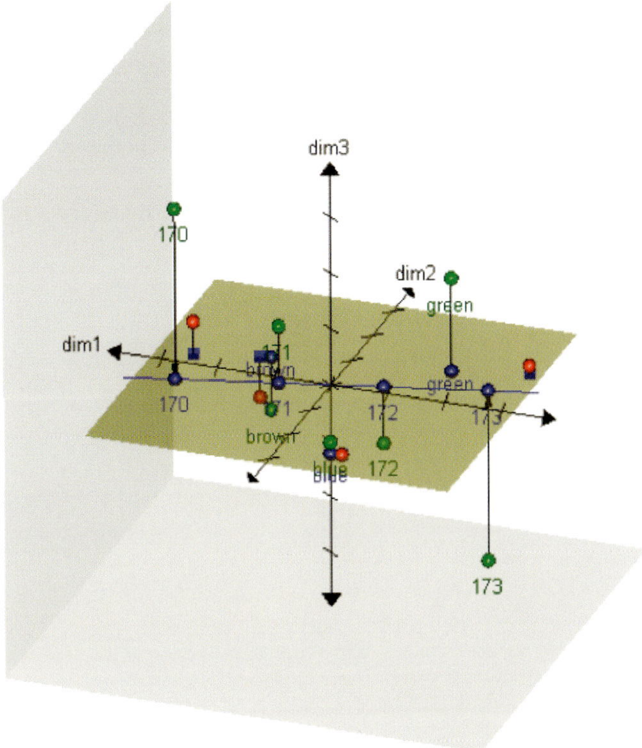

Figure 9.6 Principal axes projection for two dimensional generalized biplot.

from which it follows that

$$\sum_{k=1}^{p} \mathbf{Z}_k = \sum_{k=1}^{p} (\mathbf{I} - \mathbf{11}'/n)\mathbf{D}_k(\mathbf{I} - \mathbf{11}'/n)\mathbf{Y}\boldsymbol{\Lambda}^{-1} = \sum_{k=1}^{p} \mathbf{B}_k\mathbf{Y}\boldsymbol{\Lambda}^{-1} = \mathbf{Y},$$

where $\mathbf{B}_k$ is the double-centred version of $\mathbf{D}_k$. These are transition formulae, the first equation giving the basic points for given coordinates $\mathbf{Y}$ and the second giving $\mathbf{Y}$ as a sum of the basic points. Similar results occur for MCA which can be seen as special case of this much more general formulation (see Gower and Hand (1996), for details).

For all variables we have $\mathbf{1}'\mathbf{Z}_k = \mathbf{0}'$ and, in particular, for categorical variables, $\mathbf{Z}_k$ will have only L_k distinct values, so $\mathbf{1}'\mathbf{Z}_k = \mathbf{0}'$ implies that the weighted mean of the category points is at the centroids of the sample points.

9.6 Interpolation

Interpolation in the generalized biplot is achieved similarly to the method described in Chapter 5 for the nonlinear biplot. Of course, a new numerical observation $\mathbf{x}^*$

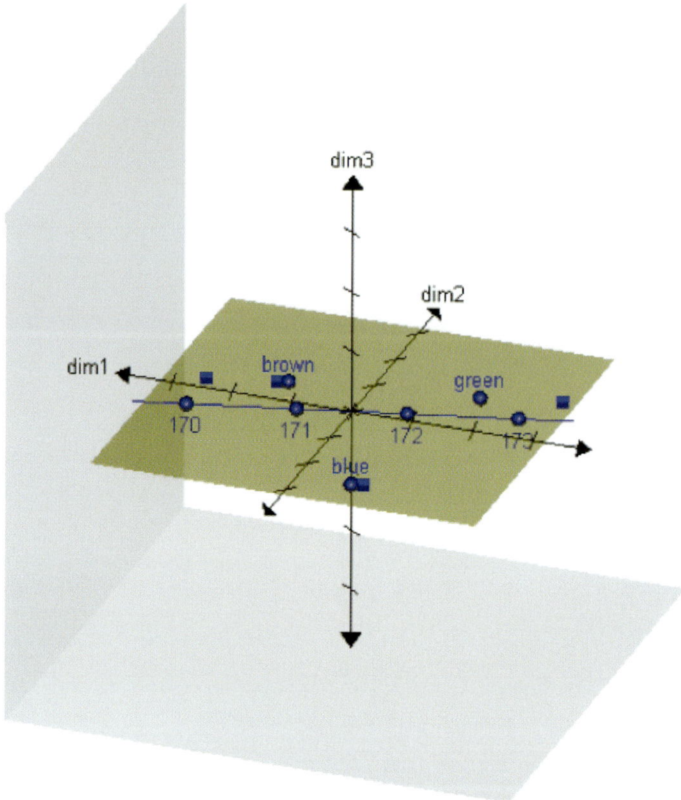

Figure 9.7 Generalized interpolative biplot in two-dimensional space.

must be scaled in the same way as were the original variables. The vector $\mathbf{d}_{n+1}$ is calculated with the chosen Euclidean embeddable distance measures, for example those described in Section 9.2. For a categorical variable, the matrix $\mathbf{G}$ can be appended with an $(n+1)$th row denoting the category levels for the new sample $\mathbf{x}^*$. For instance, when using Pythagorean distance and the EMC the ith element in the vector $\mathbf{d}_{n+1}$ is obtained by

$$
\begin{aligned}
-\frac{1}{2}d^2_{n+1,i} &= \sum_{k=1}^{p} f_k(x_{ik}, x^*_k) \\
&= \sum_{k=1}^{p_{(1)}} -\frac{1}{2}(x_{ik} - x^*_k)^2 + \sum_{k=p_{(1)}+1}^{p} \left\{ -\frac{1}{2}\sum_{h=1}^{L_k} I(g^{(k)}_{ih} = g^{*(k)}_{n+1,h}) \right\}.
\end{aligned}
$$

Several new samples can be interpolated by repeated use of this method, forming a new matrix $\mathbf{G}^*$: $m \times L$ for the m new samples, with each row corresponding to a sample and the columns corresponding to the category levels similar to the matrix $\mathbf{G}$. Combining $\mathbf{G}$

and $\mathbf{G}^*$ into the matrix

$$\tilde{\mathbf{G}} = \begin{bmatrix} \mathbf{G} \\ \mathbf{G}^* \end{bmatrix},$$

a matrix $\tilde{\mathbf{D}}_{(2)} = -\frac{1}{2}(p_{(2)}\mathbf{1}_{n+m}\mathbf{1}'_{n+m} - \tilde{\mathbf{G}}\tilde{\mathbf{G}}')$ can be calculated. The matrix $\tilde{\mathbf{D}}_{(2)}$ can be partitioned as

$$\tilde{\mathbf{D}}_{(2)} = \begin{bmatrix} \mathbf{D}_{11} : n \times n & \mathbf{D}_{12} : n \times m \\ \mathbf{D}_{21} : m \times n & \mathbf{D}_{22} : m \times m \end{bmatrix},$$

where the matrix $\mathbf{D}_{11}$ is equivalent to $\mathbf{D}_{(2)}$, while each of the m columns of $\mathbf{D}_{12}$ forms one of the $\mathbf{d}_{n+1}$ vectors necessary for the interpolation formula $\mathbf{z}^* = \mathbf{\Lambda}_r^{-1}\mathbf{Y}'_r(\mathbf{d}_{n+1} - \frac{1}{n}\mathbf{D1})$.

Graphical interpolation remains possible, but the nonconcurrency of the trajectories introduces complications concerned with the marker O_k for the mean on the kth trajectory ($k = 1, 2, \ldots, p_{(1)}$) and the centroid G of the sample points. For details, see Gower and Hand (1996).

9.7 Prediction

To predict the values of quantitative variables, we have seen that the generalized biplot pseudo-samples lead to nonconcurrent trajectories that are parallel to the trajectories of nonlinear biplots. So far as prediction is concerned, we may continue to use the concurrent nonlinear biplot trajectories themselves. Prediction for continuous variables is performed with either normal or circular projection as described in Section 5.3.3 exactly as in the case of the nonlinear biplot. The predicted category for each categorical variable is determined by the prediction regions. In $\mathscr{R}^+$ the nearest-neighbour region for the hth level of the kth categorical variable will be the subspace that contains all points that are nearer to this CLP than to any of the other CLPs of the kth variable. If this variable has L_k category levels, the space $\mathscr{R}^+$ will be divided into L_k disjoint and exhaustive neighbour regions, $\mathscr{F}_1, \mathscr{F}_2, \ldots, \mathscr{F}_{L_k}$. To predict the category level to be associated with a point $\mathbf{z}^*$ in $\mathscr{L}$ its representation $\mathbf{z}^{*+}$ in $\mathscr{R}^+$ is needed. Since $\mathscr{L}$ is a linear subspace of $\mathscr{R}$ and the nth component of any point in $\mathscr{R}$ is 0, this is given by

$$\mathbf{z}^{*+} = \begin{bmatrix} \mathbf{z}^*: & r \times 1 \\ \mathbf{0}: & (n-1-r) \times 1 \\ 0: & 1 \times 1 \end{bmatrix}.$$

The distances between $\mathbf{z}^{*+}$ and each of the CLPs must be calculated and the predicted category corresponds to the nearest CLP.

The prediction regions of categorical variables are the representation of the nearest-neighbour regions in $\mathscr{L}$, that is, the intersection $\mathscr{F}_h \cap \mathscr{L}$. It may happen that $\mathscr{L}$ does not intersect with all the neighbour regions of a categorical variable, and in that case certain category levels do not appear and are never predicted. Furthermore, projected CLPs need not lie in their own prediction regions, and certainly do not define nearest-neighbour regions in $\mathscr{L}$ that coincide with the prediction regions. Indeed, the prediction regions are not nearest-neighbour regions for any set of points in $\mathscr{L}$.

Gower and Hand (1996) discuss a detailed algorithm for computing the boundaries of the prediction regions. If the dimension of $\mathscr{L}$ is $r = 2$ they provide a simpler, but less efficient, method for obtaining the prediction regions:

- Construct a two-dimensional $m \times m$ grid of pixels in $\mathscr{L}$, say $\mathbf{E}$: $m^2 \times 2$.
- For each pixel of $\mathbf{E}$, calculate which CLP is nearest.
- Label (i.e. colour) the pixel in $\mathscr{L}$ accordingly.

Grid points can be labelled with colours corresponding to the different category levels. In all the previous biplots, the biplot trajectories of all the variables were displayed in one representation. The coloured prediction region diagram described above, however, pertains to one categorical variable and is the equivalent of one continuous biplot trajectory. Usually, to avoid confusing overlap, different prediction region diagrams will be needed for each categorical variable.

In Figure 9.8 the pixels are coloured to show the prediction regions for *blue*, *green* and *brown* eyes. The prediction biplot axis for the continuous variable ***height*** is obtained

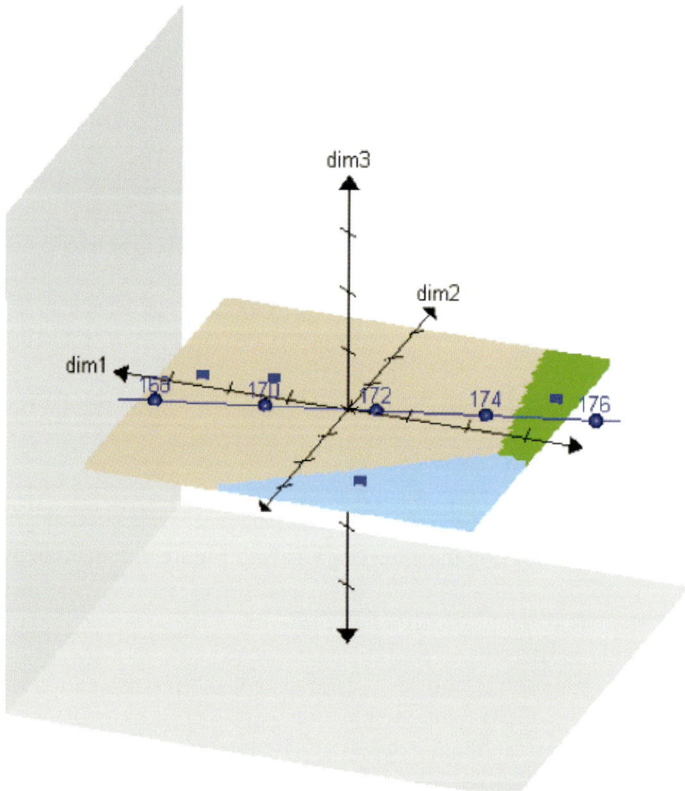

Figure 9.8 Prediction regions for the generalized biplot of the Table 9.1 data.

similarly to the prediction biplot trajectories of the nonlinear biplot described in Chapter 5. Since this example uses Pythagorean distance, all three prediction methods produce the same linear biplot trajectory shown in Figure 9.8.

9.8 An example

We now return to the remuneration data introduced in Section 4.9.1 and also used in Section 8.10 to illustrate a categorical PCA biplot where the continuous variables were categorized. In a generalized biplot, the distinction between quantitative and qualitative variables is retained. The same variables described in Section 8.10 are used in this illustration but now treating ***Remun***, ***Resrch***, ***Age*** and ***AQual*** quantitatively. We use Pythagorean distance for the continuous variables and the ECM for the categorical variables. As usual (see, for example, Section 9.2) we have to use some form of scaling for the quantitative variables. Furthermore, due to the difference in the number of categories of the qualitative variables, they too have to be normalized. The usual way to normalize the quantitative variables is to centre and then scale each to unit sum of squares. Thus we normalized each of ***Remun***, ***Resrch***, ***Age*** and ***AQual*** to unit sum of squares. Since $\mathbf{1}'\mathbf{D}_q\mathbf{1}$ for each quantitative variable is equal to $-n$ times the corrected sum of squares for that variable, this normalization process is equivalent to dividing the *ddistances* $\mathbf{D}_q$ by $-(\mathbf{1}'\mathbf{D}_q\mathbf{1})/n$. Equivalently, with the EMC, each qualitative variable was scaled so that $\mathbf{1}'\mathbf{D}_k 1 = -n$, where $\mathbf{D}_k = -\frac{1}{2}(\mathbf{1}_n\mathbf{1}'_n - \mathbf{G}_k\mathbf{G}'_k)$. This type of normalization balances the contributions of quantitative and qualitative variables to overall distance, but it is not the only possibility.

It is clear from Figure 9.9 that prediction regions for all categories of a qualitative variable are not necessarily represented in the biplot space: for academic position, prediction regions for lecturer ($R2$), senior lecturer ($R3$) and full professor ($R5$) are visible but not those for junior lecturer and associate professor, while prediction regions for only four of the nine faculties appear in the biplot space. The individual sample points are printed as solid squares. We have coloured (in the top left biplot) the squares according to gender: red squares denoting the females and green squares the males. However, the output of our function `Genbipl` provides all the necessary information for easily obtaining biplots with a different colouring scheme – for example, the different faculties or different academic positions.

That the sample points, obtained by projection, do not necessarily fall within their corresponding prediction regions, obtained by back-projection, is clear from Figure 9.9. We could show this by colouring the category levels in Figure 9.9, but this would interfere with the coloured prediction regions. Therefore we show the information numerically in Table 9.2, where the entries in bold give the numbers of correct predictions while the plain entries show the numbers of incorrect predictions. The proportion of correct predictions is closely analogous to the predictivity measure (Section 3.3) for quantitative variables. However, with categorical variables we have additional information giving the separate contributions to predictivity of each category level. Thus in Table 9.2(a) $R1$ and $R4$ are never predicted, $R2$ and $R5$ are well predicted, while $R3$ is poorly predicted. In Table 9.2(b) both genders are well predicted. Table 9.2(c) never predicts $F3, F4, F6, F7$ and $F8$, while $F1, F2, F5$ and $F9$ are all rather poorly predicted.

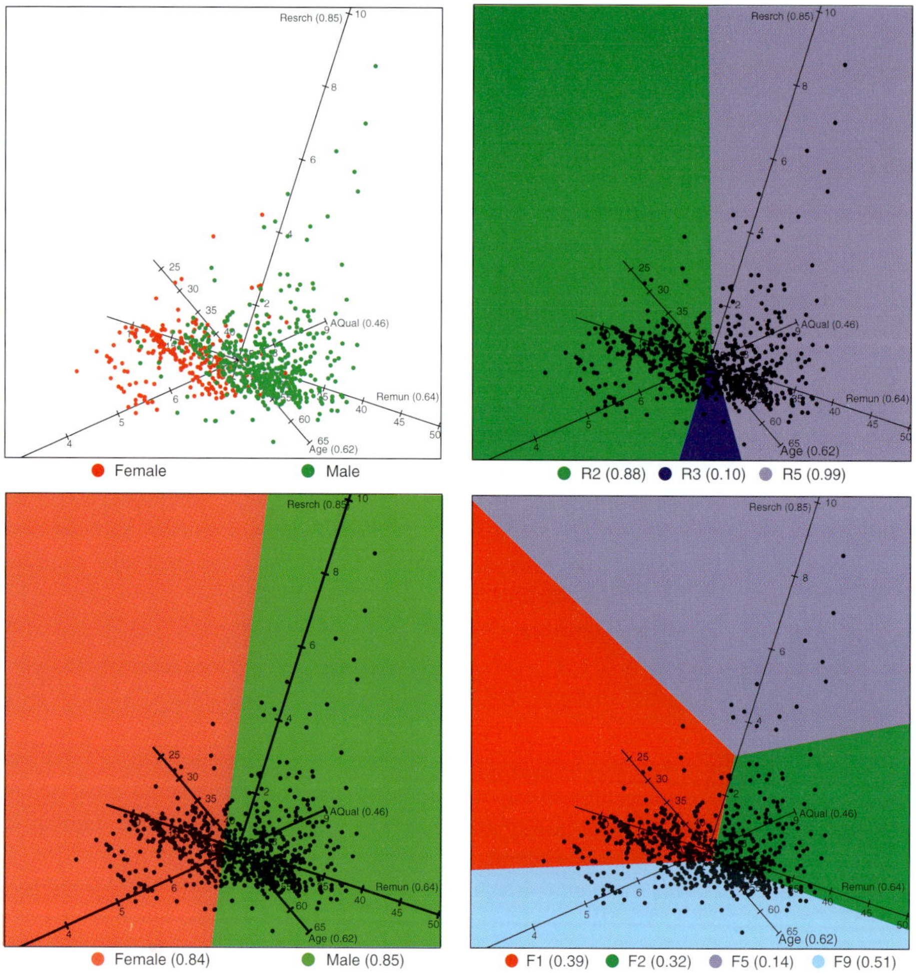

Figure 9.9 Generalized biplot of Remuneration data. Variables ***Resrch***, ***Remun***, ***Age*** and ***AQual*** are treated as quantitative variables, each centred and scaled to unit sum of squares. Variables ***Gender*** (two categories), ***Rank*** (five academic positions) and ***Faclty*** (nine different faculties) are treated as qualitative. Inter-sample distances are calculated using Pythagorean distances for the continuous variables and the EMC for the categorical variables. The quantitative variables are represented by calibrated linear axes. The calibrations are in terms of the original units using the methods described in Chapter 2. The length of each axis covers the range of the actually observed values. For the qualitative variables, prediction regions are the counterpart to biplot axes. In the top left panel we have the samples as points coloured according to gender with the biplot axes of the quantitative variables. Axis predictivities for the quantitative variables given in brackets. In the top right biplot we have the prediction regions for variable ***Rank***; in the bottom left panel the prediction regions for ***Gender*** and in the bottom right panel the prediction regions for ***Faclty***. In the biplots given in the top right panel and in both the bottom panels the proportion of correct predictions for the respective categories are given in parentheses.

Table 9.2 Predictions for the three categorical variables: (a) ***Rank***, (b) ***Gender*** and (c) ***Faclty***. Columns of Tables are given only for Prediction regions appearing in biplot space because no other prediction is possible. The 'diagonal' values giving correct predictions are shown in bold.

(a)

		Prediction			
		R2	R3	R5	Total
Observed category	R1	44	0	0	44
	R2	**180**	13	11	204
	R3	95	**20**	94	209
	R4	11	6	79	96
	R5	1	1	**173**	175

(b)

		Prediction		
		Female	Male	Total
Observed category	Female	**205**	40	245
	Male	71	**412**	483

(c)

		Prediction				
		F1	F2	F5	F9	Total
Observed Category	F1	**59**	40	3	48	150
	F2	56	**46**	1	40	143
	F3	14	7	3	17	41
	F4	24	21	0	14	59
	F5	14	9	**4**	2	29
	F6	2	8	1	1	12
	F7	53	32	0	35	120
	F8	14	32	1	16	63
	F9	32	22	0	**57**	111

Overall % correct predictions for ***Rank***:
$(180 + 20 + 173) \times 100/728 = 51.2\%$

Overall % correct predictions for ***Gender***:
$(205 + 412) \times 100/728 = 84.8\%$

Overall % correct predictions for ***Faclty***:
$(59 + 46 + 4 + 57) \times 100/728 = 22.8\%$

Several relationships are evident from Figure 9.9; we highlight the following:

- females tend to lie towards the lower end of ***Remun***, ***Age***, ***AQual*** and ***Resrch***;
- few staff (mainly male) have relatively exceptional large research output, with the majority of staff members having a value of fewer than one;
- those staff members with the highest research output come mainly from faculty $F5$, are confined to the rank of full professor and are relatively young.

Figure 9.9 can be reconstructed with the following function call:

```
X <- Remuneration.data.genbipl.2002
Genbipl(X = X, G = indmat(X[,6]), prediction.type = "normal",
   prediction.regions = TRUE, cont.scale = "unitSS", dist.cat =
   "EMC", label = F, exp.factor = 1.2, n.int=c(10,10,10,10),
   specify.classes = levels(X[,6]), x.grid = 0.005,
   y.grid = 0.005)
```

The axis predictivities given in the top left panel of Figure 9.9 for the quantitative variables are calculated from the output of the call to `Genbipl` with argument `predictions.sample = 1:nrow(X)`. Note that these predictions are expressed in terms of the original quantitative measurements. The correct way of calculating the axis predictivities is thus to first normalize these predictions exactly as the input quantitative variables were normalized using the components `centred.vec` and `scaling.vec` from the output list of the call to `Genbipl`. The sum of the squared normalized predictions for a particular variable divided by the sum of the squared normalized original observations (which is unity when normalizing to unit sum of squares) gives the required axis predictivity.

- Finally, we can compare Figure 9.9 with Figure 8.20, which gives a categorical (ordinal) PCA of the same data. It seems that the optimal ordinal scores determined by categorical PCA give very similar axes to those found by the generalized biplot using the normalized original data. In Figure 8.20 these axes have been shifted to the edges of the figure, and they have not (but could have been) in Figure 9.9. In this case, it seems that the optimal scores were similar to the original. The treatment of the fundamentally categorical variables (***Gender***, ***Rank*** and ***Faclty***) necessarily differs in the two figures because the category regions are two-dimensional, whereas the categorical PCA has represented each variable linearly. In the case of ***Gender***, with only two levels, we find that bisecting the line joining the points representing *male* and *female* in Figure 8.20 divides the space into regions that are very similar to the category regions for gender in Figure 9.9. Rather surprisingly, it seems that the EMC has given better gender separation than that given by the optimal PCA scores.

9.9 Function for constructing generalized biplots

Our main function for constructing generalized biplots is the function `Genbipl`. This function shares the following arguments with `PCAbipl`:

`X`	`c.hull.n`	`marker.size`	`pch.samples.size`
`G`	`colours`	`max.num`	`predictions.sample`
`X.new.samples`	`exp.factor`	`n.int`	`specify.bags`
`e.vects`	`label`	`offset`	`specify.classes`
`alpha`	`label.size`	`ort.lty`	`Title`
`ax`	`line.length`	`parplotmar`	`Tukey.median`
`ax.name.size`	`line.type`	`pch.means`	`zoomval`
`ax.type`	`line.width`	`pch.means.size`	
`ax.col`	`markers`	`pch.samples`	

The following arguments are specific to `Genbipl` and need special consideration.

Arguments

`CLPs.plot`	Logical value specifying if CLPs must be shown on the biplot. Defaults to FALSE.
`cont.scale`	One of `"none"`, `"unitVar"`, `"unitSS"`, `"unitRange"` for specifying normalization of quantitative variables.

`dist.cont`	The distance function for calculating that part of the **D** matrix contributed by the quantitative variables. In the current implementation, one of "`Pythagoras`", "`Clark`" or "`SqrtL1`". Can easily be extended to any qualifying distance metric as described for `Nonlinbipl`. Default is "`Pythagoras`".
`dist.cat`	The distance function for calculating that part of the **D** matrix contributed by the qualitative variables. In the current implementation, EMC.
`plot.symbol`	Plotting symbol used for colouring pixels when prediction regions are constructed.
`plot.symbol.size`	Size of `plot.symbol`.
`prediction.regions`	Logical TRUE or FALSE requesting or suppressing construction of prediction regions.
`prediction.type`	One of "`circle`", "`normal`" or "`back`". Default is "`normal`".
`straight`	If set to TRUE when `dist` = "`Pythagoras`" sample predictions are obtained as with `PCAbipl`. Should be set to FALSE for all other distance metrics.
`x.grid`	Numerical value for specifying the horizontal grid size for constructing prediction regions.
`y.grid`	Numerical value for specifying the vertical grid size for constructing prediction regions.
`colours.pred.regions`	Vector for specifying a colour for each category of a qualitative variable.

Details

In principle `Genbipl` combines the capabilities of `Nonlinbipl` and `MCAbipl`. It calls the functions `drawbipl.genbipl` and `pred.regions` for actually constructing the plots. Note that the distance from a pixel to its nearest CLP is calculated in the full space.

Value

In addition to the construction of a generalized biplot and its associated prediction regions, `Genbipl` outputs a list with the following components:

`Z`	A matrix containing the coordinates of the samples in the biplot space.
`Z.axes`	A list consisting of p_1 matrices – one matrix for each quantitative variable containing the details for constructing its biplot axis.
`e.vals`	A vector containing all the eigenvalues of the PCO of the *ddistance* matrix (9.6).
`D.cont`	The *ddistance* matrix $\mathbf{D}_{(1)}$ in (9.6).

`D.cat`	The *ddistance* matrix $\mathbf{D}_{(2)}$ in (9.6).
`normalized.X.cont`	The normalized quantitative variables in $n \times p_1$ matrix format.
`centred.vec`	A p_1-vector containing the values for centring the quantitative variables.
`scaling.vec`	A p_1-vector containing the values for scaling the quantitative variables.
`CLPs`	A matrix consisting of the coordinates of all the CLPs in the full space.
`predictions`	A $p_1 \times n$ matrix containing the predictions requested by the argument `predictions.sample`.
`category.predictions.list`	A list of p_2-vectors. Each vector contains the category predictions of all the samples for one of the qualitative variables.

10 *Monoplots*

At the outset we emphasized that the *bi-* of biplots referred not to the usual use of two dimensions but to the two types of entities exhibited – typically units and variables. Many displays exhibit only one type of entity, but represented in two or more dimensions. We term these *monoplots*. Often monoplots are concerned with symmetric matrices – for example, (dis)similarity, proximity or correlation matrices. Some are intrinsically monoplots, but others are monoplots because a deliberate choice is made to display only one type of entity; if the other were included the monoplot would become a biplot. Already, we have encountered some ambiguity in a sharp distinction between monoplots and biplots. For example, different symbols are plotted in CA (Chapter 7) to show the levels of the categorical variables labelling the rows and columns of a two-way contingency table. There, both rows and columns refer to the same type of entity. The situation is heightened in MCA (Chapter 8), where p categorical variables are presented, each with its own symbol. The same kind of entity is presented, albeit with p instances, each with its own plotting symbol – is this still a biplot? We would call it a monoplot, but it would become a biplot were the n units (samples) displayed as a set of n points or a density plot together with the information on the variables (Section 8.8). The position is aggravated when we have categorical variables because the L_k levels of the kth variable give rise to L_k points, corresponding to a single calibrated axis for a quantitative variable (Chapter 8).

10.1 Multidimensional scaling

We saw in Chapter 5, on nonlinear biplots, how a matrix may be calculated giving the distances $d_{ii'}$ between every pair of rows of a data matrix $\mathbf{X}$ and how nonlinear biplot trajectories may be found and used to predict approximations to the entries of $\mathbf{X}$. A special case is PCA, where the distance is Pythagorean – that is, $d_{ii'}^2 = (\mathbf{x}_i - \mathbf{x}_{i'})'(\mathbf{x}_i - \mathbf{x}_{i'})$ – and the trajectories are linear. In general, multidimensional scaling is concerned with finding an r-dimensional matrix $\mathbf{Z}$ whose rows generate Pythagorean distances $\delta_{ii'}$ that

Understanding Biplots John Gower, Sugnet Lubbe and Niël le Roux

approximate the observed values $d_{ii'}$ presented in a symmetric proximity matrix $\mathbf{D}$. Crucially, there is no requirement that the entries of $\mathbf{D}$ be derived from a data matrix $\mathbf{X}$; they may be directly observed in some experimental situation. Then, a monoplot of $\mathbf{Z}$ is the only possibility. We denote the distances generated by the rows of $\mathbf{Z}$ by $\boldsymbol{\Delta}$.

Different methods of multidimensional scaling depend on the criterion specified for measuring the discrepancy between observed, $\mathbf{D}$, and fitted distances, $\boldsymbol{\Delta}$, and the numerical algorithms used for optimizing the chosen criterion. These criteria may be divided into two classes – metric multidimensional scaling and nonmetric multidimensional scaling. The simplest metric multidimensional scaling method is classical scaling/principal coordinates analysis (PCO), discussed in detail in Chapter 5. PCO requires that there exists a matrix $\mathbf{Y}$ in (at most) $n - 1$ dimensions whose rows generate the observed distances exactly (the Euclidean embeddable property); in PCA, $\mathbf{Y}$ coincides with $\mathbf{X}$. Then, $\mathbf{Z}$ of rank r is obtained by minimizing $||\mathbf{Y} - \mathbf{Z}||^2$ which, as we have seen, is found from the Eckart-Young theorem on the least-squares properties of the SVD of $\mathbf{Y}$. As explained in Chapter 5, the two steps of finding $\mathbf{Y}$ and then $\mathbf{Z}$ may be subsumed into a single simple eigenvalue decomposition. In this method, $\boldsymbol{\Delta}$ plays no overt part.

When $\mathbf{D}$ is directly observed, the distances are unlikely to be generated by a real matrix $\mathbf{Y}$. Then, strictly speaking, PCO is unavailable, though it may give acceptable results when there are only a few unimportant negative eigenvalues in the eigendecomposition. Other forms of metric scaling do not require the existence of $\mathbf{Y}$ but do depend overtly on $\boldsymbol{\Delta}$. The two most popular are least-squares scaling which minimizes a criterion termed *stress*, defined by

$$\sum_{i<i'}^{n} (d_{ii'} - \delta_{ii'})^2;$$

and least-squares squared scaling which minimizes a criterion *sstress*, defined by

$$\sum_{i<i'}^{n} (d_{ii'}^2 - \delta_{ii'}^2)^2.$$

These criteria have to be minimized by available iterative methods, such as ALSCAL, KYST, or SMACOF, ... (see Cox and Cox, 2001; Borg and Groenen, 2005).

Figure 10.1 gives a geometric interpretation of the *stress* criterion. We plot the points $(d_{ii'}, \delta_{ii'})$ for all (ii') pairs. With a perfect fit, these points would lie on the line $d = \delta$. In other cases, the projection of $(d_{ii'}, \delta_{ii'})$ onto the line $d = \delta$ is $\frac{1}{2}(d_{ii'} + \delta_{ii'}, d_{ii'} + \delta_{ii'})$, giving a residual sum of squares equal to half the *stress* criterion. Note that *stress* $= \frac{1}{2}||\mathbf{D} - \boldsymbol{\Delta}||^2$. A similar representation where all distances are replaced by squared distances gives the *sstress* criterion. The representation of Figure 10.1 is not particularly interesting in itself, but it contributes to a better understanding of the corresponding nonmetric criterion, to be described next.

Nonmetric multidimensional scaling depends only on the ordinal information in $\mathbf{D}$. Then, it is required only that the fitted values $\boldsymbol{\Delta}$ approximate the same ordering as those in $\mathbf{D}$. How this is done is indicated in Figure 10.2, which is similar to Figure 10.1 except that the linear 'regression' is replaced by a monotone regression function, as shown. This monotone regression function is obtained by partitioning the points into blocks that are above and to the right of all preceding blocks. The fitted values $\hat{\delta}_{ii'}$ are determined

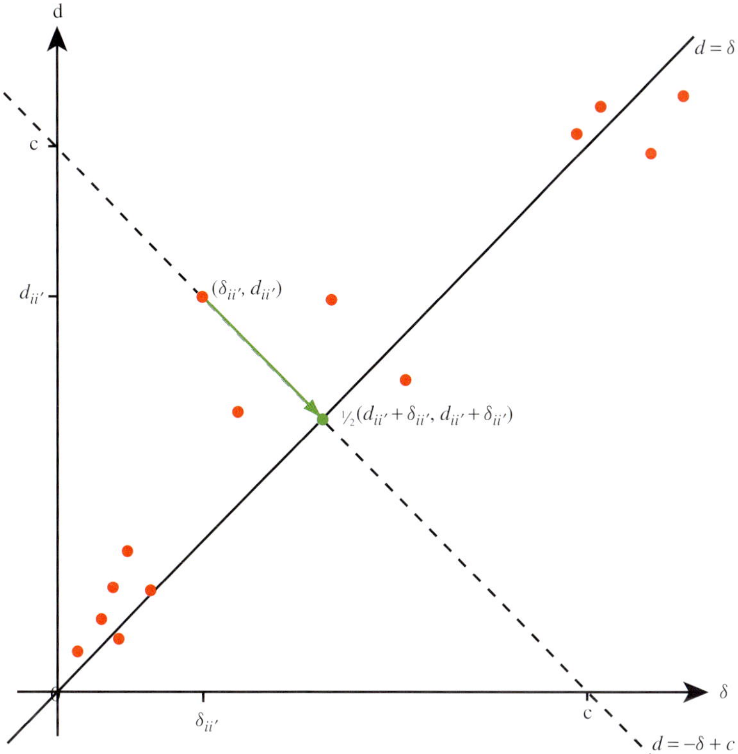

Figure 10.1 The vertical axis gives observed distances $d_{ii'}$ and the horizontal axis distances $\delta_{ii'}$ fitted in r dimensions. These determine the points ($\delta_{ii'}$, d$_{ii'}$) indicated by the red filled circles. The projection (indicated by the green arrow) of ($\delta_{ii'}$, d$_{ii'}$) onto the $d = \delta$ 'regression' line determines the point $\frac{1}{2}(d_{ii'} + \delta_{ii''}, d_{ii'} + \delta_{ii'})$ which is indicated in green. The squared residual length associated with this green point is $\frac{1}{2}(d_{ii'} - \delta_{ii'})^2$. The *stress* criterion is twice the sum of squares of these residuals.

by horizontal projection, as shown, so the residuals in this direction do not depend on the absolute values of the $d_{ii'}$, only on their ordering. Now the *stress* criterion may be modified to:

$$\sum_{i<i'}^{n} (\delta_{ii'} - \hat{\delta}_{ii'})^2 = ||\mathbf{\Delta} - \hat{\mathbf{\Delta}}||^2.$$

The above gives a very brief overview of how nonmetric MDS criteria are defined and relate to their metric counterparts. There are several variants. For example, the monotonic regression function may be expressed in terms of monotonic spline functions, thus giving a continuous monotonic regression. There is also an issue concerning what to do about tied values of $d_{ii'}$; should constraints be imposed so that they map into tied values of $\delta_{ii'}$ or not? See Borg and Groenen (2005) for a detailed discussion of this issue.

Figures 10.1 and 10.2 illustrate how the respective criteria are defined but say nothing about how to minimize them. The numerical minimization of *stress* and *sstress*, either

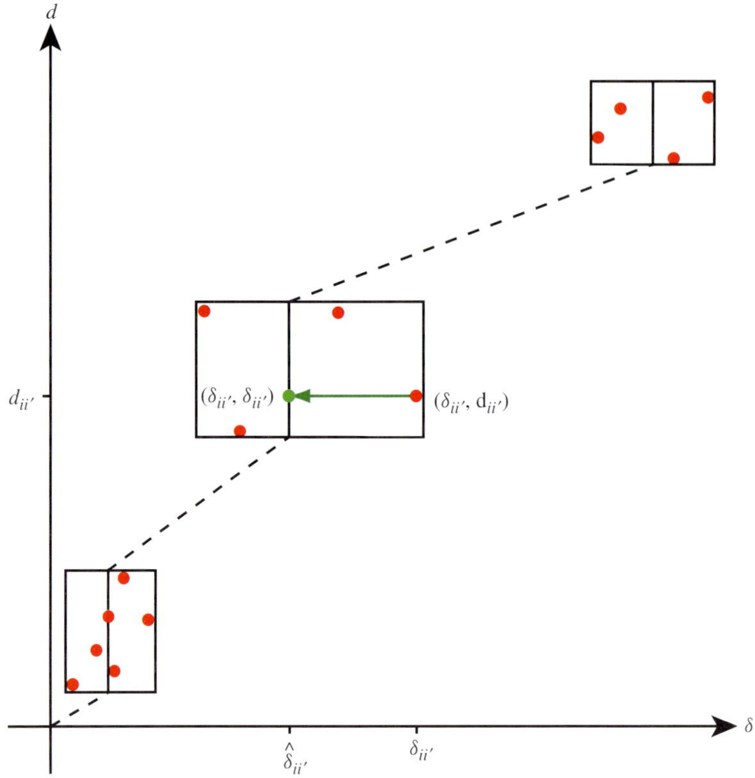

Figure 10.2 The vertical axis gives observed distances $d_{ii'}$ and the horizontal axis distances $\delta_{ii'}$ fitted in r dimensions. These determine the points $(\delta_{ii'}, d'_{ii})$ indicated by red filled circles. The rectangular boxes enclose sets of points that are monotonically related; that is, all the points in any given box have $\delta_{ii'}$ either greater or lesser than any value of the $\delta_{ii'}$ in the remaining boxes. A monotonic regression function is shown as a thick vertical line within boxes and an inclined dotted line between successive boxes. The projection (indicated by the horizontal green arrow) of $(\delta_{ii'}, d_{ii'})$ onto the monotonic regression function determines the point $(\delta_{ii'}, \hat{\delta}_{ii'})$ indicated in green, with the associated residual length of $(\delta_{ii'} - \hat{\delta}_{ii'})$. The monotonic *stress* criterion is the sum of squares of these residuals. Notice also that all the red points in any given block of points project to the same value of $\hat{\delta}_{ii'}$.

in their metric or nonmetric forms, requires iterative methods. The crucial step in the iteration is how to update a current setting of **Z** to give a convergent process. Quite sophisticated algorithms have been developed that allow different types of monotonic regression, different treatment of ties and efficient calculation. These are available in, for example, ALSCAL, KYST and SMACOF (see Cox and Cox, 2001; Borg and Groenen, 2005).

Whatever metric or nonmetric multidimensional scaling method is used, the end result is a map of n points, with coordinates given as the rows of **Z**, the distances between which approximate observed or derived distances $d_{ii'}$. These are manifestly monoplots.

However, when **D** has been derived from a data matrix **X**, we may wish to superimpose axes, possibly calibrated, to upgrade these monoplots to biplots. These axes approximate the variables, whose values are assumed to be given in the columns, of **X**. Chapter 5 has shown how nonlinear trajectories may represent variables in PCO. The regression method (see Sections 2.7, 3.5, 4.5, 5.2 and 7.3) may be used to superimpose linear axes on PCO plots and with plots derived from other methods of multidimensional scaling.

10.2 Monoplots related to the covariance matrix

10.2.1 Covariance plots

Falling into the class of monoplots where one of the entities is omitted by choice is the frequent presentation of maps of covariance and correlation matrices. These are closely related to PCA but now with interest focused on the approximation of $\mathbf{X}'\mathbf{X}$ rather than of $\mathbf{X}$ itself. With the SVD $\mathbf{X} = \mathbf{U\Sigma V}'$, in Chapter 3, PCA was based on plotting the rows of $\mathbf{U\Sigma J}$ to represent the samples and the rows of $\mathbf{VJ}$ to give directions for the variables. In monoplots we plot the rows of $\mathbf{V\Sigma J}$. The inner product $(\mathbf{V\Sigma J})(\mathbf{V\Sigma J})' = \mathbf{V\Sigma}^2\mathbf{JV}'$ approximates $\mathbf{X}'\mathbf{X}$ or, equivalently, the covariance matrix $(n-1)^{-1}\mathbf{X}'\mathbf{X}$. Now the inner product is found from pairs of points of the same kind, both representing the variables.

The inner product of $\mathbf{UJ}$ and $\mathbf{V\Sigma J}$ is unaltered so the approximation of $\mathbf{X}$ would remain valid and easily evaluated if $\mathbf{UJ}$ were added to the monoplot and the rows of $\mathbf{V\Sigma J}$ suitably calibrated, so giving a biplot. In this representation the distances between pairs of row points do not approximate the distances between the rows of $\mathbf{X}$, as was available with the PCA representations of Chapter 3. As a compromise, one may plot $\mathbf{U\Sigma J}$ for the row points and $\mathbf{V\Sigma J}$ (see CA: Chapter 7) for the variables, so producing a joint plot with no interesting inner product. It would be possible, but impracticable, to add calibrated axes based on $\mathbf{UJ}$ and $\mathbf{VJ}$ as well.

As an example we consider `Flotation.data`. This data set forms part of a larger data set collected at a base metal flotation plant. The complete data set consists of three sets of variables: digitized froth variables, operational variables and process quality variables. Six of the operational variables were chosen for `Flotation.data`, namely ***RMOF***, ***RMP***, ***BMSWA***, ***BMCFF***, ***BMCFD*** and ***EXFFBB***. It is clear from Table 10.1 that the standard deviations of these variables differ to a great extent. Furthermore, they can be considered to be measured on a ratio scale. In Figure 10.3 we give for comparative purposes the ordinary biplot of the flotation data centred but unscaled in the top panel but scaled to unit variances in the bottom panel. Figures 10.4 and 10.5

Table 10.1 Covariance matrix of the flotation data.

	RMOF	RMP	BMSWA	BMCFF	BMCFD	EXFFBB
RMOF	2695.5160	7.3482	359.0530	1987.4986	0.2112	48.4237
RMP	7.3482	0.0352	1.1955	6.9332	−0.0027	0.1348
BMSWA	359.0530	1.1955	157.0530	400.2840	−0.2119	8.2891
BMCFF	1987.4986	6.9332	400.2840	1949.6857	−0.4801	39.0583
BMCFD	0.2112	−0.0027	−0.2119	−0.4801	0.0029	0.0018
EXFFBB	48.4237	0.1348	8.2891	39.0583	0.0018	1.0498

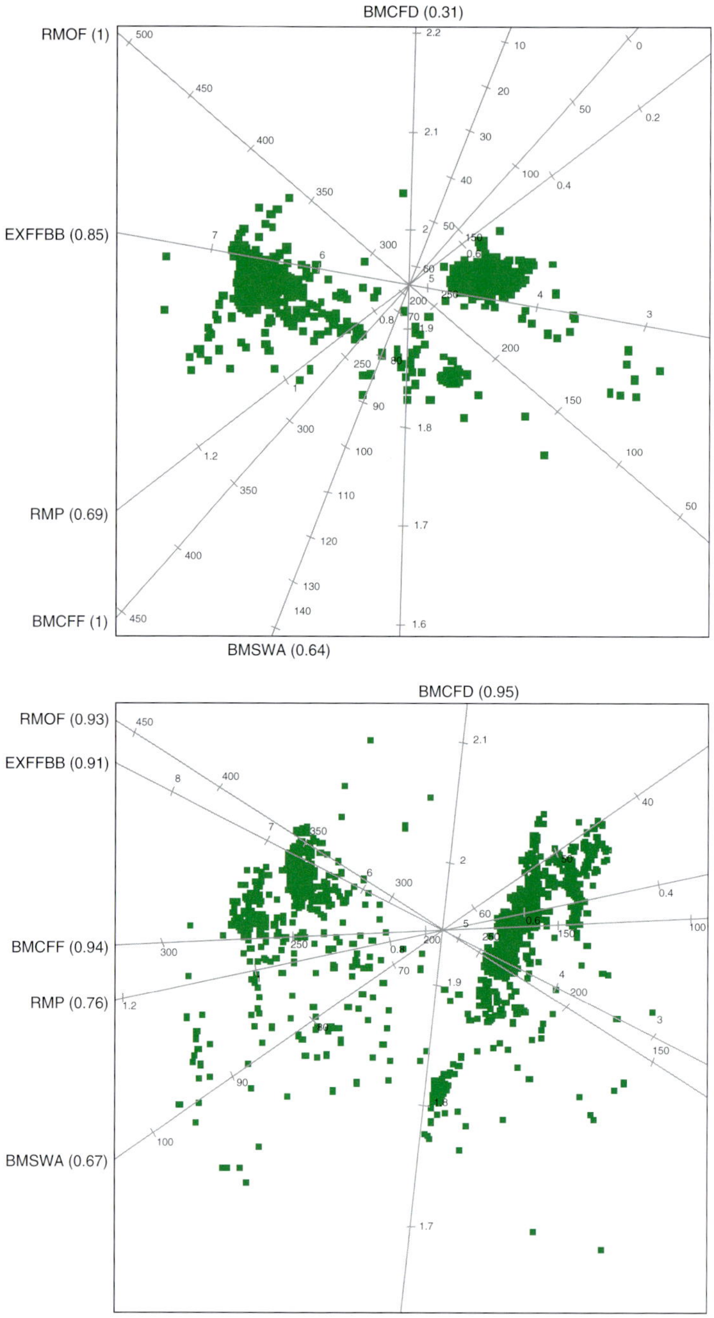

Figure 10.3 Ordinary PCA biplot of the flotation data with $\mathbf{X} = \mathbf{U\Sigma V}'$. Samples are plotted as $\mathbf{U\Sigma J}$ and variables are plotted in the form of calibrated axes where $\mathbf{VJ}$ gives directions for the axes. Axis predictivities are given in parentheses. (Top) Variables not normalized. (Bottom) Variables normalized to unit variances. We note some structure in the data but do not pursue this further.

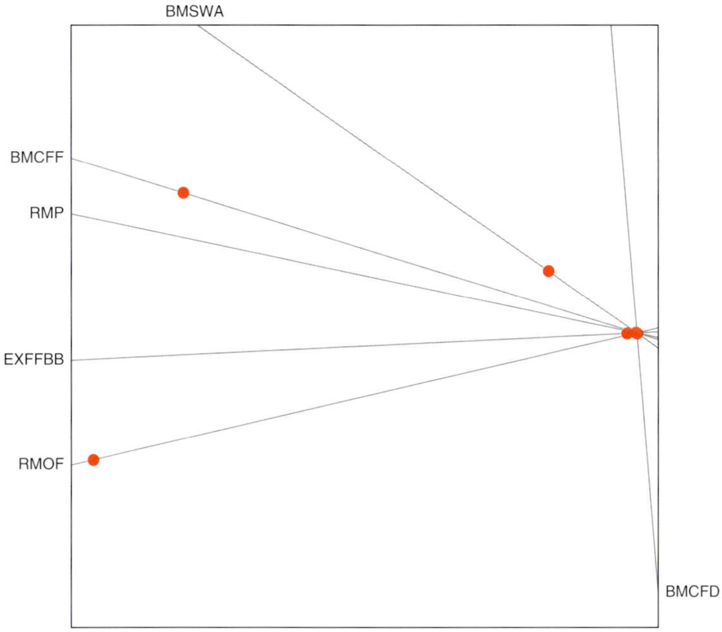

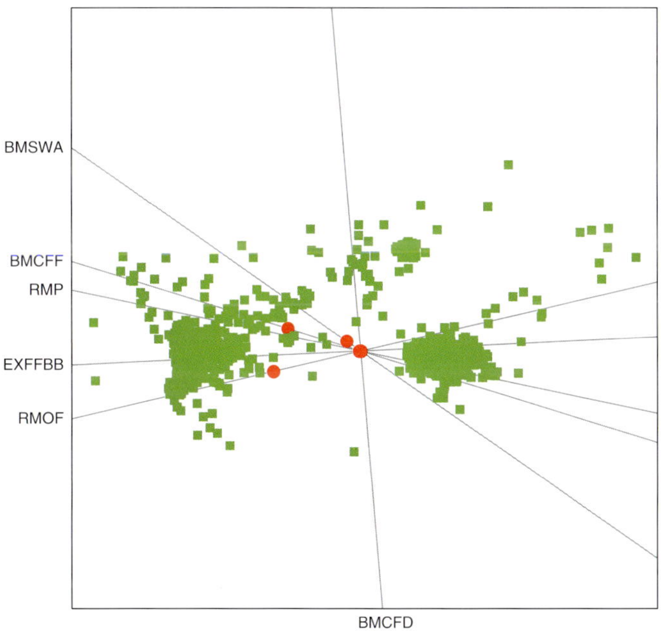

Figure 10.4 Covariance monoplot of the flotation data $(n-1)^{-1/2}\mathbf{X} = \mathbf{U\Sigma V}'$. Non-normalized $\mathbf{X}$. Points (variables) are plotted as the red filled circles using $\mathbf{V\Sigma J}$. These points defined the directions for the variables as uncalibrated axes. In the bottom panel the samples are added with coordinates given by $\mathbf{U\Sigma J}$ as in Figure 10.3. The figure in the bottom panel is what we call a *joint plot*, where row and column points are plotted but the two cannot be related by an inner product.

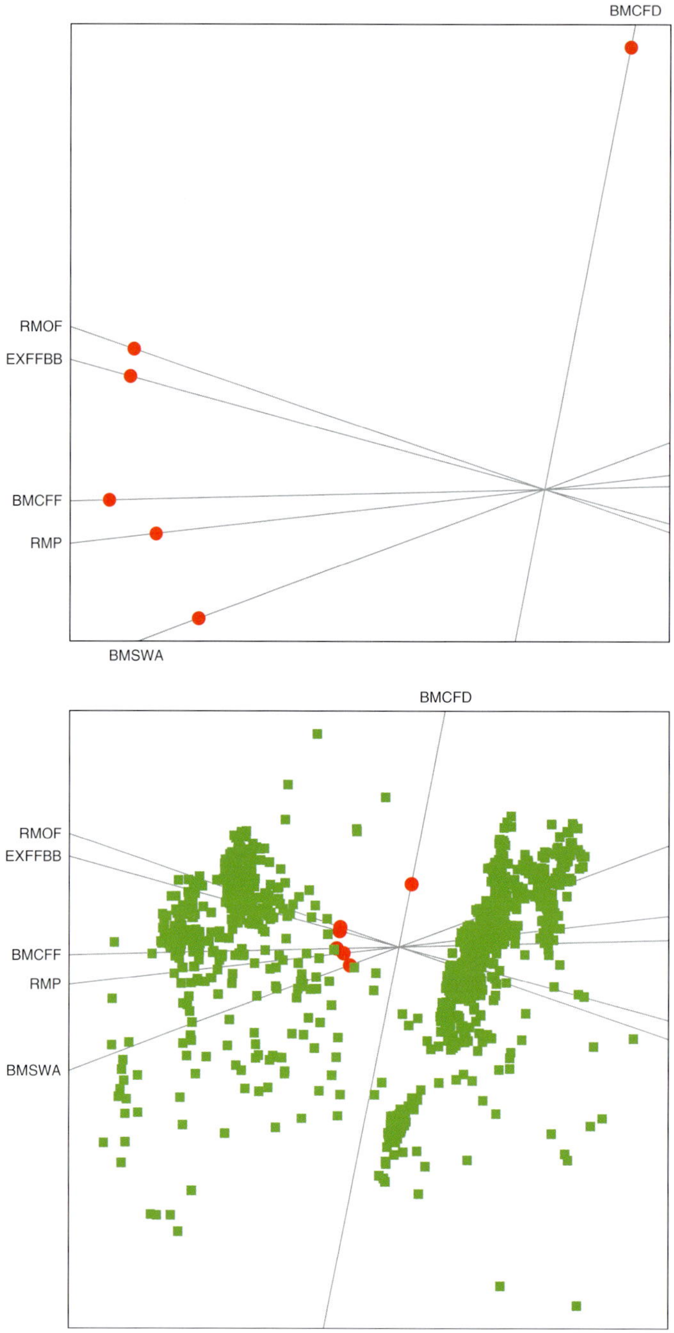

Figure 10.5 Similar to the covariance monoplot of the flotation data given in Figure 10.4 but using the normalized **X**. The top panel is a simple form of a correlation monoplot; the bottom panel a joint plot where the samples are added, similar to the bottom panel of Figure 10.4.

show the corresponding monoplots of the covariance matrices of the unscaled data and the scaled data, respectively. We note that the latter monoplot is a simple form of a correlation monoplot (see Section 10.2.2). The monoplot in the top panel of Figure 10.4 results from the function call

```
MonoPlot.cov(X = Flotation.data[,-1], samples.plot = FALSE,
   as.axes = TRUE, scaled.mat = FALSE, offset = c(0,0.3,0.2,0.1))
```

Setting argument `samples.plot = TRUE` gives the joint plot in the bottom panel. Figure 10.5 is the result of similar function calls with argument `scaled.mat = TRUE`.

10.2.2 Correlation monoplot

When $\mathbf{X}$ has been normalized to unit sums of squares, $\mathbf{X}'\mathbf{X}$ becomes a correlation matrix, $\mathbf{R}$. Then the squared distance between two points (variables) i and i' has elements

$$d_{ii'}^2 = 2(1 - r_{ii'}) \tag{10.1}$$

and so gives a direct measure of correlation that is approximated in r dimensions. In an exact representation each of the p plotted points of the map will be at unit distance from the origin. With a unit correlation the ith and i'th points coincide. A correlation of -1 generates points furthest apart. Further, (10.1) implies that if $\theta_{ii'}$ is the angle subtended at the origin then $\cos(\theta_{ii'}) = r_{ii'}$. By plotting a unit circle (see Figure 10.6) we can see the degree of approximation of each variable. This is not a geometrical representation of the adequacy criterion discussed in Section 3.3, but shows how well each variable approximates the unit correlation of exact representations, given algebraically by the square root of $\text{diag}(\mathbf{V}\boldsymbol{\Sigma}^2\mathbf{J}\mathbf{V}')$.

Figure 10.6 is a result of the function call

```
MonoPlot.cor(X = Flotation.data[,-1],
   offset = c(-0.1, 0.25, 0.1, 0.1),
   print.ax.approx = TRUE, offset.m = rep(-0.2, 6))
```

We note that the relative positions of the red solid circles in Figure 10.6 are similar to those in Figure 10.5. By considering the correlation matrix in Table 10.2, it can be concluded that the monoplot in Figure 10.6 approximates the intercorrelations quite well: in the case of the variable with the highest degree of approximation, ***BMCFD*** (0.98), for example, its position is almost orthogonal to those of ***RMOF*** and ***EXFFBB***, indicating almost zero correlations, while its position relative to those of ***BMCFF***, ***RMP*** and ***BMSWA*** shows the increasingly negative correlations with these variables.

10.2.3 Coefficient of variation monoplots

The coefficient of variation, CV, of a variable is its standard deviation divided by its mean. The CV should not be used with interval scales for then the mean depends additively

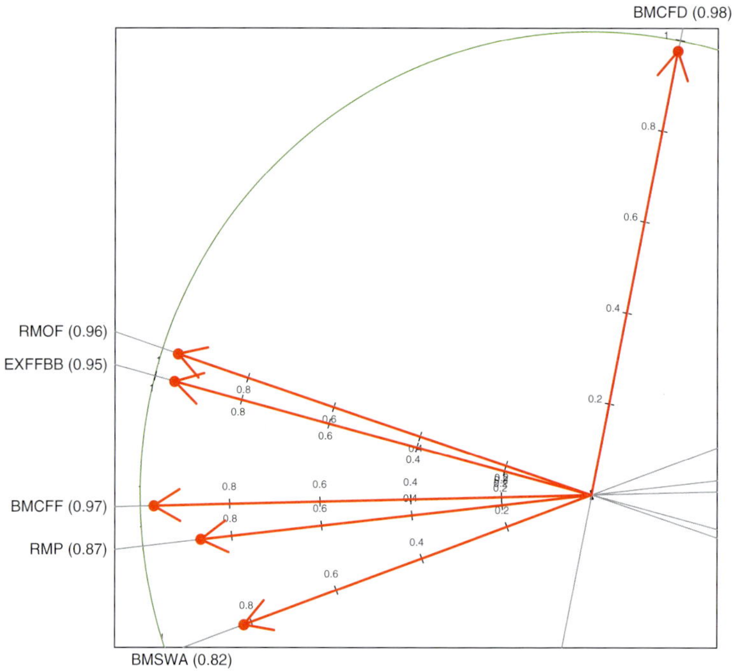

Figure 10.6 Correlation monoplot of the flotation data. The red filled circles represent the variables as points with coordinates given by $\mathbf{V\Sigma J}$. The green circle has unit radius. The tip of each arrow indicates the degree of approximation of each variable, also shown numerically with the respective labels of the monoplot axes.

Table 10.2 Correlation matrix of the flotation data.

	RMOF	RMP	BMSWA	BMCFF	BMCFD	EXFFBB
RMOF	1.0000	0.7547	0.5518	0.8670	0.0761	0.9103
RMP	0.7547	1.0000	0.5087	0.8373	−0.2645	0.7015
BMSWA	0.5518	0.5087	1.0000	0.7234	−0.3165	0.6455
BMCFF	0.8670	0.8373	0.7234	1.0000	−0.2035	0.8633
BMCFD	0.0761	−0.2645	−0.3165	−0.2035	1.0000	0.0321
EXFFBB	0.9103	0.7015	0.6455	0.8633	0.0321	1.0000

on the scale used, as when converting temperatures from Celsius to Fahrenheit, whereas the standard error changes multiplicatively. Thus, the CV is not invariant to changes in interval scales. Underhill (1990) propose a CV biplot which is based on the uncentred data matrix $\mathbf{X}$, the means of whose variables are the row vector $\mathbf{1}'\mathbf{X}/n$. These means may be assembled into a diagonal matrix $\mathbf{\Delta}$, so that $\mathbf{1}'\mathbf{\Delta} = \mathbf{1}'\mathbf{X}/n$. The matrix $\mathbf{X}$ is then replaced by

$$\mathbf{E} = \frac{1}{\sqrt{n-1}}(\mathbf{I} - \mathbf{11}'/n)\mathbf{X}\mathbf{\Delta}^{-1}. \tag{10.2}$$

We note that these quantities are dimensionless, so offer another form of scaling for PCA, but beware the caveat about interval scales. The diagonals of $\mathbf{E}'\mathbf{E}$ are the squares of the

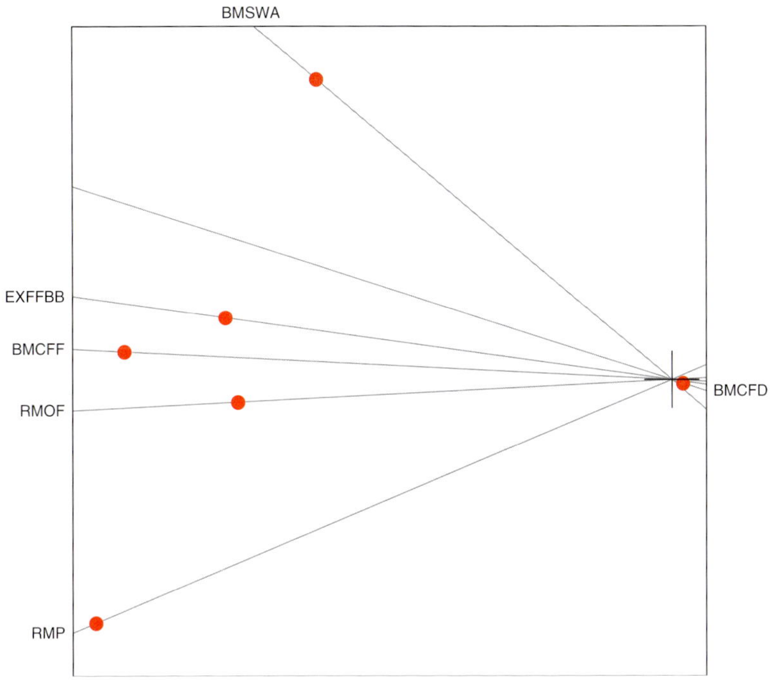

Figure 10.7 Coefficient of variation monoplot of the flotation data. This is constructed as the covariance biplot of the matrix (10.2). With $\mathbf{E}'\mathbf{E} = \mathbf{V\Sigma V}'$ the red filled circles have coordinates given by $\mathbf{V\Sigma J}$. Straight lines through the origin and each red filled circle show the variables in the form of uncalibrated monoplot axes.

coefficients of variation and if we do a covariance monoplot of **E** it is these quantities that are approximated. In Figure 10.7 we show an example of a CV monoplot obtained with the function call

```
MonoPlot.coefvar(X = Flotation.data[,-1], as.axes = TRUE, pos = "Hor",
   offset = c(0.1, 0.3, 0.2, 0.1))
```

10.2.4 Other representations of correlations

The representation of correlations, as described above, suffers from the unnecessary approximation of the known unit diagonals. It is not just that it is unnecessary to reproduce the diagonals but also that, for symmetric matrices, the Eckart-Young approximation gives twice the weight to approximating the off-diagonal terms compared with the diagonals. The unequal weighting is in the right direction but is arbitrary – see Bailey and Gower (1990) for a fuller analysis of differential weighting. This drawback is the same as that found in an even more extreme form for MCA, where whole diagonal blocks are approximated unless one uses the JCA variant of MCA (see Chapter 8). A similar solution to using JCA is available for correlation matrices, and is the basis of some of the simpler forms of factor analysis, which replace the diagonals by nonunit communalities to give a matrix of rank r. However, a simple alternative way forward was proposed by Hills (1969) who suggested that the correlation matrix be analysed by MDS and, in particular, by PCO. This is equivalent to replacing the correlation matrix $\mathbf{R}$ by a squared-distance matrix $\mathbf{11}' - \mathbf{R}$. Any MDS method fits only the off-diagonal values, so approximating the zero diagonal is not an issue. In accordance with the standard PCO procedure (Chapter 5), we require the spectral decomposition of

$$-\tfrac{1}{2}(\mathbf{I} - \mathbf{11}'/p)(\mathbf{11}' - \mathbf{R})(\mathbf{I} - \mathbf{11}'/p) \tag{10.3}$$

which is the same as $\frac{1}{2}(\mathbf{I} - \mathbf{11}'/p\ \mathbf{R}(\mathbf{I} - \mathbf{11}'/p)$. The distances approximated are given by

$$d_{ii'}^2 = (1 - r_{ii'}).$$

Now, the origin is at the centroid of the p points which, rather than unit distance, has squared distances from the centroid given by

$$\tfrac{1}{2}(1 + \mathbf{1}'\mathbf{R}\mathbf{1}/p^2)\mathbf{1} - \mathbf{1}'\mathbf{R}/p.$$

Thus, the angle properties do not apply but the distance properties do and with better approximation.

Rather than operate on $\mathbf{R}$ we may operate on $\mathbf{R}^*\mathbf{R}$, the matrix whose entries are squared correlations. Both $\mathbf{R}$ and $\mathbf{R}^*\mathbf{R}$ are positive semi-definite so PCO remains available. This gives distances defined by

$$d_{ii'}^2 = (1 - r_{ii'}^2).$$

The obvious difference is that only the absolute values of the correlations, not their signs, are approximated.

Even though we are basing these approximations on PCO, neither nonlinear biplots nor the regression method described above may be used to upgrade these monoplots to biplots. This is because the monoplot already gives information on the variables. It is likely that a method might be found for adding the n units, but we have not investigated this possibility further.

The correlation monoplot resulting from a PCO of (10.3) for the flotation data is given in the top panel of Figure 10.8, while the bottom panel is the corresponding monoplot when the $\mathbf{R}$ in (10.3) is replaced by $\mathbf{R}^*\mathbf{R}$. The distances between the points in these monoplots provide good approximations to the intercorrelations given in Table 10.2. Figure 10.8 is a result of the call

```
MonoPlot.cor2(X = Flotation.data[,-1], offset.m = rep(0.4, 6),
   pos.m = rep(2,6), rotate.degrees = -17)
```

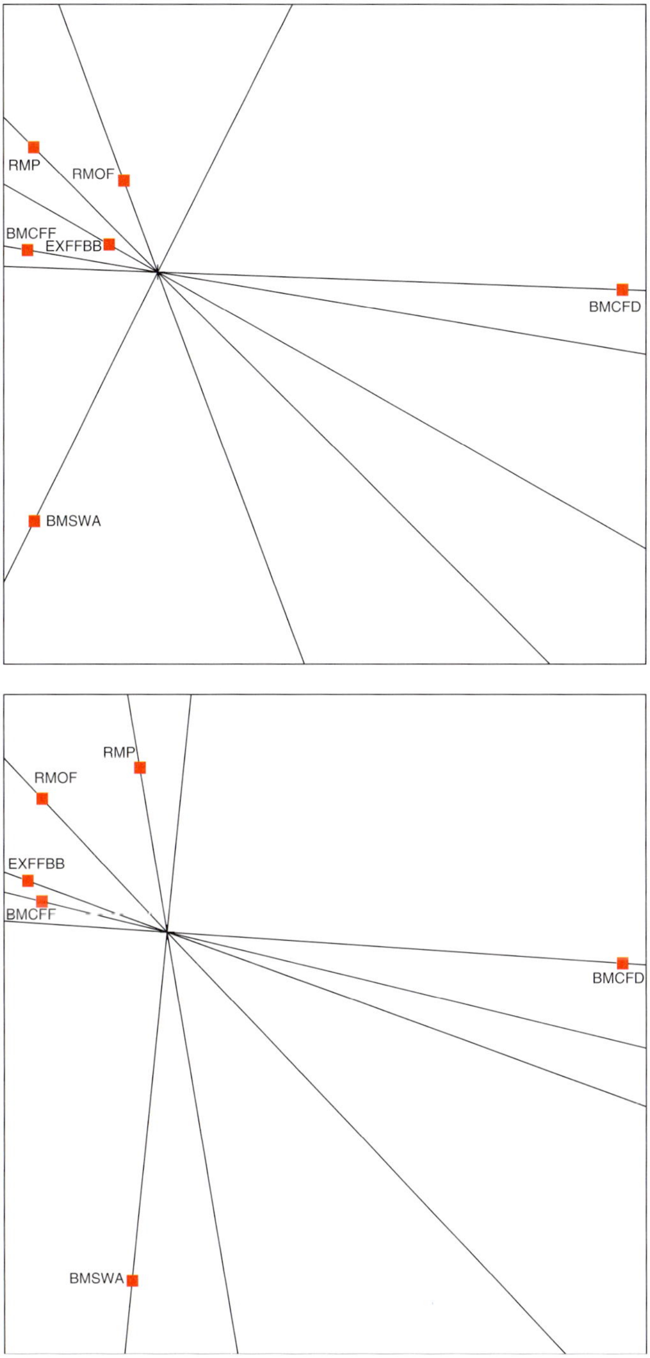

Figure 10.8 (Top) Correlation monoplot of the flotation data resulting from a PCO of $\frac{1}{2}(\mathbf{I} - \mathbf{11}'/p)\ \mathbf{R}(\mathbf{I} - \mathbf{11}'/p)$. (Bottom) Correlation monoplot of the flotation data resulting from a PCO of $\frac{1}{2}(\mathbf{I} - \mathbf{11}'/p)\ \mathbf{R}^*\mathbf{R}(\mathbf{I} - \mathbf{11}'/p)$ and rotated 17 degrees clockwise for ease of comparison with the monoplot in the top panel.

10.3 Skew-symmetry

A very different type of monoplot arises from representing the elements of a skew-symmetric matrix. Commonly, data occur in square tables **X** whose rows and columns are classified by the same type of entity but in different modes. Thus we can have import and export between the same countries, or in social science father's occupation versus son's occupation, or before–after studies. Confusion matrices offer another illustration where, in a famous example (the Morse code data of Rothkopf, 1957), the rows refer to the 26 letters and 10 digits sent in Morse code form and the columns the same letters/digits as received by 400 participants. The body of the table gives the number of times signal 'i' was perceived as signal 'j' (order taken into account), for all i, j pairs, so the diagonal gives all the correct recognitions.

Many models are available for analysing such tables, but often it is reasonable to suspect that different mechanisms govern the symmetric aspects of the data and the departures from symmetry. Gower (1977) and Constantine and Gower (1978) suggested basing analysis on the simple algebraic identity

$$\mathbf{X} = \mathbf{M} + \mathbf{N}, \tag{10.4}$$

where $\mathbf{M} = \{\frac{1}{2}(x_{ij} + x_{ji})\}$ is a symmetric matrix and $\mathbf{N} = \{\frac{1}{2}(x_{ij} - x_{ji})\}$ is a skew-symmetric matrix. We have already discussed several ways of representing symmetric matrices; here we discuss how to represent skew-symmetry. As usual, we base our plots on the singular value decomposition. For a skew-symmetric matrix, this has the special form

$$\mathbf{N} = \mathbf{U\Sigma KU}'. \tag{10.5}$$

So, like symmetric matrices, the same singular vectors are associated with the row and column singular values. Unlike symmetric matrices, the diagonal of **N** is always zero, so raises no special problems. The singular values occur in equal pairs $\mathbf{\Sigma} = \text{diag}(\sigma_1, \sigma_1, \sigma_2, \sigma_2, \ldots, \sigma_k, \sigma_k, 0)$ where the final element is zero or null according to whether the order of **X** is odd or even. Similarly, **K** is of the form

$$\mathbf{K} = \begin{bmatrix} 0 & 1 & 0 & 0 & 0 & 0 & \cdots & 0 \\ -1 & 0 & 0 & 0 & 0 & 0 & \cdots & 0 \\ 0 & 0 & 0 & 1 & 0 & 0 & \cdots & 0 \\ 0 & 0 & -1 & 0 & 0 & 0 & \cdots & 0 \\ 0 & 0 & 0 & 0 & 0 & 1 & \cdots & 0 \\ 0 & 0 & 0 & 0 & -1 & 0 & \cdots & 0 \\ \cdots & \cdots & \cdots & \cdots & \cdots & \cdots & \cdots & \cdots \\ 0 & 0 & 0 & 0 & 0 & 0 & \cdots & 1 \end{bmatrix},$$

where the final '1' is absent when the order of **X** is even.

Note that **K** is an orthogonal matrix and so $\mathbf{UK}'$ may be regarded as the equivalent of the orthogonal matrix **V** of the usual SVD. Writing $\mathbf{U} = (\mathbf{u}_{11}, \mathbf{u}_{12}, \mathbf{u}_{21}, \mathbf{u}_{22}, \ldots, \mathbf{u}_{k1}, \mathbf{u}_{k2})$, the SVD (10.5) may be expanded as

$$\mathbf{N} = \sum_{i=1}^{k} \sigma_i(\mathbf{u}_{i1}\mathbf{u}'_{i2} - \mathbf{u}_{i2}\mathbf{u}'_{i1}). \tag{10.6}$$

Thus, it turns out that it only makes sense to seek approximations in an even number of dimensions, sometimes referred to as *bimensions* but we prefer *hedron*. Previously we have often chosen to give two-dimensional maps, but now we are obliged to, and if we want better approximations we must use 4, 6, ... dimensions.

We next consider a single hedron and to simplify notation now drop the hedron identifier i and the dimension identifiers. Thus, we write $\mathbf{u}_{i1}\mathbf{u}'_{i2} - \mathbf{u}_{i2}\mathbf{u}'_{i1}$ as $\mathbf{uv}' - \mathbf{vu}'$. As usual, we may plot the rows of $(\mathbf{u}, \mathbf{v})$ as a set of p points, but what does it mean? If P_i (u_i, v_i) and P_j (u_j, v_j) are two of these points, then the area of the triangle $\mathrm{OP}_i\mathrm{P}_j$ is $\frac{1}{2}(u_i v_j - v_i u_j)$. Thus the interpretation of this kind of monoplot is in terms of area. Figure 10.9 shows the geometry. Skew-symmetry is reflected in the angular sense in which the area is measured. We take the anticlockwise sense to be positive, so that $\mathrm{OP}_i\mathrm{P}_j = -\mathrm{OP}_j\mathrm{P}_i$. Distance is not an appropriate measure, as although $\mathrm{OP}_i\mathrm{P}_j$ is zero when P_i and P_j coincide, so is it zero when O, P_i and P_j are collinear, however distant P_i may be from P_j. In conventional Euclidean plots all points equidistant from any point P_i lie on a circle; now all points P′ with equal skew-symmetry n_{ij} with P_i lie on a line parallel to OP_i as in Figure 10.9. There is another line parallel to OP_i which has skew-symmetry $-n_{ij}$, indicated by P″ in the diagram.

Axes are shown in Figures 10.9 and 10.10 although, as usual, they are merely scaffolding for plotting the points. Consequently, in Figure 10.10 the axes have been de-emphasized. From the figures, we can see that the *area* $\mathrm{P}_i\mathrm{P}_j\mathrm{P}_k = n_{ij} + n_{jk} + n_{ki}$. Clearly, this result is independent of the position of the origin. However, apart from this

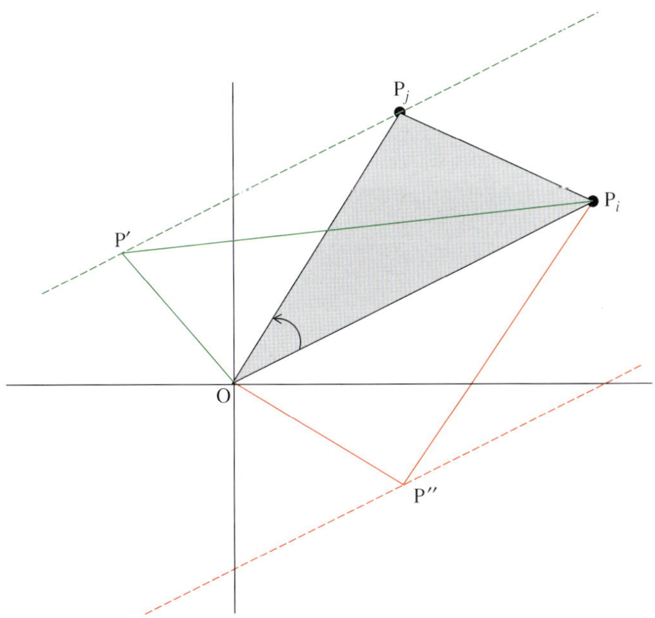

Figure 10.9 The geometry of skew-symmetry. Positive skew-symmetry n_{ij} is indicated by area measured in the anticlockwise sense $\mathrm{OP}_i\mathrm{P}_j$, negative skew-symmetry in the clockwise sense $\mathrm{OP}_j\mathrm{P}_i$. The line through P′ (P″) is the locus of all points with skew-symmetry $n_{ij}(-n_{ij})$.

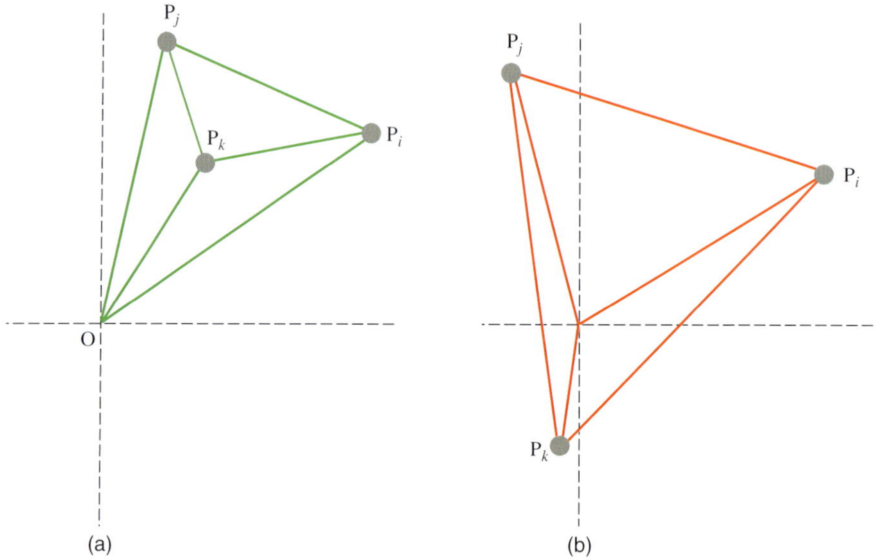

Figure 10.10 Skew-symmetry for (a) three points excluding the origin and (b) three points including the origin.

result and unlike most other plots that we show, the position of the origin is crucial and cannot be shifted as in Section 2.4. Figure 10.10 shows the importance of the origin. If we interpret $n_{ij} > 0$ as meaning that i is dominated, or preferred, by j and write this relationship as $i \succ j$, then in Figure 10.10(a) we have $i \succ j$, $i \succ k$ and $k \succ j$, so $i \succ k \succ j$ forms a consistent triad. However, in Figure 10.10(b) we have $i \succ j$, $j \succ k$ and $k \succ i$, so $k \succ i \succ j \succ k$ is inconsistent. This kind of inconsistency can occur in preference data and the area plot allows its representation according to whether or not the points representing a triad contain or do not contain the origin.

Finally, we note the important special case of linear skew-symmetry, written algebraically as

$$\mathbf{N} = \mathbf{u}\mathbf{1}' - \mathbf{1}\mathbf{u}'$$

or $n_{ij} = u_i - u_j$. This is a common feature of data and because $\mathbf{v} = \mathbf{1}$ is easily recognized by giving a linear plot in hedron representations. The line need not be vertical, as $\mathbf{u}$ and $\mathbf{v}$ in $\mathbf{uv}' - \mathbf{vu}'$ may be replaced by any pair of linear combinations of $\mathbf{u}$ and $\mathbf{v}$, as follows from the result

$$(\lambda\mathbf{u} + \mu\mathbf{v})(\alpha\mathbf{u} + \beta\mathbf{v})' - (\alpha\mathbf{u} + \beta\mathbf{v})(\lambda\mathbf{u} + \mu\mathbf{v})' = (\lambda\beta - \alpha\mu)(\mathbf{uv}' - \mathbf{vu}'). \tag{10.7}$$

This demonstrates another invariance property of hedron plots, allowing much freedom in choosing the coordinate representation. Clearly, area is not affected by rotations, neither is it by scaling in one direction when compensated by inverse scaling in the orthogonal direction; a sequence of scaling followed by rotations is permissible. All possibilities are covered by (10.7) and allow improved hedron representations, including the possibility of rotating linear features to be vertical. Even when $(\lambda\beta - \alpha\mu) \neq 1$, the only

Table 10.3 The Rothkopf Morse data for the vowels together with the **M** (symmetric) and **N** (skew-symmetric) matrices.

	Rothkopf vowels					**M** matrix					**N** matrix				
	a	e	i	o	u	a	e	i	o	u	a	e	i	o	u
a	92	3	46	6	37	92.0	4.5	55.0	6.5	25.5	0.0	−1.5	−9.0	−0.5	11.5
e	6	97	17	5	3	4.5	97.0	13.5	5.0	3.5	1.5	0.0	3.5	0.0	−0.5
i	64	10	93	7	10	55.0	13.5	93.0	4.5	15.5	9.0	−3.5	0.0	2.5	−5.5
o	7	5	2	86	11	6.5	5.0	4.5	86.0	8.5	0.5	0.0	−2.5	0.0	2.5
u	14	4	21	6	93	25.5	3.5	15.5	8.5	93.0	−11.5	0.5	5.5	−2.5	0.0

effect is to scale the hedron uniformly and so can it be absorbed into the singular value; there is no visual impact.

As an illustration of monoplots for skew-symmetric data we consider the data (in percentage form) for the vowels in the Rothkopf Morse code data (taken from the full asymmetric form of the data as available in Borg and Groenen, 2005). In Table 10.3 we show in addition to the data on the vowels the **M** and **N** matrices defined above. Figure 10.11 shows a hedron plot of the **N** matrix given in Table 10.3. The matrix $\boldsymbol{\Sigma}$ of the SVD (10.5) is diag(16.2237, 16.2337, 2.7913, 2.7913, 0). Figure 10.11 is obtained with the following code:

```
> par (mfrow = c(2,2), mar = rep(0.25,4))
> MonoPlot.skew (X = Rothkopf.vowels)
> MonoPlot.skew (X = Rothkopf.vowels, form.triangle1 = c(1,5),
     form.triangle2 = c(1,3))
> MonoPlot.skew (X = Rothkopf.vowels, form.triangle1 = c(2,3),
     form.triangle2 = c(3,4))
> MonoPlot.skew (X = Rothkopf.vowels, form.triangle1 = c(1,4),
     form.triangle2 = c(4,5))
```

In Figure 10.11, note the approximate collinearity with the origin of $\boldsymbol{e}$, $\boldsymbol{o}$ and $\boldsymbol{a}$, indicating that there is little difference in the confusion, depending on the order in which these vowels are presented. Also, $\boldsymbol{i}$, $\boldsymbol{a}$ and $\boldsymbol{u}$ have approximate linear skew-symmetry. We see that some triangles include the origin (e.g. $\boldsymbol{a}$, $\boldsymbol{u}$, $\boldsymbol{i}$) and some exclude the origin (e.g. $\boldsymbol{a}$, $\boldsymbol{u}$, $\boldsymbol{o}$).

We started with the decomposition $\mathbf{X} = \mathbf{M} + \mathbf{N}$. It is sometimes possible to combine plots derived from **M** with those derived from **N**, especially when skew-symmetry has approximate linear form. Supposing **M** has been derived from an MDS, then linear skew-symmetry may be added either as a line or as an extra dimension, possibly allowing contour plots. For example, Gower and Dijksterhuis (2004) derive a map from the symmetric part of flight times between US cities given in **X** and superimpose a line, derived from the linear skew-symmetry, giving the direction of a jet stream. Gower (2008) discusses this type of modelling that combines symmetry with departures from symmetry, so giving what may be considered a further type of biplot, not further discussed in this book.

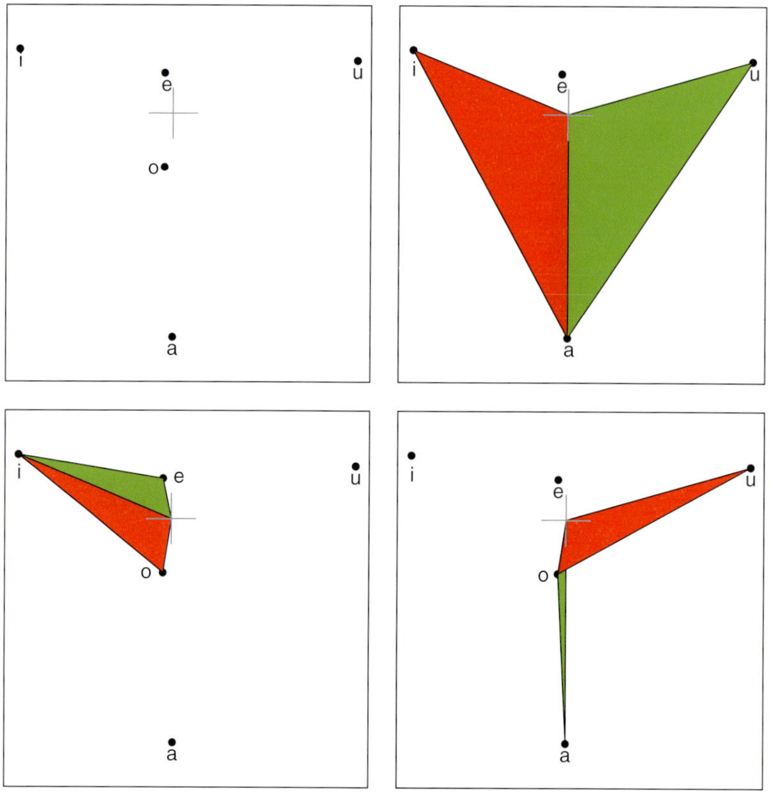

Figure 10.11 (Top left) Two-dimensional hedron plot of the vowels in the Rothkopf Morse code data. (Top right) Hedron plot with area O***au*** (O denoting the origin indicated by the grey cross) approximating $n_{15} = 11.5$ and area O***ai*** approximating $n_{13} = -9$. (Bottom left) Hedron plot with area O***ei*** approximating $n_{23} = 3.5$ and area O***oi*** approximating $n_{43} = -2.5$. (Bottom right) Hedron plot with area O***oa*** approximating $n_{41} = 0.5$ and area O***uo*** approximating $n_{43} = -2.5$.

10.4 Area biplots

The area representation of asymmetry can also be used with genuine biplots. For example, with biadditive models $\mathbf{X} = \mathbf{AB}'$ we plot $\mathbf{A}$ for rows and $\mathbf{B}$ for columns (see Chapter 6) and we have seen how the inner product can be recovered by plotting different symbols for the points. The evaluation of the inner product visually has the difficulties discussed in Section 2.3. As an alternative, we may treat either the rows or the columns as if they were variables and choose one of them to be represented by calibrated axes. This can work quite well but is an asymmetric representation of what is a symmetric data structure – symmetric in the sense that rows and columns are interchangeable, not that $\mathbf{X}$ is a symmetric matrix.

In two dimensions $\mathbf{A}$ and $\mathbf{B}$ will each have two columns. If R_i is the ith row point and C_j the jth column point, the estimate of the inner product is $r_i c_j \cos(\theta_{ij})$; see Figure 10.12 for notation. By rotating C_j through 90 degrees to C_j', we have that

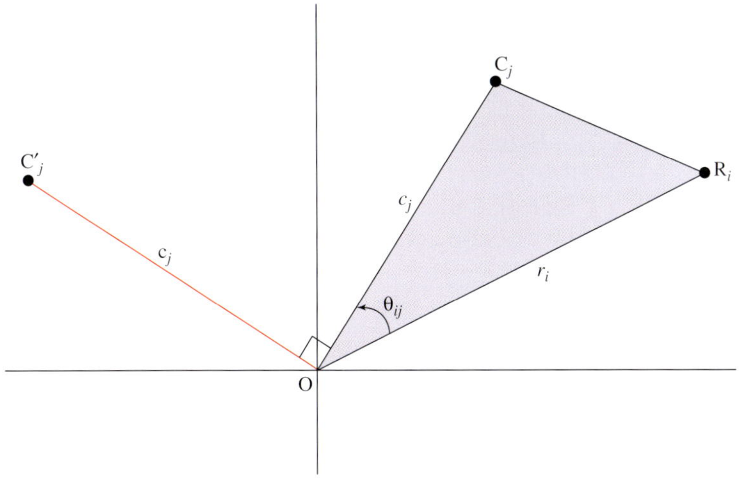

Figure 10.12 A 90 degree rotation of C_j to C'_j ensures that $r_i c_j \cos(\theta_{ij}) = r_i c_j \sin(\theta_{ij} + \frac{1}{2}\pi)$.

$r_i c_j \cos(\theta_{ij}) = r_i c_j \sin(\theta_{ij} + \frac{1}{2}\pi)$. The latter is twice the area of the triangle $OR_iC'_j$ so giving an area interpretation to inner products, for which the hedron geometry described above for representing skew-symmetry applies. If necessary, further pairs of dimensions may be added, taking care to ensure that all diagrams are on the same scale – a process termed linking diagrams that is an extension to ensuring the correct aspect ratio in a single map. See Gower *et al.* (2010) for further details and some examples.

10.5 Functions for constructing monoplots

We provide the following R functions for constructing the monoplots discussed in Sections 10.2 and 10.3: `MonoPlot.cov`, `MonoPlot.cor`, `MonoPlot.cor2`, `MonoPlot.coefvar` and `MonoPlot.skew`.

10.5.1 Function `MonoPlot.cov`

This is a function for constructing the covariance monoplot defined in Section 10.2.1.

Arguments

`X`	Data matrix of size $n \times p$.
`scaled.mat`	If TRUE a simple form of a correlation monoplot is constructed. Defaults to FALSE.
`as.axes`	If TRUE the points are represented as axes. Defaults to FALSE.
`axis.col`	Colour of an axis if `as.axes` is TRUE.
`ax.name.col`	Colour of name of an axis if `as.axes` is TRUE.

`ax.name.size`	Size of name of an axis if `as.axes` is TRUE.
`calibrate`	If set to TRUE axes are calibrated. Defaults to FALSE.
`dim.plot`	Currently only `dim.plot = 2` is implemented.
`lambda`	If set to TRUE lambda-scaling is applied. Defaults to FALSE.
`line.length`	See `PCAbipl`.
`marker.size`	Size of markers on axes.
`marker.col`	Colour of markers on axes.
`n.int`	See `PCAbipl`.
`offset`	See `PCAbipl`.
`pos`	See `PCAbipl`.
`pos.m`	See `PCAbipl`.
`samples.plot`	If set to TRUE, samples are also drawn resulting in a joint plot. Default is FALSE.
`side.label`	See `PCAbipl`.
`VJ.plot`	If set to TRUE, the points for the variables are plotted as $\mathbf{VJ}$ instead of $\mathbf{V\Sigma J}$ (the default; see Section 10.2.1 for details)
`zoomval`	See PCAbipl.zoom

Value

In addition to a covariance monoplot a list with the following two components is returned: `cov.X`, the covariance matrix of $\mathbf{X}$ and `cor.X`, the correlation matrix of $\mathbf{X}$.

10.5.2 Function `MonoPlot.cor`

This is a function for constructing the correlation monoplot defined in Section 10.2.2. It shares the following arguments with `MonoPlot.cov`:

`X`	`ax.name.col`	`lambda`	`offset.m`
`as.axes`	`ax.name.size`	`n.int`	`pos`
`axis.col`	`dim.plot`	`offset`	`pos.m`

Arguments specific to `MonoPlot.cor`

`arrows`	Defaults to TRUE, requesting arrows to be drawn from the origin to each point representing a variable.
`calib.no`	Number of calibrations on monoplot axes.
`circle`	Defaults to TRUE, requesting a circle with unit radius to be drawn.
`plot.vars.points`	Defaults to TRUE, requesting plotting the variables as points.
`print.ax.approx`	Defaults to TRUE, requesting printing as part of the axis label the measure of how well each variable approximates `unit.cor.approx`, described below.

Value

In addition to the correlation monoplot described in Section 10.2.2, a list with the following components is returned: `cor.X`, the correlation matrix of **X**; `adequacies`, the axis adequacies; `predictivities`, the axis predictivities; and `unit.cor.approx`, the measure of how well each variable approximates the unit correlation of exact representations, given algebraically by the square root of diag($\mathbf{V\Sigma}^2\mathbf{JV}'$).

10.5.3 Function `MonoPlot.cor2`

This function constructs the correlation monoplots described in Section 10.2.4 based on the PCO of (10.4) with **R** as well as with $\mathbf{R}^*\mathbf{R}$. It takes the same arguments as `MonoPlot.cor`, but arguments `exp.factor`, `rotate.degrees` and `reflect` (see `PCAbipl`) are also available.

Value

In addition to the two correlation monoplots described in Section 10.2.4, a list with the following components is returned: `cor.X`, the correlation matrix of **X**; `adequacies.R`, and `adequacies.R2`, the axis adequacies associated with **R** and $\mathbf{R}^*\mathbf{R}$, respectively.

10.5.4 Function `MonoPlot.coefvar`

This is a function for constructing the coefficient of variation monoplot defined in Section 10.2.3. Except for `scaled.mat`, it takes the same arguments as `MonoPlot.cov`. In addition to the coefficient of variation monoplot (see Figure 10.7), it returns a list with components: `cov.X`, the covariance matrix of **X**; `cor.X`, the correlation matrix of **X**; and `coefvar.vec`, the vector containing the coefficients of variation of all the variables.

10.5.5 Function `MonoPlot.skew`

This is a function for constructing the hedron plots described in Section 10.3. It takes only the following arguments:

`X`	A square matrix,
`form.triangle1`	Optional argument for constructing a triangle on the hedron plot,
`form.triangle2`	Optional argument for constructing a second triangle on the hedron plot,
`...`	Optional arguments passed to the `points` function controlling appearance of the plotted points,

In addition to the hedron plot (see Figure 10.11), it returns a list with components: `M`, the symmetric **M** matrix defined in Section 10.4; `N`, the skew-symmetric **N** matrix defined in Section 10.4; `K`, the **K** matrix defined in Section 10.4; `U` and `sigma`, the matrices **U** and $\mathbf{\Sigma}$ of the SVD (10.5).

References

Aldrich, C., Gardner, S. and le Roux, N.J. (2004) Monitoring of metallurgical process plants by using biplots. *American Institute of Chemical Engineers Journal*, **50**, 2167–2186.

Bailey, R.A. and Gower, J.C. (1990) Approximating a symmetric matrix. *Psychometrika*, **55**, 665–675.

Baraitser, A.M. and Obholzer, A. (1981) *Cape Country Furniture*. Cape Town: Struik.

Barbezat, D.A. and Hughes, J.W. (2005) Salary structure effects and the gender pay gap in academia. *Research in Higher Education*, **46**, 621–640.

Bayer, A.E. and Astin, H.E. (1975) Sex differentials in the academic reward system. *Science*, **188**, 796–802.

Benzécri, J.-P. (1973) *L'Analyse des Données* (2 volumes). Paris: Dunod.

Blackman, J.A., Bingham, J. and Davidson, J.L. (1978) Response of semi-dwarf and conventional winter wheat varieties to the application of nitrogen fertilizer. *Journal of Agricultural Science, Cambridge*, **90**, 543–550.

Blasius, J., Eilers, P.H.C. and Gower, J.C. (2009) Better biplots. *Computational Statistcs and Data Analysis*, **53**, 3145–3158.

Borg, I. and Groenen, P.J.F. (2005) *Modern Multidimensional Scaling* (2nd edition). New York: Springer.

Botha, A. (1977) *Herkoms van die Kaapse Stoel*. Cape Town: A.A. Balkema.

Bradu, D. and Gabriel, K.R. (1978) The biplot as a diagnostic tool for models of two-way tables. *Technometrics*, **20**, 47–68.

Burden, M., Gardner, S., le Roux, N.J. and Swart, J.P.J. (2001) Ou-Kaapse meubels en stinkhoutidentifikasie: moontlikhede met kanoniese veranderlike-analise en bistippings. *South African Journal of Cultural History*, **15**, 50–73.

Clarke, C.R.E., Morris, A.R., Palmer, E.R., Barnes, R.D., Baylis, W.B.H., Burley, J., Gourlay, I.D., O'Brien, E., Plumptre, R.A. and Quilter, A.K. (2003) *Effect of Environment on Wood Density and Pulp Quality of Five Pine Species Grown in Southern Africa*. Tropical Forestry Papers No 43. Oxford: Oxford Forestry Institute, Department of Plant Sciences, University of Oxford.

Constantine, A.G. and Gower, J.C. (1978) Graphical representation of asymmetry. *Applied Statistics*, **27**, 297–304.

Understanding Biplots John Gower, Sugnet Lubbe and Niël le Roux

Cook, R.D. and Weisberg, S. (1982) *Residuals and Influence in Regression*. London: Chapman & Hall.

Cox, T.F. and Cox, M.A. (2001) *Multidimensional Scaling* (2nd edition). Boca Raton, FL: Chapman & Hall/CRC.

Cramér, H. (1946) *Mathematical Methods of Statistics*. Princeton, NJ: Princeton University Press.

De Leeuw, J. (1977) Applications of convex analysis to multidimensional scaling. In J.R. Barra *et al.* (eds), *Recent Developments in Statistics*, pp. 133–145. Amsterdam: North Holland.

Eckart, C. and Young, G. (1936) The approximation of one matrix by another of lower rank. *Psychometrika*, **1**, 211–218.

Eilers, P.H.C. and Goeman, J.J. (2004) Enhancing scatterplots with smoothed densities. *Bioinformatics*, **20**, 623–628.

Fang, K.T., Kotz, S. and Ng, K.W. (1990) *Symmetric Multivariate and Related Distributions*. Boca Raton, FL: Chapman & Hall/CRC.

Flury, B. (1988) *Common Principal Components and Related Multivariate Models*. New York: John Wiley & Sons, Inc.

Flury, B. (1997) *A First Course in Multivariate Statistics*. New York: Springer.

Gabriel, K.R. (1971) The biplot graphical display of matrices with application to principal component analysis. *Biometrika*, **58**, 453–467.

Gabriel, K.R. (2002) Goodness of fit of biplots and correspondence analysis. *Biometrika*, **89**, 423–436.

Gardner, S., le Roux, N.J., Rypstra, T. and Swart, J.P.J. (2005) Extending a scatterplot for displaying group structure in multivariate data: a case study. *ORiON*, **21**, 111–124.

Gardner-Lubbe, S., le Roux, N.J. and Gower, J.C. (2008) Measures of fit in principal component and canonical variate analyses. *Journal of Applied Statistics*, **35**, 947–965.

Gifi, A. (1990) *Nonlinear Multivariate Analysis*. Chichester: John Wiley & Sons, Ltd.

Goldberg, K.M. and Iglewicz, B. (1992) Bivariate extensions of the boxplot. *Technometrics*, **34**, 307–320.

Good, P. (2000) *Permutation Tests: A Practical Guide to Resampling Methods for Testing Hypotheses* (2nd edition). Berlin: Springer-Verlag.

Gordon, N., Morton, T. and Braden, I. (1974) Is there discrimination by sex, race and discipline? *American Economic Review*, **64**, 419–427.

Gower, J.C. (1968) Adding a point to vector diagrams in multivariate analysis. *Biometrika*, **55**, 582–585.

Gower, J.C. (1977) The analysis of asymmetry and orthogonality. In J.R. Barra *et al.* (eds), *Recent Developments in Statistics*, pp. 109–123. Amsterdam: North Holland.

Gower, J.C. (1982) Euclidean distance geometry. *Mathematical Scientist*, **7**, 1–14.

Gower, J.C. (1990) Three dimensional biplots. *Biometrika*, **77**, 773–785.

Gower, J.C. (1992) Generalized biplots. *Biometrika*, **79**, 475–493.

Gower, J.C. (1993) The construction of neighbour-regions in two dimensions for prediction with multi-level categorical variables. In O. Opitz, B. Lausen and R. Klar (eds), *Information and classification: Concepts – Methods – Applications. Proceedings 16th Annual conference of the Gesellschaft für Klassifikation, Dortmund, April 1992*, pp. 174–189. Heidelberg: Springer.

Gower, J.C. (2004) The geometry of biplot scaling. *Biometrika*, **91**, 705–714.

Gower, J.C. (2006) Divided by a common language: Analyzing and visualizing two-way arrays. In M. Greenacre and J. Blasius (eds), *Multiple Correspondence Analysis and Related Methods*, pp. 77–106. Boca Raton, FL: Chapman & Hall/CRC.

Gower, J.C. (2008) Asymmetry analysis: The place of models. In K. Shigemasu *et al.* (eds), *New Trends in Psychometrics*, pp. 69–78. Tokyo: Universal Academy Press.

Gower, J.C. and Dijksterhuis, G.B. (2004) *Procrustes Problems*. Oxford: Oxford University Press.

Gower, J.C. and Hand, D.J. (1996) *Biplots*. London: Chapman & Hall.

Gower, J.C. and Harding, S. (1988) Nonlinear biplots. *Biometrika*, **75**, 445–455.

Gower, J.C. and Legendre, P. (1986) Metric and Euclidean properties of dissimilarity coefficients. *Journal of Classification*, **3**, 5–48.

Gower, J.C. and Ngouenet, R.F. (2005) Nonlinearity effects in multidimensional scaling. *Journal of Multivariate Analysis*. **94**, 344–365.

Gower, J.C., Meulman, J.J. and Arnold, G.M. (1999) Non-metric linear biplots. *Journal of Classification*, **16**, 181–196.

Gower, J.C., Groenen, P.J.F. and Van de Velden, M. (2010) Area biplots. *Journal of Computational and Graphical Statistics*, **19**, 46–61.

Green, P.J. (1985) Peeling data. In S. Kotz and N.L. Johnson (eds), *Encyclopedia of Statistical Science, Volume 6*, pp. 660 – 664. New York: John Wiley & Sons, Inc.

Greenacre, M.J. (1984) *Theory and Applications of Corresepondence Analysis*. London: Academic Press.

Greenacre, M.J. (1988) Correspondence analysis of multivariate categorical data by weighted least squares. *Biometrika*, **75**, 457–467.

Greenacre, M.J. (2007) *Correspondence Analysis in Practice* (2nd edition). Boca Raton, FL: Chapman & Hall/CRC.

Heiser, W.J. and De Leeuw, J. (1977) *How to use SMACOF-1*. Research Report. Leiden: Department of Data Theory, University of Leiden.

Hills, M. (1969) On looking at large correlation matrices. *Biometrika*, **56**, 249–253.

Hirschfeld, H.O. (1935) A connection between correlation and contingency. *Proceedings of the Cambridge Philosophical Society*, **31**, 520–524.

Hyndman, R.J. (1996) Computing and graphing highest density regions. *American Statistician*, **50**, 120–126.

Jolliffe, I.T. (2002) *Principal Component Analysis* (2nd edition). New York: Springer.

Jorion, P. (1997) *Value at Risk*. New York: McGraw-Hill.

Kempton, R.A. (1984) The use of biplots in interpreting variety by environment interactions. *Journal of Agricultural Science, Cambridge*, **103**, 123–135.

Krzanowski, W.J. (2004) Biplots for multifactorial analysis of distance. *Biometrics*, **60**, 517–524.

Lawley, D.N. and Maxwell, A.E. (1971) *Factor Analysis as a Statistical Method* (2nd edition). London: Butterworths.

Le Roux, B. and Rouanet, H. (2004) *Geometric Data Analysis: From Correspondence Analysis to Structured Data*. Dordrecht: Kluwer.

Le Roux, N.J. and Gardner, S. (2005) Analysing your multivariate data as a pictorial: a case for applying biplot methodology? *International Statistical Review*, **73**, 365–387.

Liu, R.Y., Parelius, J.M. and Singh, K. (1999) Multivariate analysis by data depth: descriptive statistics, graphics and inference. *Annals of Statistics*, **27**, 783–858.

McNabb, R. and Wass, V. (1997) Male-female salary differentials in British universities. *Oxford Economic Papers*, New Series, **49**, 328–343.

Pison, G., Struyf, A. and Rousseeuw, P.J. (1999) Displaying a clustering with CLUSPLOT. *Computational Statistics and Data Analysis*, **30**, 381–392.

R Development Core Team (2009) *R: A Language and Environment for Statistical Computing*. Vienna: R Foundation for Statistical Computing. http://www.R-project.org

Rao, C.R. (1952) *Advanced Statistical Methods in Biometric Research*. New York: John Wiley & Sons, Inc.

Rothkopf, E.Z. (1957) A measure of stimulus similarity and errors in some paired-associate learning. *Journal of Experimental Psychology*, **53**, 94–101.

Rousseeuw, P.J. and Ruts, I. (1997) The bagplot: a bivariate box-and-whiskers plot. Technical report. Antwerp: Department of Mathematics and Computer Science, Universitaire Instelling Antwerpen. http://win-www.uia.ac.be/u/statis/

Rousseeuw, P.J., Ruts, I. and Tukey, J.W. (1999) The bagplot: a bivariate boxplot. *American Statistician*, **53**, 382–387.

Ruts, I. and Rousseeuw, P.J. (1996) Computing depth contours of bivariate point clouds. *Computational Statistics and Data Analysis*, **23**, 153–168.

Scott, D.W. (1992) *Multivariate Density Estimation*. New York: John Wiley & Sons, Inc.

Shepard, R,N. and Carroll, J.D. (1966). Parametric representation of nonlinear data structures. In P.R. Krishnaiah (ed.), *Multivariate Analysis*, pp. 561–592. New York: Academic Press.

Silvey, S.D., Titterington, D.N. and Torsney, B. (1978) An algorithm for optimal designs on a finite design space. *Communications in Statistics: Theory and Methods*, **A7**, 1379–1389.

Swart, J.P.J. (1980) Non-destructive wood sampling methods from living trees: a literature survey. *IAWA Bulletin*. **1 – 2**, 42.

Swart, J.P.J. (1985) *'n Sistematies-houtanatomiese ondersoek van die Lauracea in Suidelike Afrika*. Unpublished MSc thesis, Department of Wood Science, Stellenbosch University, South Africa.

Swart, J.P.J. and Van der Merwe, J.J.M. (1980) Embuia – *Ocotea porosa* en nie *Phoebe porosa* nie. *South African Forestry Journal*, **114**, 75.

Titterington, D.N. (1976) Algorithms for computing D-optimal design on finite design spaces. In *Proceedings of the 1976 Conference on Information Science and Systems*, pp. 213–216. John Hopkins University, Baltimore, MD.

Toutkoushian, R.K. (1998) Racial and marital status differences in faculty pay. *Journal of Higher Education*, **69**, 513–541.

Toutkoushian, R.K. (1999) The status of academic women in the 1990s: no longer outsiders, but not yet equals. *The Quarterly Review of Economics and Finance*, **39**, 679–698.

Underhill, L.G. (1990) The coefficient of variation biplot. *Journal of Classification*, **7**, 41–56.

Van Blerk, S.P. (2000) *Generalising Biplots and its Applications in S-Plus*. Unpublished MComm thesis, Department of Statistics and Actuarial Science, Stellenbosch University, South Africa.

Van der Berg, S., Wood, L. and le Roux, N. (2002). Differentiation in black education. *Development Southern Africa*, **19**, 289–306.

Venables, W.N. and Ripley, B.D. (2002) *Modern Applied Statistics with S* (4th edition). New York: Springer.

Walters, I.S. and le Roux, N.J. (2008) Monitoring gender remuneration inequalities in academia using biplots. *OriON*, **24**, 49–73.

Ward, M. (2001) The gender salary gap in British academia. *Applied Economics*, **33**, 1669–1681.

Warman, C., Woolley, F. and Worswick, C. (2006) *The evolution of male-female wages differentials in Canadian universities: 1970 – 2001*. Queen's Economics Department Working Paper No. 1099. Department of Economics, Queen's University.

Wurz, S., le Roux, N.J., Gardner, S. and Deacon, H.J. (2003) Discriminating between the end products of the earlier Middle Stone Age sub-stages at Klasies River using biplot methodology. *Journal of Archaeological Science*, **30**, 1107–1126.

Zani, S., Riani, M. and Corbellini, A. (1998) Robust bivariate boxplots and multiple outlier detection. *Computational Statistics and Data Analysis*, **28**, 257–270.

Index

Understanding Biplots John Gower, Sugnet Lubbe and Niël le Roux

T